PHYSIOLOGY- LABORATORY AND CLINICAL RESEARCH SERIES

BODY TEMPERATURE REGULATION

PHYSIOLOGY- LABORATORY AND CLINICAL RESEARCH SERIES

Body Temperature Regulation
Austin B. Cisneros and Bryan L. Goins
2009 ISBN: 978-1-60741-282-3

PHYSIOLOGY- LABORATORY AND CLINICAL RESEARCH SERIES

BODY TEMPERATURE REGULATION

AUSTIN B. CISNEROS
AND
BRYAN L. GOINS
EDITORS

Nova Science Publishers, Inc.
New York

For permission to use material from this book please contact us:
Telephone 631-231-7269; Fax 631-231-8175
Web Site: http://www.novapublishers.com

Library of Congress Cataloging-in-Publication Data

Body temperature regulation / [edited by] Austin B. Cisneros and Bryan L. Goins.
p. ; cm.
Includes bibliographical references and index.
ISBN 978-1-60741-282-3 (hardcover)
1. Body temperature--Regulation. I. Cisneros, Austin B. II. Goins, Bryan L.
[DNLM: 1. Body Temperature Regulation. QT 165 B668 2009]
QP135.B625 2009
612'.01426--dc22

2009010500

Published by Nova Science Publishers, Inc. ✝ New York

Contents

Preface

Body temperature regulation refers to processes of heating and cooling that an organism uses to control its temperature. Thermoregulation is the ability of an organism to keep its body temperature within certain boundaries, even when the surrounding temperature is very different. This process is one aspect of homeostasis, a dynamic state of stability between an animal's internal environment and its external environment. The maintenance of body temperature in animals is a result of the process of metabolism, by which foodstuffs are converted into protein, carbohydrates, and fat, with the release of energy in the form of heat. Because active muscles metabolize food faster than muscles at rest, giving off more heat in the process, physical activity increases body temperature. This new book provides a great variety of research on human as well as animal body temperature regulation. The principles of thermodynamics pertinent to the exchange of heat energy between the body and its surrounding environment is reviewed. Included also are studies on the effects of diet, body weight, exercise, disease, age and drugs.

Chapter I - Thermoregulatory investigations have been an integral part of physiological research generating voluminous evidence that have illuminated latent physiological mechanisms at all levels of biological organization. Yet, substantial controversy persists on topics as critical as behavioral thermoregulation, the principal form of thermoregulation used by humans and other homoeothermic species. The main challenge in investigating behavioral thermoregulation stems from the complexity in the body's responses to different stimuli, and the difficulty to explicitly identify the system of interest and isolate it from other systems within the body, as well as external dynamics. To address these challenges I developed an experimental model where the body is supplied with an extraneous thermoregulatory system, a liquid conditioning garment, that can be accurately and objectively quantified. By situating a human in an environment that forces its use, information gained from this hybrid human-garment system provides valuable evidence on the fundamental principles that govern behavioral thermoregulation. Given that liquid conditioning garment systems have been successfully implemented to manipulate thermal balance, I monitored and modeled the fluctuations in temperature of individuals given voluntary control of their garment's temperature during prolonged cold exposure in order to detect the dependable criteria of behavioral thermoregulation based on which the conscious response to thermal input is generated. Until recently it was believed that mean skin temperature *per se* is not a dependable criterion of behavioral thermoregulation and that the main signal is of internal origin. My work thus far has shown that humans generate conscious thermoregulatory responses that are aimed towards maintaining a threshold mean body temperature, rather than temperature changes in individual body regions. Interestingly, further work in this area revealed that body heat content is linked with mental performance and that disturbances in overall thermal balance affect perceptual judgments. Mental disturbances reflected thermally-induced reductions in the threshold used to guide positive responding, which may have implications for decision-making and complex reasoning. This knowledge can contribute to our understanding of thermoregulatory responses under prolonged exposures to harsh environments and for engineering applications germane to severe thermal environments faced by athletes, military, pilots, firefighters and other emergency responders. In this light, the present chapter reviews current knowledge on behavioral thermoregulation in an attempt to understand the principles that govern it.

Chapter II - The importance of maintaining thermal balance in homoeotherms first gained recognition in the late 18th and early 19th century through the work of Lavoisier, Carnot and Mayer [1]. Since then, thermoregulatory investigations have been a vital part of physiological research providing findings that have illuminated underlying physiological mechanisms at all levels of biological organization [2-4]. Nevertheless, despite that at the cellular and molecular level our knowledge of mechanisms governing thermal balance is increasing dramatically aided by the continuous advances in technology, at the systems level (i.e., the level of architecture and organization) several fundamental questions remain unanswered [5]. The most prominent of these questions involves the mechanism by which the thermoregulatory threshold is determined which, after numerous investigations, remains a mystery [6]. And, yet, the clinical importance of the physiology of thermoregulation is immense. The prime example for this is that an exodus of body temperature from its optimal range is so suggestive of a pathological condition that body temperature is continuously monitored in all hospitalized patients and reported in every medical history, while the clinical thermometer is one of the most essential scientific instruments for the practice of medicine [7, 8]. In this light, this chapter reviews current knowledge on the functional architecture of human thermoregulation in an attempt to increase the understanding regarding the principles that govern it. The first part of this chapter reviews principles of thermodynamics pertinent to the exchange of heat energy between the human body and its surrounding environment. Thereafter the physiological principles of endothermic thermoregulation are discussed. These principles are divided into those involving thermosensors – including the location, nature and function of central thermosensitive neurons in the preoptic anterior hypothalamus and peripheral thermosensory neurons that detect shell temperatures in the skin and in the oral and urogenital mucosa – and those pertaining to thermoeffectors. The latter include the efferent neuronal pathways to thermoeffectors as well as various aspects of thermoregulatory skin vasodilation and vasoconstriction, and shivering.

Chapter III - Birds are endotherms, able to maintain their body temperature within a narrow range. However, an alteration in body temperature as a result of exposure to extreme environmental conditions may lead to a cascade of irreversible thermoregulatory events that could be lethal for the bird. To sustain thermal tolerance and avoid the deleterious consequences of thermal stresses, three direct responses are elicited: the rapid thermal stress response; acclimation; and epigenetic temperature adaptation. Although domestic fowl respond to changes in the environment with the same mechanisms as other endotherms, their intensive production causes them to face serious difficulties in coping with extreme environmental changes. Recent decades have seen significant progress in the genetic selection, on the one hand, of fast-growing meat-type broiler chickens and turkeys and, on the other hand, of egg-producing laying hens. However, the significant improvements in body and muscle growth and in egg production have not necessarily been accompanied by equivalent growth of specific visceral organs, and the probable consequence is a reduced ability to cope with extreme changes in environmental temperatures.

This chapter will focus on: a. body temperature development in chickens during embryogenesis; b. the control of body temperature during acute and chronic exposure to various environmental conditions; and c. the efficacy of controlling body temperature, as affected by thermal conditioning (epigenetic temperature adaptation) experience during perinatal period.

Chapter IV - Temperature control of the newborn infant has been a popular research topic over the last several decades and one that has significantly altered neonatal morbidity and mortality outcomes. Premature infants especially are prone to heat loss by convection, conduction, evaporation and radiation due to their thin skin, decreased body fat and immature body systems. Healthcare providers make temperature regulation one of the most important management issues for newborns, particularly in the case of premature infants. Although term and most premature infants manage their own heat production through non-shivering thermogenesis, healthcare providers need to manage the environment of those infants to minimize heat loss using special environments and procedures. Nursing and medical interventions have improved neonatal thermoregulation; however, hypothermia continues to be a problem in the extremely low birth weight (ELBW) infant population (premature infants born weighing less than 1000 grams). One intervention, placement of infants in polyurethane bags, serves to protect the smallest premature infants from evaporative heat losses in the delivery room. Once ELBW infants are admitted to neonatal intensive care units, stabilization procedures continue to expose infants to cold temperatures. ELBW infants are unable to generate heat due to immature body systems, and body temperatures can fall as low as 33° C. Therefore, nurses must pay close attention to the environment of ELBW infants, to prevent decreases in body temperature. In this way, morbidity and mortality can be decreased for ELBW infants.

Chapter V - Skin is one of the primary channels realizing the heat exchange of a human organism with its environment. When body temperature regulation is under study, the experimental methods and high-sensitive techniques allowing real-time registration of skin surface thermal patterns take on great significance. Infrared thermography supplied with two-dimensional (or focal plane array based) infrared detectors is best suited for these needs.

This chapter gives an overview of infrared thermography as a powerful research method applied to the biomedical sphere. It also embraces the material of recent IRT-assisted experimental works devoted to the studies of a human organism heat exchange with its environment under different active loadings. A high-performance infrared camera having 0.03 °C temperature sensitivity was used in the experiments. It is pictorially demonstrated how the surface thermal patterns are transformed in the course of physical and heat (hyperthermic) loads applied to the organism. Infrared thermography provides a way of visualizing the fine manifestations of vascular and perspiratory systems that is impossible to carry out using any other experimental methods. The results of quantitative analysis of two-dimensional thermograms are also presented.

It follows from the infrared observations that the typical feature of a loaded organism is its highly heterogeneous thermal pattern represented on the skin. This surface temperature non-uniformity, on the one hand, results from skin cooling due to the sweat evaporation and on the other hand from local heating due to transient dilation of separate superficial branches of vessels (the so-called "veins-collectors"). Quantitative characterization of highly heterogeneous thermograms having a direct relationship to the study of body temperature regulation is a new problem of modern biomedical infrared thermography. The original methods attacking this problem are also mentioned in this chapter.

Chapter VI - Although the idea of fever has accompanied medical practice for many centuries, the difficulty of precisely measuring the temperature has hampered its use until recent times. Initially, fever was explored by hand, and probably the idea of “heat” was related but no identical to temperature. But during the XIVth – XVIIth centuries a “measuring obsession” spread through Europe, becoming one of the most prominent features of scientific thought (at least in our western world point of view)(Crosby, 1977). Thermometry was no exception. The first clinical thermometer known was described by Santorio (Sanctorio Sanctorius) in the 17th century, although probably copying an air thermometer developed by Galileo Galilei. Santorio produced several designs but they were cumbersome and required long time for each reading, and they did not gain wide acceptance. In the 18th century Boerhaave, van Swieten, de Haen and Martine started to use the thermometer at the bedside. Diurnal variations were described, as were the relation between fever and increased heart rate or shivering. However, thermometry remained a bothersome operation, and was not widely used. In the 19th century, several improvements (notably the use of mercury instead of alcohol, and the introduction of a narrower tube to prevent the drop back of the column) made thermometers much smaller and practical. Finally, Carl Rheinhold August Wunderlichs’s seminal book *Das Verhalten der Eigenwärme in Krankenheten* (Leipzig 1868) established thermometry as an essential tool at the bedside, fixing what thereon would be considered the normal range of temperature. In his book, he reports the results of several million temperature readings on 25.000 patients. Although some of his results may be disputable (there were no formal statistics at that time, and the precision and calibration of his thermometers may be questionable)(Mackowiak et al., 1992a, Mackowiak and Worden, 1994), Wunderlich marks the starting point of temperature measuring as practiced today.

Chapter VII - In general, diagnostic imaging equipment creates images by exposing the living body to physical energy. In the living body, this applied energy can be transformed into heat. The resultant heating may have adverse effects on body temperature, which is otherwise strictly maintained within a narrow range by the thermo-regulatory system.

Physicians tend to have a general impression that ultrasonic imaging and magnetic resonance imaging (MRI) are safe diagnostic methods without radioactive exposure. However, the rise of body temperature due to these equipment has been reported as a side effect. In ultrasonic imaging, tissue temperature is increased by absorption of ultrasonic energy by the tissues. In MRI, applied radiofrequency and gradient magnetic field generate electric currents and ohmic heating in the body. The heating effect during ultrasonic imaging has been well studied, and many investigators have reported that local body temperature rise due to ultrasonic equipment may have potential adverse effects. Needless to say, the specification of ultrasonic diagnostic equipment, such as output power, is strictly regulated. Nevertheless, some use conditions, such as operating time, that still depend on the examiner. Therefore, safer use of ultrasonic diagnostic equipment cannot be accomplished unless the users of these equipment are well aware of the safety standards and potential adverse effects. This also applies to MRI. In MRI, especially conductive foreign bodies implanted inside the body and applied on the body surface can create problems of heating. Even without these objects, the human body itself also functions as a conductive object. In order to avoid thermal problems involving MRI, the examiner must pay careful attention to conductive objects including human body position.

The primary purpose of this paper is to draw attention to body temperature rise due to diagnostic imaging equipment and review the problems associated with the heating effect.

Chapter VIII - The purpose of this study was to examine the metabolic behaviour of pre-ruminant kid goats and lambs at different environmental temperatures (12, 24 and 30°C). The energy and protein retained were determined by means of a feeding experiment on different feeding levels and by the comparative slaughter method. The relationships between the total energy retained or those retained as protein or fat, and the metabolizable energy intake, were established by means of regression analyses. Energy requirements for maintenance were lower, and more closely grouped, among the kid goats than among the lambs. On the contrary, the efficiencies of energy utilization for energy retention were higher among the lambs. Under fasting conditions, the necessary thermogenesis among the kid goats was associated with the catabolism of fat. However, among the lambs, at 12°C there was found to be a high utilization of both protein and fat, while at the other two temperatures tested the corresponding utilization was mainly of fat. At satiety, the kid goats were able to regulate their intake, such that independently of the environmental temperature, their energy balance, and consequently the body composition, remained constant. On the contrary, under thermal stress induced by cold, the lambs were not capable of counteracting the greater thermal demands made by means of increasing their food intake, and thus fat was catabolized and their body composition became more lean. The temperature of 12°C was cold for both animals, located on the left hand side of their thermoneutrality zone, with the lambs being more sensitive to this cold than were the kid goats. For the goats, temperatures of 24°C and 30°C were identified as belonging to their thermoneutrality zone. However, for the lambs, 24°C could still be considered a cold temperature, while 30°C could be located within their thermoneutrality zone.

Chapter IX - Exercise in hot conditions alters the physiological response to the ensuing exercise bout and may hasten the onset of fatigue. Regardless of the environment, often pre-exercise procedures are employed in order to ensure the commencement of exercise in an optimally prepared state to perform. Two such pre-exercise procedures often proposed as being of benefit involve warming up and pre-cooling, respectively. While both have been shown to have ergogenic benefits, they are somewhat contradictory in nature. It is well documented that optimal muscular function of the contractile fibres occurs with an increased muscle temperature. Accordingly, the practice of a warm up prior to exercise commencement is endemic to most sports and athletes and is often based on the premise of increasing muscle temperature. In contrast, it is equally well documented that reducing body temperature by cooling the periphery of the body, including the musculature, is also ergogenic for exercise performance. Accordingly, the practice of pre-cooling is regularly used by athletes from a range of sports and environments. This contradiction between increasing and decreasing respective body temperatures to improve exercise performance raises interesting questions regarding the mechanisms behind the regulation of exercise in the heat. Consequently, the underlying physiological mechanisms to both warm up and pre-cooling procedures are seemingly related to thermoregulatory control. As such, this chapter will review the respective literature on both warm up and pre-cooling in relation to exercise in the heat. Included in this synthesis of relevant literature will be the physiological and performance responses to these respective pre-exercise interventions in the heat. Additionally, a comparison of the respective roles and interaction of both procedures will be discussed and finally, recommendations for the integration of both practices will be provided.

Chapter X – This chapter will examine the potential roles of absolute and relative exercise intensities during upper body exercise in able-bodied athletes and those with a spinal cord injury. The introduction will firstly provide a chronological review of studies which have examined absolute and relative workloads in relation to thermoregulatory responses and secondly provide a rationale for why these roles may be important for different populations regularly undertaking upper body exercise.

Chapter XI – The neuroleptic malignant syndrome (NMS) and the Parkinsonism-hyperpyrexia syndrome (PHS) are two clinically indistinguishable pathologies clinically characterized by hyperthermia. Their pathophysiologic mechanisms depend on decreased levels of dopamine that in NMS are due to dopamine receptor blockade by neuroleptic drugs, whereas in PHS they occur after withdrawal of dopaminergic drugs in the absence of neuroleptic medication.

This work has the aim to review the malignant syndromes depending on neuroleptic and parkinsonian drugs.

Chapter XII – Body temperature and eating behavior are two inter-dependent variables. Thermal set-point influences food intake, so that the introduction of energy can be considered as a system to reach a prefixed body temperature. Substances resetting the thermal set-point cause modifications of eating behavior. More precisely, pyrogens induce hyperphagia, while substances reducing the thermal set-point cause hypophagia. On the other hand, changes in body temperature (without changes in the thermal set-point) induce modifications of the ingestive behavior. Indeed, conditions inducing hyperthermia cause a reduction of food intake, while situations reducing body temperature are associated with an increase in food intake. The upon-exposed theory is validated by various evidences and it is in accord with the "thermoregulatory hypothesis of body weight". The present article reports our experimental observations, which corroborate this theory. The relationship between thermoregulation and body weight control are amply discussed

Chapter XIII - Perioperative hypothermia results from a combination of 1) anesthetic-induced impairment of thermoregulatory control, 2) decreased metabolism due to anesthesia, 3) cool operating room environment, and 4) factors unique to surgery that promote excessive heat loss. Perioperative hypothermia shows a characteristic three phase pattern: 1) redistribution phase, an initial rapid reduction in core temperature during the first hour resulted from an internal redistribution of body heat from the core to the peripheral shell caused by anesthetic-induced vasodilation, 2) linear phase, 2-3 hour of a slower and linear decrease at a rate determined by the difference between heat loss and heat production, 3) plateau phase shown after 2-4 hours of anesthesia, which develops due to delayed reemergence of thermoregulatory vasoconstriction restricting core to peripheral heat flow.

Perioperative Hypothermia is associated with several complications including morbid myocardial outcomes, surgical bleeding, wound infection, impaired immune function, delayeded postanesthetic recovery and prolonged hospitalization. All these data strongly suggest that anesthesiologists have to take special thermal precautions to prevent perioperative hypothermia. Diverse devices have been used to prevent hypothermia such as circulating water mattress and forced air warming device.

After the ingestion or infusion of nutrients, there is an increase in energy expenditure which has been referred to as dietary or nutrient-induced thermogenesis. This thermogenesis induced by protein or amino acids is well known to be largest and most prolonged. According to these physiological backgrounds, preoperative amino acid infusion was reported to prevent postoperative hypothermia during general anesthesia and spinal anesthesia. Also, perioperative amino acid infusion is reported to improve the outcome of the patients undergoing off-pump CABG. Fructose is also known to elicit the greatest increase in thermogenesis among various carbohydrates such as glucose. Fructose infusion prevents perioperative hypothermia during surgery. Both of amino acid infusion and fructose infusion were observed to shift upward the threshold core temperature for thermoregulatory vasoconstriction as well as to increase energy expenditure. The thermic effect of these nutrients might show the novel possibility of infusion therapy.

Chapter IVX - Observations and experiments were carried out to assess continuous thermoregulation in the nest of the Oriental hornet *Vespa orientalis* (Hymenoptera, Vespinae). The temperature of a pupating larva inside a nest without adult hornets was found to be higher by 1°C from the nest temperature in the vicinity of the brood comb. However, the temperature of larvae subjected to diethyl ether anesthesia equaled the ambient temperature. Observations showed that adult worker hornets occasionally warm the pupae from outside their cells and this by blowing warm air through the silk cap, of each cell. Upon artificial warming of combs whose adult population was intentionally removed, the temperature around these combs (at the beginning of the warming process) was lower than that of the silk-encased pupae inside the comb cells. Beyond 31.1°C, the temperature exceeded that of the pupae. The findings suggest that of the normal nest temperature is of 29°C, the pupal temperature is 30°C, which is apparently optimal for pupal development. Results of the present study prompted construction of a simple model that describes the interaction between the adults and the pupal brood which enables to maintain the latter at a constant temperature.

Chapter XV - The phylogenetic development of the febrile mechanism is not well understood because few studies have investigated the physiology of fever in species other than mammals. Considerable insight could be gained from studies in the other homeothermic group, the birds. Like mammals, birds develop fevers when exposed to bacterial endotoxin, and like in mammals, lipopolysaccharide-induced fevers are mediated by prostaglandin and nitric oxide. However, the signaling pathways that lead to the synthesis of prostaglandin and nitric oxide in birds have not been elucidated. A role for pro-inflammatory cytokines has been suggested but, as yet, there is limited evidence for the existence of such a pathway. Likewise, avian toll-like receptors have been identified and have been shown to initiate innate immune responses on pathogen exposure. It is still unknown whether such immune reactions mediate fever. Hitherto glucocorticoids have been identified as the only likely modulators of the avian febrile response. Taken together, evidence for the involvement of prostaglandin, nitric oxide and glucocorticoids in the mediation and modulation of avian fevers support the idea that the febrile mechanism in birds and mammals has a common origin. Future experiments will need to examine the physiology of fever induction and the precise roles of mediators and modulators in the avian febrile response to broaden our scientific understanding of the evolutionary and phylogenetic development of the febrile mechanism.

Chapter XVI - Aging leads to a decline in homeostasis; therefore, a decline in homeostasis might serve as a useful biomarker of aging. A homeostatic system that shows a marked decline with age in mammals is thermoregulation, which is particularly easy to monitor in mice. In this study, the authors measured the drop in mean body temperature of 34 female mice from 22 LSXSS recombinant inbred strains; the strains were divided into two cohorts that were four months apart in age. Both cohorts were monitored at the same time and showed a significant drop in mean body temperature over a period of 5 months: from 8 to 13 months of age, Cohort 1 exhibited a mean drop of 1 °C; from 4 to 9 months of age, Cohort 2 dropped 0.6 °C. Interestingly, the temperature drop in Cohort 1 (the older cohort), but not 2, exhibited significant genetic variation ($P = 0.004$). The authors discuss how his variation, occurring relatively early in the aging process, is potentially a powerful tool for elucidating the mechanisms underlying loss of thermoregulation with age.

Chapter XVII - Body temperature (BT) is strictly regulated to a constant level in human, after temperature-sensitive sensory neurons in the skin usually detect any changes of environmental temperature. Pathological conditions sometimes produce fever that results from the releases of endogenous pyretic substances after attacks such as those from bacteria and viruses. The mechanisms of BT regulation underlying these two cases are naturally different; however, in both cases the final step involves the thermoregulatory center in the central nervous system (CNS). The authors discuss whether the NO pathway is a pyretic, antipyretic or an unrelated one for BT regulation of the CNS. In conclusion, NO likely contributes to both hyperthermia and hypothermia under pathological conditions, but not under physiological conditions. The role of NO in BT regulation differs depending on when, where, and how NO is generated under pathological conditions, which is likely determined by cytokines biosynthesized after attacks of bacteria and virus.

Short Communication - This paper describes the effects of many neuroleptic drugs on the hyperthermia due to orexin A, a neuropeptide with influences on body temperature and food intake. Haloperidol reduces, while clozapine and olanzapine block the hyperthermia induced by orexin A. Risperidone enhances the elevation of body temperature due to orexin A. Quetiapine delays these hyperthermic effects. The implications on human therapeutic strategies are discussed, including the role of orexinergic system in the development of body weight gain, induced by these neuroleptic substances.

Expert Commentary A - This expert commentary reports the role exerted by cerebral cortex in the thermoregulation. Generally, the thermoregulation is described almost exclusively with the role of the hypothalamus, while the control exerted by other cerebral structures is not sufficiently underlined. Also the influences of cerebral cortex on body temperature are neglected. Indeed, a search on "PubMed" with the words "thermoregulation and cerebral cortex" reports only 261 articles, while the search of the words "thermoregulation and hypothalamus" produces 2485 papers. This commentary underlines especially the experimental evidences regarding the effects of functional decortication (obtained with cortical spreading depression) on thermogenic changes due to various conditions in different experimental models.

The hypothalamus controls body temperature. Although the hypothalamic areas exert relevant influences, the role played by the cerebral cortex on thermoregulation is not negligible. The aim of this commentary is to underline the scarcity of investigations about the control exerted by cerebral cortex on the regulation of body temperature. The preoptic-anterior and posterior hypothalamic areas are known as the playmakers of thermoregulatory mechanisms. Intracerebroventricular injection of prostaglandin E_1 acts on preoptic/anterior hypothalamic area, inducing the elevation of body temperature [11]. Although the preoptic-anterior hypothalamic area is considered the "thermostat" of body temperature regulation, other hypothalamic areas are involved in the thermoregulation.

Stimulations of the ventromedial hypothalamus causes an increase in temperature of interscapular brown adipose tissue [15, 16], that is the main effector of non-shivering thermogenesis [1]. Lesions of the ventromedial hypothalamus reduce the hyperthermia due to various stimuli. Indeed elevation of body temperature induced by an intracerebroventricular injection of prostaglandin E_1 is reduced by ibotenate damage of the ventromedial hypothalamus [9], such as post-ingestional thermogenesis is lowered by the ventromedial hypothalamic lesion [8]. Lesion of the lateral hypothalamus induces hyperthermia [14], while lysine acetylsalicylate modifies thermogenic changes induced by this lesion [10]. These experiments demonstrate a key role played by the hypothalamus in the control of body temperature.

On the other hand, the cerebral cortex exerts considerable influence on the hypothalamic areas. The functional ablation of cerebral cortex reduces the hyperthermia due to lateral hypothalamic lesions [4]. Stimulations of the frontal neocortex induces an elevation of body temperature [3], whereas cortical spreading depression blocks prostaglandin E_1 and endotoxin fever [7]. The cerebral cortex exerts significant influences on hyperthermic reaction induced by orexin A [13], which is a hypothalamic neuropeptide controlling body temperature and food intake [12].

In general, cerebral cortex controls not only voluntary behavior, but also autonomic functions. The ablation of the cerebral cortex reduces the increase of the sympathetic discharge induced by various stimuli [2, 6, 13]. Since the sympathetic activation is associated with the rise of body temperature, the influence of the cerebral cortex on the sympathetic discharge affects also body temperature. Although experimental investigations have demonstrated an evident influence of the cerebral cortex on thermoregulation, the researchers investigate scarcely on these aspects of cortical functions. Indeed, scientific papers regarding the hypothalamic influences on body temperature are much more abundant than articles about the cortical control of thermoregulation. Using a search performed on July 2008, PubMed (U.S.

National Library of Medicine) cites 2485 papers on "thermoregulation and hypothalamus", while it reports only 261 articles on "thermoregulation and cerebral cortex".

In the future, this difference should be reduced by higher investments on research about the role played by the cerebral cortex on vegetative functions, as the thermoregulation. Particular should be addressed to study regarding the influences exerted by cerebral cortex on "thermostatic hypothesis" of food intake [5]. These experimental investigations could shed further light on the controversial issue regarding the relationship between thermoregulation and control of food intake.

Expert Commentary B - For the most part, publications especially of severe antipsychotics-associated hypothermias meet criteria — in form — of uncontrolled observations and — in content — of adverse drug reactions of type B or idiosyncratic drug reactions. Both attributes may have an inhibitory effect on publication procedures in journals setting a high value on data integrity and pathomechanism. Whether publication bias is already true or not, the restrictive handling of (single) cases or uncontrolled observations hampers the constructive clinical as well as scientific discussion of this hardly known side effect.

Antipsychotics-associated hypothermias (AAH) of moderate and especially severe degree, i.e. with a core temperature of ≤ 32°C rectal, dimming of consciousness and bradyarrhythmias, respectively, turned out to be mainly published as case studies with a lower rank (IP ≤ 3). The same seems to go for reviews of this theme. In case reports clinicians often point out that on medication a hypothermia b occurred and that after discontinuation of medication a hypothermia b declined. Persons affected are often aged, rather multimorbid and polymedicated (Poncini et al., 2004).

In contrast in (prospective) studies — often higher-ranking published — rather titratable, therefore limited dose-dependent AAH of rather low degree in mammals and not multimorbid subjects, respectively, often under exposition of one substance were investigated (e.g. Van Marum et al., 2007). It is worth mentioning that antipsychotic monopharmacy and mono- / oligomorbidity correspond for example to the German guideline (Gaebel and Falkai, 2006), but not often reflects the clinical reality (Messer et al., 2006).

It is arguable to what extent the severe case reported AAH resembling type B-reactions* are etiopathologically related to the milder study-AAH, which rather correspond to type A-reactions. It seems that the rather severe forms of AAH tend to be influenced by host factors whereby environmental co-factors like changing (co-/poly-)medication play a pivotal role.

The text of a case rejection for publication in a journal with a higher impact factor could be:

"Evaluation of case reports always presents a difficult challenge. In general, such reports are intended to raise the possibility (though certainly do not prove) that Drug A is causally related to Event B. However the public media and even some aspects of the medical/scientific community do not always take it this way-not uncommonly, the case report is misinterpreted as clear evidence of cause-and-effect. Therefore all need to exercise caution when proceeding with editorial assessments of case reports. If there are plausible reasons to connect Drug A to Event B (and the connection is not already established through more rigorous study), we would be more positively inclined to accept the report for publication. On the other hand, if the clinical picture is so complex and muddy that the connection is improbable and/or uncertain, then there will be less inclined to acceptance. It may actually do more harm than good to the medical community to publish uncertain reports of this type, since they can unnecessarily raise alarm and concern among patients and physicians. The case that you describe falls more in the latter category. Temperature dysregulation associated with neuroleptics has been recognized for decades. It can be variable and unpredictable* in terms of prevalence and time-course. The patient you describe had been exposed to multiple neuroleptics, and it is not clear that exposure or non-exposure to Drug Z was plausibly related to the occurrence or non-occurrence of temperature dysregulation in either direction" (author known).

In fact it seems among others that in a single uncontrolled observation the temporal relationship between a sudden stop of the putative inducing antipsychotic (mono- or polymedication) and the recovery of temperature regulation within hours or a few days is too uncertain for assumption of a close, if not causal, connexion. In the course of this some points cause astonishment because (i) alternative hypotheses (no or other connexion) may be unlikely or absurd, (ii) cumulative evidences of many case reports of different authors as well as studies exist and (iii) the causal connexion between antipsychotics and hypothermia may be plausible for the following reasons: The relation reveals more (type A-reactions) or less (type B-reactions*) (i) dose-concentration-response properties and (ii) transient and reversible effects, respectively, i.e. basic pharmacological principles, whose closer characterization is still to be found.

Moreover, antipsychotic polypharmacy (successively, simultaneously and/ or overlapping) as well as the impossibility of reonset of the putative adverse drug for scientific reasons remain clinical requirements.

"There can be no discussion that pharmacodiligence requires prospective studies, access to regulatory filings for controlled and monitored use of drugs, and some sense of relative potency for comparison between different drug entities. However, large cohorts are needed for the detection and analysis of type B-reactions*. Since these data are lacking, case and spontaneous reports by alert clinicians must be analyzed. The outcome of such an analysis cannot be seen as solid evidence, but can help to get more insight in the adverse drug reaction. The use of measures like Reporting Odds Ratios and disproportionality can help to detect drugs with an increased association with an adverse drug reaction. A statistically significant Reporting Odds Ratio may be indicative of a higher risk for that particular event during the use of a specific medication, but is never conclusive for the actual existence of a causal relation" (Van Marum et al., 2007).

Generally speaking case reports of AAH broaden our knowledge of new constellations concerning exposition factors (medication), host factors and circumstances (life style). At least case reports remind us of the possibility as well as conditions of appearance of AAH in a way that is taken from clinical everyday life (Vandenbroucke, 2001). As always we should critically meet the fast direct naturalistic descriptive and preliminary evidence of case studies, but not strictly with refusal. This should be true for clinicians as well as scientists.

"Casual observations strike us when they are unexpected. They are therefore refutations of our previously held beliefs and will lead to new conjectures — new ideas, and new theories" (Karl Popper in Vandenbroucke, 2001).

**Explanation of type A- vs. type B-reactions:* Adverse drug reactions can be divided into at least two forms (Rawlins and Thompson, 1977; Park et al., 1992): Type A-reactions are extreme effects based on the pharmacology of a drug and therefore mostly dose-dependent and predictable. Type B-reactions or idiosyncratic drug reactions cannot be derived from the pharmacology and occasionally result often in fatalities particularly because they are hardly detectable in drug development.

Expert Commentary C - A large amount of research regarding body temperature regulation during exercise has been undertaken using lower body exercise modes such as cycling and running. However, although it may be convenient to consider arm exercise (predominantly arm cranking or arm cycling) to elicit similar physiological responses to lower body cycling, but of a lower magnitude, there are important factors to take into account which may result in different thermoregulatory responses. Indeed, upper body exercise has been suggested to pose unique thermoregulatory problems (Sawka et al, 1984). This commentary will address a number of factors that may be considered 'unique' to thermoregulation during upper body exercise.

In: Body Temperature Regulation
Editor: A. B. Cisneros and B. L. Goins
ISBN: 1-60741-282-3

Chapter I

Fundamental Principles Governing Behavioral Thermoregulation

Andreas D. Flouris*

Institute of Human Performance and Rehabilitation, Centre for Research and Technology – Thessaly. 32 Siggrou Street. Trikala, GR42100, Greece

Abstract

Thermoregulatory investigations have been an integral part of physiological research generating voluminous evidence that have illuminated latent physiological mechanisms at all levels of biological organization. Yet, substantial controversy persists on topics as critical as behavioral thermoregulation, the principal form of thermoregulation used by humans and other homoeothermic species. The main challenge in investigating behavioral thermoregulation stems from the complexity in the body's responses to different stimuli, and the difficulty to explicitly identify the system of interest and isolate it from other systems within the body, as well as external dynamics. To address these challenges I developed an experimental model where the body is supplied with an extraneous thermoregulatory system, a liquid conditioning garment, that can be accurately and objectively quantified. By situating a human in an environment that forces its use, information gained from this hybrid human-garment system provides valuable evidence on the fundamental principles that govern behavioral thermoregulation. Given that liquid conditioning garment systems have been successfully implemented to manipulate thermal balance, I monitored and modeled the fluctuations in temperature of individuals given voluntary control of their garment's temperature during prolonged cold exposure in order to detect the dependable criteria of behavioral thermoregulation based on which the conscious response to thermal input is generated. Until recently it was believed that mean skin temperature *per se* is not a dependable criterion of behavioral

* E-mail: aflouris@cereteth.gr

thermoregulation and that the main signal is of internal origin. My work thus far has shown that humans generate conscious thermoregulatory responses that are aimed towards maintaining a threshold mean body temperature, rather than temperature changes in individual body regions. Interestingly, further work in this area revealed that body heat content is linked with mental performance and that disturbances in overall thermal balance affect perceptual judgments. Mental disturbances reflected thermally-induced reductions in the threshold used to guide positive responding, which may have implications for decision-making and complex reasoning. This knowledge can contribute to our understanding of thermoregulatory responses under prolonged exposures to harsh environments and for engineering applications germane to severe thermal environments faced by athletes, military, pilots, firefighters and other emergency responders. In this light, the present chapter reviews current knowledge on behavioral thermoregulation in an attempt to understand the principles that govern it.

Introduction

Living organisms are adapted to specific temperature ranges at which they function optimally and they respond to changes in thermal load through a variety of complex mechanisms ranging from thermotaxis in bacteria to sweating in humans. Homoeothermy, the pattern of thermoregulation in which body temperature depends on a high and controlled rate of heat production [1], remains a striking adaptation in the animal kingdom that has attracted immense debate and attention to date and it is regarded as one of the most significant transitions in vertebrate evolution [2, 3]. The energetic cost of homoeothermy is substantial exceeding the corresponding values in vertebrate poikilothermal animals by a factor of 5–10 [given the same body mass and temperature [4]] and its evolution has required significant remodeling of several systems of the vertebrate body as well as substantial changes in the energetics, physiology and, particularly, the behavior of mammals and birds [3]. By using powerful behavioral, autonomic and endocrine mechanisms homoeotherms are able maintain a heat content equilibrium over a wide range of thermal loads with substantial benefits to system performance and integrity. These benefits include independence of timing daily activities, resistance to freezing, enzymatic catalysis, enhancement of aerobic capacities through greater concentration of mitochondria [3, 5], as well as rapid development of the central nervous system and acquisition of complex progressive forms of behavior [6]. Furthermore, maintaining a stable body heat content plays a major physiological role in regulating numerous molecular and chemical processes [7]. This is because the complexity of mammalian and avian vertebrates requires thermal constancy for efficient functioning because all chemical and physical reactions change their rate with changes of temperature, yet the degree of change with rising/falling temperature is different for different reactions [6].

The profound significance of maintaining thermal balance in homoeotherms first gained recognition in the late 18th and early 19th century through the work of Lavoisier, Carnot and Mayer [8]. Since then, thermoregulatory investigations have been an integral part of physiological research generating voluminous evidence that have illuminated latent physiological mechanisms at all levels of biological organization [9-11]. Yet, despite that at the cellular and molecular level our knowledge of mechanisms governing thermal balance is increasing dramatically aided by the continuous advances in technology, at the systems level (i.e., the level of architecture and organization) our understanding of homoeothermic thermoregulation remains limited. For instance, substantial controversy persists on topics as critical as behavioral thermoregulation, the principal form of homoeothermic thermoregulation. And, yet, the clinical importance of the physiology of thermoregulation is immense. The prime example for this is that an exodus of body temperature from its optimal range is so suggestive of a pathological condition that body temperature is continuously monitored in all hospitalized patients and reported in every medical history, while the clinical thermometer is one of the most essential scientific instruments for the practice of medicine [12, 13]. Increasing our knowledge on the principles that govern behavioral thermoregulation will contribute to our understanding of the functional architecture of the homoeothermic thermoregulatory system as well as the thermoregulatory responses under prolonged exposures to harsh environments and will improve current engineering applications germane to severe thermal environments faced by athletes, military, pilots, firefighters and other emergency responders [14, 15]. In this light, the present Chapter provides a review of the current knowledge on behavioral thermoregulation in an attempt to understand the principles that govern it. Thermosensation, including central and peripheral thermosensors as well as behavioral thermosensors, is reviewed first followed by behavioral thermoregulation as a thermoeffector together with experiments investigating the effectiveness of behavior in maintaining system thermal integrity.

Thermosensation

Thermoregulatory behavior has been defined as "*...an attempt to avoid what humans call thermal discomfort or displeasure and to obtain thermal pleasure*" [16] and it appears to be a reaction based on signals similar to those of autonomic and endocrine thermoregulatory reactions. Ergo, prior to reviewing current knowledge on behavioral thermoregulation from a thermoeffector standpoint, it is essential to discuss certain physiological principles of thermosensors relevant to behavioral thermoregulation – including the location, nature and function of central thermosensitive neurons in the preoptic anterior hypothalamus and peripheral thermosensory neurons that detect shell temperatures in the skin and in the oral and urogenital mucosa. Some, if not all, of the central and peripheral thermosensors discussed are also activated during autonomic and endocrine thermoregulatory adaptations, evidence of a signal overlap amongst the behavioral, autonomic and endocrine branches of the thermoregulatory system. The thermosensitive neurons dedicated only on behavioral thermoregulatory responses are also discussed, yet information on them remains limited [12].

Central Thermosensors

Thermal constancy is restricted to the core of the body, while tissues of the body's shell vary in temperature as blood flow may be increased or decreased in order to augment or limit thermal insulation, respectively, based on the internal and the ambient thermal environments. In vivo direct thermal stimulation (heat or cold) of the preoptic area and anterior hypothalamus in homoeotherms has been shown to reflexively evoke various behavioral thermoregulatory responses [12, 17-19]. Similar thermoregulatory responses have also been reported during thermal stimulation in the medulla oblongata, pons, midbrain [20-22], as well as the peridural space of the vertebral canal and of the spinal cord [12, 23-25]. These thermosensitive areas of the brain and the central nervous system have been recognized for more than a century as key components of thermoregulation based on observations from humans suffering brain damage and from the effect of experimental lesions in other mammalian models.

Peripheral Thermosensors

Beyond the superficial cold- and warm-sensitive neurons, research has shown that homoeotherms demonstrate peripheral deep-body thermosensors located in the stomach, oesophagus, large intra-abdominal veins, skeletal muscle, and other organs [12, 26]. Evidence for the significant contribution of extra-cerebral deep-body thermosensors to the control of shivering were presented for the first time by Thauer [27] who showed that, despite their distinct effect on behavioral, autonomic and endocrine thermoregulatory responses, the neuronal activities signalling the level of deep-body temperature largely remain below the level of consciousness [28]. Because of the latter as well as limitations of the thermode technique used to detect them, triggering of corresponding thermoregulatory responses to local or general changes in core body temperature (e.g., alterations in heat dissipation or generation in response to induced temperature changes) remains the only unanimously accepted criterion for the existence of functionally specific deep-body thermosensors [26, 28].

Our knowledge on the peripheral mechanisms by which we sense hot and cold temperatures has been recently updated with the discovery that ion channels of the transient receptor potential family are participating in thermosensation and are, direct or indirect, contributors to thermoregulation [12, 29]. The mammalian transient receptor potential superfamily consists of ~30 channels that are divided into six subfamilies [12, 29]. A detailed discussion of the nature and functioning of these proteins as endogenous thermosensors is beyond the scope of this chapter, yet it is important to mention that certain invertebrate transient receptor potential channels are involved in behavioral thermoregulation [12]. However, much remains to be learned regarding the involvement of these channels in the multitude of homoeothermic thermosensory functions.

Behavioral Thermosensors

It is unanimously agreed [30] that the genesis of intensive experimentation and theory regarding the functional architecture of the human thermoregulatory system is marked by the influential experiments of Barbour [31] who used the rabbit model to show that cooling a brain stem region near to what is termed today the preoptic area and anterior hypothalamus had profound effects on rectal temperature, and that the animal demonstrated marked behavioral responses. This discovery fitted exquisitely within the theory of 'brain centers' – that is, specific areas in the brain eliciting topical control of somato-sensory perception and somato-motor activity – which prevailed during those times (i.e., the beginning of the 20th century) and determined the direction of mainstream research on thermoregulation for the next 50 years. Following that time period, a number of experiments have been conducted on the underlying psychophysical mechanism(s) controlling thermoregulation, alas the results have been controversial. For instance, some of the early experiments in the area suggested that mean skin temperature *per se* is not a dependable criterion of behavioral thermoregulation [32] and that the main control signal is of internal origin [33]. This is in line with the finding that thermal stimulation of the preoptic area and anterior hypothalamus trigger behavioral thermoregulatory responses such as postural changes that increase/decrease heat loss and moving to a warmer/cooler area [34, 35]. Data from stimulation experiments adopting the rat model suggest that different thermoregulatory behaviors such as thermoregulatory grooming (i.e., spreading saliva on the skin to increase evaporative heat loss), relaxed postural extension, and locomotion use distinct neural circuitries [36], while neurons in the median preoptic nucleus are implicated in moving to a reward zone during heat exposure to trigger a breeze of cold air [37].

An important distinction between autonomic/endocrine and behavioral thermoregulatory responses is that the latter do not require an intact preoptic area and anterior hypothalamus [30, 38-40]. Indeed, the posterior hypothalamus and the medulla oblongata also appear to be involved in behavioral thermoregulation since preoptic area and anterior hypothalamus lesions may influence autonomic reflexes but leave behavioral thermoregulation intact [22, 30, 38, 39, 41, 42]. Furthermore, experiments using functional magnetic resonance imaging and positron emission tomography demonstrated that the signals for behavioral thermoregulation involved several areas in the cerebral cortex including the insular, cingulated, orbiofrontal, primary and secondary somatosensory cortex [30, 43]. In contrast, other experiments using similar techniques demonstrated the involvement of amygdala in cold-induced thermal discomfort but no activation of the thalamus, somatosensory, cingulated, or in the insular corex [44]. Indeed, the amygdala is considered vital in the control of emotional behavior that is related to maintaining body homeostasis systems such as food intake, aversive behavior, and drinking [45].

Relatively recent neurophysiological developments revealed that neurons of the dorsomedial hypothalamic nucleus and fibers passing through the paraventricular hypothalamic nucleus are involved in cold-seeking behavior [40]. Additional experiments in the same brain areas demonstrated that disinhibition of neurons in the dorsomedial hypothalamus by local microinjection of the $GABA_A$ receptor antagonist bicuculline methiodide evokes a host of behavioral, autonomic, and endocrine thermoregulatory responses [30, 46, 47]. It appears that neurons in the dorsomedial hypothalamus activate premotor sympathetic neurons in the rostral raphe pallidus that are involved in elevating body temperature, in cold defense, and in fever [47]. Based on these findings it has been proposed that neurons in the dorsomedial hypothalamus play a vital role in a variety of thermoregulatory responses, including those based on behavior [30, 47]. Indeed, the dorsomedial hypothalamus is involved in chemical signals related to food consumption and ingestive behavior both of which represent hypothalamic functions linked with thermoregulation [48]. For instance, leptin, a peptide involved in feeding behavior, increases body temperature in mice and rats [49-52] and is implicated in the human behavioral, autonomic and endocrine control of circadian body temperature rhythm [53]. Administration of two other peptides, vasopressin and oxytocin, through intracerebroventricular injection induces grooming behavior, characteristic for heat loss [54, 55].

Although several experiments have presented data in favor of a central control of behavioral thermoregulation, other authors reported that this mechanism is independent of core body temperature [56], or that thermal pleasantness sensation is a function of both mean skin temperature and core body temperature [57]. Several studies examined different deep-body compartments in order to detect extra-cerebral deep-body thermosensitivity based on the initial observation by Simon and colleagues of shivering in response to cooling of the peridural space of the vertebral canal and of the spinal cord [25]. This notion was in contrast to the popular view which favored the straightforward and, hence, convenient idea of only one region of deep-body thermosensitivity. Ergo, to prove the involvement of the spinal cord or the vertebral canal in thermoregulation it was necessary to ensure that hypothalamic temperature remained constant (or increased) when the spinal cord was cooled. This was achieved in several experiments which clearly established thermoregulatory functions of extra-cerebral deep-body compartments [58-60] demonstrating that: (i) afferent thermosensory input from the spinal cord interact with thermal inputs from the preoptic area and anterior hypothalamus, (ii) thermoregulatory reactions, including conscious behavior, influenced by hypothalamic temperature and mean skin temperature are also influenced by the temperature of the spinal cord, (iii) in the dog, thermal inputs from the spinal cord and the preoptic area and anterior hypothalamus were functionally identical and quantitatively similar. These results were confirmed by experiments adopting the rabbit model which showed that hypothalamic temperature interacts with skin and abdominal viscera temperatures in governing thermoregulation [61-63]. Additional experiments in pigs, sheep, and squirrel monkeys also supported the involvement of extra-cerebral areas including the spinal cord, the abdominal viscera, as well as the colon in behavioral thermoregulation [19, 64-66]. The properties of the behavioral thermosensors discovered to date are summarized in Table 1.

The notion of extra-cerebral deep-body thermosensitivity received support from several experiments demonstrating that behavioral thermoregulation is primarily regulated by alterations in mean skin temperature [67, 68] and that an increase in body temperature during exercise triggers behavioral responses to lower work rate in order to maintain a similar body heat storage [69]. According to this premise, mean skin temperature seems to carry important information to the brain to regulate exercise behavior – and thus metabolic rate – independently of core body temperature [69, 70]. Indeed, based on one of the most recent reviews on the topic [12], it appears that behavioral thermoregulation depends more on shell temperature signals from peripheral thermosensors than core temperature signals from central thermosensors. In contrast, deep core body temperature is comparatively more important for eliciting autonomic and endocrine responses. This arrangement mirrors the fact that behavioral responses are most often prescient in that they are aimed at preventing an anticipated thermal insult [30]. On the other hand, autonomic and endocrine thermoregulatory responses are usually employed when body heat content starts to change because behavioral responses were inadequate or could not be used due to conflicting behavioral demands.

Table 1. Properties of behavioral thermosensors discovered to date

	Location	Function	Species	Reference
Central	Preoptic area; anterior hypothalamus	Postural changes; moving to a warmer/cooler area; grooming	Rat	[34-37]
	Insular; cingulate primary/secondary somatosensory, orbiofrontal cortex	Various thermoregulatory behaviors	Human	[30, 43]
	Amygdala	Cold-induced thermal discomfort	Human	[44]
	Dorsomedial hypothalamus; paraventricular hypothalamic nucleus	Cold-seeking behavior; food consumption; ingestive behavior	Rat	[30, 40, 46, 47]
Periphera l	Peridural space of the spinal cord and the vertebral canal	Various behavioral thermoregulatory responses	Monkey	[12, 23-25]
	Spinal cord; abdominal viscera; colon	Various behavioral thermoregulatory responses	Pig; sheep; squirrel monkey	[19, 64-66]

As mentioned earlier in this section, research on the underlying psychophysical mechanism(s) controlling thermoregulation remains limited and controversial, especially regarding the neuroanatomy and dependable criteria of human behavioral thermoregulation [71]. Possible reasons for this may be the lack of appropriate techniques to measure directly thermal sensation and/or comfort in animals and humans as well as the complexity in the body's responses to different stimuli and the difficulty to explicitly identify behavioral thermoregulation and isolate it from other systems within the body, as well as from external dynamics. In a recent experiment in my laboratory we attempted to address these challenges by supplying the body with an additional thermoregulatory system that can be accurately and objectively quantified. Situating a human in an environment that forced its use, information gained from this hybrid system provided valuable evidence on the fundamental principles that govern behavioral thermoregulation. More specifically, given that liquid conditioning garment systems have been successfully implemented in the past to manipulate thermal balance [15], we monitored and modeled the fluctuations in temperature of individuals given voluntary control of the inlet temperature of their garments during prolonged passive cold exposure interspersed with a period of moderate exercise. Our purpose in this experiment was to detect the dependable criteria of behavioral thermoregulation based on which the conscious response to thermal input is generated. In order to eliminate psychological bias, the participants were informed of the specific purpose of the study only after the end of data collection.

The results of this experiment revealed that the human conscious response to thermoregulatory input during both rest and exercise in a cold environment is adjusted to changes in mean body temperature. Indeed, as body core and skin areas seemed to counteract one another, mean body temperature as well as subjective thermal comfort and thermal sensation remained comparatively stable across time. Ergo, our results showed a strong association between conscious thermoregulatory responses (i.e., changes in liquid conditioning garment inlet water temperature) and changes in mean body temperature, rather than temperature changes in individual body regions.

Previous reports in the germane literature suggested that autonomic effector response thresholds depend upon the integration of core body temperature and mean skin temperature [56, 57]. Our results support this notion demonstrating that behavioral thermoregulatory responses attempt to maintain thermoneutrality by achieving a compromise between mean skin temperature and core body temperature rather than core body temperature alone. The mid-exposure exercise bout significantly increased core body temperature and should have normally resulted in substantial heat gain [we have previously recorded a heat gain of approximately 137 W for a 75-kg individual from this exercise bout in similar settings [72]]. However, in response to the increased metabolic rate, the participants – without any psychological bias – lowered the inlet water temperature of their liquid conditioning garments and thus achieved a lower mean skin temperature, with core body temperature and mean skin temperature almost perfectly equalizing one another again (for a 75-kg individual, based on the average change in mean body temperature at the end of the exercise regimen body heat content was decreased by 1.3 W). Recent work on behavioral responses to an increase in body temperature during exercise [69] showed that individuals lower their exercise work rate in order to maintain a similar rate of heat storage. These results are in line with the findings from our experiment demonstrating that individuals lowered inlet water temperature during exercise resulting in minimal change in mean body temperature.

Conscious Behavior as a Thermoeffector

With the exception of skin vasomotor control which is constantly active, homoeothermic organisms rely primarily on behavioral thermoregulation, mediated through conscious responses, and only secondarily on autonomic and endocrine thermoregulation [56, 73-75]. This is because autonomic and endocrine responses have a limited capacity in preventing hyper/hypothermia in extreme environments. Thus, behavioral thermoregulation – which is a continuously active mechanism – is utilized to trigger conscious decisions in order to preserve thermal homeostasis.

As mentioned in the previous section, it appears that behavioral thermoregulation relies primarily on shell temperature signals from peripheral thermosensors [12, 30]. In contrast, deep core body temperature is comparatively more important for eliciting autonomic and endocrine responses. This arrangement mirrors the fact that behavioral thermoregulation is a prescient mechanism in that it is generally activated prior to any change in body heat content occurs [30]. Simple behavioral thermoregulatory responses found in a multitude of species include changes in orientation to the direction of the wind/sun, positional measures like curling up, cuddling, and huddling in order to alter the surface/mass ratio and, hence, the rate of heat exchange between the organism and the environment, as well as search for sunny, dry, warm, wind-sheltered environments and avoidance of dangerous/hostile environmental temperatures. More complex behavioral thermoregulatory responses include building a nest, spreading saliva on the skin to increase evaporative heat loss (grooming), the creation and application of clothing, or, simply, turning on or off the air conditioner. Furthermore, primary alterations of consumption behavior can significantly alter body heat content and are, thus, considered behavioral thermoregulatory responses. For instance, it has been shown that food intake increases in the cold in order to increase metabolic rate and compensate for the excessive heat loss in order to maintain thermal balance [76]. On the other end of the spectrum, febrile illnesses, seen as principally thermoregulatory phenomena, are accompanied by anorexia [76]. Unfortunately, the physiology of this behaviourally-based regulation remains to be elucidated.

Testing the Effectiveness of Behavioral Thermoeffector Mechanisms

Microclimate Liquid Conditioning Garment Systems

Behavioral thermoeffector responses have been employed as the basis of many bioengineering applications including 'smart' buildings [77] and microclimate liquid conditioning systems worn underneath protective clothing [15]. With regards to the latter, behavioral thermoregulatory responses are used for the control of temperature and cooling distribution which are very important elements for microclimate liquid conditioning systems efficiency. Optimization attempts for temperature control have been conducted since the early years of the microclimate liquid conditioning system project and examined the potential for implementing an automatic control of coolant temperature [78-83]. This is mainly due to the complexity of tasks often executed by individuals wearing a microclimate liquid conditioning system (such as pilots and astronauts). Additions to the regular tasks executed by the wearer inevitably affects his/her physical performance and has a significant impact on his/her ability to successfully accomplish primary tasks and attend to relevant input. Moreover, although behavioral thermoregulatory responses are generally very effective, results from several studies have demonstrated that the manual/subjective microclimate liquid conditioning system control, although acceptable, is not optimal [84-86]. This may be because the environment can influence perceptual sensations of heat/cold stress leading to variations between perceptual and physiological strain which may result in inappropriate behavioral thermoregulatory response [87, 88]. In addition, an individual's physical condition has been shown to significantly alter his/her perception of physiological thermal strain [89]. Ergo, the need for applying an automatic control of coolant temperature to the liquid conditioning system has been repeatedly highlighted and there has been a number of attempts to design such technology [79, 81, 82, 84-86, 90-96]. However, even the most advanced microclimate liquid conditioning systems currently used by astronauts are equipped with a three- or four-position manual control.

Hand Function

As human thermoregulation in cold environments has generated substantial interest amongst physiologists during the past 80 years with several experiments demonstrating marked and progressive impairments in fine and gross manual dexterity as early as the first five minutes of induced local cooling [97-99], along with rapid impairments of muscular function upon exposure to cold environments [100]. Proposed mechanisms to explain these events involve thermal state of the small muscles of the hand [101], between-finger sensory integration [102], as well as nerve conduction velocity and finger –synovial fluid viscosity [101], –tactile sensitivity [103], and –blood flow [104]. These mechanisms have been examined through manipulations of thermal balance during cold exposure by applying added heat either indirectly (i.e., heating another region to increase body heat content and augment vasodilation in the extremities) or directly (i.e., electrically heated gloves or auxiliary heating) [98, 99, 104-111].

In an experiment conducted recently by my research team [72], we assessed the effectiveness of behavioral thermoregulatory responses in preserving the neuromuscular function and manual dexterity of the hands during a passive cold exposure interspersed by a short period of sub-maximal exercise. We compared this behavioral maintenance of thermal balance with (i) added heat applied prior to the cold exposure and (ii) no added heat either prior to or during the cold exposure. The added heat was applied with a liquid conditioning garment [15]. Based on the premise that maintenance of hand function represented part of system integrity and would, thus, be defended through behavioral thermoregulatory responses, a secondary objective was to detect the dependable criteria of behavioral thermoregulation by quantifying and comparing the relationships between either body heat storage, or change in body heat content, or mean finger temperature and hand function.

The results from this experiments revealed that the added heat applied both prior to and throughout the cold exposure maintained twitch force, maximum voluntary contraction, hand grip, as well as finger tapping during a 130-minute cold exposure to -20°C. Decrements in Purdue Pegboard performance were prevented for approximately 54 minutes when added heat was applied prior to the exposure, while no decrements were observed when body heat content was increased throughout the cold exposure through behavioral control of the liquid conditioning garment. However, it was be noted that the benefits of added heat was applied prior to the exposure weakened over time and the overall advantage was less than that from maintaining body heat content throughout cold exposure. With regards to our secondary objective, results suggested that the best indicator of hand function was the change in body heat content followed by mean finger temperature while body heat storage was not significantly associated with any of the manual dexterity parameters. These results suggested that, under these experimental conditions, behavioral responses to maintain thermal balance were based firstly on changes in body heat content and, secondly, on local tissue temperature. On the other hand, we found no significant association between any of the thermal variables and the parameters of neuromuscular function. The latter showed a gradual decrease over time which corresponded to the general thermal body state, but this association did not reach statistical significance. Therefore, it appears that although maintenance of thermal balance does influence force generation during cold exposure, this link is not as clear as in the case of manual dexterity tasks that require precise coordination of force in both time and space, and are also influenced by factors external to thermoregulation such as vibrotactile sensitivity.

Cognitive Performance

In circumstances where behavioral responses are unable to prevent changes in thermal balance, autonomic and endocrine adaptations are triggered, albeit with some degree of risk to system integrity because these adaptations have a limited capacity in preventing hyper/hypothermia in extreme environments. It is not surprising, therefore, that cognition and mental performance deteriorate with increases in environmental temperature and disruptions in thermal homeostasis [112]. For instance, vigilance – the ability to sustain attention for prolonged periods of time – is an important component of mental performance for many contemporary workers (e.g., military, offshore workers) routinely exposed to thermal stress that is undermined by changes in thermal balance [112]. Most of the documented effects of cold ambient temperatures and/or hypothermia on vigilance suggest a delay in reaction time and an increase in the number of errors incurred during target detection; indeed, the most recent meta-analysis in this area concluded that under cold ambient conditions (≤10 °C) performance in vigilance tasks is markedly impaired [113]. Yet the magnitude of disturbance in thermal balance that generates performance decrements in vigilance has remained elusive to date. Moreover, little research has explored the possibility that vigilance decrements may be reduced through the use of thermal countermeasures.

In a recent experiment my research team examined the influence of body heat content on vigilance during a passive 130-minute cold exposure interspersed by a short period of sub-maximal exercise [14]. Body heat content was manipulated by applying added heat either prior to or throughout the cold exposure using a liquid conditioning garment system [15]. We expected that cold exposure would lead to a deteriorated vigilance performance over time, as measured by reaction time to detect visual targets embedded in the stimulus sequences based on previous research showing detrimental effects of cold on vigilance [112, 113]. Moreover, we anticipated that increases in body heat content not only throughout but also prior to the cold exposure would be effective in preserving vigilance performance based on a proposed relationship between maintenance of thermal balance and increased levels of vigilance [113, 114].

Results from this experiment revealed a decline in the number of hits as well as significantly deteriorated reaction time and reaction consistency as early as 45 minutes into the cold exposure [14]. At approximately the same time, the signal detection criterion showed marked alterations indicating a bias in favor of 'target-present' responses (i.e., that a target was detected) over time. Furthermore, we also found that that increasing body heat content not only throughout but also prior to the cold exposure was effective in preserving vigilance at baseline levels. Specifically the manipulations of thermal balance in our experiment were beneficial to reaction time, hits and false alarms. Interestingly, our analyses suggested that increasing body heat content prior to the cold exposure was generally more effective than increasing body heat content throughout the cold exposure in maintaining reaction time and reaction consistency at baseline levels [14].

Based on the data from this experiment we were unable to elucidate the underlying mechanism(s) involved in the beneficial effect of added heat on limiting the bias in favor of 'target-present' responses (i.e., that a target was detected), but our findings supported a link between thermal homeostasis and perceptual judgments. A comprehensive understanding of this link is hampered by the lack of published data on the effects of ambient and/or body temperature on signal detection measures of vigilance. To date, the results of previous research on vigilance reaction time are conflicting, but the lack of accuracy and increased false alarm rate seems to be a consistent finding in response to cold [115-118]. Our results suggested that changes in performance during vigilance tasks may reflect cold-induced reductions in the threshold used to guide positive responding, which may have implications for decision-making and complex reasoning during cold exposures. In our experiment, these adverse changes in judgmental systems were related to skin temperature and thermal comfort vote but not rectal temperature [14].

Another interesting finding generated by this experiment involved the effects of sex and body mass on vigilance. Results from non-thermal research proposed that women exhibit slower reaction time in vigilance tasks than men [119-121]. In our experiment, the baseline reaction time, consistency of reaction time and the bias in favor of 'target-present' responses partly supported these findings, yet the sex differences in our experiment did not reach statistical significance. During the cold exposure, however, significant differences were observed between sexes with reaction time, consistency of reaction time and the bias in favor of 'target-present' responses showing more pronounced signs of deterioration in women. We attributed these findings to differences in body mass, size, muscle mass, and body geometry between men and women which may have confounded heat loss/gain and, thus, the effects of cold and/or thermal manipulation on vigilance performance based on significant associations between body mass index and vigilance performance in both men and women. More importantly, given that the rate of body heat loss in women was significantly increased compared to males, we postulated that this increased heat loss from an already-decreased body heat content at baseline resulted in the more pronounced signs of vigilance deterioration in women [14].

In summary, from this experiment we concluded that: (i) vigilance is markedly deteriorated in cold conditions within 45 minutes of exposure; (ii) increasing body heat content prior to a cold exposure is efficacious in preserving vigilance performance during exposures with a maximum duration of two hours (iii) the mechanism that underlies the protective effects of increased body heat content prior to and/or during cold exposures on vigilance may be related to decision-making systems. Based these conclusions, operators working in cold environments for maximum durations of an hour may be better able to sustain vigilance/attention and thus performance if they are pre-heated prior to exposure. For longer exposures, continuous maintenance of body heat content is needed.

Regulation of Exercise Intensity

If the human organism represented a simplistic system, a command for continuing muscular effort would be executed until system collapse, whether that is of thermal, chemical, or of different origin. Yet, based on a proposed feedforward anticipatory model of behavior during exercise in the heat, humans circumvent the physiological impact of exercise-induced hyperthermia through feedforward and feedback voluntary control of effort at the start or during heat stress [122]. In this context, fatigue is viewed as a conscious or unconscious behavioral command to lower or cease work rate in order to prevent the body from reaching a thermal point of near-collapse and occurs well before the attainment of physiological impairment to exercise through neuromuscular or other physiological mechanisms [122, 123].

The 'central governor model' introduced in the previous paragraph suggests that voluntary effort during exercise is regulated by a central nervous system governor with the aim of preserving system thermal homeostasis and has received ample support thus far [69, 70, 122, 124-126]. It has been shown, for instance, that within the first 30% of a maximal self-paced time-trial in the heat, power output and skeletal muscle recruitment begin to fall reaching statistical significance after 80% of the trial is completed [70]. More importantly, these phenomena are independent of rectal temperature, perception of effort or heart rate and occur well before rectal temperature reaches 40°C, the generally accepted thermal point of near-collapse [70, 123]. Based on these results, the decreases in power output and skeletal muscle recruitment occur as part of a centrally-originated neural mechanism that predicts abnormal increases in body temperature and alters skeletal muscle recruitment to allow exercise completion and maintenance of thermal homeostasis. Similar results have been reported in an experiment showing a reduction in power output following the first 15 minutes of a 30-minunte, self-paced cycling time-trial in the heat [127]. This was despite similar rates of increase in core temperature in a hot and temperate environment. Consequently, the investigators suggested that the brain chooses a power output relative to the rate core temperature rise, based on information gained from the rate of increase in arterial blood temperature [127].

The 'central governor model' is supported, indirectly, by research in exercise science showing that performance feedback, or knowledge of results, can enhance motivation and goal-setting with beneficial effects on self-efficacy and physiological arousal [128]. This is the main reason why performance feedback is incorporated in field testing used for the evaluation of cardiorespiratory fitness [129, 130]. Furthermore, performance feedback aids test administrators by providing them with valid information on the participant's performance. In a recent experiment, we explored these phenomena in the 20m Multistage Shuttle Run Test and the 20m Square Shuttle Test, two field tests designed to predict cardiorespiratory fitness. Both tests incorporate running on a predetermined course, using performance feedback emitted from an audio-CD. The 20m Multistage Shuttle Run Test is based on shuttle running [130], while the 20m Square Shuttle Test has adopted a more continuous elliptic running-mode [129]. Performance feedback (i.e., test-stage information and sound signals determining the running pace) has been included in these testing protocols to motivate individuals maintain the prescribed pace until not physically able to continue. The results of this experiment demonstrated substantial performance decrements (e.g., >5 beats·min^{-1} in heart rate at the end of the test) when performance feedback was absent in both tests, a phenomenon indicating non-maximal effort [131]. Moreover, it appeared that the levels of performance in the cardiorespiratory fitness field tests studied depended on the provided feedback. Finally, absence of performance feedback reduced the validity and precision of the two tests in predicting laboratory-determined maximal oxygen uptake. These results provide partial support to the 'central governor model' showing that performance feedback, or knowledge of results, is a significant factor in the regulation of human exercise performance and in stimulating maximal performance.

The relatively recent 'central governor model' comes in contrast to the traditionally-held view according to which the heart's pumping capacity, not peripheral factors in the muscles, is the factor limiting effort during exercise in the heat [132-135]. Other factors, including limiting blood supply, oxygen delivery or fuel utilization – all involving a limit of the cardiovascular system – have been also proposed [136], yet it is now known that such homeostatic failures do not occur, as skeletal muscle blood flow and metabolism do not reach limiting values in the heat [136, 137]. However, the debate regarding the cause of fatigue during exercise in the heat continues strong with experiments from Brink-Elfegoun and colleagues showing that there can be an increase in skeletal muscle recruitment despite unchanged maximal oxygen consumption [134, 135]. Based on these findings, Brink-Elfegoun and colleagues believe that voluntary effort is not limited by a central governor but by the central circulation [134, 135].

From a philosophical standpoint, the 'central governor model' suggests that effort or fatigue during exercise in the heat is regulated as an active behavioral mechanism. On the other hand, the traditionally-held view holds that effort or fatigue during exercise in the heat represents a passive limit in the system's capacity to transform chemical energy into mechanical. However, it is possible that the two measures of control can harmoniously coexist within a system, the former being a prescient behavioral mechanism preventing imminent myocardial ischemia by limiting peripheral blood flow, and the latter being a safety measure employed when behavioral means are insufficient. This arrangement fits precisely the organization of the endothermic thermoregulatory system with behavioral responses, depended on shell temperature signals from peripheral thermosensors, deployed to prevent anticipated thermal insults [138], while autonomic and endocrine responses, based more on core temperature signals from central thermosensors, employed when body heat content changes because behavioral responses were inadequate or could not be used due to conflicting behavioral demands [12]. For instance, recent findings reported a centrally-originating progressive impairment in neuromuscular activation, independent of local muscle temperature, occurring prior to temperature-induced system catastrophe [139]. If this hypothesis is true, however, the 'central governor model' would be more prevalent as factor regulating effort and/or fatigue, since the mechanism described in the traditionally-held view would be employed only prior to system collapse. This would explain the increased support that the 'central governor model' has received to date [69, 70, 122, 124-126].

Conclusion

A brief review of the germane literature shows that in recent years there has been remarkable progress in the area of behavioral thermoregulation with fundamental and applied experiments providing novel evidence on the underlying psychophysical mechanism(s) controlling endothermic thermoregulation. Prominent new insights include the important distinction between autonomic/endocrine and behavioral thermoregulatory responses whereby the latter do not require an intact preoptic area and anterior hypothalamus [30, 40] and the relevant discoveries of several cerebral loci involved in behavioral thermoregulation including the insular, cingulate, and primary/secondary somatosensory cortex [30, 43], the amygdala [44], as well as the dorsomedial hypothalamus [30, 40, 46, 47]. On the other hand, applied experiments have provided strong evidence of the involvement of extra-cerebral thermosensitivity on behavioral thermoregulation and the role of skin temperature [67, 68] demonstrating that behavioral thermoregulatory responses defend changes in body heat content [14, 69, 70, 72, 138]. Ergo, there is now enough evidence to postulate that the dependable criterion of behavioral thermoregulation is whole-body thermal homeostasis and not the thermal status of certain cerebral or extra-cerebral areas.

Notwithstanding the recent attention on behavioral thermoregulation and the excitement for the new discoveries in this area, we remain largely naive to the principles governing the psychophysical mechanism(s) of endothermic thermoregulation. Many pertinent questions remain unanswered including the role of individual ion channels of the transient receptor potential family in the control of different behavioral thermoeffectors, the afferent and efferent neural circuits of behavioral thermoregulatory reactions and the neuroanatomy of behavioral thermoregulation, as well as the exact neuroanatomic substrate of specific thermoregulatory behaviors. These include potential avenues of future research in this area. Increasing our knowledge on the principles that govern behavioral thermoregulation will contribute to our understanding of the functional architecture of the endothermic thermoregulatory system as well as the endothermic thermoregulatory responses under exposures to harsh environments and will improve current bioengineering and clinical applications germane to acute or chronic deviations from thermal homeostasis.

An important aspect that must be remembered is that all thermoregulatory models rely on humans to prove true; yet humans in all their emotional, irrational glory are hard to predict. Thermoregulatory models that once enjoyed wide acceptance, have revealed a plethora of problems – we now realize [12]. They were too focused on the body's core, they didn't account for the physiological complexity of the entire organism, but perhaps most of all, they were simply too reliant on automation. In other words, they did not sufficiently account for human behavior. Fortunately, this situation has started to change. Looking ahead, one can't but help wonder what that means for the future of thermal biology.

References

[1] IUPS Thermal Commission. Glossary of terms for thermal physiology, 3rd ed. *Jap J Physiol.* 2001;51(2):245-80.

[2] Grigg GC, Beard LA, Augee ML. The evolution of endothermy and its diversity in mammals and birds. *Physiol Biochem Zool.* 2004 Nov-Dec;77(6):982-97.

[3] Bennett AF, Ruben JA. Endothermy and activity in vertebrates. *Science.* 1979 Nov 9;206(4419):649-54.

[4] Ivanov KP. The development of the concepts of homeothermy and thermoregulation. *J Therm Biol.* 2006;31:24–9.

[5] Crompton AW, Taylor CR, Jagger JA. Evolution of homeothermy in mammals. *Nature.* 1978 Mar 23;272(5651):333-6.

[6] Burton AC, Edholm OG. *Man in a cold environment.* London: Edward Arnold, Ltd.; 1955.

[7] Voets T, Droogmans G, Wissenbach U, Janssens A, Flockerzi V, Nilius B. The principle of temperature-dependent gating in cold- and heat-sensitive TRP channels. *Nature.* 2004 Aug 12;430(7001):748-54.

[8] Riggs AJ, Millecchia RJ, Riggs JE. The contributions of Lavoisier, Carnot, and Mayer to understanding heat illness. *Aviat Space Environ Med.* 2004 Oct;75(10):916-7.

[9] Greene CW. *Kirkes' Handbook of Physiology.* Revised by C. W. Greene. New York: W. Wood; 1913.

[10] Ott I. Heat center in the brain. *J Nerv Ment Dis.* 1887;14:152.

[11] Richet C. Die beziehung des gehirns zur korperwarme und zum fiever. *Arch Ges Physiol.* 1885;37:624.

[12] Romanovsky AA. Thermoregulation: some concepts have changed. Functional architecture of the thermoregulatory system. *Am J Physiol Regul Integr Comp Physiol.* 2007 Jan;292(1):R37-46.

[13] Burton AC. The clinical importance of the physiology of temperature regulation. *Can Med Assoc J.* 1956 Nov 1;75(9):715-20.

[14] Flouris AD, Westwood DA, Cheung SS. Thermal balance effects on vigilance during 2-hour exposures to −20°C. *Aviat Space Environ Med.* 2007;78(7):679-79.

[15] Flouris AD, Cheung SS. Design and control optimization of microclimate liquid cooling systems underneath protective clothing. *Ann Biomed Eng.* 2006 Feb 7;34(3):359-72.

[16] Cabanac M, Serres P. Peripheral heat as a reward for heart rate response in the curarized rat. *J Comp Physiol Psychol.* 1976 May;90(5):435-41.

[17] Hammel HT, Hardy JD, Fusco MM. Thermoregulatory responses to hypothalamic cooling in unanesthetized dogs. *Am J Physiol.* 1960 Mar;198:481-6.

[18] Satinoff E. Behavioral thermoregulation in response to local cooling of the rat brain. *Am J Physiol.* 1964 Jun;206:1389-94.

[19] Carlisle HJ, Ingram DL. The effects of heating and cooling the spinal cord and hypothalamus on thermoregulatory behaviour in the pig. *J Physiol.* 1973 Jun;231(2):353-64.

[20] Holmes RL, Newman PP, Wolstencroft JH. A heatsensitive region in the medulla. *J Physiol.* 1960 Jun;152:93-8.

[21] Newman PP, Wolstencroft JH. Cardiovascular and respiratory responses to heating the carotid blood. J Physiol. 1960 Jun;152:87-92.

[22] Lipton JM. Thermosensitivity of medulla oblongata in control of body temperature. *Am J Physiol.* 1973 Apr;224(4):890-7.

[23] Simon E. Temperature regulation: the spinal cord as a site of extrahypothalamic thermoregulatory functions. *Rev Physiol Biochem Pharmacol.* 1974(71):1-76.

[24] Chai CY, Lin MT. Effects of heating and cooling the spinal cord and medulla oblongata on thermoregulation in monkeys. *J Physiol.* 1972 Sep;225(2):297-308.

[25] Simon E, Rautenberg W, Thauer R, Iriki M. Ausloesung thermoregulatorischer Reaktionen durch lokale Kuehlung im Vertebralkanal (Stimulation of thermoregulatory reactions by local cooling within the vertebral canal). *Naturwissenschaften.* 1963;50:337.

[26] Simon E. Deep-body thermosensitivity: Another look. A tribute to Harold Theodore (Ted) Hammel (1921–2005). *J Therm Biol.* 2006;31:4–18.

[27] Thauer R. Die Ausloesungsmechanismen der chemischen Temperaturregulation (Mechanisms stimulating chemical temperature regulation). *Proceedings of the 22nd Congress of the International Union of Physiological Sciences* (IUPS); 1962; Volume 1, part 1. Leiden, Netherlands; 1962. p. 390–402.

[28] Simon E. The enigma of deep-body thermosensory specificity. *Int J Biometeorol.* 2000 Sep;44(3):105-20.

[29] Caterina MJ. Transient receptor potential ion channels as participants in thermosensation and thermoregulation. *Am J Physiol Regul Integr Comp Physiol.* 2007 Jan;292(1):R64-76.

[30] Nagashima K. Central mechanisms for thermoregulation in a hot environment. *Ind Health.* 2006 Jul;44(3):359-67.

[31] Barbour HG. The effect of direct warming and cooling of the thermoregulatory centers on body temperature (translated from the German title: Die Wirkung unmittelbarer Erwärmung und Abkühlung der Wärmezentra auf die Körpertemperatur). *Naunyn-Schmiedeberg's Arch.* 1912;70:1-26.

[32] Topliff ED, Livingstone SD. Thermal comfort in relation to mean skin temperature. *Can J Physiol Pharmacol.* 1970 Feb;48(2):98-101.

[33] Cabanac M, Cunningham DJ, Stolwijk JA. Thermorequlatory set point during exercise: a behavioral approach. *J Comp Physiol Psychol.* 1971 Jul;76(1):94-102.

[34] Hensel H. *Thermoreception and temperature regulation.* London: Academic Press; 1981.

[35] Boulant JA. Celllular mechanisms of neuronal thermosensitivity. *J Therm Biol.* 1999;24:333–38.

[36] Roberts WW. Differential thermosensor control of thermoregulatory grooming, locomotion, and relaxed postural extension. *Ann N Y Acad Sci.* 1988;525:363-74.

[37] Konishi M, Kanosue K, Kano M, Kobayashi A, Nagashima K. The median preoptic nucleus is involved in the facilitation of heat escape/cold-seeking behavior during systemic salt loading in rats. *Am J Physiol Regul Integr Comp Physiol.* 2007 Jan;292(1):R150-9.

[38] Carlisle HJ. Behavioural significance of hypothalamic temperature-sensitive cells. *Nature.* 1966 Mar 26;209(5030):1324-5.

[39] Carlisle HJ. Effect of preoptic and anterior hypothalamic lesions on behavioral thermoregulation in the cold. *J Comp Physiol Psychol.* 1969 Oct;69(2):391-402.

[40] Almeida MC, Steiner AA, Branco LG, Romanovsky AA. Neural substrate of cold-seeking behavior in endotoxin shock. *PLoS ONE.* 2006;1:e1.

[41] Mercer JB, Simon E. Lessons from the past - human and animal thermal physiology. *J Therm Biol.* 2001;26:249-53.

[42] Adair ER. Hypothalamic control of thermoregulatory behavior. Preoptic-posterior hypothalamic interaction. In: Lederis K, Cooper KE, editors. *Recent Studies of Hypothalamic Function.* Basel, Switzerland: Karger; 1974. p. 341-58.

[43] Craig AD, Bushnell MC, Zhang ET, Blomqvist A. A thalamic nucleus specific for pain and temperature sensation. *Nature.* 1994 Dec 22-29;372(6508):770-3.

[44] Kanosue K, Sadato N, Okada T, Yoda T, Nakai S, Yoshida K, et al. Brain activation during whole body cooling in humans studied with functional magnetic resonance imaging. *Neurosci Lett.* 2002 Aug 30;329(2):157-60.

[45] Rolls ET. *The brain and emotion.* Oxford, UK: Oxford University Press; 1999.

[46] Bailey TW, Dimicco JA. Chemical stimulation of the dorsomedial hypothalamus elevates plasma ACTH in conscious rats. *Am J Physiol Regul Integr Comp Physiol.* 2001 Jan;280(1):R8-15.

[47] Dimicco JA, Zaretsky DV. The dorsomedial hypothalamus: a new player in thermoregulation. *Am J Physiol Regul Integr Comp Physiol.* 2007 Jan;292(1):R47-63.

[48] Hamilton CL, Ciaccia PJ. Hypothalamus, temperature regulation, and feeding in the rat. *Am J Physiol.* 1971 Sep;221(3):800-7.

[49] Stehling O, Doring H, Ertl J, Preibisch G, Schmidt I. Leptin reduces juvenile fat stores by altering the circadian cycle of energy expenditure. *Am J Physiol.* 1996 Dec;271(6 Pt 2):R1770-4.

[50] Stehling O, Doring H, Nuesslein-Hildesheim B, Olbort M, Schmidt I. Leptin does not reduce body fat content but augments cold defense abilities in thermoneutrally reared rat pups. *Pflugers Arch.* 1997 Nov;434(6):694-7.

[51] Pelleymounter MA, Cullen MJ, Baker MB, Hecht R, Winters D, Boone T, et al. Effects of the obese gene product on body weight regulation in ob/ob mice. *Science.* 1995 Jul 28;269(5223):540-3.

[52] Campfield LA, Smith FJ, Guisez Y, Devos R, Burn P. Recombinant mouse OB protein: evidence for a peripheral signal linking adiposity and central neural networks. *Science.* 1995 Jul 28;269(5223):546-9.

[53] Simon C, Gronfier C, Schlienger JL, Brandenberger G. Circadian and ultradian variations of leptin in normal man under continuous enteral nutrition: relationship to sleep and body temperature. *J Clin Endocrinol Metab.* 1998 Jun;83(6):1893-9.

[54] Drago F, Stanciu M, Salehi S, Scapagnini U. The block of central vasopressin V1 but not V2 receptors suppresses grooming behavior and hypothermia induced by intracerebroventricular vasopressin in male rats. *Peptides.* 1997;18(9):1389-92.

[55] Olson BR, Drutarosky MD, Chow MS, Hruby VJ, Stricker EM, Verbalis JG. Oxytocin and an oxytocin agonist administered centrally decrease food intake in rats. *Peptides.* 1991 Jan-Feb;12(1):113-8.

[56] Cabanac M, Massonnet B, Belaiche R. Preferred skin temperature as a function of internal and mean skin temperature. *J Appl Physiol.* 1972 Dec;33(6):699-703.

[57] Attia M, Engel P. Thermal alliesthesial response in man is independent of skin location stimulated. *Physiol Behav.* 1981 Sep;27(3):439-44.

[58] Jessen C, Ludwig O. Spinal cord and hypothalamus as core sensors of temperature in the conscious dog. II. Addition of signals. *Pflugers Arch.* 1971;324(3):205-16.

[59] Jessen C, Mayer ET. Spinal cord and hypothalamus as core sensors of temperature in the conscious dog. I. Equivalence of responses. *Pflugers Arch.* 1971;324(3):189-204.

[60] Jessen C, Simon E. Spinal cord and hypothalamus as core sensors of temperature in the conscious dog. III. Identity of functions. *Pflugers Arch.* 1971;324(3):217-26.

[61] Stitt JT. The regulation of respiratory evaporative heat loss in the rabbit. *J Physiol.* 1976 Jun;258(1):157-71.

[62] Stitt JT, Hardy JD, Stolwijk JA. PGE1 fever: its effect on thermoregulation at different low ambient temperatures. *Am J Physiol.* 1974 Sep;227(3):622-9.

[63] Riedel W, Siaplauras G, Simon E. Intra-abdominal thermosensitivity in the rabbit as compared with spinal thermosensitivity. *Pflugers Arch.* 1973;340(1):59-70.

[64] Adair ER. Skin, preoptic, and core temperatures influence behavioral thermoregulation. J Appl Physiol. 1977 Apr;42(4):559-64.

[65] Rawson RO, Quick KP. Intraabdominal thermoreceptors and the regulation of body temperature of sheep. *Isr J Med Sci.* 1976 Sep;12(9):1040-3.

[66] Adair ER. Displacements of rectal temperature modify behavioral thermoregulation. *Physiol Behav.* 1971 Jul;7(1):21-6.

[67] Cheng C, Matsukawa T, Sessler DI, Ozaki M, Kurz A, Merrifield B, et al. Increasing mean skin temperature linearly reduces the core-temperature thresholds for vasoconstriction and shivering in humans. *Anesthesiology.* 1995 May;82(5):1160-8.

[68] Grahn D, Brock-Utne JG, Watenpaugh DE, Heller HC. Recovery from mild hypothermia can be accelerated by mechanically distending blood vessels in the hand. *J Appl Physiol.* 1998 Nov;85(5):1643-8.

[69] Tucker R, Marle T, Lambert EV, Noakes TD. The rate of heat storage mediates an anticipatory reduction in exercise intensity during cycling at a fixed rating of perceived exertion. *J Physiol.* 2006 Aug 1;574(Pt 3):905-15.

[70] Tucker R, Rauch L, Harley YX, Noakes TD. Impaired exercise performance in the heat is associated with an anticipatory reduction in skeletal muscle recruitment. *Pflugers Arch.* 2004 Jul;448(4):422-30.

[71] Nagashima K, Nakai S, Tanaka M, Kanosue K. Neuronal circuitries involved in thermoregulation. *Auton Neurosci.* 2000 Dec 20;85(1-3):18-25.

[72] Flouris AD, Cheung SS, Fowles JR, Kruisselbrink LD, Westwood DA, Carrillo AE, et al. Influence of body heat content on hand function during prolonged cold exposures. *J Appl Physiol.* 2006 Sep;101(3):802-8.

[73] Van Someren EJ, Raymann RJ, Scherder EJ, Daanen HA, Swaab DF. Circadian and age-related modulation of thermoreception and temperature regulation: mechanisms and functional implications. *Ageing Res Rev.* 2002 Sep;1(4):721-78.

[74] Attia M. Thermal pleasantness and temperature regulation in man. *Neurosci Biobehav Rev.* 1984 Fall;8(3):335-42.

[75] Hardy JD. Thermal comfort and health. *ASHRAE J.* 1971;77:43-51.

[76] Szekely M, Szelenyi Z. Regulation of energy balance by peptides: a review. *Curr Protein Pept Sci.* 2005 Aug;6(4):327-53.

[77] Cena K, de Dear R. Thermal comfort and behavioural strategies in office buildings located in a hot-arid climate. *J Thermal Biol.* 2001;26:409-14.

[78] Troutman S. J. Jr. Automatic control of water cooling. *Symposium on Individual Cooling;* 1969; Kansas State University; 1969. p. 262-80.

[79] Webb P, Annis JF, Troutman S. J. Jr. Automatic control of water cooling in space suits. Contract Rep. NASA CR-1085. *NASA Contract Rep NASA CR.* 1968 Mar:1-84.

[80] Webb P. *Automatic Cooling: Strategies, designs, and evaluations. Portable life systems.* Washington, DC: NASA report SP-234; 1970.

[81] Troutman S. J. Jr., Webb P. *Automatic controllers for the Apollo LCG: Final report on contract NASA-9778 for NASA Manned Spacecraft Center;* 1970.

[82] Nunneley SA. *Water cooled garments: a review. Space Life Sci.* 1970 Dec;2(3):335-60.

[83] Starr JB. Fluidic temperature control for liquid-cooled space suits: *Portable Life Support Systems,* NASA-SP-234; 1970.

[84] Chambers AB. Controlling Thermal Comfort in EVA. Space Suit. *ASHRAE J.* 1970;12(3):33-8.

[85] Hexamer M, Werner J. Control of liquid cooling garments: subjective versus technical control of thermal comfort. *Appl Human Sci.* 1995 Nov;14(6):271-8.

[86] Nyberg KL, Diller KR, Wissler EH. Automatic control of thermal neutrality for space suit applications using a liquid cooling garment. *Aviat Space Environ Med.* 2000 Sep;71(9):904-13.

[87] Cheung SS, Mekjavic IB. Human temperature regulation during subanesthetic levels of nitrous oxide-induced narcosis. *J Appl Physiol.* 1995 Jun;78(6):2301-8.

[88] Mekjavic IB, Tipton MJ, Eiken O. *Thermal considerations in diving.* In: Brubakk AO, Neuman TS, editors. Bennett and Elliot's Physiology of Medicine of Diving. Toronto: Saunders; 2003.

[89] Tikuisis P, McLellan TM, Selkirk G. Perceptual versus physiological heat strain during exercise-heat stress. *Med Sci Sports Exerc.* 2002 Sep;34(9):1454-61.

[90] Hexamer M, Werner J. Control of liquid cooling garments: technical control of body heat storage. *Appl Human Sci.* 1996 Jul;15(4):177-85.

[91] Hexamer M, Werner J. Control of liquid cooling garments: technical control of mean skin temperature and its adjustments to exercise. *Appl Human Sci.* 1997 Nov;16(6):237-47.

[92] Kuznetz LH. Automatic control of human thermal comfort by a liquid-cooled garment. *J Biomech Eng.* 1980 May;102(2):155-61.

[93] Nyberg KL, Diller KR, Wissler EH. Model of human/liquid cooling garment interaction for space suit automatic thermal control. *J Biomech Eng.* 2001 Feb;123(1):114-20.

[94] Troutman S. J. Jr., Webb P. *Automatic control of water cooled suits from differential temperature measurements:* Final report on contract NAS 12-682 for NASA Electronics Research Center; 1969.

[95] Webb P, Troutman SJ, Jr., Annis JF. Automatic cooling in water cooled space suits. *Aerosp Med.* 1970 Mar;41(3):269-77.

[96] Xu X, Hexamer M, Werner J. Multi-loop control of liquid cooling garment systems. *Ergonomics.* 1999 Feb;42(2):282-98.

[97] Cheung SS, Montie DL, White MD, Behm D. Changes in manual dexterity following short-term hand and forearm immersion in 10 degrees C water. *Aviat Space Environ Med.* 2003 Sep;74(9):990-3.

[98] Brajkovic D, Ducharme MB, Frim J. Influence of localized auxiliary heating on hand comfort during cold exposure. *J Appl Physiol.* 1998 Dec;85(6):2054-65.

[99] Brajkovic D, Ducharme MB. Finger dexterity, skin temperature, and blood flow during auxiliary heating in the cold. *J Appl Physiol.* 2003 Aug;95(2):758-70.

[100] Cheung SS, Sleivert GG. Lowering of skin temperature decreases isokinetic maximal force production independent of core temperature. *Eur J Appl Physiol.* 2004 May;91(5-6):723-8.

[101] Hunter J, Kerr E, Whillans M. The relation between joint stiffness upon exposure to cold and the characteristics of synovial fluid. *Can J Med Sci.* 1952;30:367-77.

[102] Rubin L, Frederick M. *Manual dexterity of the gloved and bare hand as a function of ambient temperature and duration of exposure: US Army Chemical Warfare Laboratories,* Army Chemical Center. Rep. No. CWLR 2107; 1957.

[103] Vanggaard L. Physiological reactions to wet-cold. *Aviat Space Environ Med.* 1975 Jan;46(1):33-6.

[104] Ducharme MB, Brajkovic D, Frim J. The effect of direct and indirect hand heating on finger blood flow and dexterity during cold exposure. *J Therm Biol.* 1999;24:391-6.

[105] Brajkovic D, Ducharme MB. Maintaining finger dexterity in the cold: A comparison of passive, direct and indirect hand heating methods. Dresden, Germany: NATO; 2002. *Report No.: NATO Proceedings* No. RTO-MP-076 AC/ 323(HFM-061)TP/40.

[106] Brajkovic D, Ducharme MB, Frim J. Relationship between body heat content and finger temperature during cold exposure. *J Appl Physiol.* 2001 Jun;90(6):2445-52.

[107] Brajkovic D, Frim J, Ducharme MB. The effects of clothing insulation and levels of torso heating on finger comfort and dexterity. *8th International Conference on Environmental Ergonomics;* 1998 Octover 21; San Diego, California; 1998.

[108] Koscheyev VS, Leon GR, Coca A. Finger heat flux/temperature as an indicator of thermal imbalance with application for extravehicular activity. *Acta Astronaut.* 2005 Nov;57(9):713-21.

[109] Koscheyev VS, Leon GR, Paul S, Tranchida D, Linder IV. Augmentation of blood circulation to the fingers by warming distant body areas. *Eur J Appl Physiol.* 2000 May;82(1-2):103-11.

[110] Koscheyev VS, Leon GR, Trevino RC. Efficacy of wrist/palm warming as an EVA countermeasure to maintain finger comfort in cold conditions. *Aviat Space Environ Med. 2*001 Aug;72(8):713-9.

[111] Lockhart JM, Kiess HO. Auxiliary heating of the hands during cold exposure and manual performance. *Hum Factors.* 1971;13:457-65.

[112] Poulton EC, Hitchings NB, Brooke RB. Effect of cold and rain upon the vigilance of lookouts. *Ergonomics.* 1965;8:163-8.

[113] Pilcher JJ, Nadler E, Busch C. Effects of hot and cold temperature exposure on performance: a meta-analytic review. *Ergonomics.* 2002 Aug 15;45(10):682-98.

[114] Ballard JC. Computerized assessment of sustained attention: a review of factors affecting vigilance performance. *J Clin Exp Neuropsychol.* 1996 Dec;18(6):843-63.

[115] Palinkas LA. Mental and cognitive performance in the cold. *Int J Circumpolar Health.* 2001 Aug;60(3):430-9.

[116] Enander A. Effects of moderate cold on performance of psychomotor and cognitive tasks. *Ergonomics*. 1987 Oct;30(10):1431-45.

[117] Thomas JR, Ahlers ST, House JF, Schrot J. Repeated exposure to moderate cold impairs matching-to-sample performance. *Aviat Space Environ Med.* 1989 Nov;60(11):1063-7.

[118] Van Orden KF, Ahlers ST, Thomas JR, House JF, Schrot J. Moderate cold exposure shortens evoked potential latencies in humans. *Aviat Space Environ Med.* 1990 Jul;61(7):636-9.

[119] Blatter K, Graw P, Munch M, Knoblauch V, Wirz-Justice A, Cajochen C. Gender and age differences in psychomotor vigilance performance under differential sleep pressure conditions. *Behav Brain Res.* 2006 Apr 3;168(2):312-7.

[120] Lum J, Enns JT, Pratt J. Visual orienting in college athletes: explorations of athlete type and gender. *Res Q Exerc Sport.* 2002 Jun;73(2):156-67.

[121] Versavel M, van Laack D, Evertz C, Meier F, Kuhlmann J. Effect of educational level, gender and age on the performance in a multi-user computerized psychometric test system for use in clinical pharmacological studies. *Arzneimittelforschung.* 1996 Dec;46(12):1179-85.

[122] Noakes TD. 1996 J.B. Wolffe Memorial Lecture. Challenging beliefs: ex Africa semper aliquid novi. *Med Sci Sports Exerc.* 1997 May;29(5):571-90.

[123] Cheung SS. Hyperthermia and voluntary exhaustion: integrating models and future challenges. *Appl Physiol Nutr Metab.* 2007 Aug;32(4):808-17.

[124] Dennis SC, Noakes TD. Advantages of a smaller bodymass in humans when distance-running in warm, humid conditions. *Eur J Appl Physiol Occup Physiol.* 1999 Feb;79(3):280-4.

[125] Noakes TD, St Clair Gibson A, Lambert EV. From catastrophe to complexity: a novel model of integrative central neural regulation of effort and fatigue during exercise in humans: summary and conclusions. *Br J Sports Med.* 2005 Feb;39(2):120-4.

[126] Marino FE. Anticipatory regulation and avoidance of catastrophe during exercise-induced hyperthermia. *Comp Biochem Physiol B Biochem Mol Biol.* 2004 Dec;139(4):561-9.

[127] Tatterson AJ, Hahn AG, Martin DT, Febbraio MA. Effects of heat stress on physiological responses and exercise performance in elite cyclists. *J Sci Med Sport.* 2000 Jun;3(2):186-93.

[128] Schmidt R. *Motor learning and performance: from principles to practice.* Champaign, IL: Human Kinetics; 1991.

[129] Flouris AD, Koutedakis Y, Nevill A, Metsios GS, Tsiotra G, Parasiris Y. Enhancing specificity in proxy-design for the assessment of bioenergetics. *J Sci Med Sport.* 2004 Jun;7(2):197-204.

[130] Léger L, Gadoury C. Validity of the 20 m shuttle run test with 1 min stages to predict VO2max in adults. *Can J Sports Sci.* 1989;14:21-6.

[131] Ingier F. Factors influencing assessment of maximal heart rate. *Scand J Med Sci Sports.* 1991;1:134-40.

[132] Andersen P, Saltin B. Maximal perfusion of skeletal muscle in man. *J Physiol.* 1985 Sep;366:233-49.

[133] Rowell LB. Central circulatory adjustments to dynamic exercise. *Human cardiovascular control.* New York: Oxford University Press; 1993.

[134] Brink-Elfegoun T, Holmberg HC, Ekblom MN, Ekblom B. Neuromuscular and circulatory adaptation during combined arm and leg exercise with different maximal work loads. *Eur J Appl Physiol.* 2007 Nov;101(5):603-11.

[135] Brink-Elfegoun T, Kaijser L, Gustafsson T, Ekblom B. Maximal oxygen uptake is not limited by a central nervous system governor. *J Appl Physiol.* 2007 Feb;102(2):781-6.

[136] Noakes TD. Physiological models to understand exercise fatigue and the adaptations that predict or enhance athletic performance. *Scand J Med Sci Sports.* 2000 Jun;10(3):123-45.

[137] Gonzalez-Alonso J, Teller C, Andersen S, Jensen F, Nielsen B. Influence of body temperature on the development of fatigue during prolonged exercise in the heat. *J Appl Physiol.* 1999;86:1032–9.

[138] Flouris AD, Cheung SS. Human conscious response to thermal input is adjusted to changes in mean body temperature. *Br J Sports Med.* 2008 Jan 23.

[139] Thomas MM, Cheung SS, Elder GC, Sleivert GG. Voluntary muscle activation is impaired by core temperature rather than local muscle temperature. *J Appl Physiol.* 2006 Apr;100(4):1361-9.

In: Body Temperature Regulation
Editor: A. B. Cisneros and B. L. Goins
ISBN: 1-60741-282-3

Chapter II

Reviewing the Functional Architecture of the Human Thermoregulatory System

Andreas D. Flouris[1,*] and Stephen S. Cheung[2]
[1]Institute of Human Performance and Rehabilitation, Centre for Research and Technology – Thessaly, Greece
[2]Department of Physical Education and Kinesiology, Brock University, Canada

Abstract

The importance of maintaining thermal balance in homoeotherms first gained recognition in the late 18th and early 19th century through the work of Lavoisier, Carnot and Mayer [1]. Since then, thermoregulatory investigations have been a vital part of physiological research providing findings that have illuminated underlying physiological mechanisms at all levels of biological organization [2-4]. Nevertheless, despite that at the cellular and molecular level our knowledge of mechanisms governing thermal balance is increasing dramatically aided by the continuous advances in technology, at the systems level (i.e., the level of architecture and organization) several fundamental questions remain unanswered [5]. The most prominent of these questions involves the mechanism by which the thermoregulatory threshold is determined which, after numerous investigations, remains a mystery [6]. And, yet, the clinical importance of the physiology of thermoregulation is immense. The prime example for this is that an exodus of body temperature from its optimal range is so suggestive of a pathological condition that body temperature is continuously monitored in all hospitalized patients and reported in every medical history, while the clinical thermometer is one of the most essential scientific instruments for the practice of medicine [7, 8]. In this light, this chapter reviews current

* Corresponding address. Andreas D. Flouris. Institute of Human Performance and Rehabilitation, Centre for Research and Technology – Thessaly. 32 Siggrou Street. Trikala, GR42100, Greece.

knowledge on the functional architecture of human thermoregulation in an attempt to increase the understanding regarding the principles that govern it. The first part of this chapter reviews principles of thermodynamics pertinent to the exchange of heat energy between the human body and its surrounding environment. Thereafter the physiological principles of endothermic thermoregulation are discussed. These principles are divided into those involving thermosensors – including the location, nature and function of central thermosensitive neurons in the preoptic anterior hypothalamus and peripheral thermosensory neurons that detect shell temperatures in the skin and in the oral and urogenital mucosa – and those pertaining to thermoeffectors. The latter include the efferent neuronal pathways to thermoeffectors as well as various aspects of thermoregulatory skin vasodilation and vasoconstriction, and shivering.

Introduction

Every living creature is adapted to a particular temperature range at which it functions optimally. Organisms are able to find appropriate temperature niches and/or to adapt to changes in temperature through numerous complex mechanisms ranging from thermotaxis in bacteria to sweating in humans. Yet endothermy, the pattern of thermoregulation in which body temperature (T_b) depends on a high (tachymetabolic) and controlled rate of heat production (Q_P) [9], remains a striking adaptation in the animal kingdom that has attracted immense debate and attention to date, and it is regarded as one of the most significant transitions in vertebrate evolution [10, 11].

Although endothermy appears during intense muscular activity in several otherwise ectothermic organisms, in mammals and birds it is maintained even under resting conditions. The energetic cost of this maintenance is substantial exceeding the corresponding values in vertebrate poikilothermal animals by a factor of 5–10 [given the same mass and the body temperature) [12]] and its evolution has required significant restructuring of various systems of the vertebrate body as well as profound alterations in the energetics, physiology and behavior of mammals and birds [11]. Regardless, by using powerful autonomic and behavioral mechanisms, endotherms are able maintain a heat content equilibrium over a wide range of thermal loads with substantial benefits to system performance and integrity such as enzymatic catalysis, independence of timing daily activity, resistance to freezing, enhancement of aerobic capacities through greater concentration of mitochondria [11, 13], as well as rapid development of the central nervous system and acquisition of complex progressive forms of behavior [14]. In addition, the maintenance of a comparatively stable T_b has a vast physiological significance in regulating numerous molecular and chemical processes [15]. Indeed, the complexity of mammalian and avian vertebrates requires constancy of temperature for efficient functioning as all chemical and physical reactions change their rate with changes of temperature, yet the degree of change with rising/falling temperature is different for different reactions [14].

The importance of maintaining thermal balance in endotherms first gained recognition in the late 18th and early 19th century through the work of Lavoisier, Carnot and Mayer [1]. Since then, thermoregulatory investigations have been an integral part of physiological research generating voluminous evidence that have illuminated latent physiological mechanisms at all levels of biological organization [2-4]. Nevertheless, despite that at the cellular and molecular level our knowledge of mechanisms governing thermal balance is increasing dramatically aided by the continuous advances in technology, at the systems level (i.e., the level of architecture and organization) several fundamental questions remain unanswered [5]. The most prominent of these questions involves the mechanism by which the thermoregulatory threshold is determined which, after numerous investigations, remains a mystery [6]. And, yet, the clinical importance of the physiology of thermoregulation is immense. The prime example for this is that an exodus of T_b from its optimal range is so suggestive of a pathological condition that T_b is continuously monitored in all hospitalized patients and reported in every medical history, while the clinical thermometer is one of the most essential scientific instruments for the practice of medicine [7, 8].

The fact that knowledge on the functional architecture of the endothermic thermoregulatory system is not complete even after more than two centuries of continuous study makes a great impression to all young scientists entering this very exciting area of research. They immediately feel that their efforts should focus on contributing somehow to dissolving this uncertainty. Later on, however, they come to realize that their precious question is the Holy Grail in thermal physiology and that finding an answer is not as simple as they thought it would be. Yet we sincerely hope that this will only increase their enthusiasm to study the functional architecture of the human thermoregulatory system. In our mind, thermoregulation is the most important and rewarding branch in human physiology. As A. C. Burton wrote: "...(the physiology of thermoregulation) is the most complete example of integrated activity of the organism to create homeostasis" [8].

Life and Thermodynamics

The first law of thermodynamics, an expression of the universal law of conservation of energy, identifies heat transfer as a form of energy transfer. The most common enunciation of this law is: "*The increase in the internal energy of a thermodynamic system is equal to the amount of heat energy added to the system minus the work done by the system on the surroundings*". Many researchers in the past including, amongst others, Lavoisier, Richet, Lusk, Zunst, and DuBois demonstrated, without a shadow of a doubt, that all the energy that appears ultimately as heat is quantitatively accounted for by the chemical energy of the food ingested [1, 14]. This will be discussed further in a later subsection devoted to calorimetry; first, it is necessary to present some basic principles of heat transfer.

Thermodynamic Principles of Heat Transfer

Temperature

At a molecular level, temperature can be defined as the mean kinetic energy (heat) in a body. When heat energy is lost from a body its temperature falls, and if heat energy flows into a body, its temperature rises. This is a direct application of the Second Law of

Thermodynamics which postulates that there is a net energy flow from bodies at higher temperatures to bodies of lower temperatures. The temperature of the human body, therefore, will be greatly influenced by the temperature of the fluids or solids surrounding it because these affect the heat transfer to and from the body. The exchange of heat energy between the environment and the human body occurs via radiation, conduction, convection, and evaporation.

Radiation

Radiation refers to transfer of heat energy via electromagnetic waves between the environment and the human body and is a combination of the sun's direct radiation, the ground's reflected radiation, and the diffuse radiation from the collision of atmospheric molecules. In environments with hot objects (e.g. workplaces exposed to the sun or steel mills), the radiant temperature often exceeds the skin temperature, resulting in radiant heat transfer from the environment to the skin. The radiation field surrounding an individual is dependent upon the shade and cloud cover, the time of day, time of year, terrain, location, as well as the altitude [16]. Solar radiation is a major source of heat load in hot environments but is significantly reduced by clothing from 288 W in a partially clothed individual to 110 W in a fully clothed individual [17].

Conduction

This form of heat exchange refers to the conveying of energy between surfaces in direct contact and is highly dependent on tissue conductivity [18, 19]. It has been demonstrated that, in specific areas of the body, approximately 70% of the endogenous heat can be transferred from the working muscles to the skin surface solely by conduction [18]. The majority of the remaining heat transfer to the surface is achieved through blood flow.

Convection

Convection refers to the exchange of heat energy between a surface and a gas or a liquid and is dependent on the temperature gradient, the rate of motion of the gas (e.g. air movement) or liquid (e.g. water movement), as well as its relative heat storage capacity [17, 20, 21].

Evaporation

During exposure to high dry bulb temperatures the temperature gradient between the skin and the environment is reduced and radiative and convective heat exchanges are diminished while the dependence of the human body on evaporative heat loss is growing. To prevent gains in heat energy through radiative, conductive, and convective pathways at times when the skin-environment gradient becomes negative, the human body employs evaporation which has a high capacity for heat exchange. Each liter of sweat transfers approximately 2400 kJ of heat energy from the body to the environment, but the evaporative capacity is dependent on humidity as well as air and body motion [22, 23].

Air Temperature

Air temperature is defined as "*the temperature of the air surrounding the human body which is representative of that aspect of the surroundings which determines heat flow between the human body and the air*" [24]. Naturally, the air temperature varies as heat exchange between bodies is a continuous process.

Humidity

When liquid, such as water or sweat, is heated by the human body and it evaporates to a vapor and is lost to the surrounding environment, then heat has been exchanged from the body to the environment and the body is cooled [24]. The reason for this vapor transfer is the variation in mass per unit volume of moist air: the difference in absolute humidity (i.e., the mass concentration/density of water vapor) between that at the skin surface and that in the environment. Humidity in the environment is, therefore, an important parameter. The two most commonly used forms to express humidity are relative humidity and partial vapor pressure:

Relative humidity (φ), often expressed as a percentage, represents the ratio of the prevailing partial pressure of water vapor to the saturated water vapor pressure:

$$\text{Relative Humidity } (\varphi) = \frac{P_{\text{a}}}{P_{\text{sa}}} \tag{1}$$

where the partial vapor pressure (P_a in $\text{N}\cdot\text{m}^{-2}$) is the prevailing partial pressure of water vapor in the air. The saturated vapor pressure (P_{sa} in mb) at a temperature t (°C) is given by Antoine's equation:

$$\text{Saturated Vapor Pressure } (P_{\text{sa}}) = \exp\left(18.956 - \frac{4030.18}{t + 235}\right) \tag{2}$$

The different units of pressure are related as follows: 1000 $\text{N}\cdot\text{m}^{-2}$ = 1 kP_a = 10 mb = 7.52 mmHg = 7.52 Torr.

Radiation, conduction, convection, evaporation, air temperature and humidity are some of the main variables considered in calorimetry, a technique described in detail in the following section.

Calorimetry

Calorimetry measures energy expenditure rate at which heat is transferred from the body to the environment [24]. As mentioned in the beginning of this section, the first law of thermodynamics dictates that all the energy that appears ultimately as heat is quantitatively accounted for by the chemical energy of the food ingested [1, 14]. The most accurate method for determining metabolic heat production is to measure the heat produced by the body in a whole-body calorimeter. A number of devices have been developed for this purpose ranging from controlled chambers [25], water perfused suits [26, 27], or partitional calorimetric techniques [28, 29]. The latter method is advantageous in that it allows for precise measurements of body heat storage (S) while allowing for exposure to various conditions including water immersion, extreme ambient temperatures and exercise. Regardless of the device used, calorimetric measurements require substantial exposure time due to variations in body temperature, hemodynamics, and overall thermal balance. Dry (that is, non-evaporated) heat loss (i.e., conduction, convection, radiation) can be measured either 'passively' from the temperature gradient across the walls of a poorly insulated chamber, or 'actively' as the rate at which heat must be extracted from a chamber to counteract heat loss (Q_L) through well-insulated walls [24]. Evaporative heat loss is calculated as the change of water vapor in the air of the chamber. Total heat loss/gain, also known as heat storage (S), is the sum of dry and evaporative heat loss. If precise measurements of all the heat exchange pathways are taken and careful calculations are conducted (see following section on heat balance equation), S will be equivalent to the calorific value of food ingested.

The Heat Balance Equation

The maintenance of T_b at around 37 °C dictates the existence of a heat balance between the body and its environment. This does not necessarily suggest a steady state, because a steady state suggests unchanging temperatures. In fact temperatures within the body and the avenues of heat exchange vary constantly. Yet, heat balance represents a dynamic relationship between the heat transfer into the body and the heat generation within the body with the heat outputs from the body that results in a constant temperature. The heat balance equation is:

$$\text{Heat Balance Equation: } S = M \pm W \pm E_{res} \pm C_{res} \pm E_{sk} \pm K \pm C \pm R \tag{3}$$

where M represents the metabolic Q_P or metabolic rate of the body that provides energy to allow the body to do mechanical work (W). The rates of evaporative and dry (convective) heat exchanges through the respiratory tract are E_{res} and C_{res}, respectively. As mentioned earlier in this section, heat exchange between the skin and the environment can occur via conduction (K), convection (C), radiation (R) and evaporation (E). When added together, the dynamic balance between the rates of Q_P and heat loss (Q_L) provide a rate of heat storage (S). When the human body is in a heat balance (i.e., stable temperature), the rate of heat storage is zero ($S = 0$). All the above terms are rates of Q_P or Q_L; the units of the rate of energy gain or loss are energy per unit time, therefore, kilojoules per hour ($kJ \cdot h^{-1}$). Many investigators standardize the values of the heat balance equation over individuals of different sizes by expressing them in units of energy per square meter of the total body surface area (A_D) [24]. The latter is estimated from the equation of Dubois and Dubois [30]:

$$\text{Body Surface Area } (A_D) = 0.007184 \cdot \text{Weight}^{0.425} \cdot \text{Height}^{0.725} \tag{4}$$

where the weight of the body is measured in kg and the height of the body is measured in m. This formula has been traditionally used for standardizing the results of the heat balance equation [24]. Although other body surface area equations were developed in more recent years, the DuBois equation has been validated to a greater extent and its accuracy has been confirmed more than others [24, 31].

Thermometry

Although very precise, calculations of S through the thermal balance equation require the use of sophisticated controlled chambers [25], water perfused suits [26, 27], or partitional calorimetric techniques [28, 29]. Ergo, thermometric models have been developed – requiring only temperature readings from various sites around the body – to estimate changes in S or body heat content (H_b) based on the fact that the change in body heat content (ΔH_b) can be derived by the product of the change in the mean temperature of the body's tissues ($\Delta \overline{T}_b$), the total body mass, and the mean specific heat of the body's tissues. Therefore, if body mass and mean specific heat are known, ΔH_b can be estimated by approximating $\Delta \overline{T}_b$.

The most popular model to derive $\Delta \overline{T}_b$ is the two-compartment model [32] that uses changes in rectal temperature (ΔT_{re}) – representing the body's 'core' compartment – and the changes in mean skin temperature ($\Delta \overline{T}_{sk}$) – representing the body's 'shell' compartment. The relative core:shell contribution ratios to the final $\Delta \overline{T}_b$ range from 9:1 or 4:1 for hot environments [33] to 2:1 for moderate/cold environments [34, 35]. Based on the thermometric two-compartment model [32], body heat storage (S; in kJ) for every minute is calculated as:

$$S = 3.47 \cdot m_b \cdot \Delta \overline{T}_b \tag{5}$$

where 3.47 is the average specific heat of body tissues (in kJ · kg^{-1} · °C^{-1}), m_b is the individual's body mass, and $\Delta \overline{T}_b$ is the change in mean body temperature at *time t* from the initial $\overline{T}_b$ at *time 0* (in °C^{-1} · h^{-1}), calculated as

$$\Delta \overline{T}_b = \overline{T}_b^{t} - \overline{T}_b^{0} \tag{6}$$

$\overline{T}_b$ is calculated as the weighted sum of T_{re} and $\overline{T}_{sk}$ using the previously described [36] formula:

$$\overline{T}_b = 0.67 T_{re} + 0.33 \overline{T}_{sk} \tag{7}$$

Following these calculations, the rate of change in body heat storage ($\dot{S}$, in W) for every minute is calculated as the change in S at *time t* (in minutes, S^t) from the change in S at *time t-1* [S^{t-1}] demonstrating, in essence, the magnitude of change per minute in body heat from its initial level:

$$\dot{S} = \frac{S^t - S^{t-1}}{1000} \cdot 60 \tag{8}$$

Concomitantly, body heat content (H_b; in kJ · kg^{-1} · °C^{-1}) is calculated as follows:

$$H_b = 3.47 \cdot m_b \cdot \overline{T}_b \tag{9}$$

while ΔH_b (in kJ) is calculated for every minute as the change in H_b at *time t* (in minutes, H_b^t) from the change in H_b at *time t-1* (H_b^{t-1}) demonstrating the change in total body heat per minute:

$$\Delta H_b = H_b^t - H_b^{t-1} \tag{10}$$

The aforementioned models and the weightings of the different indices have been estimated empirically by fitting experimental data to the true changes in H_b, as determined by calorimetric methods [28, 37]. Yet, several authors have reported that the thermometric two-compartment model underestimates $\Delta\overline{T}_b$ [38, 39]. The main limitation of the thermometric method is that the weighting coefficients (usually based on semi-nude young men resting or performing steady-state exercise in a hot environment) remain constant unlike temperatures within the body and the avenues of heat exchange that vary constantly. In addition, there are other external factors that play a major role in the heat exchange between the body and the environment. For instance, acclimatization and training allow the human body to develop a greater skin blood flow in response to a given heat load, bringing $\overline{T}_{sk}$ closer to core temperature (T_C), particularly in hot and humid environments [40-42]. Similarly, protective clothing curtails transfer of heat from the skin surface causing $\overline{T}_{sk}$ to increase close to T_C values [37, 43, 44]. The frequently adopted index of $\overline{T}_{sk}$ is also based upon the respective surface areas of the various regions of the body [32, 36, 45] despite that local protective devices such as gloves [44] as well as inter-individual variation in subcutaneous fat distribution [46] can influence significantly the contribution of individual readings and, in turn, the calculation of $\overline{T}_{sk}$. Another factor that has been suggested to contribute towards the limited adequacy of the two-compartment model is the lack of a separate index representing the heat stored in muscle tissues [39, 47, 48]. Therefore, a three-compartment model to estimate $\Delta\overline{T}_b$ has been proposed [39], yet it is severely limited by the need for invasive measurements of intramuscular temperature.

Physiological Principles of Endothermic Thermoregulation

Understanding the complex interactions amongst biological, behavioral, and environmental factors in endothermic organisms in a diverse range of health conditions as well as the impact of various environmental exposures on the manifestation of disease represent primary goals of thermal physiology. In this light, this discipline has served as a bridge between biology, pharmacology, and bioengineering. Since its initial years, thermal physiology has evolved from a primarily applied study of thermoregulation into a comprehensive science that combines observational methods with analytic techniques to describe the complex mechanisms by which thermal balance is maintained.

Following a phase of quasistasis, the field of thermal physiology is evolving yet again influenced by advances in biology and neuroscience, and is acquiring a more fundamental theme that is largely dependent on technological and computational infrastructure. This evolution is led by reductionists who perform exceptionally interesting research by combining thermal physiology with fields such as biochemistry, pharmacology, neuroscience, molecular biology, computer science and genetics. On the other hand, the International Union of Physiological Sciences has expressed concerns regarding the need for an integrative approach in order to appropriately understand physiology and biology [49]. In addition, many researchers worldwide have appeared in favor of integrative system physiology expressing their fears for the dramatic predominance of reductionism in current physiology and neuroscience [50-53]. Although being in favor of integrative system physiology, we see tremendous value in attempting to understand complex systems by reducing them to their components. Therefore, this section – which is the main body of this chapter – devoted to the physiological principles of endothermic thermoregulation will attempt to encapsulate the remarkable recent progress of both fundamental and integrative research. Although the functional architecture of the human thermoregulatory system is the primary concern of this section, some of its fundamental aspects have to be presented first. The principles involving thermosensors are reviewed first, including the location, nature and function of central thermosensitive neurons, and peripheral thermosensory neurons that detect shell temperatures in the skin and in the oral and urogenital mucosa. Thereafter, the efferent neuronal pathways to thermoeffectors are presented as well as various aspects of thermoregulatory skin vasodilation and vasoconstriction, control of eccrine sweat gland function, shivering and thermoregulatory behavior. The section concludes with the most recent progress from investigations on the functional architecture of the endothermic thermoregulatory system.

Thermosensors

Central Thermosensors

Thermal constancy is restricted to the core of the body, while tissues of the body's shell vary in temperature as blood flow may be increased or decreased in order to augment or limit thermal insulation, respectively, based on the internal and the ambient temperatures. This constancy of core temperature (T_C) implies its role as the regulated variable in endothermic thermoregulation. Its deviation from the physiological threshold represents a 'load error' initiating a sequence of events that lead to conservation, dissipation, or increased production of heat in order to restore T_C. Indeed, *in vivo* direct thermal stimulation (heat or cold) of the preoptic area and anterior hypothalamus (PO/AH) in endotherms has been shown to reflexively evoke various autonomic and behavioral thermoregulatory responses [7, 54-56]. Different thermoregulatory responses have also been reported during thermal stimulation in the medulla oblongata, pons, midbrain [57-59], as well as the peridural space of the vertebral canal and of the spinal cord [7, 60-62]. These thermosensitive areas of the brain and the central nervous system have been recognized for more than a century as key components of thermoregulation based on observations from humans suffering brain damage and from the effect of experimental lesions in other mammalian models.

The mechanism by which T_C and/or hypothalamic temperature (T_{hy}) is sensed remains controversial for more than five decades, with several different theories attempting to explain the nature and function of the underlying phenomena involved. The most widely accepted approach for more than half a century has been put forth by Hammel and colleagues [54, 63] who postulated that the hypothalamus – and particularly the PO/AH – is the brain region not only controlling but also perceiving deep-body temperature. The majority of PO/AH neurons are comparatively insensitive to temperature, yet approximately 30% of the neurons in these areas increase significantly their firing rates when hypothalamic temperature is increased and have, thus, been classified as 'warm-sensitive' [64]. The thermosensitivity of these cells is due to currents that determine the rate of spontaneous depolarization between successive action potentials and are therefore considered to play a 'pacemaker' role [65, 66]. Cold-sensitive neurons (i.e., neurons that increase their activity/firing rate with a decrease in brain temperature) also exist and play a role in thermoregulation, albeit a minor one since their population is quite small compared to the warm-sensitive neurons [67-69] and their activity appears to be due to inhibitory synaptic input from adjacent warm-sensitive neurons [7, 54, 63]. Indeed, research adopting thermal and chemical stimulation of PO/AH cells has demonstrated that autonomic responses to maintain thermal balance in both heat and cold are evoked by appropriate changes in the firing rate of warm sensitive neurons: increased firing rate of warm-sensitive PO/AH neurons generates heat-defense responses, while decreased firing rate in the same neurons generates cold-defense responses [70, 71].

Because of the increased attention on warm-sensitive neurons, their morphological identification has been made possible in recent years. These cells are characterized by the horizontal orientation of their dendrites in the direction of the third ventricle medially and the medial forebrain bundle laterally [72]. This orientation appears ideal [7] for receiving input via neurons conveying temperature signals from the surface of the body and the viscera through the periventricular stratum and medial forebrain bundle [73]. Indeed, the fact that the firing rate of the majority of warm-sensitive neurons is influenced not only by hypothalamic temperature, but also by skin or spinal temperatures has been known for some time [74]. In fact, whole-animal experiments showed that about 60% of warm-sensitive PO/AH neurons receive afferent input from spinal and skin thermoreceptors [74]. Based on these findings, it has been hypothesized that PO/AH warm-sensitive neurons may integrate central and peripheral thermal information by sensing changes in their own local temperature and receiving input from ascending medial and lateral pathways relaying spinal and skin thermal afferent signals [72].

While taking into account the prevailing theory, it is important to present the results from elegant studies by Kobayashi [75-77] suggesting a more active role of cold-sensitive neurons in thermoregulation. Indeed, results from this research suggested that, although the population of temperature-sensitive neurons is comprised predominantly of warm-sensitive ones – as seen by others both *in vivo* and *in vitro* [67-69] – this is only true when one is searching at the higher end of the temperature spectrum (i.e., when the PO/AH slice is at higher temperatures). In contrast, when one is searching at the lower end of the temperature spectrum by having the PO/AH slice at low temperatures, the proportions of temperature-sensitive neurons greatly changes showing a higher proportion of cold-sensitive neurons [76]. This, of course, is not surprising as cold-sensitive neurons are only active at lower tissue temperatures. Based on these results, the author proposed that the whole proportion of cold-sensitive neurons is equal to that of warm-sensitive ones [76, 77].

Peripheral Thermosensors

Peripheral thermosensory neurons detect shell temperatures in the skin as well as in the oral and urogenital mucosa and are predominantly cold-sensitive [7, 78]. The cold-sensitive neurons are unmyelinated or thinly myelinated afferent terminals located in or immediately beneath the epidermis. Their signals are transferred by thin myelinated Aδ fibers and are triggered when their temperature falls below approximately 33°C with their peak activity usually seen at a range around 25°C. If temperature falls further, the activity of the cold-sensitive neurons decreases and may cease at temperatures below 10°C [78, 79]. This bell-shaped temperature/response curve suggests that the discharge rate of cold-sensitive neurons can be equal at two different – yet cool – temperatures.

The warm-sensitive neurons are located slightly deeper than their cold-sensitive counterparts, in the dermis and their signals are transferred through unmyelinated C fibers [7, 78]. These neurons increase their discharge rate with increasing temperature, starting at a level of approximately 35°C. Their response to fast temperature changes commonly demonstrates a dynamic component besides the static response with a peak at a rather wide range [79]. This dynamic component involves increased firing rates during temperature changes and quick adaptation (or slower firing rates) when temperature stabilizes, even if the new temperature is not equal to the normal temperature [7, 78, 79]. This type of response allows the organism to quickly react to environmental changes and suggest that the regulated variable may be $\dot{S}$ and not temperature *per se*.

Beyond the superficial cold- and warm-sensitive neurons, research has shown that endotherms demonstrate peripheral deep-body thermosensors located in the stomach, esophagus, large intra-abdominal veins, skeletal muscle, and other organs [7, 52]. Evidence for the significant contribution of extra-cerebral deep-body thermosensors to the control of shivering were presented for the first time by Thauer [80] who showed that, despite their distinct effect on autonomic and behavioral thermoregulatory responses, the neuronal activities signaling the level of deep-body temperature largely remain below the level of consciousness [79]. Because of the latter as well as limitations of the thermode technique used to detect them, triggering of corresponding thermoregulatory responses to local or general changes in T_C (e.g., alterations in heat dissipation or generation in response to induced temperature changes) remains the only unanimously accepted criterion for the existence of functionally specific deep-body thermosensors [52, 79].

Assessing the change of thermoregulatory heat generation or dissipation per degree °C change in deep-body temperature has been suggested as one way to assess the integrated thermosensory input from the deep-body tissues [81]. Previous research has shown that the range of deep-body temperature changes that generate responses from a thermoregulatory effector from its threshold to maximum response is approximately 2-4°C, with a linear relationship between temperature and effector response [79, 81]. It has been suggested that this 'integrated deep-body thermosensitivity' is sufficient to account for the relative constancy of T_C in endotherms [81]. Experiments adopting the animal model generally yielded coefficients of approximately 4 to 6 $W \cdot kg^{-1} \cdot °C^{-1}$ of heat dissipation in response to cooling as well as evaporative heat loss coefficients in the order of 1 to 3 $W \cdot kg^{-1} \cdot °C^{-1}$ in response to heating [81]. It has been suggested that the contribution of deep-body thermosensitivity to the control of metabolic heat defense by shivering or of heat evaporation by sweating in humans are within the same ranges [79].

The following four subsections present information on recent developments regarding peripheral thermosensors and thermosensation in order to provide a complete understanding on the location, nature and function of peripheral thermosensitive neurons.

Transient Receptor Potential Ion Channels (TRP)

Our knowledge on the peripheral mechanisms by which we sense hot and cold temperatures has been recently updated with the discovery that ion channels of the transient receptor potential (TRP) family are participating in the thermosensation and are, direct or indirect, contributors to thermoregulation [82]. The mammalian TRP superfamily consists of ~30 channels that are divided into six subfamilies classified as the TRPA (ankyrin), TRPC (canonical), TRPML (mucolipin), TRPM (melastatin), TRPP (polycystin), and TRPV (vanilloid) [7, 82]. The demonstration that the TRPV1-V4, M2, M4, and M5 can be activated by increases in T_a and that the TRPM8 and A1 can be activated by decreases in T_a, together with the identification of their heterogeneous expression patterns and heterogeneous temperature sensitivities, resulted in the evaluation of these proteins as candidate endogenous thermosensors [82]. Concomitantly, invertebrate TRP channels have been identified that are involved in behavioral thermoregulation [7]. However, much remains to be learned regarding the involvement of these channels in the multitude of thermosensory functions that have been reported in the central and peripheral nervous systems.

Afferent Channels

Afferent channels of thermal sensation and thermoregulation begin with primary thermosensory neurons, the bodies of which are located in the dorsal root ganglia with their central axons projecting to the dorsal horn of the spinal cord (mainly lamina-I) to synapse with secondary monopolar neurons [7]. The temperature signals are carried from the lamina-I neurons to the insular cortex (i.e., the island of Reil). This occurs either with one relay in the posterior part of the ventromedial nucleus of the thalamus or with two relays in the parabrachial nucleus and the basal part of the ventromedial nucleus of the thalamus [7]. Both branches of this channel contribute to discriminative temperature sensation allowing sensation of temperature at a high spatial resolution, such as the temperature of the surface on the tip of a finger [83]. Yet, despite that this channel is crucial for daily living and for making decisions regarding interactions with the environment, it appears not to participate in T_b regulation [7, 83].

Control of Homeostasis

The structures and characteristics of afferent neuronal channels for homeostatic control have been recently reviewed [7]. Thermal stimuli that are significant enough to influence thermal homeostasis trigger appropriate thermoeffector responses. The temperature-induced signals generated from large regions of the body's shell are sent from the peripheral thermosensory neurons to the brain via the spino-reticulo-hypothalamic channel. In this channel, thermosensory inputs converge at the level of lamina-I neurons so that a single lamina-I cell can receive inputs from multiple thermosensory neurons [84] so that downstream neurons such as brain stem neurons are able to collect signals from large thermoreceptive fields [85]. Finally, the neurons of the afferent spino-reticulo-hypothalamic channel project to the hypothalamic regions including the PO/AH either through the periventricular stratum along the wall of the third ventricle, or through the medial forebrain bundle [7]. As described earlier in the section regarding central thermosensors, the dendrites of warm-sensitive PO/AH neurons are oriented ideally for collecting information from both the periventricular stratum and the medial forebrain bundle.

The Alternative 'Comparator Model'

Elegant studies by Kobayashi and colleagues during the past 25 years have challenged the hypothalamic proportional control model (described in detail in the section pertaining to central thermosensors and elsewhere in this chapter) which remains the most widely accepted approach for the past 50 years. Kobayashi opposes the dogma that receptors are sensors [i.e., the 'sensor model' [86]]. Based on the sensor model, thermoreceptors convert temperatures of the various regions of the body into impulses in the form of a neural code, conveying the information of objects to the brain where these impulses are integrated by a separate network to form a mean temperature [7]. Thereafter, this integrating network compares the mean temperature with a reference signal and through various processes, which have not been elucidated yet, orders are sent to thermoeffectors in order to evoke appropriate thermoregulatory responses [7]. The sensor model is in line with empiricism, initiated by Aristotle during the fourth century B.C. [87, 88] and explicitly formulated by John Locke in the 17th century [89].

Kobayashi argues that receptors are comparators (i.e., the 'comparator model') and opposes the view of signal transduction from the physical world to the brain [76, 77, 86, 90]. He proposes that heat- and cold sensitive neurons are comparators of temperature and evoke impulses when temperature surpasses a receptor activation threshold. In turn, these impulses are not a form of neural code (as assumed by the sensor model) but only triggers to activate target neurons in the brain. For instance, when skin temperature is above its threshold temperature, peripheral heat-sensitive neurons evoke impulses to stimulate target neurons in the brain (central thermosensors). In turn, thermal sensations are merely side-products of the activation of central thermosensory neurons connected to generate appropriate effector responses. Consequently, skin hotness "*...only occurs in the sensation world in our mind*" [86], in line with innatism formulated by Descartes during the 17th century [91], based on which stimuli to evoke sensation are innately present at the target brain. Ergo, the comparator model suggests that effectors are triggered via sensory neurons *automatically*, without the involvement/presence of a separate decision-making network or temperature code computations such as calculation of mean temperature. This concept extends to regulation of thermal balance in the entire body, but this aspect of Kobayashi's comparator model will be presented in detail in a subsequent section pertaining to models regarding the functional architecture of the human thermoregulatory system. For now, it is crucial to present the nature, location, and function of thermoeffector neurons.

Thermoeffectors

The most widely accepted model [i.e., the hypothalamic proportional control model [54, 63]] postulates that heat loss effector neurons control heat loss responses that are usually initiated when the PO/AH is heated above its normal temperature. The heat loss effectors are either spontaneously firing neurons or silent neurons that are synaptically excited by warm-sensitive neurons and synaptically inhibited by temperature-insensitive neurons. In turn, heat production effector neurons control heat production/retention responses that are generally initiated when the PO/AH is cooled below its normal temperature. According to Hammel's model, the heat production neurons are not inherently cold-sensitive, but their activity is due to inhibitory synaptic input from adjacent warm-sensitive neurons [54, 63]. In-depth information on the function of thermoeffectors neurons and their involvement in thermoregulation is presented in a subsequent section pertaining to the functional architecture of the human thermoregulatory system, where Hammel's model is presented in detail.

Although the human efferent channels to thermoeffectors have recently begun receiving attention [92], they have not been well characterized to date. In contrast, a significant proportion of these channels have been mapped during the past decade in the rat – including the tail vasculature (a specialized heat-exchange organ) as well as the brown adipose tissue and skeletal muscles (heat-production effectors) – due to advances in tracing techniques and pharmacological treatments [93]. The skin vasculature and brown adipose tissue are controlled by sympathetic ganglia, the preganglionic neurons of which are located in the intermediolateral spinal column and receive direct input from cells located mainly in the raphé/peripyramidal area of the medulla [7, 93]. In turn, these medullary sympathetic ganglia, the preganglionic neurons of which are located in the intermediolateral spinal column and receive direct input from cells located mainly in the raphé/peripyramidal area of the medulla [7, 93]. In turn, these medullary neurons are controlled by pontine (i.e., locus coeruleus), midbrain (i.e., retrorubral field, ventral tegmental area, and periaqueductal gray matter), and hypothalamic (i.e., dorsomedial and paraventricular nuclei) that receive input from warm-sensitive PO/AH neurons [7, 94-96]. As discussed in the section pertaining to thermosensors, the efferent channels for non-shivering thermogenesis and skin vasomotor tone share certain similarities, while the channel controlling shivering is substantially different. Direct and indirect inputs from the midbrain and brainstem reach α and γ motorneurons within the shivering pathway [97]. The axons of the midbrain neurons move down the spinal canal and are controlled by posterior hypothalamic neurons which, sequentially, receive inhibitory input from warm-sensitive PO/AH neurons.

The following three subsections present information on current knowledge regarding the physiology of mammalian skin vasomotor tone, sweating and shivering thermoregulatory aspects in order to provide a complete understanding on the nature and function of thermoeffectors on human thermoregulation.

Thermoregulatory Aspects of Skin Vasomotor Tone

Skin vasomotor tone is a major contributor to the maintenance of thermal homeostasis and varies considerably in response to heat and cold: thermoregulatory vasoconstriction can decrease skin blood flow (BF_{sk}) to nearly zero [98], while thermoregulatory vasodilation can increase BF_{sk} to 6-8 L·min^{-1} [99] from the 0.25 L·min^{-1} of resting BF_{sk} seen in thermoneutral environments [100, 101]. As long as T_b is regulated physically, control of skin vasomotor tone through the sympathetic nervous system prevails over the other physiological regulatory responses (i.e., shivering and sweating). During exposures to hot environments and/or exercise, increased BF_{sk} through vasodilation is essential to heat dissipation. Accordingly, decreased BF_{sk} through vasoconstriction curtails Q_L and protects against hypothermia. The tone of cutaneous vessels is controlled primarily by skin sympathetic nerve activity which is composed of vasoconstrictor and sudomotor components. Therefore, skin sympathetic nerve activity includes the control of cutaneous blood flow but also sweating [102]. The latter component will be further analyzed in a subsequent section.

Heat dissipation at resting BF_{sk} (i.e., 0.25 L·min^{-1}] in a thermoneutral environment results in approximately 85 kcal·h^{-1}, which is equal to the level of resting metabolic Q_P [100, 101]. Increased T_b from exercise and/or exposure to hot environments triggers cutaneous vasodilation which, as mentioned earlier, can increase BF_{sk} severalfold augmenting convective heat exchange from the core to the periphery [103]. These alterations are generally achieved by adaptations in cardiac output as well as blood flow redistribution from regions such as the splanchnic area that vasoconstrict without compromising oxygen supply to organs such as the heart [101]. In thermally neutral environments, ~5-10% of cardiac output is directed to the skin, while during heat stress, the proportion of cardiac output directed to the skin can reach 50-70%, approaching 8 L·min^{-1} [101]. In glabrous areas of the skin the increase in BF_{sk} during heat stress is initially achieved via withdrawal of sympathetic vasoconstrictor activity, while, in nonglabrous areas, it is achieved through an increase in sympathetic vasodilator activity. The major part of reflex thermoregulatory increase in BF_{sk} is mediated by the active vasodilator component [104] as well as nonthermoregulatory reflexes, including baroreflexes [105]. In general, the relationship between T_C and BF_{sk} is linear. In circumstances where T_C continuous to increase despite an initial rise in BF_{sk}, the latter will continue to increase until maximal responsiveness is reached or the body reaches a steady state where heat gain is equal to Q_L and, thus, T_C is constant. When T_C returns to normal resting values, BF_{sk} decreases toward normal, demonstrating a classical negative feedback loop [106].

Previous research suggested that identifiable T_C thresholds for cutaneous vasodilation exist during whole-body heating [101, 103]. These thresholds are defined as the T_C at which a rapid increase in BF_{sk} occurs, while the gain or sensitivity of the vasodilator response relative to T_C is defined as the slope of the $T_C - BF_{sk}$ relationship following the threshold. Therefore, variations in BF_{sk} at a given T_C can be explained by differences in the vasodilation threshold, the response sensitivity or both [106]. For instance, two responses with similar sensitivities but disparate thresholds result in significant differences in BF_{sk}, and, in turn, heat dissipation, for a given T_C. Factors that can influence the threshold and sensitivity of cutaneous vasodilation include circadian rhythm [107, 108], as well as exercise training and heat acclimation [41].

During exposures to cold environments BF_{sk} is decreased through cutaneous vasoconstriction resulting in curtailed convective heat exchange from the core of the body to its surface, as well as reduced heat transfer from the skin surface to the environment [100, 109]. This response is mediated by the sympathetic nervous system and it is triggered in part by stimulation of cutaneous cold receptors on the region exposed to cold, and in part by cooled blood returning to the general circulation which stimulates the thermoregulatory center in the PO/HA area [104, 110]. The activity of skin sympathetic nerves in response to cold or warm exposure demonstrates interindividual variability, which may be the determinant of cold tolerance [111].

Measurement of BF_{sk} is generally conducted via laser Doppler flowmetry or venous occlusion plethysmography [101, 112]. The former technique provides a linear index of skin blood flow from ~1mm^2 of skin area and is based on the shift of coherent laser light induced by erythrocytes moving in the cutaneous vessels and it is advantageous in that it allows for increased temporal resolution (i.e., measurements can be conducted for prolonged periods of time) and specificity to the cutaneous microcirculation [113, 114]. The two major disadvantages of the laser Doppler technique are the measurement of flow in arbitrary laser Doppler units or volts rather than measuring the absolute flow in milliliters/liters per minute, as well as the confinement to a comparatively small measurement area. To overcome the former limitation, the derived BF_{sk} is often expresses as cutaneous vascular conductance (CVC), defined as the ratio of laser Doppler flow to mean arterial pressure, representing the fraction of skin vascular conductance from the total vascular conductance. Another method used to overcome this limitation is to normalize laser Doppler flowmetry values to those measured during maximal vasodilation achieved through local warming of the skin or local sodium nitroprusside [115].

Absolute measurements of blood flow in larger areas such as the forearm, lower leg, or finger are conducted via venous occlusion plethysmography. Yet, this technique is limited in that is not specific to the skin with changes in muscle blood flow generating difficulties in data interpretation, and that it provides discontinuous measurements [generally at most 4 measurements per minute [106]]. Specifically, given that this technique measures absolute/total blood flow, it can be used to measure BF_{sk} of the forearm or the lower leg when blood flow to the underlying muscle(s) remains unchanged (e.g., passive heat stress). The contribution of muscle blood flow is limited in the finger and, therefore, venous occlusion plethysmography has been frequently adopted to assess finger skin perfusion. Yet, such a measurement includes areas of glabrous (i.e., palmar side) and nonglabrous (i.e., dorsal aspect) skin [116] which limits data interpretation.

Reflex Neural Control of BF_{sk}

The skin vasodilator and vasoconstrictor systems play a major thermoregulatory role in the maintenance of thermal balance because small alterations in BF_{sk} lead to relatively large changes in heat conservation or dissipation [106, 117]. The human cutaneous circulation is controlled by the sympathetic vasodilator nerves, activated when heat dissipation is required, and the sympathetic adrenergic vasoconstrictor nerves, activated when heat conservation is required. This control by two distinct populations of sympathetic nerves makes human cutaneous circulation unique. This system is particularly complex demonstrating significant interactions of reflex and local mechanisms in both general and local cooling or heating [98]. The reflex effects are attributed to general sympathetic vasomotor activity [118, 119] whereas the local effects are due to stimulation of local adrenergic nerves, nonadrenergic mechanisms, and inhibition of reflex vasoconstriction [98, 120, 121]. The areas of glabrous skin in the body, such as lips, soles and palms, are innervated only by the sympathetic vasoconstrictor nerves, while the areas of nonglabrous skin are innervated by both sympathetic vasodilator and vasoconstrictor nerves [101]. An additional distinction between glabrous and nonglabrous skin, which is a very important aspect of skin anatomy and physiology, pertains to the presence of arteriovenous anastomoses (AVA). These are thick-walled, low-resistance blood vessels that allow high flow rates directly from arterioles to venules and play a major role in endothermic thermoregulation [122-125]. AVA exist almost exclusively in glabrous skin where they are innervated heavily and solely by sympathetic vasoconstrictor nerves [106, 126]. Ergo, in areas of glabrous skin AVA opening or closing can result in considerable change in BF_{sk}.

The large increases in BF_{sk} are mediated to a large extent by activation of sympathetic vasodilator nerves in the skin [101]. Based on current knowledge, active vasodilation is mediated by nitric oxide to an extent of approximately 30% and, for the most part, by cotransmission from sympathetic cholinergic nerves, yet neither the cotransmitter nor the precise nature of the link between sudomotor and active vasodilator nerves have been elucidated to date [106, 127].

The central pathway regulating activity in cutaneous vasoconstrictor neurons involved in thermoregulation is distinct from that of other vasoconstrictor systems, such as those supplying the kidney or skeletal muscles [128, 129]. Decreases in BF_{sk} are mediated by activation of sympathetic vasoconstrictor nerves that release norepinephrine and one or more cotransmitters that interact with postsynaptic α1- and α2-receptors on cutaneous arterioles and AVA [130, 131].

Effects of Local Skin Warming on BF_{sk}

The control of local BF_{sk} is not only depended upon sympathetic vasodilator and vasoconstrictor systems, but also on the local T_{sk}. Local skin warming triggers a direct and significant local skin vasodilation which peaks at a sustained local T_{sk} of 42°C and is biphasic in nature [132]. A typical vasodilatory response to 30 minutes of 42°C local T_{sk} is characterized by an initial rapid BF_{sk} increase during the first 3-5 minutes, followed by a moderate decrease in BF_{sk}, and a subsequent slower vasodilation that reaches a peak following 25-30 minutes of warming [132, 133]. The vasodilation peak correlates to T_{sk}, unless a maximum response has been reached [132]. The first phase of vasodilation during local warming ranging from 29°C to 42°C is dependent to a large degree on the local activity of sensory nerves, primarily C-fiber afferents [134, 135]. The latter are stimulated by capsaicin – the pungent ingredient found in hot peppers – which binds to vanilloid receptors on nerve endings causing depolarization through opening of a cation channel, resulting in vasodilation and a local sensation of heat [136]. The second slower phase of vasodilation triggered by local warming is initiated and maintained by nitric oxide [115, 133].

Effects of Local Skin Cooling on BF_{sk}

Locally cooling the skin produces a strong localized vasoconstriction, potentially through an supraspinal axon reflex, that can decrease skin blood flow to nearly zero [98]. Local vasoconstriction is achieved solely through stimulation of local adrenergic nerves [98, 120, 121] as well as an increase in the affinity of postsynaptic α-receptors with decreasing temperature [137, 138], and does not require intact connection to the central nervous system [132]. Nevertheless, the involvement of supraspinal regions of the central nervous system has been suggested in the control of a paradoxical phenomenon termed 'cold-induced vasodilation' (CIVD). In brief, the CIVD phenomenon involves a counterintuitive acute increase in local cutaneous blood flow that develops sporadically in peripheral areas of glabrous skin of the body such as the feet in birds, ears in rabbits, tail in rats, as well as the face, hands and feet in humans when these organisms are exposed in cold environments [122, 124, 125, 139, 140]. Although local mechanisms appear to play a role in CIVD [141-144], a central component of the phenomenon has been clearly demonstrated in different mammalian models [145-150]. Further, thermal body state has been shown to influence CIVD in a way that whole body cooling results in delayed CIVD onset [142, 148, 151, 152], while increased thermogenesis [153] or body heat content [152] result in accelerated CIVD response. These findings provide support to the notion that thermal balance has a strong impact on the CIVD reaction which seems to be superimposed on the general vascular status of the peripheral blood vessels. The exact mechanism of CIVD has yet to be elucidated and is not a primary aspect of this chapter. For now, let us dwell upon the mechanisms of eccrine sweat gland function which are functionally linked and often temporally coincident with cutaneous vasodilation.

Control of Eccrine Sweat Gland Function

Heat exchange from the skin to the environment during heat stress is regulated in part by thermoregulatory control of eccrine sweat gland function. During exercise or heat exposure, increases in T_b trigger sweating [101]. The evaporation of sweat reduces T_{sk} as the amount of energy used up in the phase change that occurs when this biological fluid is removed induces a cooling effect. By reducing T_{sk}, sweating cools the blood in the dilated skin vessels prior to its return to the core of the body.

Generally, sweating increases linearly with T_b until maximal sweat secretion is reached or the body reaches a steady state where heat gain is equal to Q_L and, thus, T_b is constant. It is not surprising, therefore, that sweating and cutaneous vasodilation are functionally linked and often temporally coincident. Nevertheless, changes in one system do not always parallel changes in the other. For instance, brief exercise increases the internal temperature threshold for cutaneous vasodilation (compared with resting heat stress) but does not influence the sweating threshold [154].

Previous research has shown that the quantitative responses of human eccrine sweat glands – under any given set of conditions – are characterized by an integration of the effects of outflow of sudomotor neuronal impulses from the central nervous system and the local conditions of the microenvironment at the level of the gland [155]. Sweating secretion occurs following the release of the neurotransmitter acetylcholine from sudomotor nerves which innervates muscarinic receptors on sweat glands [156]. Therefore, acetylcholine is capable of modulating sweat gland function [157]. In addition to acetylcholine, studies in horses during exercise demonstrated an involvement of nitric oxide synthase in sweating, since its inhibition by NG-nitro-L-arginine methyl ester (coded as L-NAME) limits sweat secretion [158]. The exact mechanism(s) involved in this phenomenon remain to be elucidated, but an explanation put forth is that L-NAME has antimuscarinic action and limits sweat gland function by competitive antagonism with acetylcholine [159].

Shivering

Shivering is involuntary tremor of skeletal muscles caused most often by cold, but also by other stimuli such as strong emotions. Cold-induced shivering is an emergency mechanism for endotherms that, although can be curtailed to some extent by voluntary control, it is included in autonomic thermoregulatory effector activities as it is driven involuntarily by cold signals from peripheral and central receptors. The control of shivering by the central nervous system is quite unusual given that it is an involuntary response of skeletal muscles which are generally under voluntary control. The basic mechanisms for shivering including tremor movements are localized in the spinal cord [60]. The neuronal circuitries involved in shivering begin at the PO/AH which sends inhibitory efferent signals which descend in the medial forebrain bundle [160]. The dorsomedial region of the posterior hypothalamus is particularly important for shivering control [161, 162] as it encapsulates neurons generating excitatory signals for shivering [93]. This is also in line with the view that the Q_L center in the PO/AH suppresses shivering by inhibiting the heat production center in the posterior hypothalamus [163], although this notion has not been proven experimentally hitherto. The signals that trigger shivering leave the posterior hypothalamus and run caudally through the midbrain area, dorsolateral to the red nucleus [93, 164]. In the medulla oblongata and the pons, the channel reaches close to the ventrolateral surface and then continues to the lateral columns of the spinal cord [164].

Given that shivering is generated by rhythmic muscle contraction, the metabolic effects of shivering resemble that of light exercise. The intensity of shivering can reach approximately 40% of maximal oxygen uptake ($\dot{V}O_{2\,max}$) [or five times the resting metabolic rate) [165]]. Although the exact physiological mechanism for this limit remains elusive, several experiments have demonstrated that shivering intensity is associated with cold exposure severity and duration as well as with the morphological characteristics of individuals, including surface-to-volume ratio, percent body fat, and blood flow [165-167]. The energy required to sustain muscle contractions during shivering is provided through oxidization of carbohydrates, lipids, and proteins, although it is generally believed that the main energy sources for shivering are carbohydrates and lipids [168]. As a result, the human muscles are able to sustain shivering for several hours, either by using a variety of fuel mixtures or by changing muscle fiber recruitment [i.e., altering the recruitment of type II fibers [169]].

Functional Architecture of the Endothermic Thermoregulatory System

It is unanimously agreed [170] that the genesis of intensive experimentation and theory regarding the functional architecture of the human thermoregulatory system is marked by the influential experiments of Barbour [171] who used the rabbit model to show that cooling a brain stem region near to what is termed today the PO/AH had profound effects on rectal temperature (T_{re}), and that the animal demonstrated analogous behavioral responses. This discovery fitted exquisitely within the theory of 'brain centers' – that is, specific areas in the brain eliciting topical control of somato-sensory perception and somato-motor activity – which prevailed during those times (the beginning of the 20th century) and determined the direction of mainstream research on thermoregulation for the next 50 years. Since the 1960's, there have been four major models of human thermoregulation and they will be presented in detail in the following pages. It is important for the reader to recall that none of these models have been unanimously accepted. Each presents with inherent advantages and limitations in their attempts to explain the various phenomena of human nature and, after several decades of continuous research, the precise mechanisms of human thermoregulation remain uncertain.

Hypothalamic Proportional Control with an Adjustable Set Point

In 1960, Hammel and colleagues [54] conducted hypothalamic cooling and heating in conscious dogs using chronically implanted thermodes and found significant effects on T_C, vasoconstriction, shivering and panting. Interestingly, although alternating or sustained PO/AH heating and cooling demonstrated qualitative consistency, it revealed quantitative variation with the induced thermoregulatory response resulting in changes in T_C opposite to that of T_{hy} [54]. Based on these and other findings [172-174], Hammel's group proposed the concept of hypothalamic proportional control of T_b around an adjustable set point, summarized in the following formula:

$$\textit{Hamel's } T_b \textit{ Model: } R - R_0 = \alpha_R(T_{hy} - T_{set}) \tag{11}$$

where $R - R_0$ represents a specific net thermoregulatory response, such as shivering, panting, skin blood flow, α_R is the proportionality constant for the response (negative for heat conservation/generation and positive for heat dissipation), T_{hy} is the temperature of the hypothalamus, and T_{set} is a hypothetical set value considering that all thermal and non-thermal parameters known to influence temperature regulation do so by altering T_{set}. In this light, T_{set} is not viewed as a set point temperature, but as a complex variable reflecting the activity of neuron populations. This particular view of T_{set} was the offspring of the first experiment on the relationship between temperature and hypothalamic neurons [174]. Hammel's paper evaluated precisely the skin cold and warm receptors as thermal inputs influencing T_{set} , and demonstrated that it was possible to quantify their systematic input by $\overline{T}_{sk}$ [172] and, therefore, present T_{set} as a function (amongst other variables) of $\overline{T}_{sk}$ [175].

The eloquent model of proportional hypothalamic control around an adjustable set point quickly received attention as well as some criticism, particularly on $\overline{T}_{sk}$ and T_{hy} being the only pertinent regions generating thermal inputs. In the years to follow, Hammel's group conducted precise experiments to quantify the parameters thought to affect T_{set}. T_a and T_C [176], sleep-wakefulness cycle [172] and hibernation [177, 178] were evaluated as modulators of T_{set} with respect to Q_P. Regarding heat dissipation by thermal and non-thermal parameters, modulation of T_{set} was quantified for exercise [179, 180], T_a and T_C [176], and fever [180]. This series of experiments revealed a problem with the model of proportional hypothalamic control, namely that the slope of the regression curves linking T_{hy} to variables determined as indicators of evaporative heat dissipation were sharper in circumstances when T_{set} was shifted above its normal level (i.e., when evaporative heat loss was inhibited). Analogously, the slopes of the regression curves were less sharp in circumstances shifting T_{set} below its normal level (i.e., when evaporative heat loss was promoted). Although these limitations of the proportional hypothalamic control model were less evident in terms of cold-defense mechanisms in the dog, in other organisms deviations from an additive interaction between T_{hy} and $\overline{T}_{sk}$ or T_a could be substantial [181]. Consequently, a critical evaluation of the various experiments on the concept of hypothalamic proportional T_b control was showing an increased complexity regarding the interactions between T_{hy} and $\overline{T}_{sk}$ in the control of T_C [52]. Probable explanations put forth to explain this evident variation in the thermosensitivity of the hypothalamus included the effector saturation as well as the hypothesized exponential link between a neuron's temperature and its activity [178, 182]. This relationship between neuronal activity and temperature was based on the concept of Q_{10}–like temperature dependence, defined as a relationship for the effect of temperature on a process such that the process rate increases by the same multiple for every 10°C rise in temperature [182].

Based on the idea of Q_{10}–like temperature dependence, Hammel [178, 182] was the first to suggest the hypothesis of a non-linear relationship between T_{hy} and neuronal activity, proposing the concept of an adjustable set point. Yet, this concept required re-evaluation of the Q_{10} properties of hypothalamic neurons. Simon [52] provides the following example: "*...Q_{10} values 41 predict that the gain of any thermoregulatory response to extrahypothalamic thermal stimulation should decrease with decreasing T_{hy}, irrespective of the kind of Q_{10} distribution between warm and cold signal pathways.*" Subsequent experiments in mammalian [183, 184] and avian [185] models confirmed the idea of Q_{10}–like temperature dependence of hypothalamic neurons. As a result, the hypothalamic proportional control model at its final state postulates that, amongst hypothalamic neurons receiving warm and cold afferent signals, there is variability in the distribution of different degrees of a Q_{10}-like temperature dependence. For instance, the control of shivering can demonstrate an average Q_{10} lower in the warm signal than in the cold signal pathway. This variability depends on the species and on the thermoregulatory function controlled by a specific hypothalamic network.

The idea of a common set point in endothermic temperature regulation has been challenged by experiments on extrahypothalamic thermoregulatory functions, and particularly that of the spinal cord. Several studies examined different deep-body compartments in order to detect extra-cerebral deep-body thermosensitivity based on the initial observation by Simon and colleagues of shivering in response to cooling of the peridural space of the vertebral canal and of the spinal cord [62]. This notion was in contrast to the popular view which favored the straightforward and, hence, convenient idea of only one region of deep-body thermosensitivity. Ergo, to prove the involvement of the spinal cord or the vertebral canal in thermoregulation it was necessary to ensure that T_{hy} remained constant (or increased) when the spinal cord was cooled. This was achieved by several experiments which clearly established thermoregulatory functions of extrahypothalamic deep-body compartments [186-188] demonstrating that: (i) afferent thermosensory input from the spinal cord interact with thermal inputs from the PH/AH, (ii) thermoregulatory reactions, including conscious behavior, influenced by T_{hy} and $\overline{T}_{sk}$ are also influenced by the temperature of the spinal cord, (iii) in the dog, thermal inputs from the spinal cord and the PO/AH were functionally identical and quantitatively similar. These results lead to the recognition that extrahypothalamic body core thermal inputs are active contributors to temperature regulation. Further criticism for the proportional hypothalamic control theory came from more recent experiments demonstrating that the PO/AH generating signals for vasomotion, shivering and salivation are disparate and functionally independent [160, 189-191]. Arguments against the adjustable set point theory stemmed also from research demonstrating the widely differing inter-threshold zones between heat and cold defense activation found in circumstances of pathophysiological thermoregulatory disturbances [192].

For five decades the prevailing view on the functional architecture of the human thermoregulatory system postulated that the hypothalamus is the brain region perceiving and controlling deep-body temperature. It still remains unknown whether this is true or not. Whatever the case may be, it is crucial to recognize that the seminal experiments of Hammel during the 1960's and onwards [54, 63, 172, 173, 175, 178, 182] resulting in the concept of Q_{10}–like temperature dependence of hypothalamic neuronal thermoreception created the necessary knowledge platform that has nurtured a higher level of understanding regarding the thermoregulatory control in endothermic organisms. Indeed, evolution of the hypothalamic proportional control model through the decades brought a lot of attention on the neural process by which the stability of body temperature in endotherms is achieved. The models presented in the following three subsections, although opposing the original hypothalamic proportional control model, owe their existence to the attention and the higher level of understanding generated by the research that created it.

The Comparator Model

The prevailing approach proposed by Hammel and colleagues [54] has been challenged from elegant studies by Kobayashi and colleagues conducted during the past 25 years the results of which oppose the dogma that receptors are sensors [i.e., the 'sensor model' [86]]. Based on Hammel's hypothalamic proportional control sensory model [54, 172-174], thermoreceptors convert temperatures of the various regions of the body into impulses in the form of a neural code, conveying the information of objects to the brain where these impulses are integrated by a separate network to form a mean temperature. Thereafter, this integrating network compares the mean temperature with a reference signal – that is, a set point – and through various processes, which have not been elucidated yet, orders are sent to thermoeffectors in order to evoke appropriate thermoregulatory responses.

In his 1989 review paper [77], Kobayashi argues that for warm- and cold-sensitive neurons in the central nervous system to be sensors, their firing rate (i.e., impulses·sec^{-1}) must be a code by which these neurons convey local temperature information. For this assumption to be correct the neural code of firing rate must be directly related to temperature and the receivers (i.e., inter-neurons in the central nervous system) of these impulses must possess the ability to decipher the code of firing rate and detect the local temperature information. Yet, as demonstrated by Kobayashi [77, 86], no convincing evidence has been presented hitherto confirming either of these assumptions.

Unlike the hypothalamic proportional control sensory model, the comparator model argues that receptors are comparators and opposes the view of signal transduction from the physical world to the brain [76, 77, 86, 90]. Based on the comparator model, heat- and cold-sensitive neurons are comparators of temperature eliciting impulses when temperature surpasses a receptor activation threshold. In turn, these impulses are not a form of neural code (as assumed by the sensor model) but only triggers to activate target neurons in the brain.

Recently, Kobayashi's group provided some experimental evidence supporting the fact that receptors are not sensors but, instead, comparators by recording membrane potential of neurons exhibiting cold receptors in primarily cultured dorsal root ganglion neurons [90]. Further support to these results came from the recent identification of four types of heat receptors and two types of cold receptors as members of the transient receptor potential family – a concept covered in detail in a previous section – [82]. This provides credence to the belief that heat and cold receptors are comparators of temperature and other physical quantities.

Heat and cold receptors exist in nerve endings with the skin around the entire body surface eliciting thermoregulatory autonomic sweating/shivering and behavioral responses when local skin temperature deviates from normal. Consequently, when skin temperature is above its threshold temperature, peripheral heat-sensitive neurons evoke impulses to stimulate target neurons in the brain (central thermosensors). In essence, the comparator model suggests that effectors are triggered via sensory neurons *automatically*, without the involvement/presence of a separate decision-making network or temperature code computations such as calculation of mean temperature.

Regarding the contribution of core and shell temperatures to thermoregulation, the comparator model holds that T_C is a very stable regulated variable serving as feedback signal, while shell temperatures are highly variable, not regulated parameters serving as feed-forward signals that allow the body to respond to a thermal load in advance to avoid changes in T_C [193]. Both core and shell temperatures trigger effector responses in a similar fashion. The combination of core and shell temperature activating a specific thermoeffector depends on the central and peripheral sensory neurons wired to it.

Core Temperature Defense at a Null Zone

The main idea behind the theoretical model of core temperature defense at a null zone is routed within the definitions of the terms 'thermal homeostasis' – that is, "*a relative constancy of thermal properties of the internal environment of an organism as being maintained by regulation*" [9] – and 'homoeothermy' – that is, "*the pattern of temperature regulation in a tachymetabolic species in which the cyclic variation in core temperature, either nychthemerally or seasonally, is maintained within arbitrarily defined limits despite much larger variations in ambient temperature, i.e., homoeotherms regulate their body temperature within a narrow range*" [9] – both of which refer to a regulated system [194]. Based on Bligh's theoretical concept [194], even in the most simplistic regulatory processes there must be some degree of variation. Inevitably, this will have consequences on the activities of neural interconnections between sensor-to-effector pathways, which vary the magnitude of a particular response to a particular species of sensor [194].

The crucial difference between the warm- and cold-sensitive neurons is that they exhibit disparate activity-temperature profiles. Within a limited range of temperature variation, the activities of warm-sensitive neurons increase as temperature increases, while that of cold-sensitive neurons decrease as temperature increases. Concomitantly, the effector responses to changes in thermal balance can be distinguished into those that control Q_P within the body and those that control the exchange of heat with the surrounding environment. The basic assumption of this neuronal model is that the cold- and warm-sensitive neuron inputs are each separately used to produce two thresholds, one for Q_P effectors and the other for Q_L effectors [194]. Within the central nervous system these two sensor-to-effector channels as well as other excitatory and inhibitory influences derived from nonthermal channels are reciprocally crossed. As a result, a range of thermoneutral zone is defined – which differs depending on the organism as a consequence of the effect of nonthermal factors on the vasomotor response [193] – where there is no need for any thermoregulatory response to occur. When the capacity of the vasomotor response to maintain a constant T_C is exceeded, autonomic sweating or shivering is activated. The T_C at which these effectors are activated is classified as the thermoeffector threshold T_C [5].

The concept of T_C defense at a null zone does not support the probability of body temperature regulation at a precise level, that is, a set point. In contrast, it postulates that body temperature regulation is coarse, permitting T_C under normal physiological conditions to vary within the null zone. Additionally, proponents of this theory claim that "*...clear and unequivocal neuronal evidence supportive of the reference signal hypothesis has remained elusive, yet the notion of a stable reference signal with which the variable is being constantly compared remains a popular way of considering the physiology of homoeothermy*" [194]. Indeed, it would be fallacious not to admit that, in the parts of the central nervous system known to be involved in thermoregulation, there are vast numbers of constantly discharging neurons with a firing rate independent of local temperature. It is also important to highlight that the firing rates of a large proportion of hypothalamic neurons – the aforementioned temperature-insensitive neurons – is unaffected, or minimally affected, by T_{hy} or changes in T_{hy} [64]. Based on the null zone model, that is indicative of the fact that they are unrelated to preservation of homoeothermy [5, 194].

The null zone model accepts that temperature stability does not necessarily require a constant reference signal [5, 194], a notion also supported by others [195]. Indeed, the essential components for the maintenance of T_b at a set level are two input signals with different temperature coefficients [196]. Although the comparison of a signal varying with temperature with another that is not affected by temperature – as postulated by the set point model – does fit this definition, another arrangement may well be the differing activity-temperature coefficients of signals derived from warm- and cold-sensitive neurons, as suggested by the null zone model [194]. Yet, this remains to be confirmed as the null zone model is still more theoretical than proven.

The Model of Heat Regulation

The model of hypothalamic proportional control dealing with temperature regulation implies the existence of temperature-sensitive neurons that respond appropriately to provide the body with a fixed temperature level – a set point T_{hy} or T_C – and its control is driven in proportion to the difference between the current temperature and the set point temperature [54, 172-174]. In contrast to this view, the model of heat regulation holds that the human body maintains a heat content equilibrium over a wide range of heat loads by sensing heat flow to/from the body and subsequently defending the body heat content (H_b) through specific physiological responses [197, 198]. Therefore, the main difference between these two models pertains to the *regulated variable* based on which thermoregulatory responses are triggered. Helped by the easiness to obtain them, temperature data – and the temperature regulation theory – have dominated thermal physiology since the first studies on the topic, yet as calorimetric techniques are improving, the concept of heat regulation is receiving more attention. It is also noteworthy to comment that the heat regulation model is an attempt to explain human (endothermic) thermoregulation employing only physiological canons and thermodynamic principles, and particularly those governing the heat balance equation. Little, or nothing, is mentioned regarding the efferent neural control of thermoeffectors, thus, one is left to assume that the temperature regulation (i.e., hypothalamic proportional control) and the heat regulation models share views regarding the efferent pathways to thermoeffectors.

The main driving force behind the heat regulation model has been Paul Webb who wrote several articles in favor of this hypothesis, the most prominent of which was his 1995 review paper [197]. Nevertheless, the concept of heat regulation was proposed several times by others, including Snellen – "*...the regulating mechanism operates as if there is heat regulation instead of temperature regulation*" [199] – and Adolph – "*...in heat regulation, the most general links are between a) heat content of the body and parts (or net temperature), b) heat additions, and c) heat dissipations*" [53], as well as others who dealt directly with the concept of heat regulation [200, 201].

The central premise of heat regulation is that the body constantly generates heat and has a characteristic range of temperature. Heat generation varies constantly according to the requirements of daily living, yet a relatively constant temperature implies that Q_L is adjusted to match Q_P. Therefore, it is more likely that the controller regulates a matching of outflow to inflow of heat than a T_b or some combination of T_b's [197]. Indeed, oxidation of fuel required for internal work generates heat which must be dissipated since there is more produced than needed in order to counter inevitable surface Q_L. Therefore, as metabolic rate, and, therefore, Q_P, changes to meet the energy demands of life's activities, Q_L must adapt to match it. Consequently, "*...Q_P is an independent variable, Q_L is a dependent variable*" [197].

Based on the model of heat regulation, heat flow in the body is sensed by monitoring transcutaneous thermal gradient via temperature sensors located on the skin surface and in different subcutaneous tissue layers [197]. These sensors have, indeed, been detected [198, 202-204] and changes in the transcutaneous thermal gradient have been related to regulatory processes [204-207], yet it has not been confirmed that they allow the monitoring of heat flow.

The heat regulation model is exceptionally successful in explaining/predicting thermal responses to phenomena are that are difficult to explain by the temperature regulation model (i.e., hypothalamic proportional control). One of these phenomena is the circadian temperature rhythm which, based on the results of many experiments [27, 208-212] is a result from phase- and time-course differences in Q_P and Q_L. Possibly the strongest evidence on this regard come from a 'constant metabolism experiment', where individuals performed steady-state exercise for 24 hours on a cycle ergometer while seated in a comfortable chair wearing a suit calorimeter [197]. The results showed that the constant metabolic rate – and, thus, Q_P – was matched by a constant rate of body Q_L throughout the 24 hours. Consequently, T_{re} did not present with the normal circadian rhythm showing that for a temperature variation from day to night, metabolic rate – and, thus, Q_P – must vary. More importantly, this experiment showed that changes in metabolic rate always drive changes in T_{re}, and not the other way around [197].

Another phenomenon which the heat regulation model is exceptionally successful in explaining/predicting thermal responses to is steady-state exercise. Indeed, although Q_L is normally regulated via cutaneous vasodilation and vasoconstriction (i.e., without employing shivering and sweating), when there is added heat generation from exercise, Q_L increases via sweating to dissipate the added endogenous heat. This powerful mechanism of evaporative cooling enables the body to achieve a heat balance by matching Q_L with Q_P [197]. This achievement of body heat balance during steady-state exercise has been confirmed in many experiments [200, 203, 204, 213, 214]. Yet, as demonstrated in a seminal experiment by Nielsen [215] and confirmed several times since [204, 216, 217], T_b increases linearly with exercise intensity independently of T_a. Further, what Webb's experiments demonstrated was that heat balance occurs first and T_{re} finally stabilizes only when a steady-state Q_L equals the steady-state metabolic rate [197]. This phenomenon has been recently observed also during cold water immersion [48].

In addition to circadian rhythm and exercise, other thermal phenomena are explained more appropriately by heat regulation than by temperature regulation (i.e., hypothalamic proportional control). For instance, the seemingly abnormal combinations of shell and core temperatures following precooling before heat exposure would normally predict increased sweating rates, yet sweating is absent until H_b is restored [197, 203, 218]. Similar results arise from experiments incorporating precooling before exercise [219, 220] as well as rewarming after chilling submerged swimmers in cold water [221]. On the other end of the thermal spectrum, results from several reports suggest that during rewarming, body heat is conserved until H_b – and not necessarily T_b – is restored [35, 222, 223]. Finally, other conditions explained more successfully by heat regulation than by temperature regulation include core cooling at fixed skin temperature and negative work [197].

Notwithstanding the problems in the hypothalamic proportional control model evaded by the heat regulation model, it is necessary to observe that proponents of the latter are faced with an even more difficult problem, that is how H_b as an extensive property might be measured. Also, the monitoring of transcutaneous thermal gradient via temperature sensors located on the skin surface and in different subcutaneous tissue layers has not been confirmed, while, even if one accepts that possibility, the fact that these sensors are not distributed uniformly around the body [52] prohibits a precise assessment of H_b.

Concluding Remarks

The clinical importance of the physiology of thermoregulation is immense. The prime example for this is that an exodus of T_b from its optimal range is so suggestive of a pathological condition that T_b is continuously monitored in all hospitalized patients and reported in every medical history, while the clinical thermometer is one of the most essential scientific instruments for the practice of medicine [7, 8]. Yet, despite that at the cellular and molecular level our knowledge of mechanisms governing thermal balance is increasing dramatically aided by the continuous advances in technology, at the systems level (i.e., the level of architecture and organization) several fundamental questions remain unanswered [5]. The most prominent of these questions involves the mechanism by which the thermoregulatory threshold is determined which, after numerous investigations, remains a mystery [6].

A general aspect exemplified by the history of research in the field of thermal physiology is that its development over the decades has been characterized by putting forward theses and antitheses and, eventually, by their synthesis requested by new convincing experimental evidence, i.e., a succession of concepts essential to progress in science. Which is the true nature of endothermic thermoregulation? Is any of the existing hypotheses put forth correct? Can it be employed in order to explain the various phenomena of human thermoregulation and/or to predict/simulate human thermoregulatory responses for different engineering applications? The present chapter reviewed a plethora of attempts to provide answers to these questions in order aid in the understanding of the functional architecture of the homoeothermic thermoregulatory system.

Future research endeavors should aim in further examining the dependable criterion of conscious thermoregulatory input, a very important topic given that behavioral thermoregulation is the principal form of thermoregulation used by humans. In addition, further research on the performance of the temperature regulation and heat regulation models is necessary in order to unravel the feedback mechanism generating the responses needed to maintain thermal balance. As local effects of cold as well as skin and core temperatures may play differential roles in CIVD, future experiments should also aim in isolating the local and central components of this response. Finally, further research on the role of vasoconstrictor nerve activity in maintaining thermal balance in humans should be projected in future studies.

References

[1] Riggs AJ, Millecchia RJ, Riggs JE. The contributions of Lavoisier, Carnot, and Mayer to understanding heat illness. *Aviat Space Environ Med.* 2004 Oct;75(10):916-7.

[2] Greene CW. *Kirkes' Handbook of Physiology.* Revised by C. W. Greene. New York: W. Wood; 1913.

[3] Ott I. Heat center in the brain. *J Nerv Ment Dis.* 1887;14:152.

[4] Richet C. Die beziehung des gehirns zur korperwarme und zum fiever. *Arch Ges Physiol.* 1885;37:624.

[5] Mekjavic IB, Eiken O. Contribution of thermal and nonthermal factors to the regulation of body temperature in humans. *J Appl Physiol.* 2006 Jun;100(6):2065-72.

[6] Cooper KE. Some historical perspectives on thermoregulation. *J Appl Physiol.* 2002 Apr;92(4):1717-24.

[7] Romanovsky AA. Thermoregulation: some concepts have changed. Functional architecture of the thermoregulatory system. *Am J Physiol Regul Integr Comp Physiol.* 2007 Jan;292(1):R37-46.

[8] Burton AC. The clinical importance of the physiology of temperature regulation. *Can Med Assoc J.* 1956 Nov 1;75(9):715-20.

[9] IUPS Thermal Commission. Glossary of terms for thermal physiology, 3rd ed. Jap J *Physiol.* 2001;51(2):245-80.

[10] Grigg GC, Beard LA, Augee ML. The evolution of endothermy and its diversity in mammals and birds. *Physiol Biochem Zool.* 2004 Nov-Dec;77(6):982-97.

[11] Bennett AF, Ruben JA. Endothermy and activity in vertebrates. *Science.* 1979 Nov 9;206(4419):649-54.

[12] Ivanov KP. The development of the concepts of homeothermy and thermoregulation. *J Therm Biol.* 2006;31:24–9.

[13] Crompton AW, Taylor CR, Jagger JA. Evolution of homeothermy in mammals. *Nature.* 1978 Mar 23;272(5651):333-6.

[14] Burton AC, Edholm OG. *Man in a cold environment.* London: Edward Arnold, Ltd.; 1955.

[15] Voets T, Droogmans G, Wissenbach U, Janssens A, Flockerzi V, Nilius B. The principle of temperature-dependent gating in cold- and heat-sensitive TRP channels. *Nature.* 2004 Aug 12;430(7001):748-54.

[16] Wilson E. Advances in representing the degradative effects of chemical and biological weapons in conflict. *15th International Symposium on Military Operational Research;* 1998 1-4 September; Operational Research Society, Swindon, UK; 1998.

[17] Keim SM, Guisto JA, Sullivan JB, Jr. Environmental thermal stress. *Ann Agric Environ Med.* 2002;9(1):1-15.

[18] Xu X, Hexamer M, Werner J. Multi-loop control of liquid cooling garment systems. *Ergonomics.* 1999 Feb;42(2):282-98.

[19] Young AJ, Sawka MN, Epstein Y, Decristofano B, Pandolf KB. Cooling different body surfaces during upper and lower body exercise. *J Appl Physiol.* 1987 Sep;63(3):1218-23.

[20] Holmer I, Nilsson H, Havenith G, Parsons K. Clothing convective heat exchange--proposal for improved prediction in standards and models. *Ann Occup Hyg.* 1999 Jul;43(5):329-37.

[21] Parsons KC. Protective clothing: heat exchange and physiological objectives. *Ergonomics*. 1988 Jul;31(7):991-1007.

[22] Epstein Y, Sohar E. Fluid balance in hot climates: sweating, water intake, and prevention of dehydration. *Public Health Rev.* 1985;13(1-2):115-37.

[23] Havenith G, Holmer I, den Hartog EA, Parsons KC. Clothing evaporative heat resistance--proposal for improved representation in standards and models. *Ann Occup Hyg.* 1999 Jul;43(5):339-46.

[24] Parsons K. *Human thermal environments*. London: Third Ed. Taylor and Francis Group; 2003.

[25] Reardon FD, Leppik KE, Wegmann R, Webb P, Ducharme MB, Kenny GP. The Snellen human calorimeter revisited, re-engineered and upgraded: design and performance characteristics. *Med Biol Eng Comput.* 2006 Aug;44(8):721-8.

[26] Hambraeus L, Sjodin A, Webb P, Forslund A, Hambraeus K, Hambraeus T. A suit calorimeter for energy balance studies on humans during heavy exercise. *Eur J Appl Physiol Occup Physiol.* 1994;68(1):68-73.

[27] Webb P, Annis JF, Troutman SJ, Jr. Energy balance in man measured by direct and indirect calorimetry. *Am J Clin Nutr.* 1980 Jun;33(6):1287-98.

[28] Hardy JD, Stolwijk JA. Partitional calorimetric studies of man during exposures to thermal transients. *J Appl Physiol.* 1966 Nov;21(6):1799-806.

[29] Stolwijk JA, Hardy JD. Partitional calorimetric studies of responses of man to thermal transients. *J Appl Physiol.* 1966 May;21(3):967-77.

[30] DuBois D, DuBois EF. A formula to estimate the approximate surface area if height and weight be known. *Arch Intern Med.* 1916;17:863-71.

[31] Kouno T, Katsumata N, Mukai H, Ando M, Watanabe T. Standardization of the body surface area (BSA) formula to calculate the dose of anticancer agents in Japan. *Jpn J Clin Oncol.* 2003 Jun;33(6):309-13.

[32] Burton AD. Human calorimetry: The average temperature of the tissues of the body. *J Nutr.* 1935;9:261-80.

[33] Stolwijk JA, Hardy JD. Control of body temperature. *Handbook of Physiology Reactions to environmental agents.* Bethesda, MD: American Physiological Society; 1977. p. 45-84.

[34] Havenith G. Individualized model of human thermoregulation for the simulation of heat stress response. *J Appl Physiol.* 2001 May;90(5):1943-54.

[35] Webb P. Heat storage and body temperature during cooling and rewarming. *Eur J Appl Physiol Occup Physiol.* 1993;66(1):18-24.

[36] Hardy JD, DuBois EF. Basal metabolism, radiation, convection, and vaporization at temperatures 22°C to 35°C. *J Nutr.* 1938;15:477.

[37] Aoyagi Y, McLellan TM, Shephard RJ. Determination of body heat storage in clothing: calorimetry versus thermometry. *Eur J Appl Physiol Occup Physiol.* 1995;71(2-3):197-206.

[38] Vallerand AL, Savourey G, Hanniquet AM, Bittel JH. How should body heat storage be determined in humans: by thermometry or calorimetry? *Eur J Appl Physiol Occup Physiol.* 1992;65(3):286-94.

[39] Jay O, Gariepy LM, Reardon FD, Webb P, Ducharme MB, Ramsay T, et al. A three-compartment thermometry model for the improved estimation of changes in body heat content. *Am J Physiol Regul Integr Comp Physiol.* 2007 Jan;292(1):R167-75.

[40] Armstrong LE, Pandolf KB. Physical training, cardiorespiratory physical fitness and exercise-heat tolerance. In: Pandolf KB, Sawka MN, Gonzalez RR, editors. *Human performance physiology and environmental medicine at terrestrial extremes.* indianapolis: Benchmark Press; 1988. p. 199-226.

[41] Roberts MF, Wenger CB, Stolwijk JA, Nadel ER. Skin blood flow and sweating changes following exercise training and heat acclimation. *J Appl Physiol.* 1977 Jul;43(1):133-7.

[42] Wenger CB. Human heat acclimatization. In: Pandolf KB, Sawka MN, Gonzalez RR, editors. *Human performance physiology and environmental medicine at terrestrial extremes.* Indianapolis: Benchmark Press; 1988. p. 153-97.

[43] Aoyagi Y, McLellan TM, Shephard RJ. Effects of training and acclimation on heat tolerance in exercising men wearing protective clothing. *Eur J Appl Physiol Occup Physiol.* 1994;68(3):234-45.

[44] Goldman RF. Heat stress in industrial protective encapsulating garments. In: Levine SP, Martin WF, editors. *Protecting personnel at hazardous waste sites.* Boston: Butterworth; 1985. p. 215-66.

[45] Ramanathan NL. A new weighting system for mean surface temperature of the human body. *J Appl Physiol.* 1964 May;19:531-3.

[46] Frim J, Livingstone SD, Reed LD, Nolan RW, Limmer RE. Body composition and skin temperature variation. *J Appl Physiol.* 1990 Feb;68(2):540-3.

[47] Nadel ER, Bergh U, Saltin B. Body temperatures during negative work exercise. *J Appl Physiol. 1*972 Nov;33(5):553-8.

[48] Tikuisis P. Heat balance precedes stabilization of body temperatures during cold water immersion. *J Appl Physiol.* 2003 Jul;95(1):89-96.

[49] Noble D, Boyd CAR. The Logic of Life. *The Challenge of Integrative Physiology.* New York: The Physiological Society and Oxford University Press; 1993.

[50] Cowley A. Global manpower needs for integrative systems physiologists. *Physiologist.* 2005 Feb;48(1):1, 3-8.

[51] Cabanac M. Adjustable set point: to honor Harold T. Hammel. *J Appl Physiol.* 2006 Apr;100(4):1338-46.

[52] Simon E. Deep-body thermosensitivity: Another look. A tribute to Harold Theodore (Ted) Hammel (1921–2005). *J Therm Biol.* 2006;31:4–18.

[53] Adolph EF. Look at physiological integration. *Am J Physiol.* 1979 Nov;237(5):R255-9.

[54] Hammel HT, Hardy JD, Fusco MM. Thermoregulatory responses to hypothalamic cooling in unanesthetized dogs. *Am J Physiol.* 1960 Mar;198:481-6.

[55] Satinoff E. Behavioral thermoregulation in response to local cooling of the rat brain. *Am J Physiol.* 1964 Jun;206:1389-94.

[56] Carlisle HJ, Ingram DL. The effects of heating and cooling the spinal cord and hypothalamus on thermoregulatory behavior in the pig. *J Physiol.* 1973 Jun;231(2):353-64.

[57] Holmes RL, Newman PP, Wolstencroft JH. A heatsensitive region in the medulla. *J Physiol.* 1960 Jun;152:93-8.

[58] Newman PP, Wolstencroft JH. Cardiovascular and respiratory responses to heating the carotid blood. *J Physiol.* 1960 Jun;152:87-92.

[59] Lipton JM. Thermosensitivity of medulla oblongata in control of body temperature. *Am J Physiol.* 1973 Apr;224(4):890-7.

[60] Simon E. Temperature regulation: the spinal cord as a site of extrahypothalamic thermoregulatory functions. *Rev Physiol Biochem Pharmacol.* 1974(71):1-76.

[61] Chai CY, Lin MT. Effects of heating and cooling the spinal cord and medulla oblongata on thermoregulation in monkeys. *J Physiol.* 1972 Sep;225(2):297-308.

[62] Simon E, Rautenberg W, Thauer R, Iriki M. Ausloesung thermoregulatorischer Reaktionen durch lokale Kuehlung im Vertebralkanal (Stimulation of thermoregulatory reactions by local cooling within the vertebral canal). Naturwissenschaften. 1963;50:337.

[63] Hammel HT. Neurons and temperature regulation. In: Yamamoto WS, Brobeck JR, editors. *Physiological controls and regulations*. Philadelphia, PA: Saunders; 1965. p. 71–97.

[64] Boulant JA. Hypothalamic neurons controlling body temperature. In: Blatteis CM, Fregly MJ, editors. *Handbook of Physiology,* section 4, Environmental Physiology. New York: Oxford University Press; 1996. p. 105–26.

[65] Wechselberger M, Wright CL, Bishop GA, Boulant JA. Ionic channels and conductance-based models for hypothalamic neuronal thermosensitivity. *Am J Physiol Regul Integr Comp Physiol.* 2006 Sep;291(3):R518-29.

[66] Zhao Y, Boulant JA. Temperature effects on neuronal membrane potentials and inward currents in rat hypothalamic tissue slices. *J Physiol.* 2005 Apr 1;564(Pt 1):245-57.

[67] Hori T, Nakashima T, Hori N, Kiyohara T. Thermo-sensitive neurons in hypothalamic tissue slices in vitro. *Brain Res.* 1980 Mar 17;186(1):203-7.

[68] Hardy JD, Hellon RF, Sutherland K. Temperature-sensitive neurones in the dog's hypothalamus. *J Physiol.* 1964 Dec;175:242-53.

[69] Kelso SR, Perlmutter MN, Boulant JA. Thermosensitive single-unit activity of in vitro hypothalamic slices. *Am J Physiol.* 1982 Jan;242(1):R77-84.

[70] Zhang YH, Yanase-Fujiwara M, Hosono T, Kanosue K. Warm and cold signals from the preoptic area: which contribute more to the control of shivering in rats? *J Physiol.* 1995 May 15;485 (Pt 1):195-202.

[71] Chen XM, Hosono T, Yoda T, Fukuda Y, Kanosue K. Efferent projection from the preoptic area for the control of non-shivering thermogenesis in rats. *J Physiol.* 1998 Nov 1;512 (Pt 3):883-92.

[72] Griffin JD, Saper CB, Boulant JA. Synaptic and morphological characteristics of temperature-sensitive and -insensitive rat hypothalamic neurones. *J Physiol.* 2001 Dec 1;537(Pt 2):521-35.

[73] Cliffer KD, Burstein R, Giesler GJ, Jr. Distributions of spinothalamic, spinohypothalamic, and spinotelencephalic fibers revealed by anterograde transport of PHA-L in rats. *J Neurosci.* 1991 Mar;11(3):852-68.

[74] Boulant JA, Hardy JD. The effect of spinal and skin temperatures on the firing rate and thermosensitivity of preoptic neurones. *J Physiol.* 1974 Aug;240(3):639-60.

[75] Kobayashi S, Murakami N. Thermosensitive neurons in slice preparations of rat medulla oblongata. *Brain Res Bull.* 1982 Jun;8(6):721-6.

[76] Kobayashi S. Warm- and cold-sensitive neurons inactive at normal core temperature in rat hypothalamic slices. *Brain Res.* 1986 Jan 1;362(1):132-9.

[77] Kobayashi S. Temperature-sensitive neurons in the hypothalamus: a new hypothesis that they act as thermostats, not as transducers. *Prog Neurobiol.* 1989;32(2):103-35.

[78] Hensel H. Thermoreception and temperature regulation. London: Academic Press; 1981.

[79] Simon E. The enigma of deep-body thermosensory specificity. *Int J Biometeorol.* 2000 Sep;44(3):105-20.

[80] Thauer R. Die Ausloesungsmechanismen der chemischen Temperaturregulation (Mechanisms stimulating chemical temperature regulation). *Proceedings of the 22nd Congress of the International Union of Physiological Sciences* (IUPS); 1962; Volume 1, part 1. Leiden, Netherlands; 1962. p. 390–402.

[81] Simon E, Pierau FK, Taylor DC. Central and peripheral thermal control of effectors in homeothermic temperature regulation. *Physiol Rev.* 1986 Apr;66(2):235-300.

[82] Caterina MJ. Transient receptor potential ion channels as participants in thermosensation and thermoregulation. *Am J Physiol Regul Integr Comp Physiol.* 2007 Jan;292(1):R64-76.

[83] Hua le H, Strigo IA, Baxter LC, Johnson SC, Craig AD. Anteroposterior somatotopy of innocuous cooling activation focus in human dorsal posterior insular cortex. *Am J Physiol Regul Integr Comp Physiol.* 2005 Aug;289(2):R319-R25.

[84] Nomoto S, Shibata M, Iriki M, Riedel W. Role of afferent pathways of heat and cold in body temperature regulation. I*nt J Biometeorol.* 2004 Nov;49(2):67-85.

[85] Asami T, Hori T, Kiyohara T, Nakashima T. Convergence of thermal signals on the reticulospinal neurons in the midbrain, pons and medulla oblongata. *Brain Res Bull.* 1988 May;20(5):581-96.

[86] Kobayashi S, Okazawa M, Hori A, Matsumura K, Hosokawa H. Paradigm shift in sensory system - Animals do not have sensors. J Therm Biol. 2006;31:19-23.

[87] Sorabji R. *Aristotle on Memory.* London: Duckworth; 1972.

[88] Aristotle. Posterior Analytics (translation with commentary). *Clarendon Aristotle Series.* Oxford, 1996: Clarendon Press; 350BC.

[89] Locke J. In: Nidditch PH, editor. *An essay concerning human understanding Oxford,* 1979: Oxford University Press; 1690.

[90] Okazawa M, Takao K, Hori A, Shiraki T, Matsumura K, Kobayashi S. Ionic basis of cold receptors acting as thermostats. *J Neurosci.* 2002 May 15;22(10):3994-4001.

[91] Cottingham J, Stoothoff R, Murdoch D. *The philosophical writings of Descartes.* Cambridge: Cambridge University Press; 1985.

[92] Egan GF, Johnson J, Farrell M, McAllen R, Zamarripa F, McKinley MJ, et al. Cortical, thalamic, and hypothalamic responses to cooling and warming the skin in awake humans: a positron-emission tomography study. *Proc Natl Acad Sci U S A.* 2005 Apr 5;102(14):5262-7.

[93] Nagashima K, Nakai S, Tanaka M, Kanosue K. Neuronal circuitries involved in thermoregulation. *Auton Neurosci.* 2000 Dec 20;85(1-3):18-25.

[94] Bamshad M, Song CK, Bartness TJ. CNS origins of the sympathetic nervous system outflow to brown adipose tissue. *Am J Physiol.* 1999 Jun;276(6 Pt 2):R1569-78.

[95] Cano G, Passerin AM, Schiltz JC, Card JP, Morrison SF, Sved AF. Anatomical substrates for the central control of sympathetic outflow to interscapular adipose tissue during cold exposure. *J Comp Neurol.* 2003 Jun 2;460(3):303-26.

[96] Oldfield BJ, Giles ME, Watson A, Anderson C, Colvill LM, McKinley MJ. The neurochemical characterisation of hypothalamic pathways projecting polysynaptically to brown adipose tissue in the rat. *Neuroscience.* 2002;110(3):515-26.

[97] Tanaka M, McAllen RM. A subsidiary fever center in the medullary raphe? *Am J Physiol Regul Integr Comp Physiol.* 2005 Dec;289(6):R1592-8.

[98] Alvarez GE, Zhao K, Kosiba WA, Johnson JM. Relative roles of local and reflex components in cutaneous vasoconstriction during skin cooling in humans. *J Appl Physiol.* 2006 Feb 16;100(6):2083-8.

[99] Taylor WF, Johnson JM, O'Leary D, Park MK. Effect of high local temperature on reflex cutaneous vasodilation. *J Appl Physiol.* 1984 Jul;57(1):191-6.

[100] Johnson JM, Brengelmann GL, Hales JR, Vanhoutte PM, Wenger CB. Regulation of the cutaneous circulation. *Fed Proc.* 1986 Dec;45(13):2841-50.

[101] Johnson JM, Proppe DW. Cardiovascular adjustments to heat stress. In: Fregly MJ, Blatteis CM, editors. *Handbook of Physiology Section 4: Environmental Physiology* Vol 1. New York: Oxford University Press; 1996. p. 215-43.

[102] Bini G, Hagbarth KE, Hynninen P, Wallin BG. Regional similarities and differences in thermoregulatory vaso- and sudomotor tone. *J Physiol.* 1980 Sep;306:553-65.

[103] Johnson JM. Exercise and the cutaneous circulation. *Exerc Sport Sci Rev.* 1992;20:59-97.

[104] Roddie IC. Circulation to skin and adipose tissue. In: Shepherd JT, Abboud FM, editors. Handbook of physiology. *Bethesda: American Physiological Society;* 1983. p. 285–317.

[105] Crandall CG, Johnson JM, Kosiba WA, Kellogg DL, Jr. Baroreceptor control of the cutaneous active vasodilator system. *J Appl Physiol.* 1996 Nov;81(5):2192-8.

[106] Charkoudian N. Skin blood flow in adult human thermoregulation: how it works, when it does not, and why. *Mayo Clin Proc.* 2003 May;78(5):603-12.

[107] Stephenson LA, Kolka MA. Menstrual cycle phase and time of day alter reference signal controlling arm blood flow and sweating. *Am J Physiol.* 1985 Aug;249(2 Pt 2):R186-91.

[108] Aoki K, Stephens DP, Johnson JM. Diurnal variation in cutaneous vasodilator and vasoconstrictor systems during heat stress. *Am J Physiol Regul Integr Comp Physiol.* 2001 Aug;281(2):R591-5.

[109] Johnson JM, Park MK. Reflex control of skin blood flow by skin temperature: role of core temperature. *J Appl Physiol.* 1979 Dec;47(6):1188-93.

[110] Greenfield ADM. The circulation through the skin. In: Hamilton WF, Dow P, editors. *Handbook of physiology. Washington: American Physiological Society;* 1963. p. 1325–51.

[111] Sawasaki N, Iwase S, Mano T. Effect of skin sympathetic response to local or systemic cold exposure on thermoregulatory functions in humans. *Auton Neurosci.* 2001 Mar 23;87(2-3):274-81.

[112] Johnson JM, Taylor WF, Shepherd AP, Park MK. Laser-Doppler measurement of skin blood flow: comparison with plethysmography. *J Appl Physiol.* 1984 Mar;56(3):798-803.

[113] Saumet JL, Kellogg DL, Jr., Taylor WF, Johnson JM. Cutaneous laser-Doppler flowmetry: influence of underlying muscle blood flow. *J Appl Physiol.* 1988 Jul;65(1):478-81.

[114] Oberg PA. Laser-Doppler flowmetry. *Crit Rev Biomed Eng.* 1990;18(2):125-63.

[115] Kellogg DL, Jr., Crandall CG, Liu Y, Charkoudian N, Johnson JM. Nitric oxide and cutaneous active vasodilation during heat stress in humans. *J Appl Physiol.* 1998 Sep;85(3):824-9.

[116] Johnson JM, Pergola PE, Liao FK, Kellogg DL, Jr., Crandall CG. Skin of the dorsal aspect of human hands and fingers possesses an active vasodilator system. *J Appl Physiol.* 1995 Mar;78(3):948-54.

[117] Pérgola PE, Kellogg DL, Jr., Johnson JM, Kosiba WA. Reflex control of active cutaneous vasodilation by skin temperature in humans. *Am J Physiol.* 1994 May;266(5 Pt 2):H1979-84.

[118] Pérgola PE, Johnson JM, Kellogg DL, Jr., Kosiba WA. Control of skin blood flow by whole body and local skin cooling in exercising humans. *Am J Physiol.* 1996 Jan;270(1 Pt 2):H208-15.

[119] Stephens DP, Saad AR, Bennett LA, Kosiba WA, Johnson JM. Neuropeptide Y antagonism reduces reflex cutaneous vasoconstriction in humans. *Am J Physiol Heart Circ Physiol.* 2004 Sep;287(3):H1404-9.

[120] Chotani MA, Flavahan S, Mitra S, Daunt D, Flavahan NA. Silent alpha(2C)-adrenergic receptors enable cold-induced vasoconstriction in cutaneous arteries. *Am J Physiol Heart Circ Physiol.* 2000 Apr;278(4):H1075-83.

[121] Johnson JM, Yen TC, Zhao K, Kosiba WA. Sympathetic, sensory, and nonneuronal contributions to the cutaneous vasoconstrictor response to local cooling. *Am J Physiol Heart Circ Physiol.* 2005 Apr;288(4):H1573-9.

[122] Daanen HA. Finger cold-induced vasodilation: A review. *Eur J Appl Physiol.* 2003 Jun;89(5):411-26.

[123] Fox RH, Goldsmith R, Kidd DJ. Cutaneous vasomotor control in the human head, neck and upper chest. *J Physiol.* 1962 May;161:298-312.

[124] Grant RT, Bland E. Observations on arterio-venous anastomoses in human skin and in the bird's foot with special reference to the reaction to cold. *Heart* 1931;15:385-411.

[125] Grant RT. Observations on direct communications between arteries and veins in the rabbit's ear. *Heart.* 1930;15:281-303.

[126] Lossius K, Eriksen M, Walløe L. Fluctuations in blood flow to acral skin in humans: connection with heart rate and blood pressure variability. *J Physiol.* 1993 Jan;460:641-55.

[127] Shastry S, Minson CT, Wilson SA, Dietz NM, Joyner MJ. Effects of atropine and L-NAME on cutaneous blood flow during body heating in humans. *J Appl Physiol.* 2000 Feb;88(2):467-72.

[128] Dampney RA, McAllen RM. Differential control of sympathetic fibres supplying hindlimb skin and muscle by subretrofacial neurones in the cat. *J Physiol.* 1988 Jan;395:41-56.

[129] McAllen RM, May CN. Effects of preoptic warming on subretrofacial and cutaneous vasoconstrictor neurons in anaesthetized cats. *J Physiol.* 1994 Dec 15;481 (Pt 3):719-30.

[130] Stephens DP, Aoki K, Kosiba WA, Johnson JM. Nonnoradrenergic mechanism of reflex cutaneous vasoconstriction in men. *Am J Physiol Heart Circ Physiol.* 2001 Apr;280(4):H1496-504.

[131] Stephens DP, Bennett LA, Aoki K, Kosiba WA, Charkoudian N, Johnson JM. Sympathetic nonnoradrenergic cutaneous vasoconstriction in women is associated with reproductive hormone status. *Am J Physiol Heart Circ Physiol.* 2002 Jan;282(1):H264-72.

[132] Pergola PE, Kellogg DL, Jr., Johnson JM, Kosiba WA, Solomon DE. Role of sympathetic nerves in the vascular effects of local temperature in human forearm skin. *Am J Physiol.* 1993 Sep;265(3 Pt 2):H785-92.

[133] Minson CT, Berry LT, Joyner MJ. Nitric oxide and neurally mediated regulation of skin blood flow during local heating. *J Appl Physiol.* 2001 Oct;91(4):1619-26.

[134] Holzer P. Neurogenic vasodilatation and plasma leakage in the skin. *Gen Pharmacol.* 1998 Jan;30(1):5-11.

[135] Stephens DP, Charkoudian N, Benevento JM, Johnson JM, Saumet JL. The influence of topical capsaicin on the local thermal control of skin blood flow in humans. *Am J Physiol Regul Integr Comp Physiol.* 2001 Sep;281(3):R894-901.

[136] Caterina MJ, Schumacher MA, Tominaga M, Rosen TA, Levine JD, Julius D. The capsaicin receptor: a heat-activated ion channel in the pain pathway. *Nature.* 1997 Oct 23;389(6653):816-24.

[137] Vanhoutte PM, Cooke JP, Lindblad LE, Shepherd JT, Flavahan NA. Modulation of postjunctional alpha-adrenergic responsiveness by local changes in temperature. *Clin Sci (Lond).* 1985;68 Suppl 10:121s-3s.

[138] Flavahan NA, Lindblad LE, Verbeuren TJ, Shepherd JT, Vanhoutte PM. Cooling and alpha 1- and alpha 2-adrenergic responses in cutaneous veins: role of receptor reserve. *Am J Physiol.* 1985 Nov;249(5 Pt 2):H950-5.

[139] Lewis T. Observations upon the reactions of the vessels of the human skin to cold. *Heart.* 1930;15:177-208.

[140] Wilson O, Goldman RF. Role of air temperature and wind in the time necessary for a finger to freeze. *J Appl Physiol.* 1970;29:658–64.

[141] Keatinge WR. Effects of temperature on blood vessels. In: *"Local mechanisms controlling blood vessels".* London: Cambridge University Press; 1980.

[142] Keatinge WR. The effect of general chilling on the vasodilator response to cold. *J Physiol.* 1957 Dec 31;139(3):497-507.
[143] Cannon P, Keatinge WR. The metabolic rate and heat loss of fat and thin men in heat balance in cold and warm water. *J Physiol.* 1960 Dec;154:329-44.
[144] Keatinge WR. The effect of low temperatures on the responses of arteries to constrictor drugs. *J Physiol.* 1958 Aug 6;142(3):395-405.
[145] Sendowski I, Savourey G, Besnard Y, Bittel J. Cold induced vasodilatation and cardiovascular responses in humans during cold water immersion of various upper limb areas. *Eur J Appl Physiol Occup Physiol.* 1997;75(6):471-7.
[146] Sendowski I, Savourey G, Launay JC, Besnard Y, Cottet-Emard JM, Pequignot JM, et al. Sympathetic stimulation induced by hand cooling alters cold-induced vasodilatation in humans. *Eur J Appl Physiol.* 2000 Mar;81(4):303-9.
[147] Meyer AA, Webster AJ. Cold-induced vasodilatation in the sheep. *Can J Physiol Pharmacol.* 1971 Oct;49(10):901-8.
[148] Werner J. Influences of local and global temperature stimuli on the Lewis-reaction. *Pflugers Arch.* 1977 Jan 17;367(3):291-4.
[149] Brown RT, Baust JG. Time course of peripheral heterothermy in a homeotherm. *Am J Physiol.* 1980 Jul;239(1):R126-9.
[150] Kunimoto M. Where is the most useful point in the skin temperature curve of the fingertip immersed in the cold water to evaluate the function of the skin sympathetic function? *Biomed Thermogr.* 1987;7:254-68.
[151] Elsner RW, Nelms JD, Irving L. Circulation of heart to the hands of Arctic Indians. *J Appl Physiol.* 1960 Jul;15:662-6.
[152] Daanen HA, Van de Linde FJ, Romet TT, Ducharme MB. The effect of body temperature on the hunting response of the middle finger skin temperature. *Eur J Appl Physiol Occup Physiol.* 1997;76(6):538-43.
[153] Takano N, Kotani M. Influence of food intake on cold-induced vasodilatation of finger. *Jpn J Physiol.* 1989;39(5):755-65.
[154] Kellogg DL, Jr., Johnson JM, Kosiba WA. Control of internal temperature threshold for active cutaneous vasodilation by dynamic exercise. *J Appl Physiol.* 1991 Dec;71(6):2476-82.
[155] Elizondo RS. Local control of eccrine sweat gland function. *Fed Proc.* 1973 May;32(5):1583-7.
[156] Pappano AJ. Cholinoceptor-activating and cholinesterase-inhibiting drugs. In: Katzung BG, editor. *Basic and clinical pharmacology.* 8th Edition ed. New York: Lange Medical Books; 1998. p. 92-107.
[157] Shibasaki M, Crandall CG. Effect of local acetylcholinesterase inhibition on sweat rate in humans. *J Appl Physiol.* 2001 Mar;90(3):757-62.
[158] Mills PC, Marlin DJ, Scott CM, Smith NC. Nitric oxide and thermoregulation during exercise in the horse. *J Appl Physiol.* 1997 Apr;82(4):1035-9.
[159] Buxton IL, Cheek DJ, Eckman D, Westfall DP, Sanders KM, Keef KD. NG-nitro L-arginine methyl ester and other alkyl esters of arginine are muscarinic receptor antagonists. *Circ Res.* 1993 Feb;72(2):387-95.

[160] Kanosue K, Yanase-Fujiwara M, Hosono T. Hypothalamic network for thermoregulatory vasomotor control. *Am J Physiol.* 1994 Jul;267(1 Pt 2):R283-8.

[161] Stuart DG, Kawamura Y, Hemingway A. Activation and suppression of shivering during septal and hypothalamic stimulation. *Exp Neurol.* 1961 Dec;4:485-506.

[162] Halvorson I, Thornhill J. Posterior hypothalamic stimulation of anesthetized normothermic and hypothermic rats evokes shivering thermogenesis. *Brain Res.* 1993 May 7;610(2):208-15.

[163] Benzinger TH, Pratt AW, Kitzinger C. The thermostatic control of human metabolic heat production. *Proc Natl Acad Sci U S A.* 1961 May;47(5):730-9.

[164] Hemingway A. Shivering. Physiol Rev. 1963 Jul;43:397-422.

[165] Eyolfson DA, Tikuisis P, Xu X, Weseen G, Giesbrecht GG. Measurement and prediction of peak shivering intensity in humans. *Eur J Appl Physiol.* 2001 Jan-Feb;84(1-2):100-6.

[166] Xu X, Tikuisis P, Gonzalez R, Giesbrecht G. Thermoregulatory model for prediction of long-term cold exposure. *Comput Biol Med.* 2005 May;35(4):287-98.

[167] Tikuisis P, Giesbrecht GG. Prediction of shivering heat production from core and mean skin temperatures. *Eur J Appl Physiol Occup Physiol.* 1999 Feb;79(3):221-9.

[168] Weber JM, Haman F. Fuel selection in shivering humans. *Acta Physiol Scand.* 2005 Aug;184(4):319-29.

[169] Haman F. Shivering in the cold: from mechanisms of fuel selection to survival. *J Appl Physiol.* 2006 May;100(5):1702-8.

[170] Nagashima K. Central mechanisms for thermoregulation in a hot environment. *Ind Health.* 2006 Jul;44(3):359-67.

[171] Barbour HG. The effect of direct warming and cooling of the thermoregulatory centers on body temperature (translated from the German title: Die Wirkung unmittelbarer Erwärmung und Abkühlung der Wärmezentra auf die Körpertemperatur). *Naunyn-Schmiedeberg's Arch.* 1912;70:1-26.

[172] Hammel HT, Jackson DC, Stolwijk JA, Hardy JD, Stromme SB. Temperature regulation by hypothalamic proportional control with an adjustable set point. *J Appl Physiol.* 1963 Nov;18:1146-54.

[173] Hammel HT, Stromme S, Cornew RW. Proportionality constant for hypothalamic proportional control of metabolism in unanesthetized dog. *Life Sci.* 1963 Dec;12:933-47.

[174] Nakayama T, Hammel HT, Hardy JD, Eisenman JS. Thermal stimulation of electrical activity of single units of the preoptic region. *Am J Physiol.* 1963;204:1122-26.

[175] Hammel HT. Concept of the adjustable set temperature. In: Hardy JD, Gagge AP, Stolwijk JAJ, editors. *Physiological and behavioral temperature regulation.* Springfield: Charles C. Thomas; 1970. p. 676–83.

[176] Hellstrøm B, Hammel HT. Some characteristics of temperature regulation in unanesthetized dog. *Am J Physiol.* 1967;213:547–56.

[177] Heller HC, Hammel HT. CNS control of body temperature during hibernation. *Comp Biochem Physiol A.* 1972 Feb 1;41(2):349-59.

[178] Hammel HT, Heller HC, Sharp FR. Probing the rostral brainstem of anesthetized, unanesthetized, and exercising dogs and of hibernating and euthermic ground squirrels. *Fed Proc.* 1973 May;32(5):1588-97.

[179] Jackson DC, Hammel HT. Hypothalamic "set" temperature decreased in exercising dog. *Life Sci.* 1963 Aug;8:554-63.

[180] Sharp FR, Hammel HT. Effects of fever on salivation response in the resting and exercising dog. *Am J Physiol.* 1972 Jul;223(1):77-82.

[181] Brueck K, Schwennicke HP. Interaction of superficial and hypothalamic thermosensitive structures in the control of nonshivering thermogenesis. *Int J Biometeorol.* 1971;15:156-61.

[182] Hammel HT. The set-point in temperature regulation: analogy or reality. In: Bligh J, Moore R, editors. *Essays on temperature regulation.* Amsterdam: North-Holland; 1972. p. 121–37.

[183] Inomoto T, Mercer JB, Simon E. Opposing effects of hypothalamic cooling on threshold and sensitivity of metabolic response to body cooling in rabbits. *J Physiol.* 1982 Jan;322:139-50.

[184] Inomoto T, Mercer JB, Simon E. Interaction between hypothalamic and extrahypothalamic body temperatures in the control of panting in rabbits. *Pflugers Arch.* 1983 Jul;398(2):142-6.

[185] Inomoto T, Simon E. Extracerebral deep-body cold sensitivity in the Pekin duck. *Am J Physiol.* 1981 Sep;241(3):R136-45.

[186] Jessen C, Ludwig O. Spinal cord and hypothalamus as core sensors of temperature in the conscious dog. II. Addition of signals. *Pflugers Arch.* 1971;324(3):205-16.

[187] Jessen C, Mayer ET. Spinal cord and hypothalamus as core sensors of temperature in the conscious dog. I. Equivalence of responses. *Pflugers Arch.* 1971;324(3):189-204.

[188] Jessen C, Simon E. Spinal cord and hypothalamus as core sensors of temperature in the conscious dog. III. Identity of functions. *Pflugers Arch.* 1971;324(3):217-26.

[189] Kanosue K, Nakayama T, Tanaka H, Yanase M, Yasuda H. Modes of action of local hypothalamic and skin thermal stimulation on salivary secretion in rats. *J Physiol.* 1990 May;424:459-71.

[190] Kanosue K, Zhang YH, Yanase-Fujiwara M, Hosono T. Hypothalamic network for thermoregulatory shivering. *Am J Physiol.* 1994 Jul;267(1 Pt 2):R275-82.

[191] Kanosue K, Yoshida K, Maruyama M, Nagashima K. The central organization of the thermoregulatory system. In: Kosaka M, Sugahara T, Schmidt KL, Simon E, editors. *Thermotherapy for neoplasia, inflammation, and Pain.* Tokyo Springer; 2001. p. 2–11.

[192] Romanovsky AA. Do fever and anapyrexia exist? Analysis of set point-based definitions. *Am J Physiol Regul Integr Comp Physiol.* 2004 Oct;287(4):R992-5.

[193] Romanovsky AA, Ivanov AI, Shimansky YP. Selected contribution: ambient temperature for experiments in rats: a new method for determining the zone of thermal neutrality. *J Appl Physiol.* 2002 Jun;92(6):2667-79.

[194] Bligh J. A theoretical consideration of the means whereby the mammalian core temperature is defended at a null zone. *J Appl Physiol.* 2006 Apr;100(4):1332-7.

[195] Werner J. The concept of thermoregulation for human body temperature. *J Therm Biol.* 1980;5:75-82.

[196] Hensel H. Neural processes in thermoregulation. *Physiol Rev.* 1973 Oct;53(4):948-1017.

[197] Webb P. The physiology of heat regulation. *Am J Physiol.* 1995 Apr;268(4 Pt 2):R838-50.

[198] Bazett HC. Theory of reflex controls to explain regulation of body temperature at rest and during exercise. *J Appl Physiol.* 1951 Oct;4(4):245-62.

[199] Snellen JW. Set point and exercise. In: Bligh J, Moore RE, editors. *Essays on temperature regulation.* Amsterdam: North Holland; 1972. p. 139-48.

[200] Grucza R. Body heat balance in man subjected to endogenous and exogenous heat load. *Eur J Appl Physiol Occup Physiol.* 1983;51(3):419-33.

[201] Houdas Y, Sauvage A, Bonaventure M, Ledru C, BGuieu JD. Thermal control in man: Regulation of central temperature or adjustments of heat exchanges by servomechanisms? *J Dynamic Syst Meas Control.* 1973;95:331-5.

[202] Ebbecke U. Uber die Temperaturempfindungen in ihrer Abhängigkeit von der Hautdurchblutung und von den Reflexzentren. *Pfluegers Arch.* 1917;169:385-462.

[203] Kerslake DM. *The stress of hot environments.* Cambridge: Cambridge University Press; 1972.

[204] Nielsen B. Thermoregulation in rest and exercise. *Acta Physiol Scand Suppl.* 1969;323:1-74.

[205] Webb P. Temperatures of skin, subcutaneous tissue, muscle and core in resting men in cold, comfortable and hot conditions. *Eur J Appl Physiol Occup Physiol. 1992;64(5):471-6.*

[206] Ivanov KP. The location and function of different skin thermoreceptors. In: Bligh J, Voigt K, editors. *Thermoreception and temperature regulation.* Berlin: Springer-Verlag; 1990. p. 37-43.

[207] Ivanov KP, Dymnikova L, Danilova N. The hypothalamus neuron's response to the signals from surface and deep skin thermoreceptors. *J Therm Biol.* 1987;12:289-93.

[208] Aschoff J, Heise A. Thermal conductance in man: its dependence on time of day and on ambient temperature. In: Itoh S, Ogata K, Yoshimura H, editors. *Advances in climatic physiology.* Tokyo: Igaku Shoin; 1972. p. 334-48.

[209] Krauchi K, Wirz-Justice A. Circadian rhythm of heat production, heart rate, and skin and core temperature under unmasking conditions in men. *Am J Physiol.* 1994 Sep;267(3 Pt 2):R819-29.

[210] Robinson EL, Demaria-Pesce VH, Fuller CA. Circadian rhythms of thermoregulation in the squirrel monkey (Saimiri sciureus). *Am J Physiol.* 1993 Oct;265(4 Pt 2):R781-5.

[211] Aschoff J, Pohl H. Thermal conductance in man: its dependence on time of day and on ambient temperature. In: Itoh S, Ogata K, Yoshimura H, editors. *Advances in climatic physiology.* Tokyo: Igaku Shoin; 1972. p. 334-48.

[212] Webb P. Continuous thermal comfort in a suit calorimeter. In: Durand J, Raynaud J, editors. *Thermal comfort: physiological and psychological bases.* Paris: INSERM; 1979. p. 177-85.

[213] Davies CT. The effects of different levels of heat production induced by diathermy and eccentric work on thermoregulation during exercise at a given skin temperature. *Eur J Appl Physiol Occup Physiol.* 1979 Feb 15;40(3):171-80.

[214] Kerslake DM. Factors concerned in the regulation of sweat production in man. *J Physiol.* 1955 Feb 28;127(2):280-96.

[215] Nielsen M. Die regulation der körpertemperatur bei muskelarbeit. *Scand Arch Physiol.* 1938;79:193-230.

[216] Lind AR. A physiological criterion for setting thermal environmental limits for everyday work. *J Appl Physiol.* 1963 Jan;18:51-6.

[217] Wyndham CH, Bouwer WM, Devine MG, Paterson HE. Physiological responses of African laborers at various saturated air temperatures, wind velocities and rates of energy expenditure. *J Appl Physiol.* 1952 Dec;5(6):290-8.

[218] Veghte JH, Webb P. Body cooling and response to heat. *J Appl Physiol.* 1961 Mar;16:235-8.

[219] Olschewski H, Bruck K. Thermoregulatory, cardiovascular, and muscular factors related to exercise after precooling. *J Appl Physiol.* 1988 Feb;64(2):803-11.

[220] Schmidt V, Bruck K. Effect of a precooling maneuver on body temperature and exercise performance. *J Appl Physiol.* 1981 Apr;50(4):772-8.

[221] Webb P. Rewarming after diving in cold water. *Aerosp Med.* 1973 Oct;44(10):1152-7.

[222] Ereth MH, Lennon RL, Sessler DI. Limited heat transfer between thermal compartments during rewarming in vasoconstricted patients. *Aviat Space Environ Med.* 1992 Dec;63(12):1065-9.

[223] Kruk B, Pekkarinen H, Harri M, Manninen K, Hanninen O. Thermoregulatory responses to exercise at low ambient temperature performed after precooling or preheating procedures. *Eur J Appl Physiol Occup Physiol.* 1990;59(6):416-20.

In: Body Temperature Regulation
Editor: A. B. Cisneros and B. L. Goins
ISBN: 1-60741-282-3

Chapter III

Controlling Body Temperature – The Opportunities for Highly Productive Domestic Fowl

S. Yahav*[*]*, D. Shinder, M. Ruzal, M. Giloh and Y. Piestun
Department of Poultry and Aquaculture Science, ARO, the Volcani Center, P.O. Box 6, Bet Dagan 50250, Israel

Abstract

Birds are endotherms, able to maintain their body temperature within a narrow range. However, an alteration in body temperature as a result of exposure to extreme environmental conditions may lead to a cascade of irreversible thermoregulatory events that could be lethal for the bird. To sustain thermal tolerance and avoid the deleterious consequences of thermal stresses, three direct responses are elicited: the rapid thermal stress response; acclimation; and epigenetic temperature adaptation. Although domestic fowl respond to changes in the environment with the same mechanisms as other endotherms, their intensive production causes them to face serious difficulties in coping with extreme environmental changes. Recent decades have seen significant progress in the genetic selection, on the one hand, of fast-growing meat-type broiler chickens and turkeys and, on the other hand, of egg-producing laying hens. However, the significant improvements in body and muscle growth and in egg production have not necessarily been accompanied by equivalent growth of specific visceral organs, and the probable consequence is a reduced ability to cope with extreme changes in environmental temperatures.

This chapter will focus on: a. body temperature development in chickens during embryogenesis; b. the control of body temperature during acute and chronic exposure to various environmental conditions; and c. the efficacy of controlling body temperature, as affected by thermal conditioning (epigenetic temperature adaptation) experience during perinatal period.

Introduction

In endothermic animals (mammals and birds) body temperature (T_b) is the most physiologically guarded parameter of the body, therefore the thermoregulatory system in these animals operates at a very high gain, in order to control T_b within a relatively narrow range, despite moderate to extreme changes in the environmental conditions. The ability to maintain a stable T_b springs from the mechanisms that control heat production and heat loss; mechanisms that changed in the course of evolution, to enable endothermia instead of ectothermia. The evolutionary changes from ectothermia to endothermia were achieved because the developmental regulatory mechanisms maintained a balance between heat production and heat loss. Both mechanisms, especially heat production, are probably older than endothermy, but both are permanently activated and regulated by both neuronal and hormonal signals (Silva, 2006; Morrison et al., 2008; Richards and Proszkowiec-Weglarz, 2008).

The center for T_b regulation is neuroanatomically located in the preoptic/anterior hypothalamus (PO/AH; Boulant, 1996; DiMicco and Zaretsky, 2007). This area plays a dual role: it monitors local temperature changes, and it integrates temperature information from the periphery (Hellon and Taylor, 1982; Patapoutain et al., 2003; Wechselberger et al., 2006). Temperature is monitored by temperature-sensitive neurons, which change their firing rate in accordance with the hypothalamic temperature (Griffin et al., 2001). These neurons comprise 40% of the PO/AH neurons; about 75% of them are warmth sensitive – their firing rate increases or decreases in parallel with the hypothalamus temperature – and about 25% of them are cold sensitive with a firing rate that changes in the opposite sense to the hypothalamus temperature (Kelso and Boulant, 1982, Dean and Boulant 1989, Tzschentke and Basta, 2002; Tzschentke et al., 2004). The T_b control site in the hypothalamus controls the metabolic capacity via the hypothalamus-hypophysis-thyroid axis and the adrenal axis (DiMicco and Zaretsky, 2007; Freake and Oppenheimer, 1995).

Thermogenesis (heat production) can be divided into obligatory and facultative thermogenesis (Silva, 2006). Obligatory thermogenesis refers to the energy required to maintain T_b while the ambient temperature (T_a) lies in the thermoneutral zone, a range in which the body is in thermal equilibrium with the environment and produces energy at a level termed the resting metabolic rate (RMR) (Gordon, 1993). Facultative thermogenesis refers to prompted energy production required when the T_a deviates below or, to some extent, above the thermoneutral zone. The facultative thermogenesis is based on the short-term shivering thermogenesis (ST) and a long-term mechanism, the non-shivering thermogenesis (NST). In birds NST occurs in the skeletal muscle (Duchamp and Barré, 1993; Dridi et al., 2008). Thermogenesis is regulated by the thyroid hormones (McNabb and King, 1993; Silva, 2006). T_a influences the rate of metabolic activity and, thereby, the amount of oxygen required by

* Corresponding author: yahavs@agri.huji.ac.il

the bird (Freeman, 1964; Kuhn et al., 1984; Buys et al., 1999). The metabolic rate is associated with the secretion of the hormone, thyroxine (T_4), which is secreted by the thyroid gland, and is deiodinated to triiodothyronine (T_3) in the periphery, in all tissues. Triiodothyronine is the main metabolism-stimulating hormone (McNabb and King, 1993; Gabarrou et al., 1997) that was found to be associated with temperature regulation (McNabb and King, 1993; Carew et al., 1998; Gonzales et al., 1999; Yahav, 2000a).

The other parameter of the thermoregulatory equation is heat loss. In birds, heat is dissipated through respiratory-evaporative mechanisms (Richards, 1968, 1970, 1976; Seymour, 1972; Marder and Arad, 1989), and cutaneous evaporative mechanism (Webster and King, 1987; Ophir et al., 2002) and sensible heat loss via radiation, convection (Yahav et al., 2005) and conduction (Wolfenson et al., 2001). The main avian hormone controlling heat dissipation is arginine vasotocin (AVT, Seth et al., 2004).

Domestic fowls especially the highly productive agricultural fowls (broilers, turkeys and laying hens) differ to some extent from other endotherms in their ability to maintain a sufficiently dynamic steady state despite severe changes in the environment (Yahav, 2000a). This can be attributed to their enhanced production.

This chapter will focus on the development of T_b in broiler chickens and turkeys and its control under fluctuating environments, by means of various strategies.

Body Temperature Development in Chickens during Embryogenesis – The Transition from Ecto to Endotherm Phase

The incubation temperature is the most important vital parameter during embryogenesis, and for several decades the practice in breeding domestic fowl was and still is to keep this parameter stable and to optimize it during incubation, to achieve the best hatchability (Swann and Brake, 1990c; French, 1997) and chick quality (Wilson, 1991; Decuypere and Michels, 1992), while avoiding possible deleterious effects on the development of the chicken embryo (Peterka et al., 1996; Krausova and Peterka, 2007). The temperature of the embryo does, indeed, depend on incubation temperature, but also on other environmental parameters such as relative humidity, air speed and air flow, which are basically the parameters of the physical microenvironment around the egg (Meir and Ar, 1987; Swann and Brake, 1990a,b,c). These, coupled with the exchange of heat between the egg and its microenvironment and the capacity of the embryo to produce heat, will determinate the embryo's T_b (Van Brecht et al., 2005). However, the embryo's T_b is usually measured by a non-invasive technique based on observation of the egg-shell temperature T_{egg}, which is well known to be an accurate indicator of the embryo T_b (Southerland et al, 1987; Joseph et al., 2006; Leksprisompong et al., 2007).

The egg shell temperature is considered to be affected by the embryo's levels of heat production and heat loss, of which the latter is affected mainly by changes in the blood flow in the chorioallantoic membrane (Reizis et al., 2005).

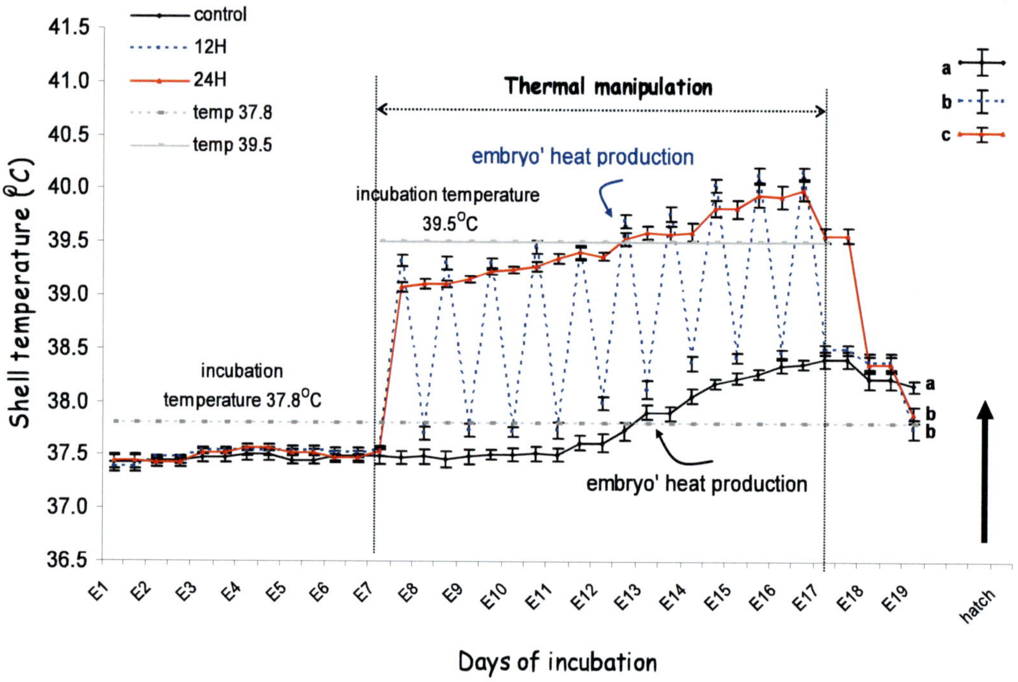

Figure 1. The effects on egg shell temperature during incubation, of thermal manipulation (TM) at 39.5°C and 65% relative humidity, applied for 12 h/day (12H) or continuously (24H) from day 7 to day 16 of incubation, i.e., E7 to E16 (inclusive). Embryo's heat production represents the period during which the embryos metabolic rate increases to a level that allows T_{egg} to rise above the incubation temperature. Within each embryogenesis day, values designated by different letters differ significantly ($P \leq 0.05$).

The broiler's T_{egg} is known to increase as the embryo develops and its pattern of change can be divided into three phases (Figure 1; control treatment). During the first phase T_{egg} is lower than the incubation temperature. This phase lasts from the beginning of incubation until E10 to E13, depending on the incubation temperature, which may range between 37 and 38°C. Whittow and Tazawa (1991) made it clear that even a small deviation in incubation temperature from 37.5 to 37.8°C would lead to changes in the total duration of incubation, because of changes in the embryo metabolic rate, which is directly related to the incubation temperature prior to piping. The 2nd phase is the one in which T_{egg} increases with advancing embryogenesis, and rises to about 1°C above the incubation temperature. During this phase the egg thermal conductance remains relatively constant (Van-Brecht et al., 2005), suggesting that at a certain point (related to incubation temperature), T_{egg} will rise above the incubation temperature because heat production exceeds heat loss. The third phase, which extends from E19 onwards, is when internal piping begins and T_{egg} drops but remains higher than the incubation temperature.

Despite the conservative idea that the incubation temperature of domestic fowl must remain relatively constant in order to maximize hatchability and synchronization of the hatching time, several recent studies demonstrated the beneficial effects of elevating the incubation temperature in order to improve thermotolerance acquisition without impairing embryo development or the post-hatch performance of the fast-growing broiler (Yahav et al., 2004a,b; Collin et al., 2007; Piestun et al., 2008a; Shinder et al., 2008). These studies further demonstrated that it is possible to alter the incubation temperature in order to improve acquisition of thermotolerance with respect to extreme environmental conditions, in case of either hot or cold spells (Piestun et al., 2008a; Shinder et al., 2008). This was based on the fact that in nature incubation temperature varies between 30 and 40°C (Webb, 1987) as a result of the necessity for parent birds to search for food or to avoid predators, and those changes may contribute to acquisition of enhanced thermotolerance for coping with changes in the environment. However, the fact that improvement of thermotolerance acquisition impairs production in domestic fowl elicited a certain caution, with limits to considered alterations of the incubation temperature.

To achieve the double goal of thermotolerance improvement without impairment of the genetic potential to produce, two strategies were used: long-term, moderate temperature alteration; and short-term, extreme temperature changes. Both strategies were based on the hypothesis that the beneficial effects of elevating the incubation temperature may be achieved through "imprinting" of physiological control systems, such as the thermoregulatory system (Nichelmann et al., 1994, 1999; Tzschentke and Basta, 2002; Tzschentke et al., 2004; Tzschentke and Plagemann, 2006), which depends on the possibility of inducing long-term temperature adaptation of an organism to acute changes in climatic conditions (Yahav and Hurwitz, 1996; Shinder et al., 2008; Piestun et al., 2008b). The differences among the strategies are related to the different stages of embryo development at which they were applied. The long-term moderate temperature change was based on the critical period of development and maturation of the hypothalamus-hypophysis-thyroid axis and the adrenal axis, from days E7 to E16 (inclusive) and was used to improve thermotolerance under hot conditions (Piestun et al., 2008a), whereas the short-term, extreme temperature change was used during the final phase of incubation, when the embryo switches from ecto- to endothermic behavior, i.e., on E18 and E19, and was used to improve thermotolerance under cold conditions (Shinder et al., 2008).

Under the first strategy, broiler embryos were intermittently exposed to 39.5°C and 65% relative humidity, either for 12 h/day (12H, Figure 1) from E7 to E16 inclusive, or continuously (24H, Figure 1). The effect of increasing incubation temperature was clearly shown in an approximately 1-day reduction of the time needed for the embryo to increase T_{egg} above incubation temperature in both 12H and 24H treatments, as compared with the control treatment. This suggests: a. an improvement of embryogenesis with increasing temperature; and b. that there was an ectothermic response of the embryos to elevated temperature. The stable and significantly lower T_{egg} that was observed from E18.5 onwards suggests that the embryos had reached their endothermic phase; it also indicated a lower metabolic rate of the thermal manipulated (TM) embryos as a result of decreased heat production. This phenomenon was maintained at hatch and post hatch, and will be discussed later in this chapter.

The second strategy – short-term exposure to extreme temperature – was used to elucidate the effect on post-hatch cold tolerance, of cold exposure episodes of 15°C lasting 30 or 60 min during E18 and E19 (Shinder et al., 2008). Figure 2 illustrates the dramatic effect of cold exposure on T_{egg}. This was characterized by a fall of T_{egg} from approximately 39.5°C to approximately 30 or 25°C after 30 or 60 min of exposure, respectively. This can be related to the exponential dynamics of cooling (Tazawa et al., 1988a, b, 2001) resulting from a dramatic increase in sensible heat loss via radiative and free-convective heat transfer from the egg (Shinder et al., 2008), following an abrupt decrease in incubation temperature. Following termination of cold exposure T_{egg} and the subsequent T_b of the hatched chick were higher than those of the controls, suggesting an increase in the metabolic rate of the cold-exposed embryos (Josef et al., 2006). Also, as a result of the improved thermotolerance efficiency under cold conditions, the number of incidences of ascites syndrome diminished (Shinder et al., 2008).

There is no doubt that the incubation temperature plays a crucial role through its effect on the embryo body temperature. The conservative practice of keeping a constant incubation temperature must be changed carefully, to ensure that adoption of the practice of altering the incubation temperature according to the above two strategies, ensures, on the one hand, improved acquisition of thermotolerance with respect to the environment, and, on the other hand, sustained production and performance.

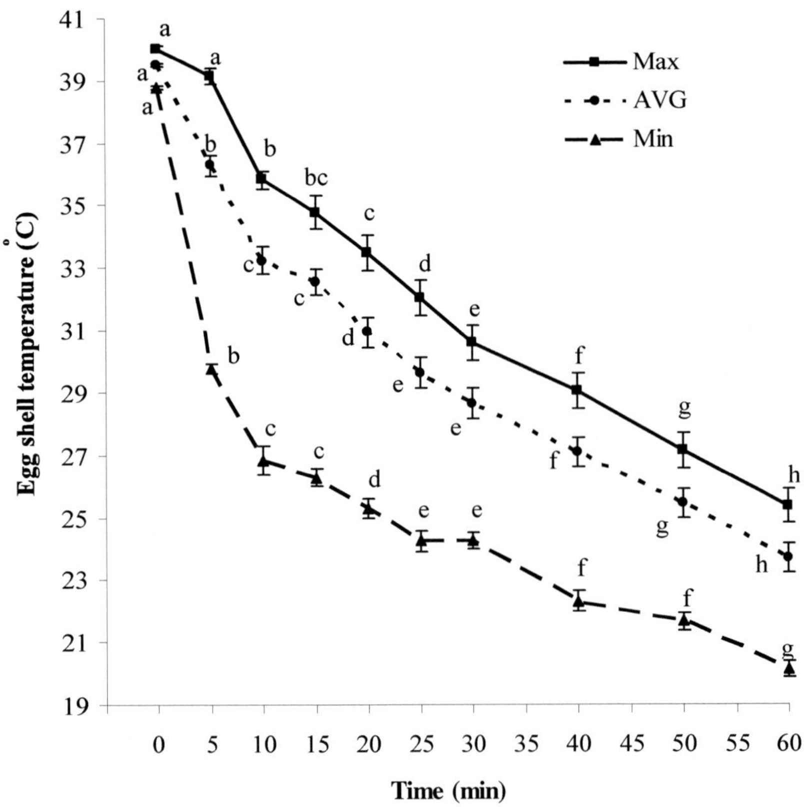

Figure 2. Egg shell temperatures (T_{egg}) (measured by thermography); at the middle of the egg (embryo) – Max; an average over the whole T_{egg} – AVG; and at the top of the egg (air cell) – Min, during 60 min of cold (15°C) exposure on d 19 of incubation (E19). Within each regression line, values designated by different letters differed significantly (Mean ± SE; n = 8; $P < 0.05$). (According to Shinder et al. 2008).

Controlling T_b Post Hatch – Three Differnt Strategies

Domestic fowl, especially the highly productive species such as broiler chickens, turkeys and laying hens, are able to maintain their T_b post-hatch within a narrow range. However, an increase or decrease in T_b above or below the regulated range, as a result of exposure to extreme environmental conditions, may lead to a cascade of irreversible thermoregulatory events that could be lethal for the bird. This irreversible situation was found to be particularly important for domestic fowl, because of the last decade's improvement in the genetic selection of these birds. Development in genetic selection of the meat-type fowl, i.e., broilers (Havenstein et al., 1994, 2003a; Cunningham, 2006) and turkeys (Havenstein et al., 2007) led to rapid growth, with increased feed efficiency and metabolic rate (Janke et al., 2004), that, as requested, provides the poultry industry with heavy domestic fowl in relatively short growth periods. Such developments logically necessitate parallel increases in the size of the cardiovascular and respiratory systems, as well as enhancements of their functional efficiency. However, inferior development of such major systems (Havenstein et al., 2003b) has led to relatively low capability to balance energy expenditure and body water balance, which has resulted in a relatively low capability to efficiently control T_b under extreme environmental conditions.

Similarly to other endotherms, domestic fowl can sustain thermal tolerance and avoid the deleterious consequences of thermal stresses by using three direct responses: the rapid thermal stress response – RTSR (Parsell and Lindquist, 1994; Yahav et al., 1997a); acclimation/acclimatization (Yahav et al., 1997b); and embryonic or post-hatch TM, based on epigenetic adaptation during the perinatal period (Janke et al., 2002; Nichelmann et al., 1999; Piestun et al., 2008a,b; Tzschentke et al., 2001; Tzschentke and Basta, 2002; Yahav and Hurwitz, 1996).

The first two of these, RTSR and acclimation, depend on exposure duration, whereas TM depends on timing. The RTSR is an immediate response to thermal stress exposure; it initiates the recruitment of bodily resources to maintain the animal in a dynamic steady state. This response cannot be maintained for a long period, therefore, under continuous exposure to severe heat or cold stress a cascade of events may lead to heat or cold stroke.

Acclimation is defined as physiological or behavioral changes occurring within the lifetime of an organism, which reduces the strain or enhances tolerance of strain caused by experimentally induced stressful changes, in particular, climatic factors (IUPS Thermal Commission, 2001). Whereas acclimation refers to experimental procedures, acclimatization refers to natural processes. Acclimation involves an array of autonomically controlled physiological mechanisms working in concert to enhance heat or cold tolerance (Horowitz, 2002). Thermal acclimation/acclimatization can be characterized as an expansion of the dynamic thermoregulatory range of the bird. Basically, it involves the expansion of the body core temperature safety margins; alteration of heat loss and heat production, which is achieved through concerted procedures at all levels of the body, and induces a shift in the level of homeostasis so as to efficiently improve thermotolerance in the new environmental conditions.

Inducing shifts in the level of homeostasis is based on the plasticity of the thermoregulatory control elements located in the preoptic anterior hypothalamus (PO/AH). This site is composed of thermo-sensitive neurons that are able to sense core T_b and to control physiological, endocrinological, and behavioral responses, so as to keep core T_b relatively constant in mammals (Boulant, 1996; Boulant and Dean, 1986) and birds (Tzschentke and Basta, 2000; Tzschentke et al., 2004). In mammals (Pierau et al., 1998) as well as in birds (Tzschentke and Basta, 2002) changing the sensitivity of the warm- and/or cold-sensitive neurons located in the PO/AH may change the threshold for heat production and/or heat loss of the animal, and thereby ensure better capability to cope with changes in the environment.

Acclimation/acclimatization is initiated in response to persistent or repeated challenge of the thermoregulatory system by environmental temperatures, and has a long-acting effect. In rats a biphasic model for acclimation to heat was suggested (Horowitz, 1998); it comprises a short first phase (1-5 d) and a second phase that lasts for over 3 wk. Although, there is no evidence for two phases of acclimation in birds, it seems that 4-7 days may be sufficient to control T_b efficiently in broilers (Yahav et al., 1997b), turkeys (Yahav, 1999) and laying hens (Yahav et al., 2000) exposed to different T_a. However, changes in the cardiovascular system can be detected only after approximately 2-3 wk in broilers (Yahav et al., 1997b) and turkeys (Yahav, 1999).

On the other hand, the TM response is timing-dependent and it basically occurs during the critical time-segment of embryogenesis or post-hatch, in which changes in the environmental conditions can induce a long-lasting physiological memory.

Rapid Thermal Stress Response – The Development of Hyper- or Hypothermia

One of the immediate responses to extreme elevated or reduced T_a is the development of hyperthermia or hypothermia, respectively (Etches et al., 1995; Hillman et al., 1985). In both cases this response initiates redistribution of the blood flow in such a way that heat will be dissipated from or conserved in the body, respectively. The level and time course of induction of hyper-/hypothermia is related to the level of T_a to which the animal is exposed. They are also age dependent, which is a reflection of the changing surface-to-volume ratio.

Hyperthermia Development

Exposure to elevated T_a will increase blood flow to the skin, especially to the non-feathered areas of the fowl (Wolfenson et al., 1981; Wolfenson, 1986; Yahav et al., 1998a, 2004c, 2005), as well as to the upper respiratory passageways (Wolfenson et al., 1981), in order to transport heat from the viscera to the periphery.

Increased blood flow to the periphery – vasodilatation – will follow the development of hyperthermia that results from acute heat exposure. The development of hyperthermia followed by increase in surface temperature, as indicated by facial surface temperature, is illustrated in Figure 3. These graphs show the level of hyperthermia developed as a result of acute heat exposure. Irrespective of the broiler age, hyperthermia was developed in a short time, and T_b reached values of 42.9-44.9, 42.6-43.7, 42.6-43.3, and 43.1-44.6°C at the ages of 9, 16, 22 and 36 days, respectively; the increased facial surface temperatures that developed in parallel reached 40.9-43.0, 40.9-42.3, 41.2-41.7, and 42.5-44.0°C, respectively. Whereas before the acute exposure to heat stress a difference of approximately 2.0°C between T_b and facial surface temperature was monitored, this difference was significantly reduced during the development of hyperthermia, suggesting a limited capacity for vasodilatation despite the progressive development of hyperthermia.

The level of hyperthermia from which a sufficient recovery can occur was found to be between 44.0 and 44.5°C for broilers and turkeys (Yahav et al., unpublished data).

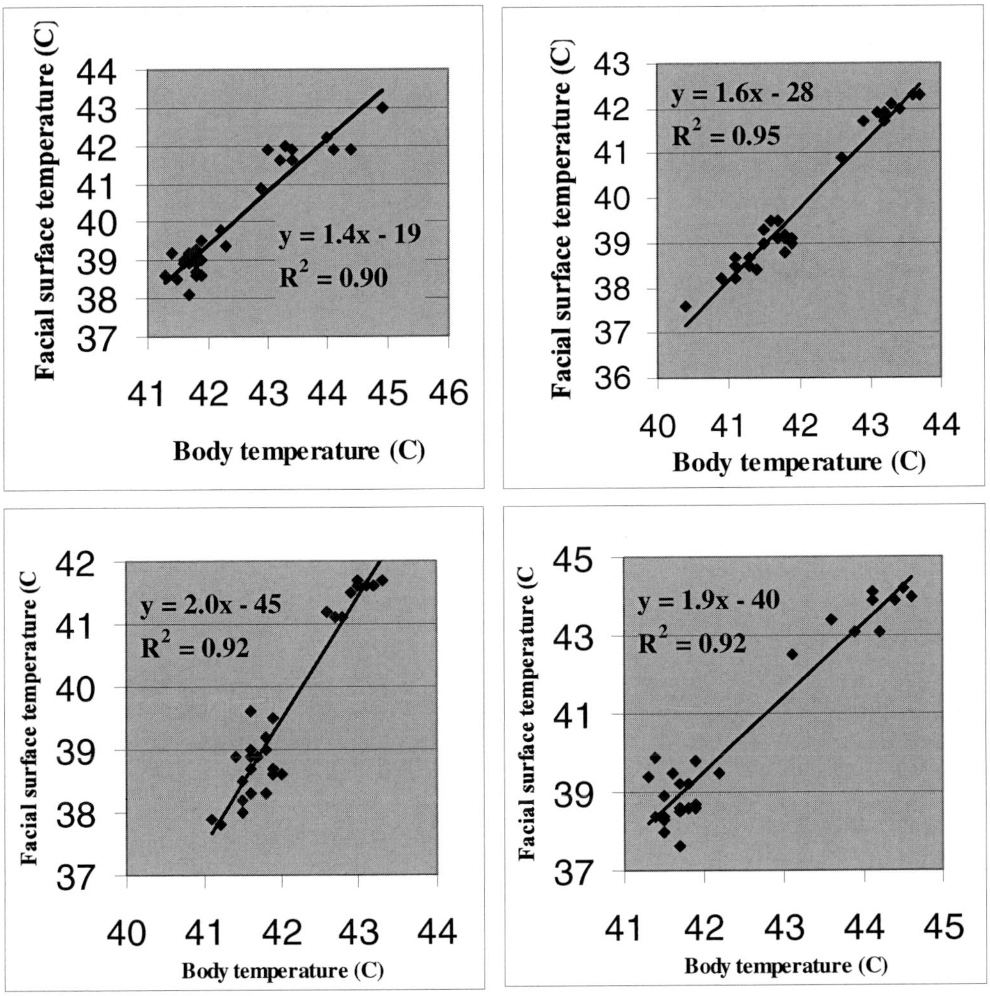

Figure 3. Hyperthermia development as correlated with the age of broiler chickens, and its correlation with vasodilatation development, as indicated by facial surface temperature. The graphs represent body and facial surface temperature of broilers exposed to 41 (upper, left), 37 (upper, right), 35 (lower, left), and 35°C (Lower, right) for 4 h, at 9, 16, 23, and 36 days of age, respectively. In each graph the lower group of squares represents the pre-exposure temperatures, and those measured after recovery from heat exposure.

It can be speculated that above this level, heat stroke is developed. Heat stroke is characterized by detrimental changes at the cellular and molecular levels that lead to the fatal cascade of events of decreased blood pressure, brain hypoxia, neuronal dysfunction, cell fatigue, etc. (Hales et al., 1996). Thus, heat stroke may be the reason for insufficient recovery that is characterized by neuronal disorders, or for complete inability to recover.

The consequence of skin temperature elevation is the enhancement of the temperature gradient between the body surface and the ambient environment, which is the main driving force of sensible heat loss (SHL) via the mechanisms of radiation, convection (Yahav et al., 2004c, 2005) and conduction (Wolfenson et al., 2001). The elevated blood flow to the skin is also responsible for cutaneous water evaporation (CWE) from the skin, a process that is observed mainly in desert-dwelling avian species (Smith and Suthers, 1969, Arieli et al., 2002; Ophir et al., 2002) – desert pigeons, doves and sand grouse – which use this evaporative route as their main cooling mechanism. CWE can be 2.5 times as great as the overall evaporative mechanism (Ophir et al., 2002) and is attributed to the microstructure of the skin tissue. Although broiler chickens have no association with desert adaptations, Yahav et al. (2005) hypothesized that there is a mechanism of passive water loss from the cutaneous tissue of broiler chickens exposed to high T_a and a high ventilation rate.

Although SHL plays an important role in balancing heat accumulation in the body, evaporative cooling through panting still has a pivotal role (Hillman et al., 1985). During acute heat exposure chickens suffer from dehydration as a result of a high rate of panting (Etches et al., 1995), and such dehydration may dramatically affect the functioning of the blood system, leading to heat stroke. During the first stages of acute heat stress, before the development of dehydration although elevation of the blood supply to the skin and the upper respiratory passageways enhanced, there is no impact on the blood supply to other important tissues. This is mainly because of the increase in cardiac output that results from the increases in heart rate and stroke volume, and from the redistribution of blood flow among tissues, when blood flow is diverted from non-vital tissues. The resulting increases in heart rate and venous return support an overall increase in cardiac output that refills the arterial pressure reservoir more rapidly and prevents further reduction in arterial pressure (Whittow et al., 1964; Sturkie, 1967; Darre and Harrison, 1987; Zhou, 2000). However, with the development of dehydration a depletion of the blood volume will cause a reduction in venous pressure which in turn diminishes blood flow to the skin and the upper respiratory tract, thereby impairing the efficacy of the evaporative heat loss routes. This will lead to lethal levels of hyperthermia, with T_b that could reach 46-47°C.

An additional combined mechanism involves the modulation of heat production. Heat production will decline with increasing duration and level of heat stress, to prevent excessive accumulation of heat in the body. The metabolic rate is associated with secretion of the hormone thyroxine (T_4) by the thyroid gland, and T_4 is deiodinated to triiodothyronine (T_3) in all peripheral tissues. Triiodothyronine is the main metabolism-stimulating hormone (McNabb and King, 1993; Gabarrou et al., 1997) that was found to be associated with temperature regulation (McNabb and King, 1993; Carew et al., 1998; Gonzales et al., 1999; Yahav, 2000a). Under acute heat stress exposure, the main and immediate mechanism to reduce T_3 and, thereby, heat production is via reduction of the process of deiodination of peripheral T_4 to T_3.

It can be concluded that recruitment of the physiological mechanisms that lead to enhancement of heat loss via every route, coupled with reduced heat production will contribute to minimizing the deleterious effect of heat stress. However, a long-term (hours) incidence of extreme heat stress may eventually lead to the cascade of heat stroke events which eventually will end in death.

Hypothermia Development

Body mass strongly affects the functioning of both adult and juvenile animals (McNab, 2002). In juvenile fowl the large surface-to-volume ratios result in high loss of heat during cold exposure, high energy expenditure, and proportionally low capacity for fat accumulation that provides insulation and energy storage. Consequently, in juveniles, the energy cost of maintaining T_b may become prohibitively high, leading to inability to maintain T_b under acutely cold conditions.

The negative effect of acute cold exposure in the adult domestic fowl is limited and is likely to cause a reduction in T_b that would be limited to the lower range of normothermia. The factor that limits changes in T_b is the massive insulation provided by skin fat deposits and feather coverage, together with the small surface-to-volume ratio. This response is completely different from that in juveniles, in which lack of fat deposits and high surface-to-volume ratio predominate over the highly effective insulation provided by plumage coverage, leading to major difficulties in controlling T_b under acute cold exposure. Furthermore, during the first 10 days post hatch the thermoregulatory mechanism of the chick suffers from immaturity (Dunnington and Siegel, 1984; Modrey and Nichelmann, 1992) with respect to the development of integration of thermal information in the hypothalamus (Rothwell, 1992), and buildup of the body-to-brain temperature difference (Arad and Itsaki-Glucklish, 1991).

Therefore, this period is a critical one from the thermoregulatory perspective. In order to minimize the deleterious effect of unsuitable T_a on domestic fowl during the juvenile period, strict controls of T_a are imposed during the first 3 weeks post hatch.

To withstand cold exposure, the bird minimizes heat loss by restricting blood flow to the skin by means of vasoconstriction that is initiated via the sympathetic nervous system. This occurs in parallel with redistribution of the blood flow to the viscera, in order to enhance heat production. In order to compensate for heat loss which, under extremely cold conditions may predominate over heat production, both shivering thermogenesis (ST) (Hillman et al., 1985) and non-shivering thermogenesis (NST) are employed (Duchamp et al., 1999). The latter is based on the involvement of avian uncoupling protein (avUCP) in the skeletal muscle of chickens (Duchamp et al., 1999; Collin et al., 2003) and ducklings (Rambault et al., 2001).

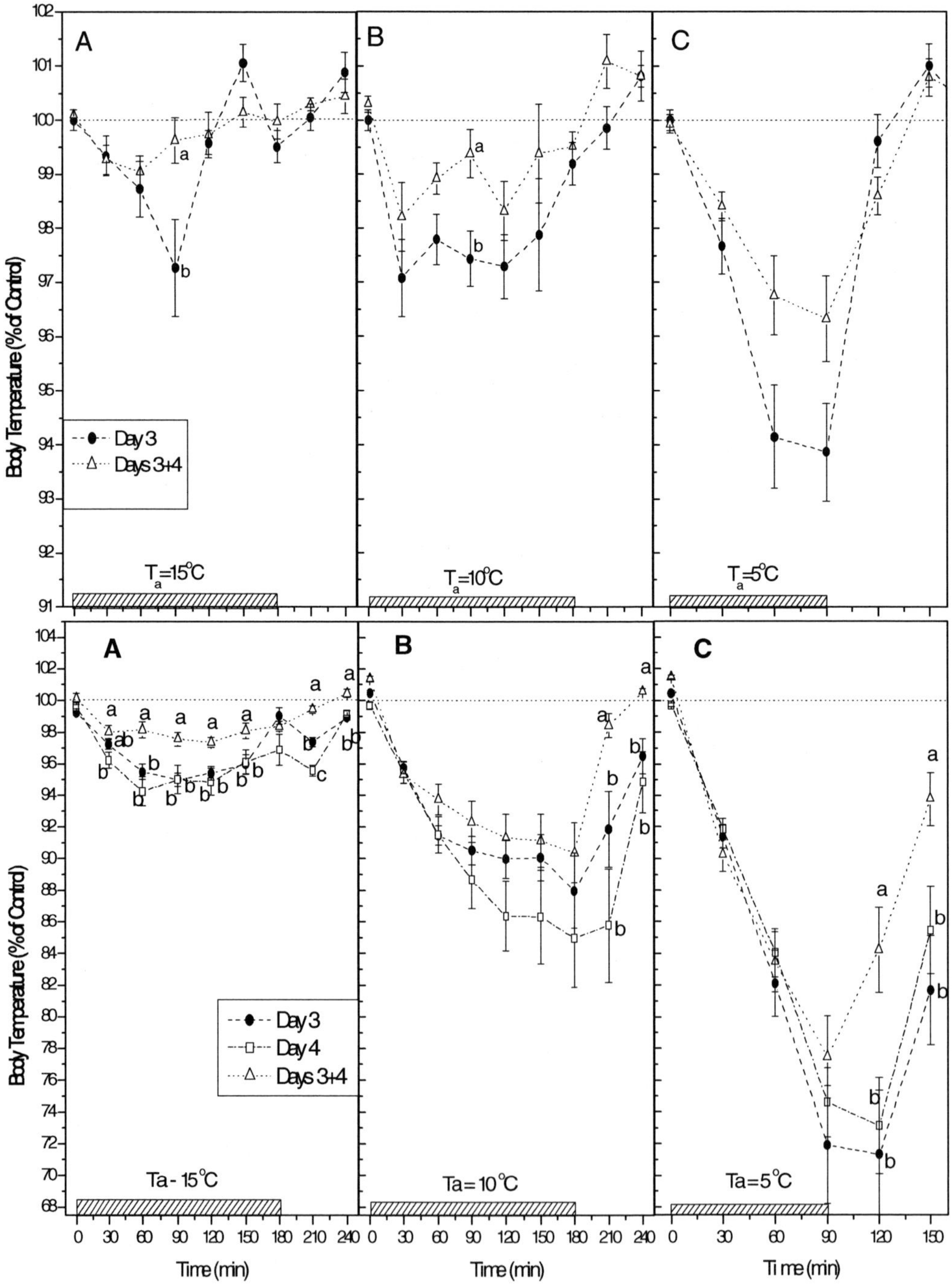

Figure 4. The effects of exposure to various low T_a – 15°C (A) and 10°C (B) for 3 h day^{-1}, and 5°C (C) for 1.5 h day^{-1} – at chick ages of 3 days (circle), 4 days (white square) and 3 + 4 days (triangle), presented as percentages of control (dotted line). At any given time, values designated by different letters differed significantly (n = 8; $P < 0.05$). Hatched bar indicates cold exposure. (According to Shinder et al., 2007).

Taking into consideration the limitations imposed by the immaturity of the juvenile chick's thermoregulatory system on the bird's capacity to cope with extremely acute cold conditions, Shinder et al. (2007) exposed broiler chicks at the age of 3 and/or 4 days to 15, 10 and 5°C for 3 or 1.5 h/day, respectively, to elucidate the effects of intermittent exposures to low T_a on the chick T_b (Figure 4), on the total SHL via convection and radiation, and on the distribution of SHL between the two major non-feathered areas, i.e., the face and legs. The three T_a that were chosen in the study imposed heavy demands on the chicks as they tried to maintain their T_b. Reduction of T_a from 15 to 5°C caused energy demands to increase to a level such that chicks at either 3 or 4 days of age could not maintain an adequate T_b level for longer than 90 min. However, it was obvious that a second exposure enhanced the chicks' capacity to maintain T_b.

In general, hypothermia caused by early-post-natal cold exposure of chicks had been documented previously (Wekstein and Zolman, 1971; Freeman and Manning, 1984; Arad, 1991; Shinder et al., 2002). However, repetitive cold exposure of post-natal chicks had been shown to improve the ability to control T_b. This was, most probably, a result of alteration in the metabolic rate, coupled with an efficient vasomotor response. Shinder et al. (2007) also demonstrated the dynamics of SHL from the two non-feathered areas, the face and the legs; they concluded that the legs differed from the face, and suggested that the legs are a major organ for vasomotor responses, whereas the face is a conservative organ.

Contrary to the situation presented by hyperthermia, in hypothermia the challenge is to increase thermogenesis. This coupled with a significant reduction in heat loss will contribute to the maintenance of T_b.

Acclimation to the Environment – The Control of Body Temperature

Acclimation to Ambient Temperature

T_a is considered to be a prominent environmental factor in thermal acclimation. Thermal acclimation involves an array of autonomically controlled physiological mechanisms that work in concert to enhance thermal tolerance (Horowitz, 2002), and in which the cardiovascular system plays a major role. Thermal acclimation can be characterized as an expansion of the dynamic thermoregulatory range of the bird, which involves: expansion of the safety margins of the body core temperature; increased heat loss; and decreased heat production. The overall process results from concerted procedures that operate at all levels of the body to induce a new level of homeostasis which efficiently improves thermotolerance in the new thermal environment. Thermal acclimation is initiated in response to persistent or repeated thermal challenge of the thermoregulatory system and has a long-acting effect.

Thermal acclimation causes a moderate level of hyperthermia in broilers, compared with the level developed during acute thermal stress (Table 1). However, whether the exposure to cold is acute or via acclimation makes no difference to the development of hypothermia, which was found to be moderate in both cases. In turkeys, however, none of the T_a levels used, even the highest, caused any hyperthermic response. The hyperthermic response of turkeys was found to differ from that of broiler chickens exposed to similar environmental conditions, although both species belong to the same order, i.e., *Galliforme*. A remarkable thermoregulatory capacity was also demonstrated in non-acclimated 11-week-old turkeys exposed to acute heat (36°C, R.H. 20-30%): they developed only a moderate hyperthermia (T_b = 42.5°C) (Yahav, unpublished data). The ability of turkeys to control T_b efficiently may result from efficient cooling mechanisms (i.e., evaporative and sensible heat losses) and/or from sufficient reduction in heat production. It can be speculated that the relatively large comb and wattles that characterize turkeys increase the heat loss via convection and radiation and thereby contribute to better control of T_b. The increase in T_b at the low T_a (10°C, Table 1) may be related to the increase in heat production. At the age of 6 weeks the amount of skin fat insulation is much lower in turkeys than in broilers, which could be the reason for their higher heat production rate.

Table 1. Comparison between the effects on broiler body temperature, of acute exposure and of broiler and turkey body temperature to acclimation to low, moderate, and high ambient temperatures. (According to Yahav et al., 1997b; Yahav, 1999, 2002)

	Ambient temperature (°C)		
	10	25	35
	Broiler body temperature (°C)		
Acute exposure	40.6 ± 0.2^{c}	41.1 ± 0.1^{b}	44.7 ± 0.4^{a}
Acclimation	40.9 ± 0.2^{c}	41.6 ± 0.1^{b}	42.5 ± 0.1^{a}
	Turkey body temperature (°C)		
Acclimation	41.9 ± 0.1^{a}	40.7 ± 0.1^{c}	41.1 ± 0.1^{b}

In rows, values designated by different letters, differ significantly ($P \leq 0.05$).

During heat acclimation the vasomotor response, which includes vasodilatation and changes in skin temperature, is also moderate, in parallel with the moderate response in T_b. Broiler chickens that had been acclimated to 35°C exhibited the following pattern of body and facial surface temperatures: at 22 days of age T_b was 42.3 ± 0.2°C, whereas facial surface temperature was 40.6 ± 0.3°C, at 36 days of age T_b and the facial surface temperature were 42.8 ± 0.3 and 40.8 ± 0.3°C, respectively. These values are significantly different from those monitored in broilers under acute heat exposure (Figure 3).

The vasodilatation response during heat acclimation coincided with vasoconstriction affecting the visceral organs, but it further differed from the RTSR in that, despite this vasoconstriction the blood flow to these organs could continue as during pre-acclimation, as a consequence of systemic blood pressure adjustment. In such a situation better adjustment of the visceral organs temperature will contribute to better heat tolerance.

Heat acclimation induces further adaptations in the circulatory system; they are manifested as lower heart rate (Forrest et al., 1968; Sturkie et al., 1970) and increased stroke volume, which suggests enhanced cardiac efficiency. In mammals, including humans, the enhancement of the cardiac output was found to be related to increased venous blood return, which allows cardiac filling pressure to reach values that can lead to elevated stroke volume (Senay et al., 1976). These changes were attributed to changes in plasma volume expansion, a phenomenon that has been mainly attributed to a buildup of higher plasma protein concentration, which alters the balance between capillary hydrostatic and colloid osmotic pressure. Expansion of plasma volume in chickens as a result of acclimation to high T_a was demonstrated by Whittow et al. (1964), May et al. (1971), and Yahav et al. (1997b). The significant plasma expansion during acclimation to constant or diurnally cyclic high T_a in broilers coincided with a significant increase in plasma protein concentration at high temperatures that could be important as a regulatory mechanism for plasma volume expansion (Yahav et al., 1997a). This expansion supports both the cardiovascular changes and the need to provide fluid for heat dissipation by panting (Marder and Arad, 1989).

In cold acclimation, changes in the cardiovascular system act mainly to accommodate oxygen needs. These changes were observed in birds adapted to low T_a (Olsen, 1937; Deaton et al., 1969a, b; May et al., 1971; May and Deaton, 1974; Julian et al., 1989; Shlosberg et al., 1992, 1996; Yahav et al., 1997b, Yahav, 1999, 2000a). The changes, which included peripheral blood flow vasoconstriction, increased blood volume, mainly as a result of expansion of the packed cell volume (PCV), and increased hematocrit/hemoglobin concentration, were often accompanied by compensatory changes in heart muscle weight (Deaton et al., 1969a, b; May et al., 1971; May and Deaton, 1974; Carey and Morton, 1976; Palomeque and Planas, 1978; Yahav et al., 1997b). The enhancement of red blood cell volume (the major component of PCV) to accommodate the oxygen needs causes an increase in blood viscosity (Fedde and Wideman, 1996) that contributes to the increases in cardiac workload (Owen et al., 1995, Wideman and Tackett, 2000), blood pressure (Owen et al., 1995), and blood flow resistance (Wideman et al., 1998).

In domestic fowl, and especially in broilers, acclimation to cold may fail and, as a result, the ascites syndrome may develop, as a result of impaired ability to regulate energy balance under low T_a or high altitude (Julian, 1993; Wideman, 2000; Luger et al., 2003; Baghbanzaden and Decuypere, 2008). These environmental conditions trigger the development of the syndrome, which is characterized mainly by hypoxemia, increased workload of the cardiopulmonary system, central venous congestion, fluid exudation, mainly into the peritoneal cavity, and, finally, death (Julian, 1993; Maxwell et al., 1995; Olkowski and Classen, 1998). Surprisingly, broilers suffering from ascites did not exhibit a level of hypothermia that could account for the impaired ability to regulate the energy balance (Yahav, unpublished data).

During acclimation to low or high T_a, modulation of heat production is crucial. Indeed, low T_a will induce an increase in heat production whereas high T_a will depress it. The main hormonal modulators are the thyroid hormones, T_4 as a pre-hormone and T_3, the active hormone, which are known to have a wide range of effects on growth, development and metabolism (McNabb and King, 1993; Tsukada et al., 1998; Vasilatos-Younken et al., 2000). There is a negative linear correlation between T_a and T_3 in domestic broiler chickens (Yahav et al., 1996) and turkeys (Yahav, 2000a), therefore, at low T_a the T_3 concentration is higher in order to meet the increased metabolic demands. Likewise, T_3 concentration was positively correlated with the chicken's nutritional status (Gabarrou et al., 1997; Gonzales et al., 1999; Buyse et al., 2000). In broilers and turkeys the adjustment of T_3 production during the short and intensive growth period is crucial. However, it must be noted that acclimation to cold or hot conditions will cause control of T_b to predominate over promotion of growth and development, so that acclimation to extreme temperatures will cause a decline in performance and production of domestic fowl.

Controlling Body Temperature in a Complex Environment – Ambient Temperature and Relative Humidity

Only a few studies have examined the effect of RH on domestic fowls, and most of them focused on its effect on performance. Prince et al. (1965) found that RH in the range of 50-90% had no effect on growth rate or food intake, and Winn and Godfrey (1967) found that humidity in the range of 25-90% did not affect growth rate at a T_a of 32.2°C. On the other hand, Adams and Rogler (1968) demonstrated better performance at RH of 40% than at 80%, when chickens were exposed to 29°C. Canton et al. (1983) concluded that the deep body temperature was affected only by the dry-bulb temperature. Yahav et al. (1995) and Yahav (2000b) found that the effects of acclimation to high T_a at various RH levels led to maximal growth rate and feed intake in male and female broiler chickens at RH 60-65%. A similar pattern was observed in broilers exposed to T_a of 30 or 28°C at the same RH values.

The reason for the response of chicken growth to RH is not clear. However, at high T_a, differences in RH might affect the amount of energy expenditure for maintenance (i.e., energy needed to control body temperature). This energy expenditure can be assessed by plotting energy intake against weight gain, and the intercept of such a plot provides an estimate of maintenance energy expenditure. This calculation assumes that when a balanced diet is provided, birds consume feed according to their needs for maintenance and for growth, and that there is no interaction between the needs for the two functions (Hurwitz et al., 1980). The energy for maintenance was lowest in chickens exposed to RH 60-65%, at all T_a, and it declined as a percentage of total energy expenditure as T_a declined (Yahav, 2000a). These results lead to the possibility that less energy is invested in thermoregulation by chickens exposed to RH 60-65%, because of better thermoregulatory efficiency. Indeed, broiler chickens developed hyperthermia during exposure to high (70-75%) or low (40-55%) RH (Yahav et al., 1995); only at 60-65% RH were they able to maintain T_b in the upper part of the normothermic range.

In turkeys the response of performance to RH has been found to be age-dependent (Yahav et al., 1995, 1998b; Yahav, 2000a). At T_a of 35°C and RH ranging from 40 to 75% maximal growth rate and feed intake in male turkeys aged 4-8 weeks were obtained at RH of 40-45%, but at 30°C no effect of RH on performance was observed. Turkeys aged 10-19 weeks, exposed to 35°C showed a bell-shaped pattern of response to RH, with maximum body weight and feed intake at RH 70-75%. Despite these wide differences in the effects of T_a and RH on turkey performance, its effect on turkey T_b was found to differ from the hyperthermic response exhibited by broilers.

Controlling Body Temperature in a Complex Environment – High Ambient Temperature and Ventilation

Ventilation is a powerful determinant for the major component of SHL, i.e., convection. It has been assumed that sensible heat loss does not play an important role for the domestic fowl when T_a is above the upper limit of the thermoneutral zone (for review see Hillman et al., 1985) and, therefore, that its influence on T_b may be neglected. This assumption was based on: a. the small differences between the body surface temperature (T_S) and T_a; and b. the fact that in fully feathered birds only limited areas (i.e., legs, head, wattle and comb) are unfeathered. Therefore, convection at high T_a could be neglected (Tzschentke et al., 1996). However, sensible heat loss has been reported to be important in desert mammals (Mohler and Heath, 1988; Klir et al., 1990; Klir and Heath, 1992; Philips and Heath, 1992), and in domestic fowls several studies demonstrated the effect of ventilation on performance (Drury, 1966; Wathes and Clark, 1981; Mitchell, 1985; Lacy and Czarick, 1992; Timmons and Hillman, 1993; Phillip and Sanborn, 1994; Tzschentke et al., 1996; Simmons et al., 1997, 2003; Lott et al., 1998; May et al., 2000; Czarick et al., 2000; Tzschentke and Nichelmann, 2000; Yahav et al., 1998a, 2001, 2004c, 2005, 2007).

Table 2. The effects of ventilation rate at different ambient temperatures on body temperature of 6-week-old broilers and turkeys

	Ventilation rate (m/s)			
	0.8	1.5	2.0	2.5
	Broiler body temperature (°C)			
T_a = 35°C	43.9±0.10[a]	42.9±0.08[c]	42.8±0.09[c]	*43.2±0.10[b]
T_a = 30°C	40.5±0.10[b]	40.4±0.06[b]	40.9±0.09[a]	41.0±0.09[a]
T_a = 25°C	40.7±0.06[b]	41.1±0.07[a]	41.1±0.06[a]	41.0±0.05[a]
	Turkey body temperature (°C)			
T_a = 35°C	41.5±0.06[a]	41.2±0.01[ab]	40.6±0.09[a]	41.2±0.07[ab]
T_a = 30°C	40.9±0.12[b]	41.1±0.08a[b]	41.4±0.07[a]	41.3±0.08[a]
T_a = 25°C	40.6±0.06	40.8±0.04	40.7±0.09	40.7±0.13
T_a = 20°C	41.4±0.06	41.0±0.09	41.3±0.10	41.2±0.09

* this value was recorded at ventilation rate of 3.0 m/s. In rows, values designated by different letters differ significantly ($P \leq 0.05$).

In broilers and turkeys it seems that by changing high T_a values and varying ventilation rates, thermoregulatory and performance responses can be altered. In broilers (Yahav et al., 2005) maximal performance was recorded at a ventilation air velocity of 2.0 m/s when T_a was held at 35°C. At T_a of 30°C the maximal performance was obtained at 2.5 m/s, and at 25°C it shifted dramatically to 0.8 m/s. These results suggest that there was a turning point in the performance response of broiler chickens to ventilation when T_a fell below 30°C, with increased air velocity probably causing chilling, which led to increased energy expenditure on maintenance, and, therefore, reduced growth performance.

In turkeys a similar effect of ventilation rate on performance was monitored at 35°C. However, at 30°C optimal performance was achieved at ventilation rates between 1.5 and 2.5 m/s. At 25°C there was no effect of ventilation rate on body weight, but a significant increase in feed intake was taken to indicate that turkeys suffer from ventilation irrespective of its rate (Yahav et al., 2007), and this effect was more pronounced at 20°C (Yahav et al., unpublished data). It seems, however, that a significant effect of ventilation rate on T_b was observed only in broilers exposed to 35°C (Table 2): they developed hyperthermia. Under these conditions a bell shaped response of T_b to air velocity was recorded, with the trough between 1.5 and 2.0 m/s. Although ventilation rate did not affect radiative heat loss over a wide range of air velocities (Table 3), a significant linear ($r^2 = 0.998$) increase of convective heat loss, Q_c with increasing ventilation rate was observed. The increase had a dramatic effect on the ratio of sensible heat loss to maintenance energy. While the ratio was only 29% in broilers exposed to air velocities of 0.8 and 1.5 m/s, it exceeded 44% of the energy expended for maintenance in those exposed to 3.0 m/s (Table 3). This demonstrates that sensible heat loss fills a major role in heat loss, whose importance depends on ventilation rate, despite the small temperature gradients between the bird surface and the environment.

Table 3. The effects of air velocity at high ambient temperature on heat loss by radiation (Qr), by convection (Qc), and radiation plus convection (Qt), expressed as percentages of energy expended for maintenance. (According to Yahav et al., 2004c)

Variable	Air velocity ($m\ s^{-1}$)			
	0.8	1.5	2.0	3.0
Qr (kcal/d)	21.7 ± 2.9	21.9 ± 3.1	27.4 ± 2.9	27.2 ± 2.9
Qc (kcal/d)	47.9 ± 6.0^{d}	67.3 ± 6.4^{c}	90.2 ± 6.0^{b}	114.0 ± 6.0^{a}
Qt (kcal/d)	69.6 ± 8.8^{b}	90.2 ± 9.4^{b}	117.6 ± 8.8^{a}	141.2 ± 8.8^{a}
Qt (% of energy expended for maintenance)	24.1 ± 3.7^{b}	29.1 ± 4.4^{b}	36.8 ± 2.2^{ab}	44.7 ± 4.7^{a}

In rows, values designated by different letters differ significantly ($P \leq 0.05$; n = 8).

Under all other environmental conditions (below 35°C), although statistically significant differences in T_b were recorded from time to time, T_b values were in the normothermia range, therefore, the physiological effects of changes in air velocity may be of small concern.

Controlling Body Temperature in a Complex Environment – High Ambient Temperature and Ammonia Pollution

The environment maintained in broiler and turkey houses has a major effect on the health, performance and thermoregulation of the birds. Ammonia gas, one of the poultry house products that are detrimental to domestic fowl health, is formed from uric acid and undigested proteins in the manure, by aerobic or anaerobic bacteria (Groot Koerkamp, 1994; Kristensen and Wathes, 2000). The degradation of uric acid and undigested proteins to form ammonia is influenced mainly by the temperature, pH and moisture content of the manure (Elliot and Collins, 1982).

Table 4. The effects of various concentrations of ammonia in air on body temperature (T_b), and arterial pH and pCO_2 of broilers raised at T_a = 32°C. (According to Yahav, 2004)

Variable	**Ammonia concentration (ppm ± SE)**				**SEM**
	16 ± 2	**28 ± 3**	**39 ± 4**	**54 ± 5**	
T_b (°C)	41.50^{c}	41.91^{b}	42.27^{a}	42.26^{ab}	0.18
pH	7.51^{b}	7.53^{ab}	7.56^{a}	7.56^{ab}	0.02
pCO_2(mmHg)	35^{a}	32^{ab}	29^{b}	29^{b}	1.43

In rows, values designated by different letters differ significantly.

Exposure to ammonia has been found to cause changes in the pulmonary ultrastructure that supports the mechanical defense components of the respiratory system (Oyetunde et al., 1978) especially the dead space, which plays a major role in the ability to control body temperature at high T_a, by panting (Hillmann et al., 1985; Marder and Arad, 1989).

The effect of increased ammonia concentration on the ability of broiler chickens to control T_b was studied by Yahav (2004), who found that body temperature was significantly affected by the concentration of ammonia in the air (Table 4). Chickens exposed to the lowest ammonia concentrations held T_b at significantly lower values than those exposed to the highest concentrations. Arterial pH increased with increasing ammonia concentration, in parallel with a decline in partial pressure of arterial CO_2, and was similar in the birds exposed to the two highest ammonia concentrations. These results suggest a higher panting rate in broilers exposed to the higher concentrations of the pollutant, but with lower efficacy in controlling T_b. Under the low ammonia concentrations, the broilers maintained T_b within the known normothermic range for domestic fowl, whereas exposure to 39 ppm of ammonia or more caused significant increases in T_b, which reached values typical of mild hyperthermia. This suggests a negative effect of ammonia concentration on the ability to control T_b, most probably through it effect on the efficacy of panting as a mechanism of heat loss.

In sum, it can be concluded that thermal acclimation may serve as an efficient tool to improve the acquisition of thermotolerance. However, the process of thermal acclimation, at least in broilers, is rendered rather impracticable by the early marketing age of broilers, coupled with: (a) the necessity to keep the environmental temperature controlled up to the age of 21 d for brooding; (b) the deleterious effect of thermal acclimation on broiler performance; and (c) the enormous cost of temperature-controlled poultry houses. The relatively efficient thermoregulatory capacity of turkeys that is induced during thermal acclimation has a transient negative effect on growth, but the long period of production may ensure that there is sufficient time to use thermal acclimation with this species.

The limitations to acclimation outlined above have stimulated studies of how to efficiently adopt the use of epigenetic temperature adaptation in broilers.

The Epigenetic Body Temperature Adaptation Approch

Epigenetic temperature adaptation is based on the assumption that environmental factors, especially T_a, have a strong role in determination of the “set-point” for physiological control systems, during “critical developmental phases.” Dörner (1974) first described this as the "determination rule." Thermosensitive neurons in the preoptic anterior hypothalamus (PO/AH) integrate afferent temperature signals from various areas of the body, in order to elicit appropriate thermoregulatory responses via control of physiological, endocrinological, and behavioral responses, and thereby to keep the core T_b relatively constant (Boulant, 1996).

Epigenetic adaptation, which has been defined as a lifelong adaptation that occurs during prenatal (embryogenesis) or early post-hatching ontogeny, takes place within critical developmental phases that affect gene expression (Nichelmann and Tzschentke, 2002; Tzschentke and Basta, 2002; Tzschentke et al., 2004), and seems to offer a suitable opportunity to reach the goal of improved acquisition of thermotolerance in broilers. During early development, most functional systems evolve from an open-loop type, without feedback, to a closed-loop control type, with feedback, according to the "transformation rule" (Dörner, 1974). Application of TM during the critical phases of this development process may induce alterations in the thermoregulatory control system.

Post-hatching Thermal Manipulations

Broiler chicks complete development of their brain and body temperature regulation systems by the age of 10 days post-hatching (Arad and Itsaki-Glucklish, 1991). During this period, body and brain temperatures are maintained at a lower level than in adult chickens. Subsequently, as age increases, the difference between body and brain temperatures increases linearly and significantly. The epigenetic response has been successfully modulated by early-age TM of post-hatching chicks, by exploiting the incomplete maturation of the thermoregulatory system. TM, involving exposure of 3-d-old broilers to 37-38°C at 60-80% RH for 24 h, was found to improve acquisition of thermotolerance. The improvement achieved was manifested in the ability of the TM chicks to reduce heat production efficiently during exposure to acute thermal challenge at marketing age (Yahav and Hurwitz, 1996). This was accompanied by: 1. an alteration in sensible heat loss through convection and radiation (Yahav et al., 2005); 2. a significant reduction in the stress level of the TM chickens, as indicated by their plasma corticosterone concentrations (Yahav, 2007); and 3) pronounced increases in the 27-, 70- and 90-kDa heat-shock proteins (HSPs) in the heart muscle and lung tissue of the control chickens during thermal challenge (Yahav et al., 1997a). It has been suggested that the induction of HSPs was correlated with T_b, and that the HSP response was not part of the long-term mechanism elicited by the early-age TM. The reduction in heat production, coupled with increased SHL, enabled relatively slow development of hyperthermia and thus dramatically reduced mortality caused by heat challenge. Application of TM at 3 d of age also induced compensatory growth that led to improved performance and muscle growth, because of the proliferation of skeletal muscle satellite cells during TM (Halevy et al., 2001). It was further found to positively affect the post-TM development of the gastrointestinal tract (Uni et al., 2001). The effect of TM on the chicks' PO/AH was studied with reference to two genes: the R-Ras3 and the brain-derived neurotrophic factor (BDNH). The R-Ras3 belongs to the small GTP-binding protein subfamily and is activated by multiple extracellular stimuli. The Ras subfamily controls a variety of cellular events that culminate in gene transcription; it plays an essential role in cell proliferation, differentiation and survival, and in modulation of synaptic functions. The BDNF binds specifically to the TrK-B receptor and initiates tyrosine phosphorylation, which activates the phosphotyrosine-binding site. This initiates the internal cellular pathway of RAS, and results in transduction of genes involved in neuronal growth and maintenance. In the chick PO/AH significant increases in the expression of R-Ras3 (Labunskay and Meiri, 2006), brain-derived neurotrophic factor (BDNF) (Katz and Meiri, 2006), and 14-3-3ε (Meiri, 2008) were detected during TM, which suggests that these genes are involved in TM.

However, uniform post-hatch temperature manipulation is difficult to achieve, whereas the use of such manipulations during incubation could be more efficient and uniform, if the interaction between the mechanisms of incubation and embryological development could be adequately understood.

Pre Hatching Thermal Manipulations

The hypotheses underlying these studies are that: 1. during embryogenesis, it is possible to induce long-lasting physiological memory, based on epigenetic adaptation; 2. long-lasting memory can be defined, most probably, as alteration in the hypothalamic threshold response to changes in the environment; and 3. TM by means of specific levels and durations of heat exposure, during sensitive periods within embryogenesis will impart improved thermotolerance during the bird's entire life span.

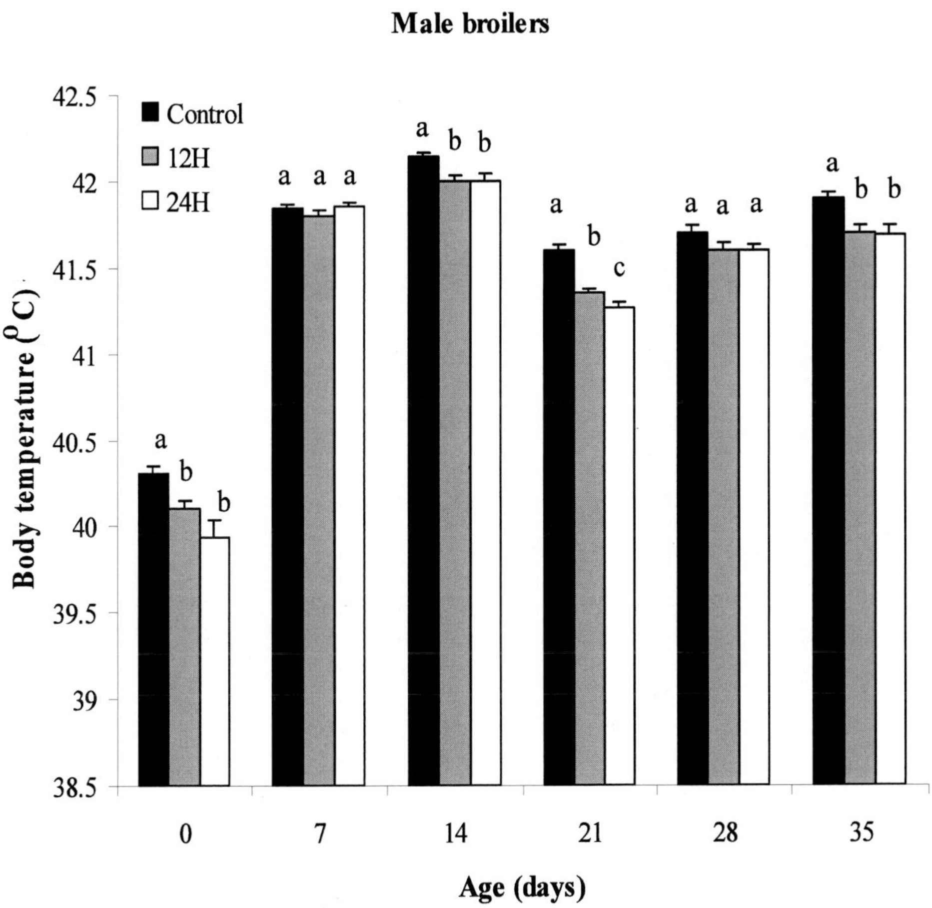

Figure 5. Body temperatures of male broiler chickens from hatching until 35 d of age after they had been thermally manipulated (TM) at 39.5°C and 65% RH from incubation days E7 to E16 for 12 h/d (12H) or continuously (24H). Different letters indicate significant ($P \leq 0.05$) differences among treatments. (According to Piestun et al., 2008a).

In contrast to the uniform temperature of commercial incubation, in nature, incubation conditions are non-uniform, because of the need to search for food and escape from predators, and because of non-uniform nest insulation (Webb, 1987). This may be one of the reasons why birds in the wild are quite capable of coping with extreme environmental temperatures.

It was previously reported that exposing embryos to high or low temperatures during incubation improved their capacity to adapt to hot or cold environments, respectively, in the post-hatching phase (Nichelmann et al., 1994; Tzschentke and Basta, 2002; Moraes et al., 2003; Yahav et al., 2004a).

Three critical parameters have to be considered in the approach to TM during chick embryogenesis: 1. the timing of the critical phase; 2. the temperature level; and 3. the duration of exposure.

Determination of the critical phase of embryogenesis that would enable successful application of TM to improve the acquisition of thermotolerance was based on the hypothesis that the "set point" or "response threshold" of controlling systems could be altered most efficiently during the development/maturation of the hypothalamus-hypophyseal-thyroid axis (thermoregulation) and/or the hypothalamus-hypophyseal-adrenal axis (stress).

In recent experiments (Piestun et al., 2008a,b) TM at 39.5°C and 65% RH was applied continuously (24H treatment) or intermittently for 12 h per day (12H) to broiler embryos from days E7 to E16 (inclusive). After hatching, chicks were raised under standard conditions to 35 d of age and then subjected to thermal challenge (35°C for 4.0 h).

A significant decline in hatchability was recorded in the 24H TM broiler eggs. This was mainly because of embryos that initiated external piping but were not able to complete it, most probably as a result of insufficient accumulation of glycogen in the liver and the *Musculus complexus* (Christensen et al., 2001). The hatched chicks of the 24H TM treatment were further characterized by lower chick quality and significantly lower body weight. However, the 12H treatment did not significantly affect hatchability or chick quality (Piestun et al., 2008b).

Application of TM, either continuously in treatment 24H or intermittently in treatment 12H, caused a significant decline in the metabolic rate of the hatched chicks (Figure 1 and 5). This was manifested in a significantly reduced T_b, accompanied by significantly reduced plasma T_4 and T_3 concentrations (Piestun et al., 2008a). Although T_3 is the most potent hormone with regard to the chick's metabolic rate, the fact that plasma T_4 concentration was also significantly reduced in the hatched TM chicks suggests that the activity of the thyroid gland was reduced upon hatching, which emphasizes the significant effect of TM in reducing the metabolic rate (Yahav et al., 2004b). Most studies have demonstrated an improvement in thermotolerance during the first 10 d post-hatching. However, chicks that had been exposed to TM during embryogenesis and then raised to marketing age did not exhibit a long-lasting improvement in thermotolerance. Although thermally treated chicks had significantly lower T_b immediately post-hatching, it persisted only until 4 or 5 weeks of age, after which the beneficial effect diminished. Furthermore, heat challenge of TM chickens at 42 d of age did not reveal any thermal advantages (Collin et al., 2007; Tona et al., 2008).

Piestun et al. (2008b) demonstrated for the first time that adopting E7 to E16 as the "critical phase" for TM of chick embryos significantly enhanced thermotolerance. In this study, T_b of the treated chickens remained lower to significantly lower throughout the growth period of 35 days (Figure 5), suggesting that the metabolic rate was lower. Moreover, thermal challenge of 35-d-old broilers at 35°C for 4 h resulted in a significant improvement in the acquisition of thermotolerance in both the 12H and 24H TM treatments; an improvement that was characterized by a significantly lower level of metabolic rate and stress, coincided with significant increase of SHL as indicated by a mortality rate half that of the control chickens.

This accumulating evidence shows that the epigenetic adaptation approach, and its association with changes in the incubation environment, with emphasis on fine-tuning the level and duration of stress to coincide with the "critical phase," can elicit efficient epigenetic temperature adaptation in broiler chickens.

General Conclusions

Collectively, domestic fowls specially the highly productive broiler chickens and turkeys differ to some extent from other endotherms, in their ability to maintain a sufficiently dynamic steady state under severe changes in the environmental conditions. This ability is the result of improved genetic selection for growth and development. The three direct responses, i.e., the rapid thermal shock response, acclimation, and epigenetic temperature adaptation, that are used to withstand thermal stress can be divided into time-dependent (RTSR, acclimation) and timing-dependent (epigenetic temperature adaptation) responses.

The major concern with broilers and, to a much smaller extent, with turkeys is with high T_a, which elicits severe (under RTSR) or moderate (acclimation) hyperthermia in broilers, and moderate or shallow (if any) hyperthermia, respectively, in turkeys. Low T_a causes shallow, if any, hypothermia in both broilers and turkeys, but it affects performance. Epigenetic temperature adaptation was significantly developed during the last decade, and may serve as an efficient strategy to improve thermotolerance acquisition without impairing performance.

References

Adams, R.L. & Rogler, J.C. (1968). The effects of dietary aspirin and humidity on the performance of light and heavy breed chicks. *Poultry Science,* 47,1344-1348.

Arad, Z. (1991). Ontogeny of brain temperature regulation in chicks (*Gallus gallus domesticus*). *British Poultry Science,* 32,203-210.

Arad, Z. & Itsaki-Glucklish, S. (1991). Ontogeny of brain temperature in quail chicks (*Coturnix coturnix japonica*). *Physiology Zoology,* 64,1356-1370.

Arieli, Y., Peltonen, L. & Ophir, E. (2002). Coolong by cutaneous water evaporation in the heat acclimated rock pigeon (*Columba livia*). *Comparative Biochemistry and Physiology A*, 131,497-504.

Baghbanzaden, A. & Decuypere, E. (2008). Ascites syndrome in broilers: physiological and nutritional perspectives. *Avian Pathology*, 37,117-126.

Boulant, J.A. (1996). Environmental physiology: Hypothalamic neurons regulating body temperature. In: M.J. Fregly, & C.M. Blatteis, (Eds). *Handbook of physiology: Section 4* (pp 105-126). APS, Oxford University Press; New York.

Boulant, J.A. & Dean, J.B. (1986). Temperature receptors in the central nervous system. *Annual Review of Physiology,* 48,639-654.

Buyse, J., Decuypere, E., Darras, V.M., Vleurick, L.M., Kühn, E.R. & Veldhuis, J.D. (2000). Food deprivation and feeding of broiler chickens is associated with rapid and interdependent changes in the somatotrophic and thyrotrophic axes. *British Poultry Science,* 41,107-116.

Buyse, J., Scheele, C.W., Kwakernaak, C. & Decuypere, E. (1999). Performance and physiological variables in broiler chicken lines differing in susceptibility to the ascites syndrome: 2. Effect of ambient temperature on partial efficiencies of protein and fat retention and plasma hormone concentrations. *British Poultry Science*, 40,140-144.

Canton, G. H., Buffington, D. E. & Mather, F. B. (1983). Evaporative cooling effects on mature, male broiler breeders. *American Society of Agriculture and Engineering*, 1794.

Carew, L.B., Evarts, K.G. & Alster, F.A. (1998). Growth, feed intake, and plasma thyroid hormone levels in chicks fed dietary excesses of essential amino acids. *Poultry Science*, 77,295-298.

Cary, C. & Morton, M. (1976). Aspects of circulatory physiology of montane and lowland birds. *Comparative Biochemistry and Physiology,* 54A,61-74.

Christensen, V.L., Wineland, M.J., Fasenko, G.M. & Donaldson, E. (2001). Egg storage effects on plasma glucose and supply and demand tissue glycogen concentrations of broiler embryos. *Poultry Science*, 80,1729-1735.

Collin, A., Berri, C., Tesseraud, S., Rodón, F.E., Skiba-Cassy, S., Crochet, S., Duclos, M.J., Rideau, N., Tona, K., Buyse, J., Bruggeman, V., Decuypere, E., Picard, M. & Yahav, S. (2007). Effects of thermal manipulation during early and late embryogenesis on thermotolerance and breast muscle characteristics in broiler chickens. *Poultry Science*, 86,795-800.

Collin, A., Malheiros, R.D., Moraes, V.M.B., Van As, P. Darras, V.M., Taouis, M., Decuypere, E. & Buyse, J. (2003). Effects of dietary macronutrients on energy metabolism and uncoupling protein mRNA expression in broiler chickens. *British Journal of Nutrition,* 90,261-269.

Cunningham, D.L. (2006). Cash flow estimates for contract broiler production in Georgia: A 20 year analysis. http://pubs.caes.uga.edu/caespubs/pubcd/B1228.htm. *Accessed December.*

Czarick, M., Lott, B. & Lacy, M. (2000). Poultry Housing Tips: Is an air speed of 600 ft/min in a tunnel house harmful? Vol. 12 No. 6. (pp. 1-3) University of Georgia, Cooperative Extension Service.

Darre, M.J. & Harrison, P.C. (1987). Heart rate, blood pressure, cardiac output and total peripheral resistance of single comb white leghorn hens during an acute exposure to 35C ambient temperature. *Poultry Science*, 66,541-547.

Dean, J.B. & Boulant, J.A. (1989). Effects of synaptic blockade on thermosensitive neurons in rat diecephalon in vitro. *American Journal of Physiology*, 257,R65-R73.

Deaton, J.W., Reece, F.N., McNally, E.H. & Tarver, W.J. (1969a). Liver, heart and adrenal weight of broilers reared under constant temperatures. *Poultry Science,* 48,283-288.

Deaton, J.W., Reece, F.N. & Tarver, W.J. (1969b). Hematocrit, hemoglobin and plasma protein levels of broilers reared under constant temperatures. *Poultry Science,* 48,1993-1996.

Decuypere, E. & Michels, H. (1992). Incubation temperature as a management tool: A review. *World's Poultry Science Journal*, 48,28-38.

DiMicco, J.A. & Zaretsky, D.V. (2007). The dorsomedial hypothalamus: a new player in thermoregulation. *American Journal of Physiology – Regulation and Integration Comparative Physiology,* 292,R47-R63.

Dörner, G. (1974). Environment-dependent brain differentiation and fundamental processes of life. *Acta Biologica Medica Germanica*, 33,129-148.

Dridi, S., Temim, S., Derouet, M., Tesseraud, S., & Taouis, M. (2008). Acute cold- and chronic heat-exposure upregulate hepatic leptin and muscle uncoupling protein (UCP) gene expression in broiler chickens. *Journal of Experimental Zoology – A,* 309,381-388.

Drury, L.N. (1966). Air velocity and broiler growth in diurnally cycled hot environment. Trans. *ASAE*, 9,329-332.

Duchamp, C. & Barré H. (1999). Skeletal muscle as the major site of non-shivering thermogenesis in cold adapted ducklings. *American Journal of Physiology, Regulation integration and Comparative Physiology,* 265,R5469-R5479.

Dunnington, E.A. & Siegel, P.B. (1984). Thermoregulation in newly hatched chicks. *Poultry Science*, 63,1303-1313.

Elliot, H.A. & Collins, N.E. (1982). Factors affecting ammonia release in broiler houses. *Transactions ASAE,* 25,413-424.

Etches, R.J., John, T.M. & Verrinder-Gibbins, G.A.M. (1995). Behavioral, physiological, neuroendocrine and molecular responses to heat stress. in: N.J. Daghir, (Ed). *Poultry production in hot climate* (pp. 31-66). Cambridge, UK, CAB International.

Fedde, M.R. & Wideman, R.F. Jr. (1996). Blood viscosity in broilers: influence on pulmonary hypertension syndrome. *Poultry Science*, 75,1261-1267.

Forrest, J.C., Will, J.A., Schmidt, G.R., Judge, M.D. & Briskey, E.J. (1968). Homeostasis in animals (*Sus domesticus*) during exposure to warm environment. *Journal of Applied Physiology,* 24,33-39.

Freake, H.C. & Oppenheimer, J.H. (1995). Thermogenesis and thyroid function. *Annual Review of Nutrition,* 15,263-291.

Freeman, B.M. (1964). The emergence of the homeothermic-metabolic response in the fowl (*Gallus domesticus*). *Comparative Biochemistry and Physiology,* 13,413-422.

Freeman, B.M. & Manning, A.C. (1984). Re-establishment of the stress response in *Gallus domesticus* after hatching. *Comparative Biochemistry and Physiology*, 78,267-270.

French, N.A. (1997). Modeling incubation temperature: the effect of incubator design, embryonic development, and egg size. *Poultry Science,* 76,124-133.

Gabarrou, J.F., Duchamp, C., Williams, J. & Géraert, P.A. (1997). A role for thyroid hormones in the regulation of diet-induced thermogenesis in birds. British Journal of Nutrition, 78,963-73.

Gonzales, E., Buyse, J., Sartori, J.R., Loddi, M. & Decuypere, E. (1999). Metabolic disturbances in male broilers of different strains. 2. Relationship between the thyroid and somatotropic axes with growth and mortality. *Poultry Science*, 78,516-521.

Gordon, C.J. (1993). Temperature regulation in laboratory rodents. New-York: Cambridge University Press.

Griffin, J.D., Saper, C.B. & Boulant, J.A. (2001). Synaptic and morphological characteristic of temperature sensitive and –insensitive rat hypothalamic neurons. *Journal of Physiology,* 537,521-535.

Groot Koerkamp, P.W.G., (1994). Review on emissions of ammonia from housing systems for laying hens in relation to sources, processes, building design and manure handling. *Journal of Agriculture Engineering Research,* 59,73-87.

Hales, J.R.S., Hubbard, R.W. & Gaffin, S.L. (1996). Limitation of heat tolerance. In: M.J. FREGLY, & C.M. BLATTRIS, (Eds), *Handbook of physiology, section 4, environmental physiology, Vol. 1. (Chapter 15).* (pp. 285-359). Oxford, Oxford University Press.

Halevy, O., Krispin, A., Leshem, Y., McMurtry, J.F. & Yahav, S. (2001). Early age heat stress accelerates skeletal muscle satellite cell proliferation and differentiation in chicks. *American Journal of Physiology*, 281,R302-R317.

Havenstein, G.B., Ferket, P.R. & Qureshi, M.A. (2003a). Growth, livability, and feed conversion of 1957 versus 2001 broilers when fed representative 1957 and 2001 broiler diets. *Poultry Science,* 82,1500-1508.

Havenstein, G.B., Ferket, P.R. & Qureshi, M.A. (2003b). Carcass composition and yield of 1957 versus 2001 broilers when fed representative 1957 and 2001 broiler diets. *Poultry Science*, 82,1509-1518.

Havenstein, G.B., Ferket, P.R., Grimes, J.L., Qureshi, M.A. & Nestor, K.E. (2007). Comparison of the performance of 1966- versus 2003-type turkeys when fed representative 1966 and 2003 turkey diet: growth rate, livability, and feed conversion. *Poultry Science*, 86,232-240.

Havenstein, G.B., Ferket, P.R., Scheideler, S.E. & Larson, B.T. (1994). Growth, livability, and feed conversion of 1991 versus 1957 broilers when fed "typical" 1957 and 1991 broiler diets. *Poultry Science*, 73,1785-1794.

Hellon, R.F. & Taylor, D.C.M. (1982). An analysis of a thermal afferent pathway in the rat. *Journal of Physiology*, 326,319-328.

Hillman, P.E., Scott, N.R. & Van Tienhoven, A. (1985). Physiological responses and adaptations to hot and cold environments. In: M.K. YOUSEF, (Ed), *Stress physiology in livestock (vol. 3, Poultry)* (pp. 27-71). Boca Raton, FL, CRC Press, Inc.

Horowitz, M. (1998). Do cellular heat acclimation responses modulate central thermoregulatory activity? *News in Physiological Science,* 13,218-225.

Horowitz, M. (2002). From molecular and cellular to integrative heat defense during exposure to chronic heat. *Comparative Physiology and Biochemistry*, 131,475-483.

Hurwitz S., Weiselberg, M., Eisner, U., Bartov, I., Riesenfeld, G., Sharvit, M., Niv, A. & Bornstein, S. (1980). The energy requirements and performance of growing chickens and turkeys, as affected by environmental temperature. *Poultry Science,* 59,2290-2299.

IUPS Thermal Commission (2001). Glossary of terms for thermal physiology. *The Japanese Journal of Physiology,* 51,245-280.

Janke, O., Tzschentke, B. & Boerjan, B. (2004). Comparative investigations of heat production and body temperature in modern chicken breeds. *Avian and Poultry Biological Reviews*, 15,191-196.

Janke, O., Tzschentke, B., Höchel, J. & Nichelmann, M. (2002). Metabolic responses of chicken and muscovy duck embryos to high incubation temperatures. *Comparative Biochemistry and Physiology*, 131(A),741-750.

Joseph, N.S., Lourens, A. & Moran E.T. Jr. (2006). The effects of suboptimal eggshell temperature during incubation on broiler chick quality, live performance, and further processing yield. *Poultry Science*, 85,932–938.

Julian, R. J. (1993). Ascites in poultry. *Avian Pathology*, 22,419-454.

Julian, R.J., McMillan, I. & Quinton, M. (1989). The effect of cold and dietary energy on right ventricular hypertrophy, right ventricular failure and ascites in meat type chickens. *Avian Pathology,* 18,675-684.

Katz, A. & Meiri, N. (2006). Brain-derived neurotrophic factor is critically involved in thermal-experience-dependent developmental plasticity. *Journal of Neuroscience*, 12,3899-3907.

Kelso, S.R. & Boulant, J.A. (1982). Effect of synaptic blockade on thermosensitive neurons in hypothalamic tissue slices. *American Journal of Physiology,* 243,R480-R490.

Klir, J.J. & Heath, J.E. (1992). An infrared thermographic study of surface temperature in relation to thermal stress in three species of foxes: the red fox (*Vulpes vulpes),* arctic fox *(Alopex lagopus),* and kit fox (*Vulpes macrotis). Physiological Zoology*, 65,1011-1021.

Klir J.J., Heath, J.E. & Benanni, N. (1990). An infrared thermographic study of surface temperature in relation to Thermal stress in the Mongolian Gerbil, *Meriones unguiculatus. Comparative Biochemistry and Physiology*, 96A,141-146.

Krausova, T. & Peterka, M. (2007). Teratogenic and lethal effects of 2–24 h hyperthermia episodes on chick embryos. *Journal of Thermal Biology*, 32,193–203.

Kristensen, H.H. & Wathes, C.M. (2000). Ammonia and poultry welfare: a review. *World's Poultry Science Journal,* 56,235-245.

Kühn, E.R., Decuypere, E. & Rudas, P. (1984). Hormonal and environmental interactions on thyroid function in the chick embryo and post-hatching chickens. *Journal of Experimental Zoology,* 232,653-658.

Labunskay, G. & Meiri, N. (2006). R-Ras3/(M-Ras) is involved in thermal adaptation in the critical period of thermal control establishment. *Journal of Neurobiology*, 66,56-70.

Lacy, M.P. & Czarick, M. (1992). Tunnel ventilated broiler houses: broiler performance and operating cost. *Journal of Applied Poultry Research*, 1,104-109.

Leksrisompong, N., Romero-Sanchez, H., Plumstead, P.W., Brannan, K.E. & Brake, J. (2007). Broiler incubation. 1. Effect of elevated temperature during late incubation on body weight and organs of chicks. *Poultry Science*, 86,2685-2691.

Lott, B.D., Simmons, J.D. & May, J.D. (1998). Air velocity and high temperature effects on broiler performance. *Poultry Science*, 77,391-393.

Luger, D., Shinder, D., Wolfenson, D. & Yahav, S. (2003). Disorders in erythropoiesis regulation during the development of ascites syndrome – a possible role of corticosterone. *Journal of Animal Science,* 81(3),784-790.

Marder, J. & Arad, Z. (1989). Panting and acid-base regulation in heat stressed birds. *Comparative Biochemistry and Physiology*, 94A,395-400.

Maxwell, M.H., Alexander, I.A., Robertson, G.W., Mitchell, M.A. & McCorquodale, C.C. (1995). Identification of tissue hypoxia in the livers of ascitic and hypoxia-induced broilers using trypan blue. *British Poultry Science*, 36,791-798.

May, J.D. & Deaton, J.W. (1974). Environmental temperature effect on heart weight of chickens. *Journal of Biometeorology,* 18,295-300.

May, J.D., Deaton, J.W., Reece, F.N., Mitlin, N. & Kubena, L.F. (1971). The effect of environmental temperature on blood volume. *Poultry Science,* 50,1867-1870.

May, J.D., Lott, B.D. & Simmons, J.D. (2000). The effect of air velocity on broiler performance and feed and water consumption. *Poultry Science,* 179,1396-1400.

McNabb, B.K. (2002). *The physiological ecology of vertebrates*. Cornell University Press, Ithaca.

McNabb, F. & KING, D.B. (1993). Thyroid hormones effects on growth development and metabolism. In: M.P. Schreibman, C.G. SCANES, & P.K.T. PANG, (Eds), *The Endocrinology of Growth Development and Metabolism in Vertebrates* (pp. 393-417). Academic Press, New York, NY, USA.

Meir, M. & Ar A. (1987). Improving turkey poult quality by correcting incubator humidity to match eggshell conductance. *British Poultry Science,* 28,337-342

Meiri, N. (2008). 14-3-3epsilon expression is induced during the critical period of thermal control establishment. *Developmental Neurobiology*, 68,62-72.

Mitchell, M. A. (1985). Effects of air velocity on convective and radiant heat transfer from domestic fowls at environmental temperatures of 20 and 30°C. *British Poultry Science*, 26,413-423.

Modrey, P. & Nichelmann, M. (1992). Development of autonomic and behavioral thermoregulation in turkeys (*Meleagris gallopavo*). *Journal of Thermal Biology*, 17,287-292.

Mohler, F.S., & Heath, J.E. (1988). Comparison of IR thermography and thermocouple measurement of heat loss from rabbit pinna. *American Journal of Physiology*, 254,389-395.

Moraes, V.M.B., Malehiros, D., Bruggmann, V., Collin, A., Tona, K., Van As P., Onagbesan, O.M., Buyse, J., Decuypere, E. & Macari, M. (2003). Effect of thermal conditioning during embryonic development on aspects of physiological responses of broilers to heat stress. *Journal of Thermal Biology,* 28,133-140.

Morrison, S.F., Nakamura, K. & Madden, C.J. (2008). Central control of thermogenesis in mammals. *Experimental Physiology*, 93,773-797.

Nichelmann, M., Höchel, J. & Tzscentke, B. (1999). Biological rhythms in birds – development, insights and perspectives. *Comparative Biochemistry and Physiology*, 124A,437-439.

Nichelmann, M., Lange, B., Pirow, R., Langbein, J. & Herrmann, S. (1994). Thermal balance in health and disease. *Advances in pharmacological science*. Birkhäuser Verlag, Basel.

Nichelmann, M. & Tzschentke, B. (2002). Ontogeny of thermoregulation in precocial birds. *Comparative Biochemistry and Physiology*, 131(A),751-763

Olkowski, A. A. & Classen, H. L. (1998). Progressive bradycardia, a possible factor in the pathogenesis of ascites in fast growing broiler chickens raised at low altitude. *British Poultry Science*, 39,139-146.

Olsen, C. (1937). Variations in the cells and hemoglobin content in the blood of normal domestic chickens. *Cornell Veterinary,* 27,235-263.

Ophir, E., Arieli, Y., Marder, J. & Horowitz, M. (2002). Cutaneous blood flow in the pigeon *Columba livia*: its possible relevance to cutaneous water evaporation. *Journal of Experimental Biology,* 205,2627-2636.

Owen, R.L., Wideman, R.F. Jr, & Cowen, B.S. (1995). Changes in pulmonary arterial and femoral arterial blood pressure in broiler chickens. *Poultry Science*, 74,708-715.

Oyetunde, O.O.F., Thomson, R.G. & Carlson, H.C. (1978). Aerosol exposure of ammonia, dust and *Escerichia coli* in broiler chickens. *Canadian Veterinary Journal*, 19,187-193.

Palomeque, J. & Planas, J. (1978). Blood volume in domestic pigeon. *Comparative Biochemistry and Physiology,* 59A,413-417.

Parsell, D.A. & Lindquist, S. (1994). Heat shock proteins and stress tolerance. In: R.I. Morimoto, A. Tissieres, & C. Georgopoulos, (Eds), *Biology of heat shock proteins and molecular chaperones* (pp. 457-494). Cold Spring Harbor, NY, Cold Spring Harbor Laboratory Press.

Patapoutain, A., peier, AM., Story, G.M. & Viswanath, V. (2003). ThermoTRP cahnnels and beyond: mechanisms of temperature sensation. *Nature Review Neuroscience,* 4,529-539.

Peterka, M., Peterkova, R. & Likovsky, Z. (1996). Teratogenic and lethal effects of long-term hyperthermia and hypothermia in the chick embryo. *Reproduction Toxicology*, 10,327–332.

Phillips, P. K. & Heath, J.E. (1992). Heat loss by the pinna of the African Elephant (*Loxodonta africana*). *Comparative Biochemistry and Physiology*, 101A,693-699.

Phillips P. K. & Sanborn, A. F. (1994). An infrared, thermographic study of surface temperature in three ratites: Ostrich, Emu and Double-Wattled Cassowary. *Journal of Thermal Biology*, 19,423-430.

Pierau, F.K., Sann, H., Yakimova, K.s & Haug, P. (1998). Plasticity of hypothalamic temperature-sensitive neurons. *Progress in Brain Research,* 115,63-84.

Piestun, Y., Shinder, D., Ruzal, M., Halevy, O., Brake, J. & Yahav, S. (2008a). Thermal manipulations during broiler's embryogenesis – its effect on the improvement of thermotolerance acquisition. *Poultry Science*, 87,1516-1525.

Piestun, Y., Shinder, D., Halevy, O. & Yahav, S. (2008b). The effect of thermal manipulations during development of the thyroid and adrenal axes on in-hatch and post-hatch thermoregulation. *Journal of Thermal Biology*, 33,413-418.

Prince R. R., Whitaker J. H., Matterson L. D. & Luginbuhl, R. E. (1965). Response of chickens to temperature and relative humidity environments. *Poultry Science*, 44, 73-77.

Raimbault, S., Dridi, S., Denjean, F., Lachuer, J., Couplan, E., Bouillaud, F., Bordas, A., Duchamp, C., Taouis, M. & Ricquier, D. (2001). An uncoupling protein homologue putatively involved in facultative muscle thermogenesis in birds. *Biochemistry Journal,* 353,441-444.

Reizis, A., Hammel, I. & Ar, A. (2005). Regional and developmental variations of blood vessel morphometry in the chick embryo chorioallantoic membrane. *Journal of Experimental Biology,* 13,2483-2488.

Richards, S.A. (1968). Vagal control of thermal panting in mammals and birds. *Journal of Physiology*, 199,89-101.

Richards, S.A. (1970). The biology and comparative physiology of thermal panting. *Biological Review,* 45,223-264.

Richards, S.A. (1976). Evaporative water loss in domestic fowls and its partition in relation to ambient temperature. *Journal of Agricutural Sciences,* 87,527-532.

Richards, M.P. & Proszkowiec-Weglarz, M. (2007). Mechanisms regulating feed intake, energy expenditure, and body weight in poultry. *Poultry Science*, 86,1478-1490.

Rothwell, N.J. (1992). Hypothalamus and thermogenesis. In: M.J. Kinney, & H.N. Tucker, (Eds), *Energy Metabolism: Tissue Determinants and Cellular Corollaries* (pp. 229-245). Raven Press, New York, NY.

Senay, L.C., Mitchell, D. & Wyndham, C.H. (1976). Acclimatization in a hot humid environment: body fluid adjustment. *Journal of Applied Physiology,* 40,786-796.

Seth, R., Köhler, A., Grossmann, R., & Chaturvedi, C.M. (2004). Expression of hypothalamic arginine vasotocin in response to water deprivation and sex steroid administration in female Japanese quail. *Journal of Experimental Biology,* 207,3025-3033.

Seymour, S.R. (1972). Convective heat transfer in the respiratory systems of panting animals. *Comparative Biochemistry and Physiology,* 35,119-127.

Shinder, D., Luger, D., Rusal, M., Rzepakovsky, V., Bresler, V. & Yahav, S. (2002). Early age cold conditioning in broiler chickens (*Gallus domesticus*): Thermotolerance and growth responses. *Journal of Thermal Biology*, 27,517–523.

Shinder, D., Rusal, M., Giloh, M. & Yahav, S. (2008). The effect of repetitive acute cold exposures at the latest phase of embryogenesis of broilers on cold resistance during life span. *Poultry Science*, (in press).

Shinder, D., Rusal, M., Tanny, J., Druyan, S. & Yahav, S. (2007). Thermoregulatory responses of chicks (*Gallus domesticus*) to low ambient temperatures at an early age. *Poultry Science*, 86,2200-2209.

Shlosberg, A., Bellaiche, M., Zeitlin, M., Ya'acobi, M. & Cahaner, A. (1996). Hematocrit values and mortality from ascites in cold stressed broilers from parents selected by hematocrit. *Poultry Science,* 75,1-5.

Shlosberg, A., Zadikov, I., Bendheim, U., Handji, V. & Berman, E. (1992). The effect of poor ventilation, low temperature, type of feed and sex of bird on development of ascites in broilers. Physiopathological factors. *Avian Pathology,* 21,369-382.

Silva, J. E. (2006). Thermogeneic mechanisms and their hormonal regulation. *Physiological Review,* 86,435-464.

Simmons, J.D., Lott, B.D. & May, J.D. (1997). Heat loss from broiler chickens subjected to various wind speeds and ambient temperatures. *Applied Engineering and Agriculture*, 13,665-669.

Simmons, J.D., Lott, B.D. & Miles, D.M. (2003). The effects of high air velocity on broiler performance. *Poultry Science*, 82,232-234.

Smith, R.M. & Suthers, R. (1969). Cutaneous water loss as a significant contribution to temperature regulation in heat stressed pigeon. *Physiologist,* 12,358-364.

Sotherland, P.R., Spotila, J.R. & Paganelli, C.V. (1987). Avian eggs: barriers to the exchange of heat and mass. *Journal of Experimental Zoology*, 1 (Suppl.),81-86.

Sturkie, P.D. (1967). Cardiovascular effects of acclimatization to heat and cold in chickens. *Journal of Applied Physiology,* 22,13-15.

Sturkie, P.D., Lin, Y.C. & Ossorio, N. (1970). Effects of acclimatization to heat and cold on heart rate in chickens. *American Journal of Physiology,* 219,34-36.

Swann, G.S. & Brake, J. (1990a). Effect of dry-bulb temperature, relative humidity and egg shell conductance during the first three days of incubation on egg weight loss and chick weight. *Poultry Science*, 69,535-544.

Swann, G.S. & Brake, J. (1990b). Effect of dry-bulb temperature, relative humidity and egg shell conductance during days seventeen to twenty one of incubation on egg weight loss and chick weight. *Poultry Science*, 69,45-553.

Swann, G.S. & Brake, J. (1990c). Effect of incubation dry-bulb and wet-bulb temperatures, on time of hatch and chick weight at hatch. *Poultry Science*, 69,887-897.

Tazawa H., Moriya, K., Tamura, A., Komoro, T. & Akiyama, R. (2001). Ontogenetic study of thermoregulation in birds. *Journal of Thermal Biology*, 26,281–286.

Tazawa, H., Turner, J.S. & Paganelli, C.V. (1988a). Cooling rates of living and killed chicken and quail eggs in air and in helium-oxygen gas mixture. *Comparative Biochemistry and Physiology*, 90A,99–102.

Tazawa, H., Wakayama, H., Turner, J.S. & Paganelli, C.V. (1988b). Metabolic compensation for gradual cooling in developing chick embryos. *Comparative Biochemistry and Physiology*, 89A,125–129.

Timmons, M. B. & Hillman, P. E. (1993). Partitional heat losses in heat stressed poultry as affected by wind speed. *4th International Livestock Environment Symposium* London England. ASAE Special Publication, St. Joseph, MI.

Tona, K., Onagbesan, O., Bruggmann, V., Collin, A., Berri, C., Duclos, M.J., Tesseraud, S., Buyse, J., Decuypere, E. and Yahav, S. (2008). Effects of heat conditioning at d 16 to 18 of incubation or during early broiler rearing on embryo physiology, post-hatch growth performance and heat tolerance. *Archive Geflügelk*, 72,75-83.

Tsukada, A., Ohkubo, T., Sakaguchi, K., Tanaka, M., Nakashima, K., Hayashida, Y., Wakita, M. & Hoshino, S. (1998). Thyroid hormones are involved in insulin-like growth factor-I (IGF-I) production by stimulating hepatic growth hormone receptor (GHR) gene expression in the chicken. *Growth Hormone IGF Research,* 8,235-242.

Tzschentke, B. & Basta, D. (2000). Development of hypothalamic neuronal thermosensitivity in birds during the prenatal period. *Journal of Thermal Biology,* 25,119-123.

Tzschentke, B. & Basta, D. (2002). Early development of neuronal hypothalamic thermosensitivity in birds: influence of epigenetic temperature adaptation. *Comparative Biochemistry and Physiology – A,* 131,825-832.

Tzschentke, B., Basta, D., Janke, O. & Maier, I. (2004). Characteristics of early development of body functions and epigenetic adaptation to the environment in poultry: focused on development of central nervous mechanisms. *Avian and Poultry Biology Reviews*, 15,107-118.

Tzschentke, B., Basta, D. & Nichelmann, M. (2001). Epigenetic temperature adaptation in birds: peculiarities and similarities in comparison to acclimation. *News in Biomedical Science,* 1,26-31.

Tzschentke, B. & Nichelmann, M. (2000). Influence of age and wind speed on total effective ambient temperature in three poultry species (*Gallus domesticus, Cairina moschata, Meleagris gallapavo*). *Archive Geflügel,* 64,1-8.

Tzschentke, B., Nichelmann, M. & Postel, T. (1996). Effects of ambient temperature, age and wind speed on the thermal balance of layer-strain fowl. *British Poultry Science,* 37,501-520.

Tzschentke, B, & Plagemann, A. (2006). Imprinting and critical periods in early development. *World's Poultry Science Journal,* 62,626-637.

Van Brecht, A., Hens, H., Lemaire, J.L., Aerts, J.M., Degraeve, P. & Berckmans, D. (2005). Quantification of the heat exchange of chicken eggs. *Poultry Science,* 84,353-361.

Vasilatos-Younken, R., Zhou, Y., Wang, X., McMurtry, J.P., Rosebrough, R.W., Decuypere, E., Buys, N., Darras, V.M., Van Der Geyten, S. & Tomas, F. (2000). Altered chicken thyroid hormone metabolism with chronic GH enhancement in vivo: consequences for skeletal muscle growth. *Journal of Endocrinology,* 166,609-620.

Uni, Z., Gal-Garber, O., Geyra, A., Sklan, D. & Yahav, S. (2001). Changes in growth and function of chick small intestine epithelium due to early thermal conditioning. *Poultry Science,* 80,438-445.

Wathes, C.M. & Clark, J.A. (1981). Sensible heat transfer from the fowl: radiative and convective heat losses from a flock of broiler chickens. *British Poultry Science,* 22,185-196.

Webb, D.R. (1987). Thermal tolerance of avian embryos: a review. *The Condor,* 89, 874-898.

Webster, M.D. & King, J.R. (1987). Temperature and humidity dynamics of cutaneous and respiratory evaporation in pigeons, *Columba livia. Journal of Comparative Physiology,* 157,253-260.

Wechselberger, M., Wright, C.L., Bishop, G.A. & Boulant, J.A. (2006). Ionic channels and conductance-based models for hypothalamic neuronal thermosensitivity. *American Journal of Physiology, Regulation Integration and Comparative Physiology,* 291,R518-R529.

Wekstein, D.R. & Zolman, J.F. (1971). Cold stress regulation in young chickens. *Poultry Science,* 50,56-61.

Whittow, G.C., Sturkie, P.D. & Stein, G. Jr. (1964). Cardiovascular changes associated with thermal polypnea in the chicken. *American Journal of Physiology,* 207,1349-1353.

Whittow, G.C. & Tazawa, H. (1991). The early development of thermoregulation in birds. *Physiological Zoology,* 64,1371-1390.

Wideman, R. F. (2000). Cardio-pulmonary hemodynamics and ascites in broiler chickens. *Avian and Poultry Biology Review,* 11,21-43.

Wideman R.F., Forman, M.F., Hughes, J.D. Jr, Kirby, Y.K., Marson, N. & Anthony, N.B. (1998). Flow-dependent pulmonary vasodilation during acute unilateral pulmonary artery occlusion in Jungle Fowl. *Poultry Science,* 77,615-626.

Wideman, R.F. & Tackett, C.D. (2000). Cardio-pulmonary function in broilers reared at warm or cool temperatures: effect of acute inhalation of 100% oxygen. *Poultry Science*, 79,257-264.

Wilson H.R. (1991). Physiological requirements of the developing embryo: temperature and turning. In: S.G. Tullet (ed.), *Avian Incubation* (pp. 145-156). Butterworth-Heinemann, Oxford, UK

Winn, P. N. & Godfrey, E.F. (1967). The effect of humidity on growth and feed conversion of broiler chickens. *International Journal of Biometeorology*, 11,39-50.

Wolfenson, D. (1986). The effect of acclimatization on blood flow and its distribution in normothermic and hyperthermic domestic fowl. *Comparative Biochemistry and Physiology A*, 85,739-742.

Wolfenson, D., Bachrach, D., Maman, M., Graber, Y. & Rozenboim, I. (2001). Evaporative cooling of ventral regions of the skin in heat-stressed laying hens. *Poultry Science*, 80,958-964.

Wolfenson, D., Frei, Y.F., Snapir, N. & Berman, A. (1981). Heat stress effects on capillary blood flow and its redistribution in the laying hen. *Pflugers Archives*, 390,86-93.

Yahav, S. (1999). The effect of constant and diurnal cyclic temperatures on performance and blood system of young turkeys. *Journal of Thermal Biology*, 24, 71-78.

Yahav, S. (2000a). Domestic fowl—strategies to confront environmental conditions. *Poultry and Avian Biology Reviews*, 11,81-95.

Yahav, S. (2000b). Relative humidity at moderate temperatures, its effect on male broiler chickens and turkeys. *British Poultry Science*, 41,94-100.

Yahav, S. (2002). Limitation in energy consumption affects the ability of young turkeys to cope with low ambient temperature. *Journal of Thermal Biology*, 27,103-108.

Yahav, S. (2004). Ammonia affects performance and thermoregualtion of male broiler chickens. *Animal Research*, 53,289-293.

Yahav, S. (2007). Thermal manipulations during perinatal period – does it improve thermotolerance and performance of broiler chickens? *19th Annual Australian Poultry Science Symposium.* pp. 14-18.

Yahav, S., Collin, A., Shinder, D. & Picard, M. (2004a). Thermal manipulations during broiler chick embryogenesis: Effects of timing and temperature. *Poultry Science*, 83,1959–1963.

Yahav, S., Goldfeld, S., Plavnik, I. & Hurwitz, S. (1995). Physiological responses of chickens and turkeys to relative humidity during exposure to high ambient temperature. *Journal of Thermal Biology*, 20,245-253.

Yahav, S. & Hurwitz, S. (1996). Induction of thermotolerance in male broiler chickens by temperature conditioning at an early age. *Poultry Science*, 75,402-406.

Yahav, S., Luger, D., Cahaner, A., Dotan, M., Ruzal, M. & Hurwitz, S. (1998a). Thermoregulation in naked neck chickens subjected to different ambient temperatures. *British Poultry Science*, 39,133-138.

Yahav, S., Plavnik, I., Rusal, M. & Hurwitz, S. (1998b). The response of turkeys to relative humidity at high ambient temperature. *British Poultry Science*, 39,340-345.

Yahav, S., Ruzal, M. & Shinder, D. (2007). The effect of ventilation on performance, body and surface temperature of young turkeys. *Poultry Science*, 87,133-137.

Yahav, S., Sasson-Rath, R. & Shinder, D. (2004b). The effect of thermal manipulations during embryogenesis of broiler chicks (*Gallus domesticus*) on hatchability, body weight and thermoregulation after hatch. *Journal of Thermal Biology*, 29,245-250.

Yahav, S. Shamay, A., Horev, G., Bar-Ilan, D., Genina, O. & Friedman-Einat, M. (1997a). Effect of acquisition of improved thermotolerance on the induction of heat shock proteins in broiler chickens. *Poultry Science,* 76,1428-1434.

Yahav, S., Shinder, D., Razpakovski, V., Rusal, M. & Bar, A. (2000). Lack of response of laying hens to relative humidity at high ambient temperature. *British Poultry Science,* 41,660-663.

Yahav, S., Shinder, D. Tanny, J. & Cohen, S. (2005). Sensible heat loss – the broiler's paradox. *World's Poultry Science Journal,* 61,419-435.

Yahav, S., Straschnow, A. Luger, D., Shinder, D., Tanny, J. & Cohen, S. (2004c). Ventilation, sensible heat loss, broiler energy, and water balance under harsh environmental conditions. *Poultry Science*, 83,253-258.

Yahav, S., Straschnow, A., Plavnik, I. & Hurwitz, S. (1997b). Blood system response of chickens to changes in environmental temperature. *Poultry Science,* 76,627-633.

Yahav, S., Straschnow, A., Plavnik, I. & Hurwitz, S. (1996). Effect of diurnal cyclic versus constant temperatures on chicken growth and food intake. *British Poultry Science,* 37,43-54.

Yahav, S., Straschnow, A., Vax, E., Razpakovski, V. & Shinder, D. (2001). Air velocity alters broiler performance under harsh environmental conditions. *Poultry Science,* 80,724-726.

Zhou, W. (2000). Physiological significance of the change in blood viscosity of broiler chickens under high ambient temperature. *Japanese Poultry Science,* 37,201-211.

In: Body Temperature Regulation
Editor: A. B. Cisneros and B. L. Goins
ISBN: 1-60741-282-3

Chapter IV

Neonatal Thermoregulation

Robin B. Knobel*
Duke University, School of Nursing, 307 Trent Drive, Durham, NC 27710, USA

Abstract

Temperature control of the newborn infant has been a popular research topic over the last several decades and one that has significantly altered neonatal morbidity and mortality outcomes. Premature infants especially are prone to heat loss by convection, conduction, evaporation and radiation due to their thin skin, decreased body fat and immature body systems. Healthcare providers make temperature regulation one of the most important management issues for newborns, particularly in the case of premature infants. Although term and most premature infants manage their own heat production through non-shivering thermogenesis, healthcare providers need to manage the environment of those infants to minimize heat loss using special environments and procedures. Nursing and medical interventions have improved neonatal thermoregulation; however, hypothermia continues to be a problem in the extremely low birth weight (ELBW) infant population (premature infants born weighing less than 1000 grams). One intervention, placement of infants in polyurethane bags, serves to protect the smallest premature infants from evaporative heat losses in the delivery room. Once ELBW infants are admitted to neonatal intensive care units, stabilization procedures continue to expose infants to cold temperatures. ELBW infants are unable to generate heat due to immature body systems, and body temperatures can fall as low as 33° C. Therefore, nurses must pay close attention to the environment of ELBW infants, to prevent decreases in body temperature. In this way, morbidity and mortality can be decreased for ELBW infants.

* Corresponding author. Robin B. Knobel* PhD, RNC, NNP, Assistant Professor. Tel: 919-684-9295; E-mail: Robin.knobel@duke.edu

Neonatal Thermoregulation

Temperature control of the newborn infant has been a popular research topic over the last several decades and one that has significantly altered neonatal morbidity and mortality outcomes. Healthcare providers make temperature regulation one of the most important management issues for newborns, particularly in the case of premature infants. Although term and most premature infants manage their own heat production through non-shivering thermogenesis, healthcare providers make it a top priority to manage the environment of those infants to minimize heat loss using special environments and procedures. The earliest and smallest of premature infants are not able to produce heat adequately; therefore, these infants are the most vulnerable to increased morbidity and mortality from cold stress.

Full-term infants produce heat by non-shivering thermogenesis or brown fat metabolism and if they become cold during admission procedures or nursing care, they can generate their own heat to keep their bodies warm. Premature infants, between 37 and 32 weeks gestation have the ability to generate heat by non-shivering thermogenesis, but lack the body fat that insulates term infants. Therefore, these infants need to be cared for in heated environments such as incubators or radiant warmers, preventing heat loss. Preterm extremely low birth weight (ELBW) infants have little to no ability to generate heat by non-shivering thermogenesis due to immature systems (Houstek et al., 1993).

ELBW infants are infants weighing less than 1000 grams, which corresponds to less than about 27-28 weeks gestation at birth. The earliest that an ELBW infant can be delivered and stand a chance of survival is 23 weeks gestation, with a birth weight as low as 400 grams (Kattwinkle, 2000). Altogether, about 33,000 ELBW infants are born each year in the United States (Mathews & MacDorman, 2006). Technology has advanced to the point that most ELBW infants survive (O'Shea, Klinepeter, Goldstein, Jackson, & Dillard, 1997; Subramanian, Yoon, & Toral, 2002). Survivability is dependent on birth weight corresponding with the gestational age; therefore, survival rates are approximately 11% for infants weighing less than 500 grams survive, 51% for infants weighing 500-749 grams, and 84% for infants weighing 750-1000 grams (Subramanian et al., 2002).

Premature infants are born then stabilized in a delivery room or stabilization room, typically on a radiant warmer. Stabilization after birth can take a few minutes to an hour or more, during which the infant is exposed to the cold surrounding environment. Once stabilization is complete, the infant is then transferred to a neonatal intensive care unit, where further stabilization and admission procedures take place. Most stabilization procedures take place while the infant is cared for under radiant heat; however, the infant continues to lose heat by conduction, convection, radiation and evaporation. The infant will usually be transferred into an incubator, which should provide the infant with a warm, humid environment and minimize heat loss. This will be the infant's home until the infant can generate heat and retain heat without this external source, usually around 32-34 weeks gestation.

Heat production is accomplished through non-shivering thermogenesis. Full-term infants rely on brown fat metabolism or non-shivering thermogenesis for heat production (Houstek, et al, 1993). The metabolism of brown fat is inefficient in the ELBW infant due to extreme immaturity (Bruck, 1998; Houstek et al., 1993) and may not produce enough heat to prevent body temperature from falling. Because ELBW infants are vulnerable to cold stress, knowing the optimal range for their temperature is important in order to protect the infant from cold stress and organ system insult. However, very little research has examined this range in ELBW infants. Most of the research on optimal body and environmental temperatures for premature infants occurred 20-40 years ago, before significant numbers of ELBW infants or ventilated infants survived (Adams, Fujiwara, Spears, & Hodgman, 1964; Bruck, 1961; Buetow & Klein, 1964; Day, Caliguiri, Kamenski, & Ehrlich, 1964; Silverman, Fertig, & Berger, 1958; Yashiro, Adams, Emmanouilides, & Mickey, 1973).

Many studies have linked cold stress to increased mortality and morbidity (Buetow & Klein, 1964; Day et al., 1964; Hazan, Maag, & Chessex, 1991; Vohra, Grent, Campbell, Abbott, & Whyte, 1999) and shown the importance of a warm environment in caring for preterm infants (Baumgart, Engle, Langman, Fox, & Polin, 1980; LeBlanc, 1982; Meyer, Payton, Salmont, Hutchinson, & de Klerk, 2001; Miller & Oliver, 1966; Yashiro et al., 1973) to prevent cold stress. Many ELBW infants exhibit low body temperatures even with the advances in themoregulation of the premature infants and with current technology. This chapter will focus on neonatal body temperature regulation in premature and sick infants during the first 30 days of life.

Human Temperature Regulation

Humans are homeothermic in that they can regulate their core body temperature within a narrow limit (Nadel, 2003; Widmaier, Raff, & Strang, 2005). Core (internal) body temperature is maintained at approximately 37° C, but ranges from 36° C to 37.5° C (Nadel, 2003; Widmaier et a., 2005). Humans require precisely regulated body temperature because large elevations in temperature can cause nerve malfunction and protein denaturation, and the brain ceases to control temperature after large increases (Widmaier et al., 2005). Convulsions may occur at 41° C and 43° C is the absolute limit of survival (Widmaier et al., 2005). Alternatively, severe hypothermia will cause the heart to slow down. Once the internal body temperature reaches 25° C humans will suffer cardiac standstill or fibrillation (Guyton & Hall, 2006).

Temperature is regulated by balancing heat production against heat loss (Guyton & Hall, 2006), a balance that is continuously being disturbed by changes in metabolic rate or external environment (Widmaier et al., 2005). The thermoregulatory system in a human consists of thermal sensors, afferent pathways, an integration system in the central nervous system, efferent pathways, and target organs that control heat generation and transfer (Nadel, 2003).

Central and peripheral thermoreceptors sense the alteration in temperature on the skin and internally. Peripheral thermoreceptors are free nerve endings that are distributed over the entire skin surface and are thermosensitive (Nadel, 2003). These peripheral thermoreceptors provide information that is sent forward through the temperature control pathway and have the ability to detect warm or cold areas on the skin (Widmaier et al., 2005). Receptors increase their firing rate from a steady state with a change in the skin's normal temperature (Nadel, 2003). The skin thermoreceptors provide, by way of afferent nerve fibers that carry sensory information to the hypothalamic regulatory center, an early warning when there is a change in the ambient temperature of the skin. This information also travels by thalamic pathways to the cerebral cortex to cause a conscious perception of the thermal environment and behavioral adjustments (Nadel, 2003). Deep body structures including the hypothalamus, spinal cord, and abdominal organs contain central thermoreceptors (Widmaier et al., 2005). Once thermoreceptors detect a core temperature that is colder or warmer than normal, they modify heat transfer rates to restore core temperature to its regulated level (Nadel, 2003).

Peripheral and central thermoreceptors send information to the brain control center, the hypothalamus, which enables signals to be sent through neuronal pathways. The principal area of thermoregulatory regulation is the preoptic and anterior hypothalamic nuclei of the hypothalamus (Guyton & Hall, 2006). The hypothalamus monitors the prevailing thermal status with a normal set of thermal conditions and then sends efferent commands to alter the rate of heat generation and modify the rate of heat transfer within and from the body (Nadel, 2003).

Neuronal effector mechanisms attempt to increase or decrease body temperature through sending signals by way of sympathetic nerves going to the sweat glands, adjusting smooth muscle tone of cutaneous arterioles to control blood flow to the skin surface, activating motor neurons to the skeletal muscles (Guyton & Hall, 2006; Nadel, 2003; Widmaier et al., 2005) or activating chemical thermogenesis. Changes in muscle activity constitute the major control for temperature regulation. In the case of cold exposure, the hypothalamus causes motor neurons to the skeletal muscles to produce shivering which is oscillating rhythmical muscle contractions and relaxations occurring at a rapid rate (Widmaier et al., 2005). The rapid muscle contractions produce heat and can be called shivering thermogenesis. When there is too much heat, basal muscle contractions are reflexively decreased and voluntary movement slows. The autonomic system controls cutaneous blood flow over most of the skin, the body's largest organ. Active vasodilation can increase blood flow up to 10 times the resting level (Nadel, 2003), thereby releasing heat from the skin to the environment. A minor reduction in cutaneous blood flow is caused by cutaneous vasoconstriction, mediated by sympathetic nerves because of changes in core body temperature, which causes heat to be conserved. If the heat load increases, the autonomic nervous system activates the eccrine sweat glands, which cause sweat to be secreted onto the skin's surface. The sweat increases the partial pressure of water vapor on the skin and causes evaporation and loss of heat.

Chemical thermogenesis provides heat without muscle activity by increased sympathetic stimulation causing increased norepinephrine and epinephrine circulation in the blood leading to an immediate increase in the rate of cellular metabolism (Guyton & Hall, 2006; Widmaier et al., 2005). Chemical thermogenesis, also called brown fat metabolism or non-shivering thermogenesis, is the primary method of heat production for the infant up to 1 year of age (Hull & Smales, 1978). Chemical thermogenesis, if any, is minimal in adults. The degree of chemical thermogenesis that occurs in an animal is directly proportional to the amount of brown fat that exists in that animal (Guyton & Hall, 2006).

Fetal Temperature Regulation in the Fetal Period through Transition to Extra-Uterine Life

The fetal ambient environmental temperature is the maternal body temperature. Heat builds up in the fetus until the temperature gradient between the fetus and the mother causes heat to dissipate from the fetus to the mother (Power, 1998). Under normal conditions, the temperature of the fetus remains about 0.5°C higher than that of the mother, or about 37.6° C to 37.8° C (Blackburn, 2003; Power, 1998). This tight linkage with the mother prevents the fetus from regulating its own temperature independently. The heat generated by the fetal metabolism offloads to the mother through the amniotic fluid to the uterine wall or via the umbilical cord and placenta to the maternal blood stream (Blackburn, 2003). Cooling of fetal sheep in utero does not elicit non-shivering thermogenesis (Gunn & Gluckman, 1995); therefore, it is likely that the human fetus cannot independently produce heat in response to cooling. Evidence indicates that placental inhibitors, primarily prostaglandin E_2 and adenosine, prevent non-shivering thermogenesis (Gluckman, Sizonenko, & Bassett, 1999; Gunn & Gluckman, 1995; Sawa, Asakura, & Power, 1991). The inhibitors of non-shivering thermogenesis rapidly decrease once the umbilical cord is clamped upon birth. Non-shivering thermogenesis can then be activated by stimulating the infant's cold receptors by virtue of the change into a colder environment, increasing oxygenation consumption of the infant (Gunn & Gluckman, 1995; Power, 1998).

Infant Temperature Regulation

Once the infant is delivered and the umbilical cord is clamped in the relatively cold delivery room, exposure to a cold environment will trigger non-shivering thermogenesis. The infant's body temperature decreases and peripheral and central thermoreceptors detect change (Widmaier et al., 2005). The immediate response of the peripheral skin receptors in adults is to cause vasoconstriction of the peripheral vessels to diminish heat loss and inhibit sweating and shivering with a resultant increase in body heat production. The neonate lacks the ability to shiver and sweat; therefore, vasoconstriction is the predominant result of activation of peripheral skin receptors (Guyton & Hall, 2006). Central thermoreceptors send messages to areas of the hypothalamus located bilaterally in the posterior hypothalamus near the level of the mamillary bodies (Guyton & Hall, 2006). The anterior hypothalamic-preoptic area also transmits temperature signals into this posterior hypothalamic area. The hypothalamus serves as the primary overall integrator of hormonal and system responses because it is the single most important control area for homeostatic regulation and has different pathways that form the master command center for neural and endocrine coordination. A peptide, neuropeptide Y, is abundant in the hypothalamus and at normal body temperatures may inhibit thermogenesis to maintain a normal body temperature and regulate energy homeostasis (Wilding, Widdowson, & Williams, 1997).

Because the effector mechanisms of skeletal muscle stimulation are minimal in the infant, the infant will not shiver in response to cold stress. Non-shivering thermogenesis is the predominant effector mechanism. Norepinephrine is released from the nerve endings that terminate on the surface of brown adipocytes, while simultaneously causing an increase in thyroid stimulating hormone (TSH), due to an increase in sympathetic activity. The catecholamine, norepinephrine, acts via adrenergic receptors, primarily type β3 but also via α1 receptors (Power, 1998). Cold stress also causes inhibition of synthesis of Neuropeptide Y, which allows thermogenesis (Billington, Briggs, Harker, Grace, & Levine, 1994). TSH stimulates the release of thyroid hormones, mostly thyroxine ($T_{4)}$. Working simultaneously, norepinephrine activates type II iodothyronine 5'deiodinase, which causes T_4 to convert to T_3 (Barrett, 2003). In brown adipose tissue, triiodothyrone (T_3) generated locally acts to upregulate an uncoupling protein referred to UCP or Thermogenin. T_3 acts locally to uncouple mitochondrial oxidation from phosphorylation in the brown adipose tissue, causing heat production. Norepinephrine mediates the production of the second messenger, cAMP that activates cAMP-dependent protein kinase (Voet et al., 2002). The kinase activates the hormone-sensitive triacylglycerol lipase by phosphorylating it. The activated lipase hydrolyzes tricylglycerols to yield free fatty acids and glycerol (Power, 1998; Voet, Voet & Pratt, 2002). Free fatty acid presence causes an inhibitory effect of purine nucleotides (ADP, ATP, GDP, GTP) on thermogenin.

The major components of non-shivering thermogeneisis in infants are brown adipose tissue, thermogenin, and type II iodothyronine. Hatai (1902) first described brown adipose tissue in humans in 1902. Brown fat is a tissue with thermogenic capabilities, rich in mitochondria and blood supply innervated with sympathetic nerve fibers. Brown adipocytes arise from mesenchymal cells, which are associated with blood vessels and proliferate rapidly simultaneously with the blood vessels developing inside the brown fat (Nechad, 1986). Mitochondria develop and increase in size and number depending on the metabolic activity of oxidation in the tissue. During this time of development, sympathetic nerve fibers grow into the tissue and become densely innervated on the nerve endings of each blood vessel and each adipocyte (Sauer, 1995). Protein content increases considerably, as does mass of mitochondria in the tissue. The inner mitochondria membrane becomes specifically enriched with thermogenin, the rate-limiting enzyme in the process of heat production (Sauer, 1995).

Thermogenin is a 306-residue polypeptide chain protein (Bing et al., 1996) that makes up approximately 15% of the protein in the inner mitochondrial membranes of brown adipose tissue (Voet et al., 2002). Thermogenin mRNA transcription probably only occurs in brown adipose tissue (Ricquier & Bouillaud, 2000). Signals activated after norepinephrine stimulation of brown adipose tissue strictly control thermogenin gene expression at the transcription level (Voet et al., 2002). Thermogenin is an H^+ channel protein located in the inner mitochondrial membrane of brown adipose tissue (Jones & DeCherney, 2003). Free fatty acids relieve the inhibition of H^+ channel and allow conduction of protons. This action allows the protons generated by electron transport to enter the mitochondrion through Thermogenin that dissipates the H^+ gradient needed by the H^+ translocating ATP synthase. Under physiological conditions, heat is produced by the dissipation of an electrochemical H^+ gradient, which is generated by electron transport and uncoupled from ATP synthesis (Voet et al., 2002). Therefore, the mitochondria in brown adipose tissue can produce heat without storage of energy through ATP (Jones & DeCherney, 2003).

Brown adipose tissue is very vascular allowing stimulation by norepinephrine to cause a tremendous increase in blood flow to the tissue. It occurs as discrete lobes, found predominantly in the interscapular, thoracic, and cervical regions, as well as beside the great vessels and contiguous with the spinal cord (Huttunen, 2002). It has a brown red color because of the rich vascularization and the presence of mitochondrial cytochromes. Brown adipose tissue quickly exports the heat to critical organs through the massive vascularization. Most brown adipose tissue deposits are deep, located next to internal organs, and supply heat by direct conductance. Some interscapular brown fat causes heat by vascular convection because of its superficial location. A high rate of brown adipose tissue thermogenesis requires high rates of blood flow to provide oxygen for metabolism and to remove heat.

In humans, T_4 is bound more tightly to plasma proteins than T_3, causing a net effect of two times the amount of free T_4 than T_3 in circulation (Barrett, 2003). Exposure to a cold environment triggers an increase in the levels of norepinephrine and TSH. TSH stimulates the release of T_4, which is converted to T_3 mediated by type II iodothyronine in the brown adipose tissue. The conversion of T_4 to T_3 is crucial for the thermogenic response (Sauer, 1995) because T_3 locally acts to upregulate thermogenin leading to heat production (Voet et al., 2002).

All infants retain the capacity for non-shivering thermogenesis up to 6-12 months of age. However, preterm infants may be unable to meet heat loss demands from exposure to a cold environment quickly enough to prevent a fall in body temperature. Low birth weight (less than 2500 grams) and preterm infants are more susceptible to stress from a cold environment than term infants because of greater heat exchange with the environment and because non-shivering thermogenesis becomes inefficient before about 30-32 weeks gestational age due to less thermogenin and type II iodothyronine content.

Developmental Properties Related to Non-shivering Thermogenesis

Brown adipose tissue begins to develop as early as the 75 mm fetal stage (Hatai, 1902). Only 20-30 grams of brown adipose tissue are necessary to handle all the observed non-shivering thermogenesis of a newborn baby as calculated by Hull in 1977. Studies have shown that the structure of brown adipose tissue is well developed in the preterm infant as early as 25 weeks gestational age (Sauer, 1995), with brown fat comprising about 1-2% of body weight (Nechad, 1986). However, brown adipose tissue is not the only essential element for non-shivering thermogenesis. The specific content of thermogenin in infants increases from 29.4 ± 3.3 pmol/mg at 25 weeks gestational age to 62.5 ± 10.2 pmol/mg at 40 weeks gestational age (Houstek et al., 1993). A major increase in thermogenin does not occur until 32-weeks gestational age, which is approximately the time when a neonate can effectively use non-shivering thermogenesis to generate heat adequately. The enzyme type II iodothyronine is active by 25 weeks gestational age and increases fourfold by term. A major increase in this enzyme also occurs at 32 weeks gestational age (Houstek et al., 1993). The low levels of thermogenin content and type II iodothyronine content prior to 32 weeks gestational age are the likely causes of inefficient non-shivering thermogenesis.

Because non-shivering thermogenesis is inefficient, preterm infants less than 30-32 weeks gestational age are more susceptible to cold stress from cold environments. Cold stress is more likely during transition, when the infant is delivered from the warm intrauterine environment to the cold, drafty environment of the delivery room. The newly born infant is covered in amniotic fluid; therefore much heat is lost quickly by evaporation. Hammarlunc and Sedin (1979) found transepidermal water loss in infants to be inversely correlated with gestational age possibly because more premature infants have thinner skin. Transepidermal water loss is 15 times higher in infants born at 25 weeks gestational age than in term infants; therefore, heat loss would be proportionate to this evaporation of water through the skin. The high evaporative heat losses in preterm infants during the first few hours of life and even first few days will gradually decrease with advancing postnatal age (Sedin, 1995). Radiative heat losses are initially low immediately after birth in preterm infants 25-27 weeks gestational age but become the most important route of heat transfer after the first postnatal week (Sedin, 1995). In infants greater than 28 weeks gestational age, heat loss from radiation is most important route of heat transfer from birth onward. The heat loss through radiation gradually increases with age. For an infant 25-27 weeks gestational age, evaporative heat loss transfers the most heat during the first 10 days of life, if the infant is in a dry environment. If the infant is kept in an environment of 60% humidity, heat loss will be much lower and a greater heat loss will result from radiation from the fifth day of life onward (Sedin, 1995).

Heat Transfer

Heat is a by-product of metabolism. The metabolic rate of the body (rate of heat production) is determined by the basal rate of metabolism of all body cells and extra metabolism caused by muscle activity, thyroxine on the cells, sympathetic stimulation of the cells, and an increase in chemical activity in the cells themselves (Guyton & Hall, 2006).

Human thermoregulation attempts to keep body temperature in a steady state, in which heat production equals heat loss. The rate at which heat is lost depends on how rapidly heat can be conducted from where it originates within the body and how fast heat can be transferred from the skin to the environment. The body is insulated from heat loss through the protection of the skin along with subcutaneous tissues and fat. Fat conducts heat only one third as readily as other tissues in the body (Guyton & Hall, 2006). The skin transfers heat to the environment by way of radiation, conduction, convection and evaporation.

Heat decreases or increases by way of radiation, which is the process by which all body surfaces emit heat in the form of electromagnetic waves (Guyton & Hall, 2006; Nadel, 2003; Widmaier et al., 2005). The infrared portion of the electromagnetic energy spectrum carries this energy. The temperature difference between the skin and the radiating body determines the rate of heat loss. For example, heat may be transferred from the infant's body to a colder wall near the infant. Alternatively, heat may be absorbed into the skin from a radiant heat source near the infant.

Moreover, heat transfers to or from the skin surface by way of conduction when the skin touches another solid object of different temperature (Guyton & Hall, 2006; Nadel, 2003; Widmaier et al., 2005). In the NICU, this heat gain or loss is minimal because infants usually rest on prewarmed surfaces. Heat moves from molecule to molecule as the molecules from the skin surface and another surface collide. This object might be air, water, or a solid surface on which the infant rests.

Heat is transferred by convection when a moving fluid such as air or water carries heat between the body surface and the environment (Guyton & Hall, 2006; Nadel, 2003; Widmaier et al., 2005). If the body surface is warmer than the surrounding environmental air, heat is first conducted into the air, then swept away by convective air currents. Warm molecules are less dense and therefore rise into the air from the skin.

Heat loss by evaporation is a major route of heat transfer when water is lost from the skin and membrane linings of the respiratory tract. The evaporative rate is independent of the temperature gradient between the skin and the environment and is proportional to the water vapor-pressure gradient between the skin and the environment (Nadel, 2003). Evaporation causes 0.6 kcal of heat to be lost for every 1 gram of water lost from the body (Guyton & Hall, 2006; Nadel, 2003; Widmaier et al., 2005).

A neutral thermal environment is a set of environmental conditions in which the temperature of the naked body does not change when the subject is at rest and there is no muscle activity (Nadel, 2003). These environmental conditions of air temperature, airflow, humidity, and temperature of surrounding radiating surfaces will minimize heat loss or gain by radiation, conduction, convection and evaporation to keep the infant in a steady metabolic state related to thermal balance.

Heat transfer from the body causes problems for the thermoregulation of infants. Because infants have less subcutaneous fat and relatively larger surface areas proportioned to weight exposed to the environment than adults and because they rely solely on non-shivering thermogenesis for heat production, we must examine thermoregulation in infants, and in particular in preterm infants.

Current Recommendations: Thermal Care of the Preterm Infant

Current recommendations for thermal care for the preterm infant have improved as researchers have found new ways to guard against heat loss. Research has determined that caring for preterm infants under radiant warmers and in incubators can reduce heat loss (Baumgart, 1987; Bell & Rios, 1983; LeBlanc, 1982; Yashiro et al., 1973), with some studies using computerized servo control and setting the desired skin temperature (Buetow & Klein, 1964; Day et al., 1964; Malin & Baumgart, 1987; Meyer et al., 2001).

Yashiro et al. (1973) conducted an early study providing warmth using convective and radiant heat to improve the body temperature of premature infants and found that both types of heat would increase the body temperatures of infants to above 36° C. Baumgart (1987) found that infants nursed in an incubator with convective heat had lower convective heat losses compared to infants nursed under radiant warmers which provide radiant heat. LeBLanc (1982) also found that infants' oxygen consumption was less when nursed in incubators rather than under radiant warmers. Bell and Rios (1983) found that manual air temperature control (increasing the environmental temperature through increase in heat) could be just as effective in maintaining a warm environment and body temperature for infants as using the servo temperature control whereby an infant's body temperature is controlled at a set temperature by a computer.

Buetow and Klein (1964) found that infants controlled at a body temperature of 36° C or more for the first 96 hours had a significantly greater chance of surviving. Day et al. (1964) found lower mortality in infants heated to maintain an abdominal temperature at 36° C compared to infants given constant heat at 31.8° C regardless of their temperature. Malin and Baumgart (1987) found that an abdominal temperature of at least 36.5° C, rather than lower temperatures, kept oxygen consumption to a minimum for preterm infants. Meyer et al. (2001) found maintaining abdominal temperatures at a target temperature of at least 36.8° C under a radiant warmer was easier than in an incubator during the first 24 hours of life.

These researchers contributed very valuable changes to the thermal care of preterms, greatly reducing radiative, evaporative, conductive, and convective heat loss. Caring for preterms less than 32-34 weeks gestational age or less than approximately 1500 grams, in an incubator or on a radiant warmer is the current standard of care. However, researchers have given no clear indication about how they selected their control points for environmental temperature or body temperature, and most of these early studies were done on infants weighing more than 1,000 grams. None of the studies evaluated the physiological effects of differing body temperatures specifically on ELBW infants, the very group that needs the most thermal support.

Thus, neonatal nurses have little information to apply to the thermal care of ELBW infants. A current neonatal nursing text recommends trying to achieve thermoneutrality for normal growth and to meet the infants' energy needs (Kenner, 2003). The thermal neutral environment is generally 32-33.5° C for term infants, 34-35 ° C for preterm infants greater than 30 weeks gestational age, and unspecified higher temperatures for infants less than 30 weeks gestational age (Blackburn, 2003). One nursing text states that what constitutes a normal range for preterm infants is not precisely known and needs further research (Kenner, 2003). Because an optimal body temperature range has not been specified for preterm infants of different gestational ages determining the appropriate temperature during transition for ELBW infants is difficult (Kenner, 2003).

Nursing and medical texts recommend using radiant warmers and incubators to decrease evaporative losses and promote conductive gain (Bell & Glatzl-Hawlik, 1998; Blackburn, 2003; Kenner, 2003), however, no text gives precise control points. Various researchers have recommended plastic wrap as a way to reduce evaporative and convective heat losses in the NICU (Baumgart, 1984; Baumgart et al.,1982; Knauth, Gordin, McNelis & Baumgart, 1989) and in the delivery room (Besch, Perlstein, Edwards, Keenan, Sutherland, 1971; Bjorklund & Hellstrom-Westas, 2000; Knobel, Wimmer & Holbert, 2005; Vohra et al., 1999; Vohra, Roberts, Zhang, Janes, & Schmidt, 2004). Placing ELBW infants in plastic bags just after delivery has been especially helpful in reducing initial hypothermia in this gestational age group (Bjorklund & Hellstrom-Westas, 2000; Cramer, Wiebe, Hartling, Crumley, & Vohra, 2005; Knobel et al., 2005; Vohra et al., 1999; Vohra et al., 2004); implementation of this intervention is slowly increasing nationwide (Knobel, Vohra, & Lehmann, 2005).

Specific thermal guidelines for ELBW infants do not exist as standards of care due to the slow translation of research to the bedside. This lack of information can be especially detrimental during the transitional period when cold stress is likely following delivery and stabilization of the infant (Kent & Williams, 2008; Mathew, Lakshminrusimha, Cominsky, Schroder, & Carrion, 2007). Unless heat loss can be prevented, attempting to produce heat could potentially stress system homeostasis (balance of normal oxygenation, acid-base, glucose, and heart rate) for the ELBW infant in transition. Guidelines are needed for a specific body temperature range for the ELBW infant during transition so that cold stress does not cause harm to physiological homeostasis. This range probably varies with gestation similar to the ability to generate non-shivering thermogenesis (Houstek et al., 1993).

Hypothermia, or a low body temperature, can lead to a syndrome in infants characterized by slowness of movement, refusal to eat, lethargy, edema, and often sclerema (Sinclair, 1992) as described by Mann (1955) and Elliot (Mann & Elliot, 1957). The syndrome results from prolonged exposure to a cold environment, especially during the transitional period. Hypothermia has been associated with a fall in systemic arterial pressure, a decrease in plasma volume, decreased cardiac output, increased peripheral resistance, and metabolic acidosis (Sinclair, 1992).

No studies have specified the body temperature that defines hypothermia for infants. Some early researchers exposed preterm infants (1120- 2270 grams) at less than 24 hours of age to environmental temperatures of 21-23° C and found an increase in oxygen consumption (Adams et al., 1964; Bruck, 1961). Bruck (1961) also noted that body temperature fell more rapidly for preterms than for term infants when they were in an environment with a temperature of 28° C. Research in which infants were deliberately exposed to colder temperatures was only done in the 1950-1960's. Studies at that time linked exposure to cold to increased mortality (Buetow & Klein, 1964; Day et al., 1964; Silverman et al., 1958), and it is no longer ethical to deliberately expose preterm infants to cold environmental temperatures. Yet cold body temperature, or hypothermia, continues to be a problem for ELBW infants (Bhatt et al., 2007; Horns, 2002; Kenty & Williams, 2008; Knobel et al., 2005; Loughead, Loughead, & Reinhart 1997; Mathew et al., 2007; Thomas, 2003).

Researchers have shown that ELBW infants continue to have cold stress in today's technologically advanced NICUs (Bhatt et al., 2007; Horns, 2002; Knobel et al., 2005; Loughead et al., 1997; Thomas, 2003). Knobel et al. (2005) and Loughead et al. (1997) found 66-93% of ELBW infants were admitted to the NICU with hypothermic temperatures. Horns (2002) found 90% of her ELBW sample had extremely cold peripheral temperatures when cared for in the controlled environment of an incubator. Thomas (2003) found temperatures of preterm infants as low as 33.1° C associated with caregiver interventions occurring while the infant remained in an incubator.

Neonatal Hyperthermia

An infant with a body temperature greater than 37.5° C is said to be hyperthermic (LeBlanc, 2002). Because ELBW infants have immature thermoregulation, thin skin and a large body surface area to weight ratio, they can also take on too much heat from a heat-enriched environment and become hyperthermic (too high a body temperature). Preterm infants can be prone to overheating due to large surface area, limited insulation, and limited sweating ability (Jones & DeCherney, 2003; Guyton & Hall, 2006; Thomas, 1994). Full term infants and larger preterms (35-37 weeks gestation) are able to sweat and raise their evaporative losses; however, the maximal sweating rate is related to gestational age (Sulyok, Jequier, & Prod'hom, 1973). The set-point temperature for sweating tends to be lower with increasing gestational age, and Sulyok et al. (1973) found sweating rates for larger preterm infants were found to be 1.43 W/kg compared to full term infants who had a rate of 1.81 W/kg. They measured sweating rates in watts (W) with 1 W = 14.334 calories/minute = 0.860 Cal/hr = 2.975 ml of oxygen consumed per minute.

Hypothalamic thermoreceptors sense an increase in body core temperature and compare this temperature with a reference signal, then direct neural output to activate heat dissipation (Nadel, 2003). Skin blood flow and sweating will increase as core temperature increases, prompting an increase in the rate of heat transfer to the environment. The capacity to elevate skin blood flow is an essential defense against hyperthermia. ELBW infants have little ability to vasoconstrict peripherally (Horns, 2002; Lyon, Pikaar, Badger & McIntosh, 1997); therefore, it is likely that their ability to vasodilate is also inefficient.

Hyperthermia may be associated with fever in term infants related to infection; however, early onset infection is likely to coincide with normal temperatures after birth for term infants and with hypothermia in preterms (Weisman, Stoll, & Cruess, 1992). Preterm infants with hyperthermia are most likely overheated due to iatrogenic heat transfer. Horns (2002) found temperatures as high as 38.7° C associated with caregiver interventions with infants in incubators, with some incubator temperatures recorded as being high as 40.49°C.

Prevention of Hypothermia in the Delivery Room

As a neonatal nurse practitioner, I have admitted ELBW infants into the NICU with body temperatures as low as 32.8° C. Many others have documented low body temperatures on admission to the NICU after stabilization in the delivery room (Bhatt et al., 2007; Loughead et al., 1997; Hazan et al., 1991; Vohra et al., 1999). The risk of cold stress is greatest at birth during the transition from the warm, wet, well-insulated intrauterine environment to the cool, drafty delivery room. To reduce this risk, the Neonatal Resuscitation Program (NRP) has made the provision of warmth the first step in resuscitation of the newborn (Kattwinkle, 2006).

Previously reported efforts to reduce heat loss and therefore prevent hypothermia have included the use of various types of occlusive coverings (Baumgart et al., 1980; Baumgart, 1984; Bjorklund & Hellstrom-Westas, 2000; Besch et al., 1971;). These presumably work by reducing evaporation, which is a particularly important route of heat loss in the first few minutes after birth (Hammarlund, Nilsson, Oberg, & Sedin, 1980). Such barriers may also reduce convective losses.

To examine the extent of this problem, I conducted a retrospective chart review of 100 infants admitted to our NICU. ELBW infants had a 93% incidence of hypothermia upon admission according to the traditional definition of hypothermia (body temperature < 36.4°) (Knobel et al., 2005). After an extensive literature review, I located a study that had tested a promising intervention to decrease cold stress. Vohra et al. (1999) used a plastic lunch bag to wrap infants less than 32-weeks gestation, immediately after delivery. Although this study had a small sample, the intervention significantly improved admission body temperatures of infants of less than 28-week gestational age but not those of older infants.

As principal investigator, I replicated the study in 2002 with a larger sample of 88 infants, all less than 29-weeks gestation. The infants were placed into a polyurethane bag, already used for infants born with bowel defects in our unit, immediately after delivery and while still wet with amniotic fluid, up to their necks, then resuscitated normally. The infants placed in the polyurethane bags were less likely to be hypothermic on admission (44% vs 70%, Chi Square (1)=6.22, $p<.01$) and had higher mean admission temperatures than those not placed in the bag (36.5° C vs 36° C, t (86)=3.066, $p<.003$) (Knobel et al., 2005). The analyses also revealed that colder delivery room temperatures led to colder body temperatures, independent of the intervention. Once the bag was removed and stabilization procedures began, the infants were further exposed to cold environments. As a result, many of the ELBW infants did not maintain their body temperatures in the NICU. Admission temperatures for the control group ranged from 34.6° C to 37.5° C and from 34.9° C to 38.3° C for the intervention group. The very smallest of the study patients remained with extremely low temperatures on admission to the NICU, regardless of the study intervention.

We found that wrapping infants less than 29-weeks gestational age in polyurethane bags immediately after birth reduced the occurrence hypothermia upon NICU admission. This finding supports that of Vohra et al (1999) who found higher admission temperatures in infants less than 28 weeks gestation who were wrapped in polyethylene immediately after delivery. In the Vohra et al. (1999) study, all 5 of the infants who died were in the control (non-wrapped) group; the investigators postulated that reducing hypothermia might have reduced mortality. This study was small, with only 18 patients in the subgroup less than 28 weeks gestation, but it was provocative in its implications for the importance of thermal management. Since that time, many other researchers have used similar interventions to prevent heat loss during delivery room stabilization (McCall, Alderdice, Halliday, Jenkins & Vohra, 2008).

Another finding of our study was the association between the environmental temperature in the delivery room and the infant's NICU admission temperature (Knobel et al., 2005). Since this study, Kent & Williams (2008) verify that increasing delivery room temperature increases body temperature for preterm infants upon admission to the NICU. Delivery room temperatures in our study ranged from 18.9°C to 31.1°C, although we requested that the delivery room be maintained at an environmental temperature of 26°C for all premature births.We found that warmer delivery room temperatures were indeed associated with higher admission temperatures, but that only the subgroup of infants who were both delivered in warm delivery rooms and placed in polyurethane bags had a mean admission temperature > 36.4°C. Hyperthermia was seen in only one patient and was not associated with any recognized adverse effects. Neonatal Resuscitation Program standards now dictate that if the premature infant is born less than 28-weeks gestation, that the resuscitators should consider placing him, below the neck in a polyethylene bag (Kattwinkle, 2006).

Body Temperature Control in the First Few Days of Life

ELBW infants (300-1000 grams at birth and usually with a corresponding gestational age of 23-28 weeks) are exposed to many stabilization procedures and cool room temperatures, cold infusions, and cold drapes from birth through the first 12 hours of life. Because ELBW infants are similar to poikilothermic (cold-blooded animals that vary their internal temperature) beings, their temperature will continue to fall unless intervention is used to warm them. Without the ability to generate heat, these infants can experience morbidity and mortality from such low body temperatures (Buetow & Klein, 1964; Day et al., 1964; Hazan et al., 1991).

Our last study examined heart rate, abdominal and peripheral temperature, peripheral vasoconstriction and oxygenation in ten ELBW infants (510-960 grams birth weight) from 2004-2006 in a within subjects, multiple case design. Infants were followed over their first 12-hours in the NICU as this is the time of transition from the mother's uterus to stabilization in the NICU and the infants are exposed to birth, stabilization procedures and nursing care exposing these infants to cold environmental temperatures. Central and peripheral temperatures were measured using a thermistor attached to the infant on the surface of the abdomen and the sole of the foot every minute for the first 12-hours in the neonatal intensive care unit. Temperatures were recorded by a data logger and then downloaded into a laptop computer. Heart rate was measured using a SpaceLabs cardiopulmonary monitor continuously for 12 hours and downloaded into a laptop computer using a Windows hyperterminal program. Pulse oximeter readings were measured using a Nelcor Pulse Oximeter and recorded continuously at 1-minute intervals for 12-hours, then downloaded into a laptop computer. Peripheral vasoconstriction was monitored (adominal temperature greater than foot temperature by at least 2° C) for the 12-hours.

Results: Low Temperatures

Infants in this study averaged very low abdominal temperatures (35.17°-36.68° C) through out the entire 12-hour study period. Abdominal temperature was closely and generally positively related to peripheral temperature in all 10 infants. Minimum abdominal temperatures for the 10 subjects ranged from 28.96° to 33.95° C and their maximums ranged from 36.91° C to 38.25° C. The range of mean abdominal temperatures over the 12-hour study periods for each infant ranged from 35.17° to 36.68° C. Seven of ten subjects had a mean abdominal temperature for the study period less than the 36.4° C defined cut point for hypothermia.

Peripheral temperatures ran slightly higher. The minimum peripheral temperature for each infant ranged from 32.13° to 35.12° C and the maximum ranged from 36.8° to 39.6° C. The mean peripheral temperature range for all 10 study infants was 35.10° to 37.06° C. Five of the ten infants had mean peripheral temperatures that were less than the cut point for hypothermia.

I examined the plotted trends of abdominal and peripheral temperature for all study subjects; 7 of 10 infants had peripheral temperatures greater than abdominal temperatures for most of the study period. The mean abdominal temperature for all 10 study subjects was 35.93° C which is 0.38° C lower than the mean peripheral temperature for all 10 study subjects (36.31° C).

Infants did not Exhibit Peripheral Vasoconstriction

Only one infant exhibited peripheral vasoconstriction with 9% of his observations representing peripheral vasoconstriction. This was an 880-gram, 26-gestation week male. The largest infant in the study (960-gram, 26-gestational week female) showed an attempt at peripheral vasoconstriction with 63% of her measures with abdominal temperature greater than foot temperature by at least 1° C. Eight infants (500-710 grams, 24-25 weeks GA) showed a pattern where at least 50% of their peripheral temperature observations were higher than their matched abdominal temperature observations. Our results suggested that infants born less than approximately 800 grams do not exhibit peripheral vasoconstriction even though they had hypothermic body temperatures during the first 12-hours of life; moreover, they exhibit a pattern whereby their peripheral temperatures are greater than their central temperature.

Temperature and Caregiver Interactions

Graphs of all primary variables (abdominal and peripheral temperature, heart rate and oxygen saturation) over time were printed for each infant with a scale of 10 minutes equaling one inch. Each graph spanned 8-9 sheets of 8 x 11 inch paper. Clinical observation events were marked on the graphs. The curves were visually inspected to identify relationships between the physiological variable and the clinical events happening over the study period that were recorded by the observer as stimulation to the infant.

Temperatures decreased as low as 34.9° C when portholes or doorways were opened for procedures such as routine nursing care, repositioning, suctioning, drawing lab work, reintubating, or taking chest Xrays. One infant received two blood transfusions at 2 and 4 hours of age. Both transfusions were associated with a simultaneous decrease in temperature. In addition, the infant received two infusions of normal saline given through umibilical catheters using a syringe and pump, both of which were associated with decreases in abdominal temperature, heart rate and oxygen saturation.

Hypothermia

Traditionally, hypothermia was established as body temperature less than 36.4° C per the 1988 American Academy of Pediatrics guidelines. Analyses were conducted to determine at which temperature range (>36.4° C or ≤ 36.4° C) the heart rate for each infant was most stable, that is in the normal range. Because most textbooks define normal heart rate range for the neonate as 100-180 (Blackburn, 2003; Kenner, 2003) and heart rates varied largely across all 10 infants in this study, normal heart rate was defined as between the 25th and 75th percentile heart rate values for each infant.

Chi square analyses were used to compare the proportions of observations in the normal and abnormal heart rate range for each infant when their abdominal temperature was greater than 36.4° C and less than or equal to 36.4° C. Six out of the ten study subjects had a statistically significant ($p < .05$) relationship between normal heart rate observations and whether the abdominal temperature was hypothermic or not. Of these six significant relationships, a larger percentage of observations in the normal heart rate range always occurred when temperatures were above 36.4° C. Therefore, having non-hypothermic temperature was associated with normal heart rate in this sample.

Optimal Control Temperature for ELBW Infants

Heart rate data for each subject were used to determine the optimal body temperature to minimize heart rate observations in the abnormal range and maximize heart rate observations in the normal range. The optimal abdominal temperature range for infants in this study appears to be between 36.8°C and 36.9° C. In this range, observations for most infants were in the normal range (> 50% of observations for 8 of 10 infants) while minimizing (< 25% of observations in 8 of 10 infants) heart rate observations in the below normal range and above normal range (< 25% of observations in 8 of 10 infants). Visual examination of the graphs gives evidence that the optimal abdominal temperature range for infants in this study was around 36.8° C to 37.0° C in order to minimize abnormal heart rate observations and to maximize normal heart rate observations.

Conclusions

The findings of this study indicated that ELBW infants have high rates of hypothermia (less than 36.4° C by traditional definition; American Academy of Pediatrics, 1988). Infants in this study averaged very low abdominal temperatures throughout the 12-hour study period (35.17°-36.68° C). Axillary temperatures recorded as low as 33.0° C and were too low to register at times. Temperatures through initial parts of stabilization in the NICU were very low for most infants and 7 of 10 infants averaged hypothermic temperatures throughout the entire 12-hour study period. ELBW infants are exposed to many stabilization procedures and cool room temperatures, cold infusions, and cold drapes from birth through the first 12 hours of life. These infants are likely to become very cold because of this exposure. Because ELBW infants have little to no ability to generate heat by non-shivering thermogenesis (Houstek et al., 1993; Lyon et al., 1997), their body temperatures can become very low in response to environmental cold stress. Without the ability to generate heat, these infants can experience morbidity and mortality from such low body temperatures (Buetow & Klein, 1964; Day et al., 1964; Hazan et al., 1991; Vohra et al., 1999).

The temperatures of ELBW infants during stabilization easily can decrease as low as 33° C even though the infants are placed on radiant warmers. Because ELBW infants are vulnerable to cold stress, knowing the optimal range for the ELBW infant's temperature is important in order to protect the infant from cold stress and organ system insult. However, very little research has examined this range in ELBW infants. Most of the research on optimal body and environmental temperatures for premature infants was conducted 20-40 years ago, before significant numbers of ELBW infants or ventilated infants survived (Adams et al., 1964; Bruck, 1961; Buetow & Klein, 1964; Day et al., 1964; Silverman et al., 1958; Yashiro, Adams et al., 1973). Results from the current study give evidence on how to optimize temperature control in ELBW infants weighing 550-960 grams during the first 12 hours of life.

Study results also indicated that ELBW infant's abdominal temperature should be kept higher than 36.4° C. Previous research is not specific as to the point at which normal temperature ends and hypothermia begins in the neonates. The American Academy of Pediatrics specified hypothermia for neonates as 36.4° C in their 1988 guidelines (American Academy of Pediatrics, 1988). Researchers have defined hypothermic cut points to evaluate interventions to reduce hypothermia as 35.0°C (Costeloe, Hennessy, Gibson, Marlow & Wilkinson, 2000), 35.5° C (Thomas, 2003), 35.6° C (Bredemeyer, Reid, & Wallace, 2005), and 36.4° C (Knobel et al., 2005;). Thus, there is no clear definition of the hypothermic limit in ELBW infants.

We used the Academy of Pediatrics traditional hypothermia cut point of 36.4° C and found that 6 of 10 infants had normal heart rate observations when abdominal temperature was greater than 36.4° C. Six of ten infants had a significant correlation between the degree the abdominal temperature was below 36.4° C and the extent of the heart rate was above or below the normal range. Nine of ten infants had a significant relationship between the abdominal-peripheral temperature difference and whether temperature was above or below the 36.4° C cut point. Seven of ten infants had a significant correlation between the abdominal-peripheral temperature difference and the extent of abdominal temperature decreased below 36.4° C. Therefore, a cut point of at least 36.4° should define hypothermia for the ELBW infant. However the cut point may need to be even higher.

Abdominal temperature clearly needs to be kept above 36.4° C to minimize heart rate abnormality and decrease the abdominal-peripheral temperature difference. This study found that when abdominal temperature was greater than 36.4° C, heart rate was more apt to be in the normal range. The abdominal-peripheral temperature difference was also smaller when the abdominal temperature was greater than 36.4° C. In 8 of 10 infants, the abdominal-peripheral temperature difference was larger when abdominal temperature was below 36.4° C than when the abdominal temperature was above 36.4° C.

However, this traditional cut point of hypothermia may not be enough to minimize abnormalities in physiological variables. This study found the optimal abdominal temperature to minimize abnormal heart rates and maximize normal heart rates in ELBW infants was between 36.8° and 37° C. Based on early research with infants approximately 29 weeks gestation (Sauer, Dane, & Visser, 1984), current standard of care is that servo control for abdominal temperature in ELBW infants be set at 36.5° C. However, I believe the range should be increased to 36.8°-37° C. In the current study, heart rate increased above each infant's 75th percentile when abdominal temperature decreased to between 34.4°-36.4° C range and heart rate decreased below the 25th percentile when abdominal temperature decreased lower than approximately 34.4° C.

Previous researchers (Costeloe et al., 2000; Horns, 2002; Lyon et al., 1997) have found that ELBW infants exhibited extremely low temperatures during the first few hours of life due to stabilization in the NICU, and this current study confirms that this problem still exists. Thus all nursing and medical care providers need to be cognizant of the ELBW infants' inability to generate heat when exposed to cold environmental temperatures and the propensity of those infants to have abdominal temperatures that decrease to extremely low temperatures during stabilization procedures. We found that many times care was not clustered; one infant was suctioned nine different times in the 12-hour study period along with many other medical and nursing procedures. Providers should take precautions to augment the environmental air to provide warmth during stabilization procedures and attempt to adhere to a minimal stimulation policy for at least the first 12 hours of life. By observing an infant over the 12-hour study period, I noted that routine interventions to augment heat such as keeping porthole covers over portholes, providing humidity as soon as possible to the infant, minimizing time the incubator is open were many times delayed for the entire 12-hour study period. Additionally, most infants had very low abdominal temperatures while umbilical lines were being inserted and this procedure took up to 2 hours. Therefore, I recommend augmenting the environment around the ELBW infant during umibilical line insertion with heat lamps or a warming gel mattress. Providers should also expedite stabilization procedures to minimize exposure to cold environmental temperatures by placing lines as quickly as possible. Abdominal temperature should be controlled minimally at 36.5°C and results of the current study suggest the control point should be increased to between 36.8° and 37° C. Because abdominal temperature was seen to decrease in association with fluid infusion through the umbilical catheter, nurses may want to warm intravenous infusions prior to infusing through the umbilical venous or arterial catheters during the first 12 hours of life. This could be achieved by keeping syringes or bags of fluid for a new ELBW infant in the incubator while the incubator is being prewarmed for that infant.

Directions for Future Research

Future research should be directed at finding the optimal thermal control point for body temperature for ELBW infants. Studies should be designed to assess the impact of nursing and medical procedures on heat loss in the ELBW infant. Interventions need to be assessed in minimizing heat loss in this population, specifically by warming intravenous fluids and preventing heat loss when performing such procedures such as suctioning the endotracheal tube.

Our last study should be replicated with a larger sample population of ELBW infants. Each infant needs to be followed through 5 days of life for data collection and mortality outcomes for these infants through hospitalization should be determined. The first 12 hours are a time of instability, therefore at least an additional 12 hours of data collection would allow the infant to become stable and normalize physiological variables. The additional 4 days would allow for aging effects to be determined. Some infants in this current study did not become warm until the last 1-2 hours of the study period. Therefore, at least 12 additional hours of data collection would examine ELBW infants during a time of normalization and warmth. Collecting data over 5 days will give valuable information as to the development of vasomotor control to invoke peripheral vasoconstriction in infants weighing 500-700 grams.

Because morbidity and mortality are high in ELBW infants, especially infants weighing 400-700 grams, further research should examine the relationship of cerebral perfusion to low abdominal temperature and poor vasomotor control. Intraventricular hemorrhage is a possibility in the first week of life for these extremely premature infants (Gardner, 2005) and poor vasomotor control may possibly be linked to an increased incidence of intraventricular hemorrhage. Preterm infants have difficulty in autoregulating their cerebral blood flow and changes in venous and arterial blood flow can translate into increased cerebral blood pressure resulting in rupture of vessels in the brain and hemorrhage (Goddard-Finegold. & Mizrahi, 1987). Poor vasomotor control exemplified by an increased peripheral-abdominal temperature difference may be linked to the inability of the ELBW infant to protect the brain from stressful clinical events. Therefore, future studies should concentrate on development of vasomotor tone in these infants, correlating vasomotor tone with body temperature. Examining vasomotor especially in relationship to low abdominal temperature and abdominal-peripheral temperature difference measures should yield valuable information towards decreasing brain injury and mortality in ELBW infants.

References

Adams, F., Fujiwara, T., Spears, R., & Hodgman, J. (1964). Temperature regulation in premature infants. *Pediatrics, 33*, 487-495.

American Academy of Pediatrics & College of Obstetrics and Gynecologists. (1988). *Guidelines for perinatal care* (second ed.). Elk Grove Village, IL: American Academy of Pediatrics.

Barrett, E. (2003). The thyroid gland. In W. Boron & E. Boulpaep (Eds.), *Medical physiology: A cellular and molecular approach* (pp. 1035-1048). Philadelphia: Saunders.

Baumgart, S. (1984). Reduction of oxygen consumption, insensible water loss, and radiant heat demand with use of plastic blanket for low-birth-weight infants under radiant warmers. *Pediatrics_74*, 1022-1028.

Baumgart, S. (1987). Current concepts and clinical strategies for managing low-birth-weight infants under radiant warmers. *Medical Instrumentation, 21*(1), 23-28.

Baumgart, S., Engle, W., Langman, C., Fox, W., & Polin, R. (1980). Monitoring radiant power in the critically ill newborn under a radiant warmer. *Critical Care Medicine, 8*(12), 721-724.

Baumgart, S., Fox, W., & Polin, R. (1982). Physiologic implications of two different heat shields for infants under radiant warmers. *The Journal of Pediatrics, 100*, 787-790.

Bell, E., & Glatzl-Hawlik, M. (1998). Environmental temperature control. R. Polin, & W. Fox (Eds.), *Fetal and neonatal physiology* (2nd ed., vol. 1, pp. 716-727). Philadelphia: W. B. Saunders Company.

Bell, E., & Rios, G. (1983). Air versus skin temperature servocontrol of infant incubators. *The Journal of Pediatrics, 103*, 954-958.

Besch, N. J., Perlstein, P. H., Edwards, N. K., Keenan, W. J., & Sutherland, J. M. (1971). The transparent baby bag. *The New England Journal of Medicine, 284,* 121-124.

Billington, C., Briggs, J., Harker, S., Grace, M., & Levine, S. (1994). Neuropeptide Y in hypothalamic paraventricular nucleus: a center coordinating energy metabolism. *American Journal of Physiology, 266*, R1765-R1770.

Bing, C., Frankish, H., Wang, Q., Hopkins, D., Keith, D., Trayhurn, P., & Williams, G. (1996). Dissociation of hypothalamic NPY from BAT uncoupling protein mRNA in rats exposed to 24 h thermoneutrality. *American Journal of Physiology, 270*, R111-R117.

Bhatt, D., White, R., Martin, G., Van Marter, L., Finer, N., Goldsmith, J., et al. (2007). Transitional hypothermia in preterm newborns. *Journal of Perinatology, 27* Suppl 2: S45-7.

Bjorklund, L., & Hellstrom-Westas, L. (2000). Reducing heat loss at birth in very preterm infants (letter). *The Journal of Pediatrics, 137*, 739-740.

Buetow, K., & Klein, S. (1964). Effect of maintenance of "normal" skin temperature on survival of infants of low birth weight. *Pediatrics, 34*, 163-170.

Blackburn, S. T. (2003). *Maternal, fetal & neonatal physiology, 2nd Edition.* St. Louis, MO: Saunders.

Bredemeyer, S., Reid, S., & Wallace, M. (2005). Thermal management for premature births. *Journal of Advanced Nursing, 52*(5), 482-489.

Bruck, K. (1961). Temperature regulation in the newborn infant. *Biologia Neonatorum, 3*(2/3), 65-119.

Bruck, K. (1998). Neonatal thermal regulation. R. Polin & W. Fox (Eds.), *Fetal and neonatal physiology* (second ed., vol. 1, pp. 676-701). Philadelphia: W.B. Saunders Company.

Costeloe, K., Hennessy, E., Gibson, A., Marlow, N., & Wilkinson, A. (2000). The EPICure Study: Outcomes to discharge from hospital for infants born at the threshold of viability. *Pediatrics, 106*, 659-671.

Cramer, K., Wiebe, N., Hartling, L., Crumley, E., & Vohra, S. (2005). Heat loss prevention: A systematic review of occlusive skin wrap for premature neonates. *Journal of Perinatology, 25*, 763-769.

Day, R., Caliguiri, L., Kamenski, C., & Ehrlich, F. (1964). Body temperature and survival of premature infants. *Pediatrics, 34,* 171-181.

Gardner, M. (2005). Outcomes in children experiencing neurologic insults as preterm neonates. *Pediatric Nursing, 31*, 451-456.

Goddard-Finegold, J. & Mizrahi, F. (1987). Understanding and preventing perinatal, intracerebral, peri- and intraventricular hemorrhage. *Journal of Child Neurology, 2*(3), 170-185.

Gluckman, P., Sizonenko, S., & Bassett, N. (1999). The transition from fetus to neonate-an endocrine perspective. *Acta Paediatrica , 428 supplement*, 7-11.

Gunn, T., & Gluckman, P. (1995). Perinatal thermogenesis. *Early Human Development, 42,* 169-183.

Guyton, A. & Hall, J. (2006). Textbook of Medical Physiology (11th ed.). Philadelphia: W.B. Saunders.

Hammarlund K, Nilsson GE, Oberg PA, Sedin G. Transepidermal water loss in newborn infants: V. Evaporation from the skin and heat exchange during the first hours of life. *Acta Paediatr Scand.* 1980; 69:385-392.

Hatai, S. (1902). On the presence in human embryos of an interscapula gland corresponding to the so-called hibernating gland of lower mammals. *Anatomy Anzeiger, 21*, 369.

Hazan, J., Maag, U., & Chessex, P. (1991). Association between hypothermia and mortality rate of premature infants-Revisited. *American Journal of Obstetrics and Gynecology, 164,* 111-112.

Horns, K. (2002). Comparison of two microenvironments and nurse caregiving on thermal stability of ELBW infants. *Advances in Neonatal Care, 2*(3), 149-160.

Houstek, J., Vizek, K., Pavelka, S., Kopecky, J., Krejcova, E., & Hermanska, J. (1993). Type II iodothyronine 5'-deiodinase and uncoupling protein in brown adipose tissue of human newborns. *Journal of Clinical Endocrinology Metabolism, 77*, 382-387.

Hull, D. (1977). Brown adipose tissue and the newborn infant's response to cold. E. Phill, J. Barnes, & M. Newton (Eds.), *Scientific foundations of obstetrics and gynaecology* (2nd ed.). London: M. Heinemann Medical Books.

Hull, D., & Smales, O. (1978). Heat production in the newborn. J. Sinclair (Ed.), *Temperature regulation and energy metabolism in the newborn* (pp. 129-56). New York, New York: Grune & Stratton.

Huttunen, P. (2002, November). *Brown adipose tissue.* Oulu, Finland: University of Oulu, Department of Forensic Medicine.

Jones, E., & DeCherney, A. (2003). Fetal and neonatal physiology. W. Boron & E. Boulpaep (Eds.), *Medical physiology: A cellular and molecular approach* (pp. 1190-1208). Philadelphia: Saunders.

Kattwinkle, J. (2006). *Textbook of Neonatal Resuscitation.* (5th ed.) United States of America: American Academy of Pediatrics and American Heart Association.

Kenner, C. (2003). Resuscitation and stabilization of the newborn. C. Kenner & J. Lott (Eds.), *Comprehensive neonatal nursing: A physiologic perspective* (3rd ed., pp. 210-327). Philadelphia: Saunders.

Kent, A. & Williams, J. (2008). Increasing ambient operating theatre temperature and wrapping polyethylene improves admission temperature in premature infants. *Journal of Paediatric Child Health, 44,* 325-331.

Knauth, A., Gordin, M., McNelis, W., & Baumgart, S. (1989). Semipermeable polyurethane membrane as an artificial skin for the premature neonate. *Pediatrics, 83,* 945-950.

Knobel, R., Vohra, S., & Lehmann, C. (2005). Heat loss prevention in the delivery room for preterm infants: A national survey of newborn intensive care units. *Journal of Perinatology, 25*, 514-518.

Knobel, R., Wimmer, J., & Holbert, D. (2005). Heat loss prevention for preterm infants in the delivery room. *Journal of Perinatology, 25*, 304-309.

LeBlanc, M. (1982). Relative efficacy of an incubator and an open warmer in producing thermoneutrality for the small premature infant. *Pediatrics, 69*(4), 439-45.

Loughead, M., Loughead, J., & Reinhart, M. J. (1997). Incidence and physiologic characteristics of hypothermia in the very low birth weight infant. *Pediatric Nursing, 23*(1), 11-15.

Lyon, A., Pikaar, M., Badger, P., & McIntosh, N. (1997). Temperature control in very low birthweight infants during first five days of life. *Archives of Diseases in Childhood, 76*, F47-F50.

Malin, S., & Baumgart, S. (1987). Optimal thermal management for low birth weight infants nursed under high-powered radiant warmers. *Pediatrics, 79*(1), 47-54.

Mann, T. P. (1955). Hypothermia in the newborn: A new syndrome? *Lancet, 1,* 613-614.

Mann, T. P., & Elliot, R. (1957). Neonatal cold injury due to accidental exposure to cold. *Lancet, 272*, 229-234.

Mathew, B., Lakshminrusimha, S., Cominsky, K., Schroder, E., & Carrion, V. (2007). Vinyl bags prevent hypothermia at birth in preterm infants. *Indian Journal of Pediatrics, 74*, 249-253.

Mathews, T. & MacDorman, M. (2006). Infant mortality statistics from the 2003 period linked birth/infant death data set. *National Vital Statistics Reports, 54*, (16), 1-30.

McCall, E., Alderdice, F., Halliday, H., Jenkins, J., & Vohra, S. (2008). Interventions to prevent hypothermia at birth in preterm and/or low birthweight infants. *Cochrane Database System Reviews, 1*: CD004210.

Meyer, M., Payton, M., Salmont, A., Hutchinson, C., & de Klerk, A. (2001). A clinical comparison of radiant warmer and incubator care for preterm infants from birth to 1800 grams. *Pediatrics, 108*(2), 395-7.

Miller, D., & Oliver, T. (1966). Body temperature in the immediate neonatal period: The effect of reducing thermal losses. *American Journal of Obstetrics and Gynecology, 94*(7), 964-969.

Nadel, E. (2003). Regulation of body temperature. W. Boron & E. Boulpaep (Eds.), *Medical physiology* (pp. 1231-1241). Philadelphia: Saunders.

O'Shea, T., Klinepeter, K., Goldstein, D., Jackson, B., & Dillard, R. (1997). Survival and developmental disability in infants with birth weights of 501 to 800 grams, born between 1979 and 1994. *Pediatrics, 100(*6), 982-986.

Power, G. (1998). Perinatal thermal physiology. W. Polin & W. Fox (Eds.), *Fetal and neonatal physiology* (2nd ed., vol. 1, pp. 671-675). Philadelphia: W.B. Saunders Company.

Ricquier, D., & Bouillaud, F. (2000). Mitochondrial uncoupling proteins: from mitochondria to the regulation of energy balance. *Journal of Physiology, 529*(1), 3-10.

Sauer, P. (1995). Metabolic background of neonatal heat production, energy balance, metabolic response to heat and cold. A. Okken & J. Koch (Eds.), *Thermoregulation of sick and low birth weight neonates* (pp. 9-20). Germany: Springer-Verlag Berlin.

Sawa, R., Asakura, H., & Power, G. (1991). Changes in plasma adenosine during simulated birth of fetal sheep. *Journal of Applied Physiology, 70,* 1524-1528.

Sedin, G. (1995). Neonatal heat transfer, routes of heat loss and heat gain. A. Okken & J. Koch (Eds.), *Thermoregulation of sick and low birth weight neonates* (pp. 21-36). Berlin: Springer.

Silverman, W., Fertig, J., & Berger, A. (1958). The influence of the thermal environment upon the survival of newly born premature infants. *Pediatrics, 22*, 876-886.

Sinclair, J. C. (1992). Management of the thermal environment. J. C. Sinclair & M. B. Bracken (Eds.), *Effective care of the newborn infant (*pp. 40-58). Oxford: Oxford University Press.

Subramanian, S., Yoon, H., & Toral, J. (2002) *Extremely low birth weight infant* Retrieved November 1, 2003 from www.emedicine.com/ped/topic2784.htm.

Sulyok, E., Jequier, E., & Prod'hom, L. (1973). Thermal balance of the newborn infant in a heat-gaining environment. *Pediatric Research, 7*, 888-900.

Thomas, K. (1994). Thermoregulation in neonates. *Neonatal Network, 13*(2), 15-22.

Thomas, K. (2003). Preterm infant thermal responses to caregiver differ by incubator control mode. *Jounal of Perinatology, 23,* 640-645.

Voet, D., Voet, J., & Pratt, C. (2002). Electron transport and oxidative phosphorylation. D. Voet, J. Voet & C. Pratt (Eds.) *Fundamentals of biochemistry* (Upgrade ed., pp. 492-528). New York: John Wiley & Sons, Inc.

Vohra, S., Grent, G., Campbell, V., Abbott, M., & Whyte, R. (1999). Effect of polyethylene occlusive skin wrapping on heat loss in very low birth weight infants at delivery: A randomized trial. *Journal of Pediatrics, 134*(547-551).

Vohra, S., Roberts, R., Zhang, B., Janes, M., & Schmidt, B. (2004). Heat loss prevention (HeLP) in the delivery room: A randomized controlled trial of polyethylene occlusive skin wrapping in very preterm infants. *Journal of Pediatrics, 145* (750-753).

Weisman, L., Stoll, B., Cruess, D., Hall, R., Merenstein, G., Hemming, V. & Fischer, G. (1992). Early onset group B streptococcal sepsis: A current assessment. *Journal of Pediatrics, 121*, 428.

Widmaicr, J., Raff, H., & Strang, K. (2005). Vander's Human Physiology: The Mechanics of Body Function (9th ed.). New York: McGraw-Hill Companies.

Wilding, J., Widdowson, P., & Williams, G. (1997). Neurobiology. *British Medical Bulletin, 53,* 286-306.

Yashiro, K., Adams, F., Emmanouilides, G., & Mickey, M. (1973). Preliminary studies on the thermal environment of low-birth-weight infants. *The Journal of Pediatrics, 82*, 991-994.

In: Body Temperature Regulation
Editor: A. B. Cisneros and B. L. Goins
ISBN: 1-60741-282-3

Chapter V

The Use of Infrared Thermography for the Investigation of Thermoregulation in Humans

Boris G. Vainer
Rzhanov Institute of Semiconductor Physics, Siberian Branch of the Russian Academy of Sciences; Novosibirsk State University, Russia

Abstract

Skin is one of the primary channels realizing the heat exchange of a human organism with its environment. When body temperature regulation is under study, the experimental methods and high-sensitive techniques allowing real-time registration of skin surface thermal patterns take on great significance. Infrared thermography supplied with two-dimensional (or focal plane array based) infrared detectors is best suited for these needs.

This chapter gives an overview of infrared thermography as a powerful research method applied to the biomedical sphere. It also embraces the material of recent IRT-assisted experimental works devoted to the studies of a human organism heat exchange with its environment under different active loadings. A high-performance infrared camera having 0.03 °C temperature sensitivity was used in the experiments. It is pictorially demonstrated how the surface thermal patterns are transformed in the course of physical and heat (hyperthermic) loads applied to the organism. Infrared thermography provides a way of visualizing the fine manifestations of vascular and perspiratory systems that is impossible to carry out using any other experimental methods. The results of quantitative analysis of two-dimensional thermograms are also presented.

It follows from the infrared observations that the typical feature of a loaded organism is its highly heterogeneous thermal pattern represented on the skin. This surface temperature non-uniformity, on the one hand, results from skin cooling due to the sweat evaporation and on the other hand from local heating due to transient dilation of separate superficial branches of vessels (the so-called "veins-collectors"). Quantitative characterization of highly heterogeneous thermograms having a direct relationship to the study of body temperature regulation is a new problem of modern biomedical infrared thermography. The original methods attacking this problem are also mentioned in this chapter.

1. Introduction

To maintain the temperature of a human body to be stable, the balance between heat production and heat loss should be maintained. The intensity of heat loss is determined by several channels realizing the heat exchange of the organism with its environment. Here, skin is one of the primary channels. Heat removal is governed, with the participation of skin, by sweat evaporation, heat transfer to air molecules, and thermal radiation. An instant as well as a sustained surface distribution of skin temperature performs intensity and particular traits of the organism's heat production. Owing to the blood flow and heat transfer in superficial tissues, the skin thermal pattern reflects thermal processes proceeding deep in the body. On the other hand, the effectiveness of heat removal is dependent in many respects on the temperature values and on the thermal pattern of the surface of a body. In this regard, the peculiarities of surface thermal pattern and the variation of such patterns with time provide relevant information for studying human temperature regulation.

Development of adequate experimental techniques and procedures for remote quantitative control of physiological external manifestations of thermoregulatory processes and mechanisms is currently the issue of the day. One of these techniques is infrared thermography (IRT). IRT is widely used in engineering, industry, and many other technical and scientific applications [1–3]. In the studies that enter into the investigation of humans, there are a great many problems requiring temperature measurement at many points on human skin [4–7]. In such studies, an IR camera could serve as an adequate measuring instrument in most cases. At the same time, IRT is still used in physiology rather rarely [8–14]. Previously, one of the reasons for a limited interest in IRT was the lack of highly sensitive, convenient in use, and fast-operating IR cameras. To date, the situation has been improved drastically. With the advent of new instrumentation, it is now possible to identify many nuances in thermal patterns which were undetectable earlier.

In this chapter, primary attention is focused on the use of new-generation IRT as a method aimed at executing effective remote control of the body temperature regulation processes.

2. IRT in Scientific Investigations

2.1. IRT and Three Generations of Infrared Camera

To those researchers who are not familiar with the basics of IRT, let us give a short introduction into the method we are dealing with in this chapter. Thermal imaging as well as IRT is based on the use of a fundamental natural phenomenon: each body of outward things is heated, at least a bit, and, thanks to this fact, emits electromagnetic radiation into the environment. The greatest amount of this radiation belongs to the infrared (IR) part of the spectrum. The prime (parent) device equipping IRT is the IR camera, a special optical system, which receives, and processes thermal (infrared) radiation emitted by the objects under study [15–17]. A two-dimensional thermal field measured by an IR camera is then represented as a black-white or/and color thermogram to be analyzed. Normally, dark and light zones in black-white thermograms represent low-temperature and high-temperature regions, respectively. The color pallet is usually not strictly regulated. Special software is capable of reading temperature values at any point (pixel) in the thermogram.

The primary purpose of practically all IRT systems is common: to convert invisible to human eyes thermal radiation into visible images. IR detectors, most kinds of which are based on the use of semiconductors, serve in the IR cameras for this purpose. A sensitive part of each detector consists of a certain number of separate photosensitive cells. According to number and placement of cells within the detector, IR cameras are conventionally subdivided into three generations. First one covers the IR devices containing just a single photosensitive cell. In the devices of a later, second, generation, a string (line array) of normally several hundred cells is used. A detector of the third, newest, generation IR camera contains a two-dimensional array of synchronously working cells, arranged in the matrix composed of hundreds of rows and hundreds of columns. The generally accepted name of the third-generation IR detector is focal-plane array, or FPA.

Construction of IR cameras belonging to the early generations is complicated by the presence of an optical-mechanical scanner incorporated into the system. In contrast, the IR image scanning realized in third-generation cameras is executed solely in an electronic way. It gives priority to the latter IR systems, as opposed to the mechanically-scanned cameras, in temperature sensitivity, speed of operation, and spatial resolution in total. The temperature sensitivity of standard FPA-based IR cameras is now of the order of a hundredth of a degree at the frame rate of tens of frames per second. It is proved that FPA-based systems are more efficient in medicine [18–22] than the systems previously devised.

Quantitative analysis of thermograms is carried out on the basis of digitized data of IR measurements. To acquire digital data, a signal produced by each photosensitive cell (pixel) is converted by a 12-16-bit analog-digital converter. High temperature sensitivity and spatial resolution as well as a wide dynamic range of modern FPA-based IR cameras provide thermal images with photographic quality. As this takes place, electronic data processing of digitally represented thermal images permits automatic recognition and analysis of subtle temperature gradients along an investigated surface.

High performance FPA-based IRT system makes possible the real-time saving of continuous thermal movies in computer memory. Thanks to this function, the possibility exists of monitoring and study of high-rate thermal processes accompanying physiological and functional changes in the organism in the course of various diagnostic tests.

Original materials represented in this chapter were obtained with the help of FPA-based IR systems TKVr-IFP working in the narrow spectral diapason located near wavelength $\lambda = 3$ μm [23–25]. Two-dimensional array of 128×128 photosensitive cells made on the basis of indium arsenide was served as a photosensitive unit of the IR detector. IR camera had a temperature sensitivity of about 0.03 °C at a target temperature 30 °C. Maximal frame rate of the camera was equal to 50 frames per second. IR system was capable to save thermograms in computer memory with this frequency.

2.2. Human Being as a Test Subject in Biomedical IRT

The test subjects examined in this chapter are human beings; a registered physical characteristic here is the surface distribution of their skin temperature. Unfortunately, human tissues are virtually non-transparent for IR radiation owing to a high concentration of water in their content. As a consequence, radiation emitted by heated deep organs and tissues is immediately absorbed by other tissues and can not come out. As a result, information about processes developing in the interior of the organism is gleaned on the outside in the form of thermal radiation emitted by inhomogeneously heated skin. Correctly solving the problems devoted to identification of biophysical mechanisms responsible for forming a skin thermal pattern is important not only because it gives a better insight into the mechanisms and the processes of human thermoregulation, but also because the particular features of such a pattern (shape, time behavior, etc.) can be a manifestation of organism disorders, thereby serving the purpose of medical diagnostics.

It is well known that animals are used often enough as the test subjects in biomedical investigations relating to human life and health. Clinical and biomedical experimentation with humans is strictly regulated by law. However, IRT belongs to one of the few biomedical areas where replacing a human with an animal is practically inadmissible if we are going to transfer the results obtained with animals to the humans. The reason for this assertion is that the IRT offers registration just of a surface temperature, but physical characteristics of the interface layer separating the human's and animal's bodies from outside ambient are appreciably different.

Indeed, hair growth that is common to a wide variety of haematothermal mammals extensively used in experiments shields and misrepresents the real skin thermal pattern because the fur is opaque, it disperses appreciably IR radiation, and it is cooled by air. Shaving the hair off is not preferable because it results in physiological disorder of investigated living tissue and can not ensure the trustworthiness and reliability of final conclusions. A difference between human and animal sweating highly influencing surface temperature also takes place.

In this connection it only remains to conclude that biomedical IRT experiments aimed at human needs should be carried out by means of examination of humans themselves. A favorable circumstance here is the fact that in standard IRT investigations, unlike most other diagnostic means (X-rays, magnetic resonance tomography, etc.), electromagnetic radiation is received from a human body, but not applied to it. In this respect, IRT is a non-invasive and absolutely harmless instrument, allowing multiple reiterations of tests without prejudice to the health of being tested. A great advantage of IRT is its remotability, namely, the absence of any direct contact with a test subject. Such a contact is able to influence the natural behavior of a living object under investigation. In spite of the undoubted attractiveness and potentialities of the IRT method as applied to examination of humans, the use of an IR camera in physiology still remains a rather rare event.

2.3. Typical Objects under Observation at IRT-based Investigation of Body Temperature Regulation

Several types of informative objects accessible to IRT for the observation and quantitative analysis of thermoregulatory manifestations in humans can be specified. They are indicated below.

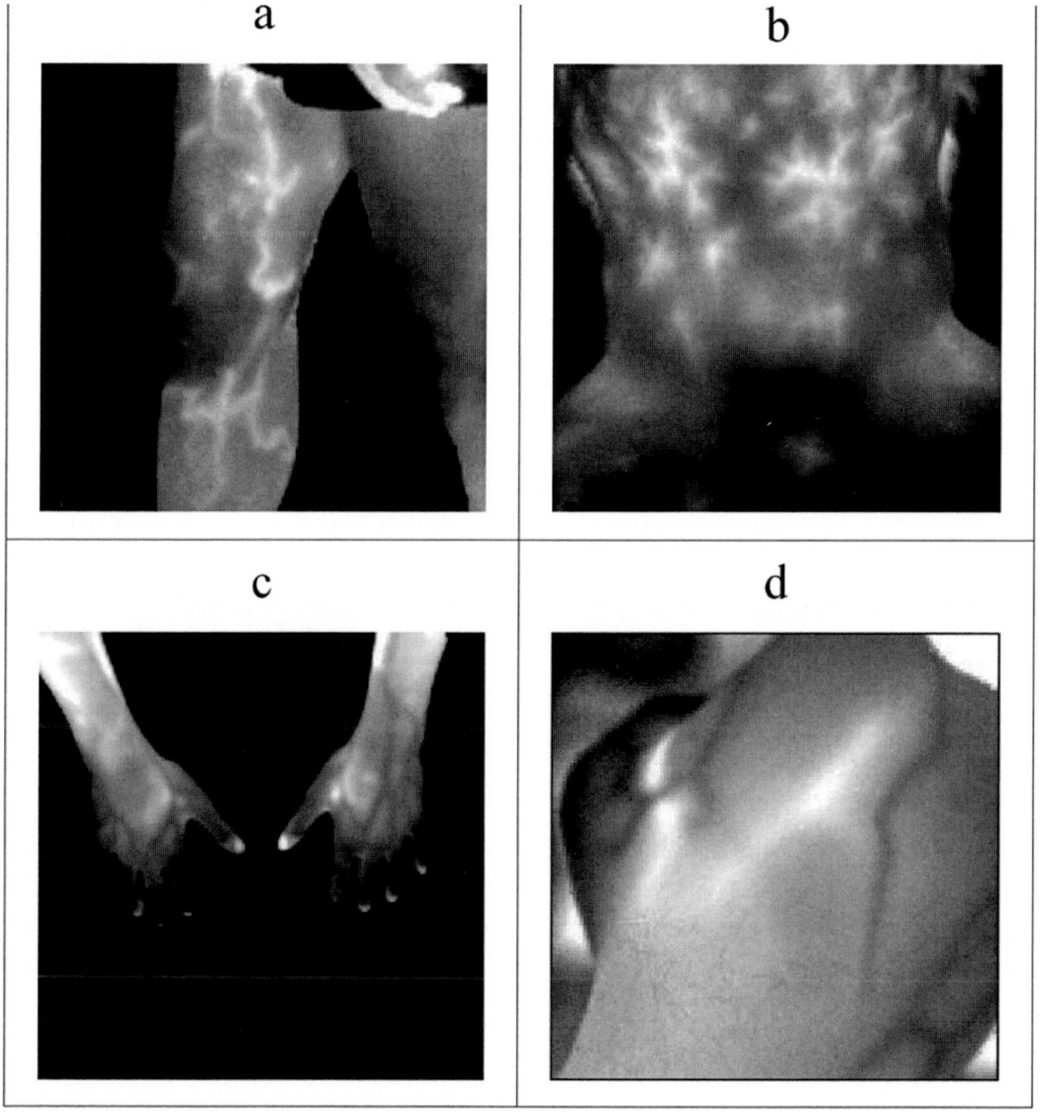

Figure 1. Representation of major surface vessels in thermograms. a – dilated veins of shin and thigh; b – branch-shaped vessels at submaxillary region; c – hand vessels cooled by the blood carried away from cold fingers; d – cooled and heated vessels registered simultaneously on the hand.

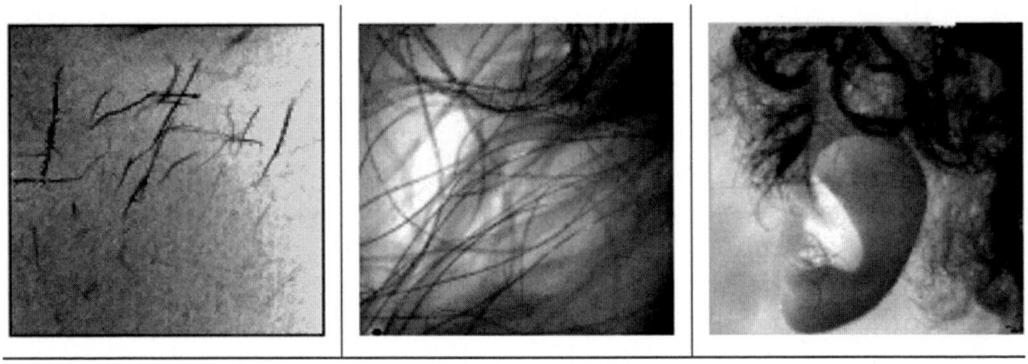

Figure 2. Three examples of IR images containing human hair. From left to right: sparse hair on leg skin; separate hair on head; dense hair layer on head.

Major Surface Vessels

"Thermal portraits" of separate major surface vessels look in the black-white thermograms as light or dark cords or branches with well-marked outlines as it is demonstrated in figure 1. Light and dark tones are defined here with reference to the tone of surrounding tissues. For the most part, major surface vessels registered by IR camera belong to venous system because arteries are usually hidden deep in the body and their representative pattern on skin is rather blurred.

When a healthy human organism is at rest, thermal image of major vessels is more particularly not come in sight. Its contrast increases essentially at abrupt increase of tissual blood flow that can appear for instance as a consequence of physical load of the whole organism or its part. As the latter takes place, major surface vessels become cooler than at rest if they drain the cool (usually – peripheral and distal) parts of body (figure 1 c, d). And vice versa, their temperature is increased if the vessels drain tissues with increased heat production (figure 1 a, b). The latter effect can take place under different loadings.

Dry Zones between Major Vessels

When skin is not wetted, superficial tissues located between major surface vessels show up in the thermogram as rather light and uniform areas. Mean temperature calculated on these areas depends essentially on activity of metabolic processes developed under the associated parts of skin both nearby body surface and deep in the tissues. Hairs at such areas are usually considered as artifacts. They are well blown by fresh atmospheric air and therefore occur darker than warm skin (figure 2). Right thermogram in figure 2 visually reminds that a dense layer of hair with the enclosed still air serves as excellent heat insulator of skin.

Regions of Profuse Sweating

Surface zones with profuse sweating located between major surface vessels show up in the thermogram as rather dark and uniform areas. Temperature decrease characterizing such parts of skin is due to the transfer of heat from body to secreta of sweat glands that is necessary to evaporate the sweat (the heat of phase transition). Owing to the fact that specific heat of vaporization of water is quite great (about 540 calories per gram), this physical mechanism results in quite great heat loss followed by noticeable decrease skin in temperature.

Separate Sweat Glands

Even if the profuse sweating is not appeared, nevertheless, separate sweat glands are able to ooze sweat quite intensively. These highly active sweat glands show up in the thermogram as black points. In figure 3, the dense group of separate sweat glands around mouth can be easily distinguished. In the thermograms of finger cushions, these points are usually grouped along papillary lines. Quite detailed IRT study of separate sweat glands is presented in [26–27]. A new terms “gland lattice” and “gland-lattice pitch” are also introduced in [26–27] on the basis of thermographically revealed features of active secretory ducts. It should be noted that not only does IRT allow bringing into sight the sweat glands, but it also allows making quantitative analysis of the gland functioning that can be seen from the graph presented in figure 4. In figure 4, the distribution of temperature in the vicinity of three finger-cushion active sweat glands showing different excretory efficiency is presented.

Regions with High Thermal Heterogeneity

Highly heterogeneous surface thermal pattern is a new subject of investigation aroused interest in it in the course of experimental study of thermoregulation in humans exposed to physical and thermal loads [26–27]. When highly heterogeneous pattern, like those shown in figure 5, is observed on skin, it is difficult to believe that separate light or dark spots here can be of scientific interest or can bring any valuable information about the organism status. In the situation exhibited in figure 5, of special interest are the spacious regions covered by multitude of light and dark spots. Appearance of such a “leopard hide” skin in thermal images is resulted largely from dilatation of the superficial so-called “veins-collectors” [26–27].

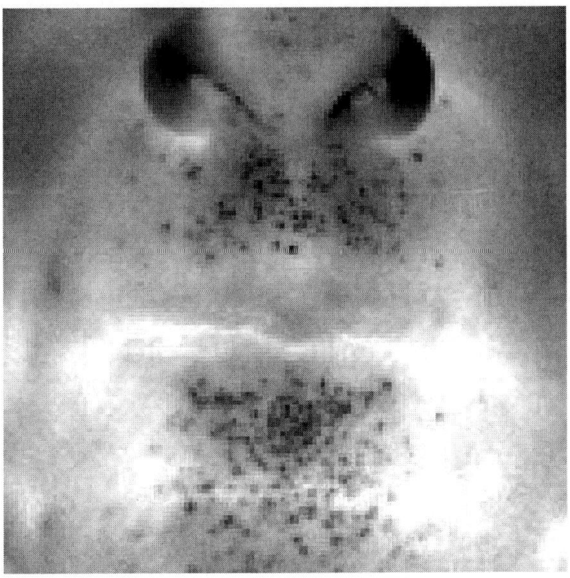

Figure 3. Thermogram showing the work of separate groups of active sweat glands located above and below mouth.

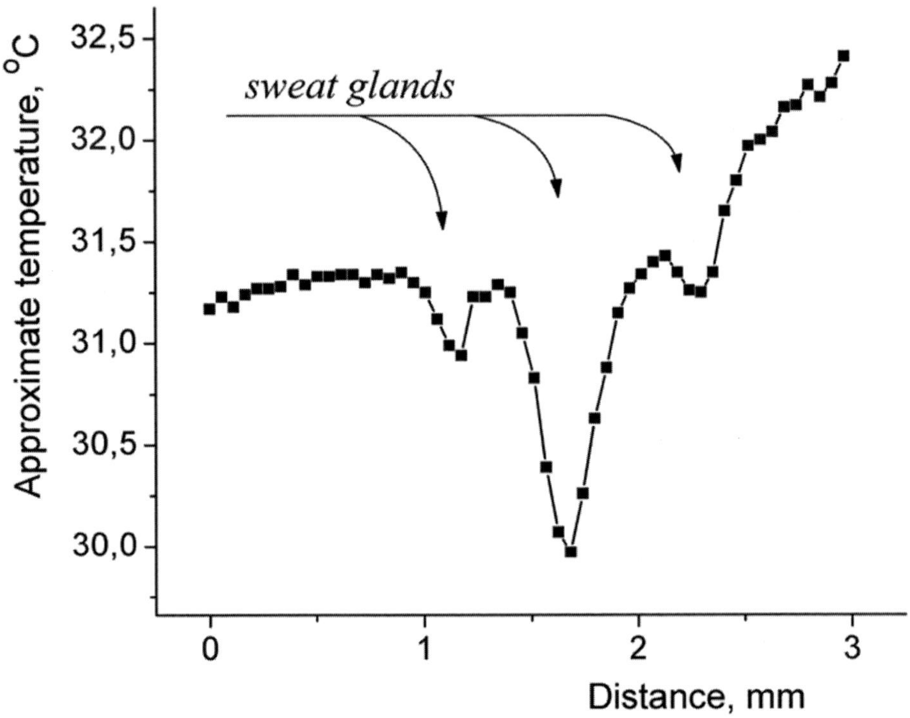

Figure 4. Temperature profile measured across the finger-cushion three active sweat glands attributed with different productivities. The data were obtained with the use of additional lens.

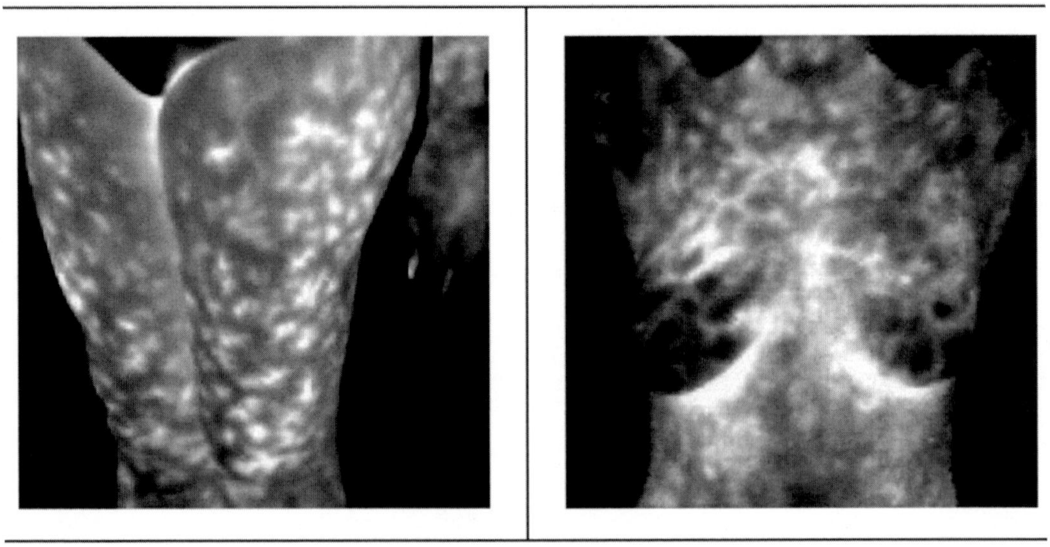

Figure 5. Heterogeneous thermal patterns: at the left – woman's thighs; at the right – front upper part of woman's body.

2.4. Physiological Grounds Justifying the Use of IRT for Examination of Humans

Energy change of any physical system during a certain time interval can be represented in the form:

$$E_{\mathrm{RESULT}} - E_{\mathrm{INIT}} = A_{\mathrm{EXTERN}} + Q_{\mathrm{INPUT}} \quad ,$$

where E_{INIT} and E_{RESULT} are initial and final energies of the system respectively, A_{EXTERN} – the work of external forces, Q_{INPUT} – quantity of heat supplied to the system. Q_{INPUT} is negative in the case of heat dissipation; A_{EXTERN} is also negative if the system itself works.

In a living organism, life-support processes (mechanical work, ion transfer, chemical synthesis, etc.) are followed by the change of energy and by heat evolution [28]. These processes are usually unthrifty. For instance, a typical efficiency of adenosine triphosphate energy consumption at protein synthesis is just of the order of 20% [28]. The rest of energy is finally lost in the form of heat dissipation. Because of a human being is the homoitherm (keeping constant his core temperature), two competiting synchronous processes progress uninterruptedly in his body: heat production and heat loss. The principal mechanism of heat removal to environment is realized by means of skin.

Circulating blood is heated in active deepened tissues cooling them, and then it is cooled down in superficial tissues, including skin, simultaneously heating the latter. At that, heat production in the organism undergoes material change when the organism or its organ or another part becomes working abnormally. It does often happen in this case that metabolism also undergoes change and the affected part begins consuming more (or less) energy than it takes place when the organism is functioning properly. Consequently more (or less) heat, as against in the normality, is released within the locality of a pathological area. On the other hand, physical mechanisms of heat release (heat transfer, convective heat exchange, etc.) remain practically unchanged, especially in the beginning of the illness. This has the effect of a skin temperature increase (decrease) within the affected part, and we can reveal an unusual thermal pattern of skin with the help of IR camera. Typical examples of abnormal (increased) intensity of heat production at inflammatory processes are given in figure 6. It should be noted that the observed surface temperature anomalies not infrequently can be associated also with vascular or neuro-vascular reactions, but not with metabolism.

2.5. Quantitative Data Acquired by IR Camera

At early stage of development of IRT systems, it was possible to observe thermal image just on the screen of cathode-ray tube and, if necessary, to take a photograph from this screen. Nevertheless, the mere opportunity to look at and to see invisible radiation meant a lot, and science made great strides with invention of IRT. Visual (qualitative) analysis of thermograms extended the capabilities of medical diagnostics that was documented by many papers and monographs [29, 30]. At the same time, a qualitative analysis alone is not enough if IRT is intended to be put to scientific use. The state of the art in development of IR systems allows radical alteration of the situation. As it was already mentioned above, computer support of the modern IR cameras makes possible the real-time data processing of measured thermal patterns and allows saving the thermograms in digital form. This allows quantitative investigation of surface temperature fields to be made, the final results of which are not dictated by subjective perception of the thermograms.

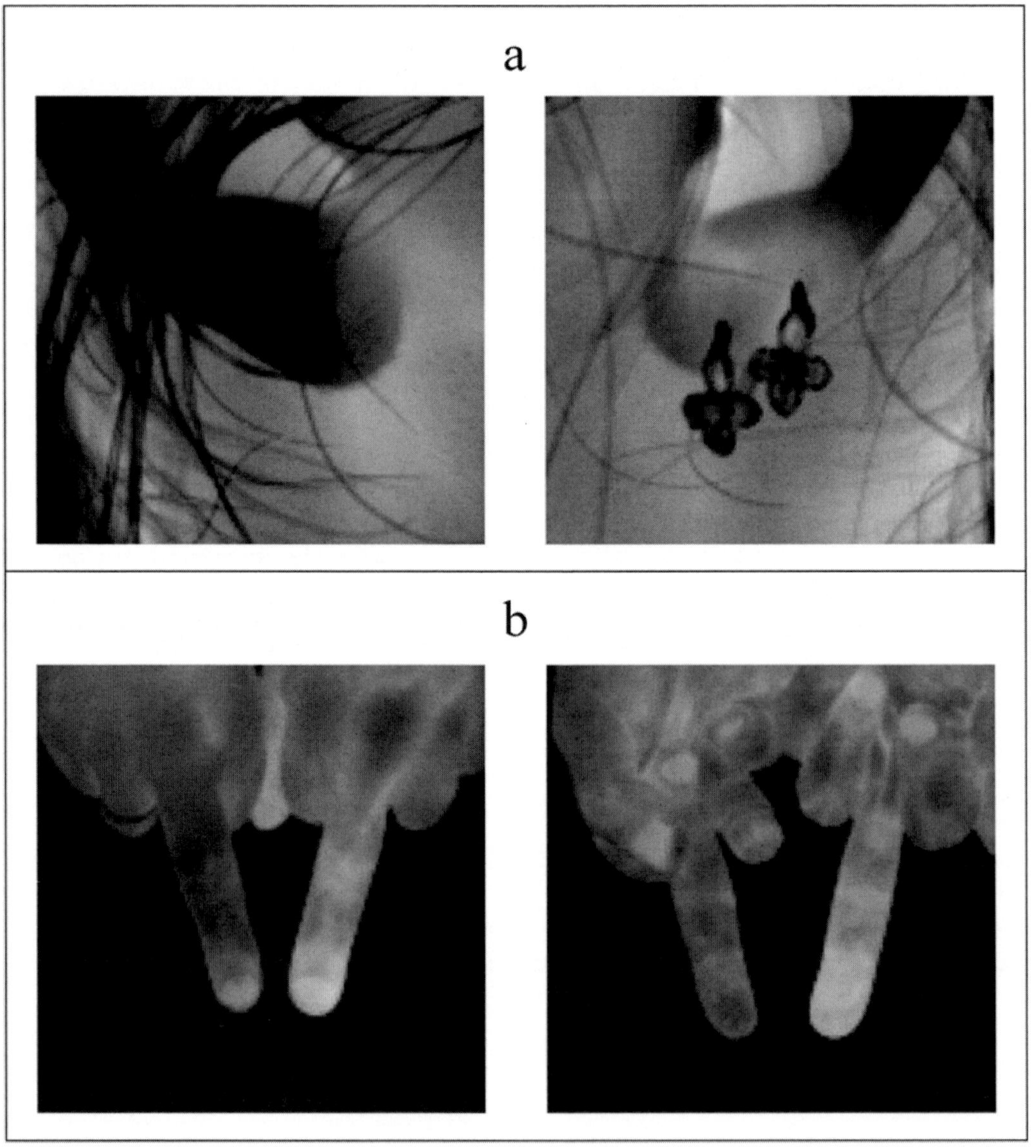

Figure 6. a – inflammation of the left ear lobe caused by ear piercing; b – inflammation of the left third finger caused by trauma.

When a human is under IRT investigation, the next quantitative characteristics acquired from the obtained thermal images are in common use:

1) *Local temperature*. It is determined at the point commensurate to one pixel of thermal image or averaged within small region covering several pixels in the vicinity of a point.

2) *Mean temperature*. It is determined within a rather spacious region in the thermogram by calculation of arithmetic average of temperature values obtained from all pixels of this region.

3) *Thermo-asymmetry*. This term defines the difference of local or mean temperatures measured at equivalent regions of symmetric parts of body. These parts can be exemplified by right and left (upper or lower) extremities, projection of kidneys to skin, etc.

4) *Thermal contrast*. It is the difference of local temperatures measured at two points of body surface. This characteristic is normally of interest when some not great zone on skin looks too hot or too cold as compared to the temperature of surrounding tissues.

In addition, measurement, calculation, and plotting of the next thermogram characteristics are of sizeable benefit to scientific research:

5) *Cross-sectional temperature profile*. It is the temperature distribution along a selected section of the thermogram (usually vertical or horizontal sections). Plotting of such distributions helps, for example, to find out that the profuse sweating increases the contrast of thermal pattern (figure 7) [31]. Another example of profitably used cross-sectional temperature profile is presented in figure 4.

6) *Three-dimensional (3D) temperature profile*. XY coordinate plane is used here to define coordinates of thermal-image pixels, and the applicate is the temperature value corresponding to each pixel. 3D-representation of thermal images was advantageously used in the method developed for quantitative analysis of heterogeneous thermograms [32]. Besides, 3D-surface thus plotted often facilitates visualization of thermal relief and makes possible to recognize much more details than it can be found in plane thermograms. In the IR studies of cutaneous perspiration [26–27] the reverse 3D temperature profile was plotted when separate sweat glands were examined. In that case, 3D-plot was presented with the temperature decreasing upwards, and the decrease of temperature associated with secreta evaporation gave an accurate account of excretory efficiency of each separate sweat gland.

7) *Temperature histogram*. Human's eyes and brain have no the capability to catch and process all information that is abundant in the thermogram measured by modern IR camera. Temperature histogram calculated and plotted for a given surface area of interest has considerable utility in helping us to pick up as more as possible information from IR data. To plot the histogram, one should count the number of pixels N which temperature values fall within the narrow temperature interval ΔT located in the vicinity of given temperature T. After that, a horizontal line segment is plotted on T–N coordinate plane between $(T - \Delta T/2)$ and $(T + \Delta T/2)$ at the ordinate value T. The next value T is taken to be equal to $(T + \Delta T)$. This process is repeated for all informative temperatures scattered within the chosen surface area. Artifact-induced temperatures fallen within the investigated area (for instance, the temperature of rather cold clothes, accessories, etc.) are normally truncated. As it was shown in [26–27], the shape of temperature histogram is very sensitive to predominant mechanisms of human organism heat exchange with environment.

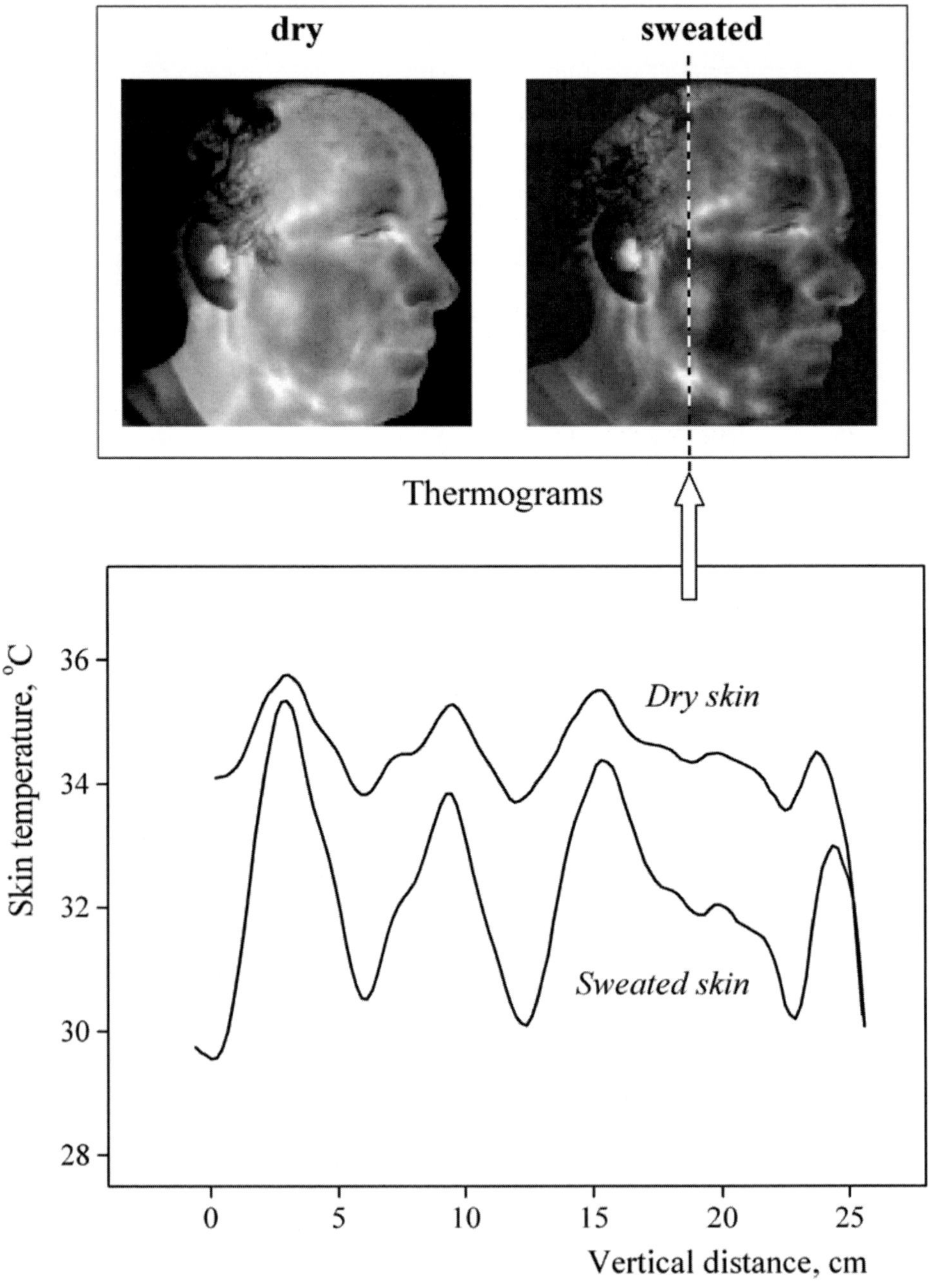

Figure 7. Thermograms of dry and sweated faces of the same person and the corresponding temperature profiles taken along the shown vertical section of both thermograms.

8) *Standard deviation of temperature (SDT)*. A shape of temperature histogram obtained from human body surface normally looks like a bell. This fact was elicited in [26–27]. It was also shown in [26–27] that the typical shape of such a bell plotted for healthy subject at rest is rather symmetric and well approximated by the normal-distribution function. Symmetry breakdown and variation of the shape of this characteristic may serve as indicators of

homeostasis shift and can be used for elaboration of quantitative criteria for the latter. Effective width of the bell is characterized by SDT, which value can be calculated by the overwhelming majority of softwares intended for statistical data manipulation. It is shown in [26–27] that the study of temperature-histogram SDT at different experimental conditions can bring much new information to the receptacle of knowledges about human organism.

9) *The degree of heterogeneity*. The degree of spatial uniformity of thermal pattern can be artificially changed by rearrangement of "warm" and "cold" pixels. Keeping in mind this operation, it can be easily comprehended that both sharply heterogeneous and non-heterogeneous thermograms can give out identical temperature histograms and, hence, identical values of SDT. Thus, the degree of heterogeneity and the temperature distribution function (histogram) generally offer two independent quantitative characteristics of the thermal pattern. This deduction was first made in [26–27]. Just one question remained: how the degree of heterogeneity could be quantitatively expressed?

One algorithm [33] recently proposed for quantitative description of heterogeneous thermograms includes spatial differentiation of a discrete image and a subsequent Fourier analysis. The use of this algorithm is based on the general view of heterogeneous image, which resembles a quasi-periodic structure. This structure, in turn, is associated with the notion of a texture. A reciprocal of the texture period was used in [33] as a quantitative measure of heterogeneity.

Another approach devoid of some bottlenecks of the first one was based on using the Haar wavelet transform. It was also described in [33]. This algorithm allowed the quantitative characteristics of heterogeneity to be better reconciled with the visual patterns of the actually measured test thermograms arranged in the order of increasing inhomogeneity.

A new method [32, 34, 35] attacking this problem includes direct analysis of multiple heat spots (3D-"hills") distributed throughout (X,Y,T)-surface plotted from the measured thermogram. The original principle of "thermal relief flooding" for searching and identification of hyperthermia foci was here developed and applied. The method was implemented with the help of original software code. This code allowed one to detect both high-temperature and weak-intensity spots on the thermogram, to analyze their shape in a three-dimensional (X,Y,T) representation, to estimate the cross-sectional area of such three-dimensional "hills", to construct the spectral characteristics reflecting the distribution of heat spots in terms of their areas, etc.

Figure 8 shows the thermograms of lower extremities taken by examination of one person at different states of his vascular system responsible for obtaining either a fairly homogeneous thermogram (on the left) or a highly heterogeneous thermal pattern (on the right) [32]. The thermograms were measured by IR camera containing 128×128-pixel detector. For research purposes, each image was digitally increased to 256×256 pixels, being simultaneously smoothed by the Lagrangian polynomial interpolation over four points (to simplify interpolation, the real size was reduced to 252×252 pixels). The software analyzed both these dimensional cases. The results of computational (automatic) analysis of the IR images plotted on the thermograms are shown in the lower row of figure 8. The shaded areas mark the fields found on the thermograms by method mentioned in the previous paragraph. It is clearly seen that the results of automatic identification of hyperthermic foci provide a quite satisfactory characterization of the true spatial distribution and approximate sizes of visually identified inhomoheneities. As the array dimension is increased, the number of detected spots becomes appreciably greater. It increases from 27 to 59 on thermograms measured before the human organism was subjected to whole-body thermal loading and from 36 to 143 after the thermal loading which is responsible for a drastic increase in heterogeneity of the temperature pattern. Moreover, digital smoothing allows the shape of some comparatively large spots to be identified more clearly.

Note that the spots were sought on both the original and zoomed-in thermograms with a minimum area of 10 pixels allocated for the spot. This means that the number of small spots detected on the 256×256 zoomed-in thermogram drastically increases. Figure 9 shows the spectral distributions of the warm spot size for data shown in figure 8. In biomedical applications, automatic sorting of heterogeneity spots in terms of their size (figure 9) is important because it can bring, in particular, the next valuable information: the state of what kind of vessels (small or large) has been changed; what group of vessels is coming back to the norm and what characteristic time is needed for that. It is seen from figure 9 that the method of recognition and direct analysis of multiple heat spots [32] provides adequate statistical treatment of the areas of thermal spots.

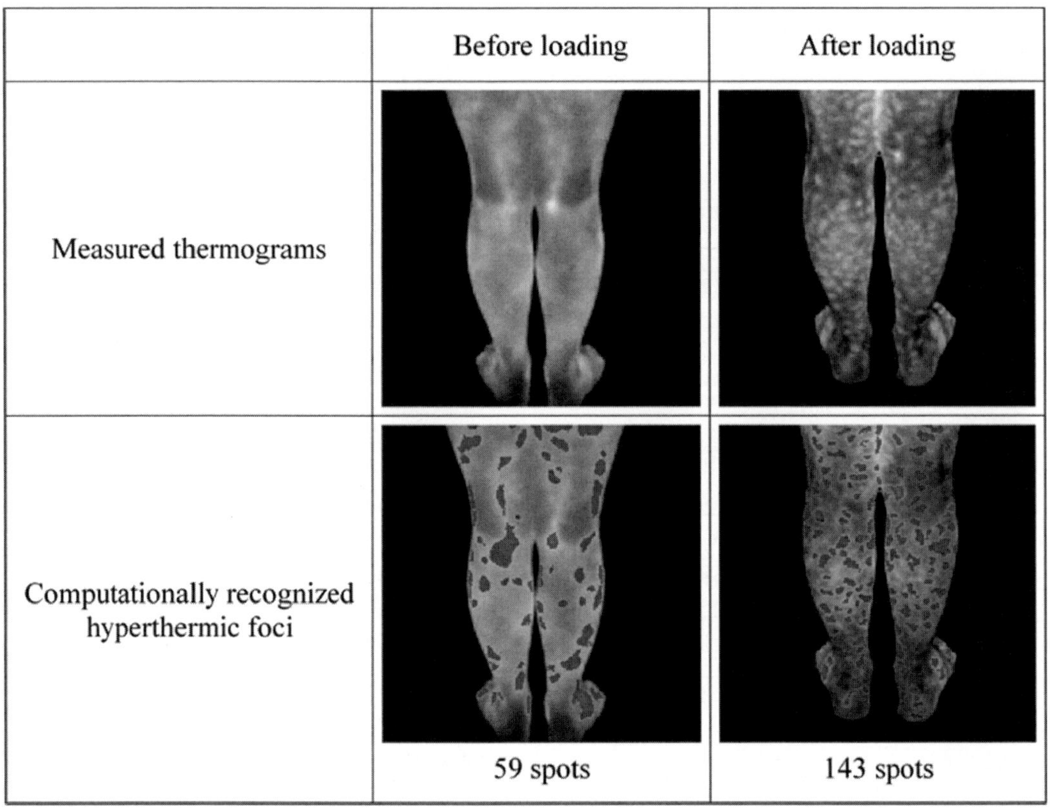

Figure 8. Thermograms of the same person measured before (left) and after (right) whole-body thermal loading.

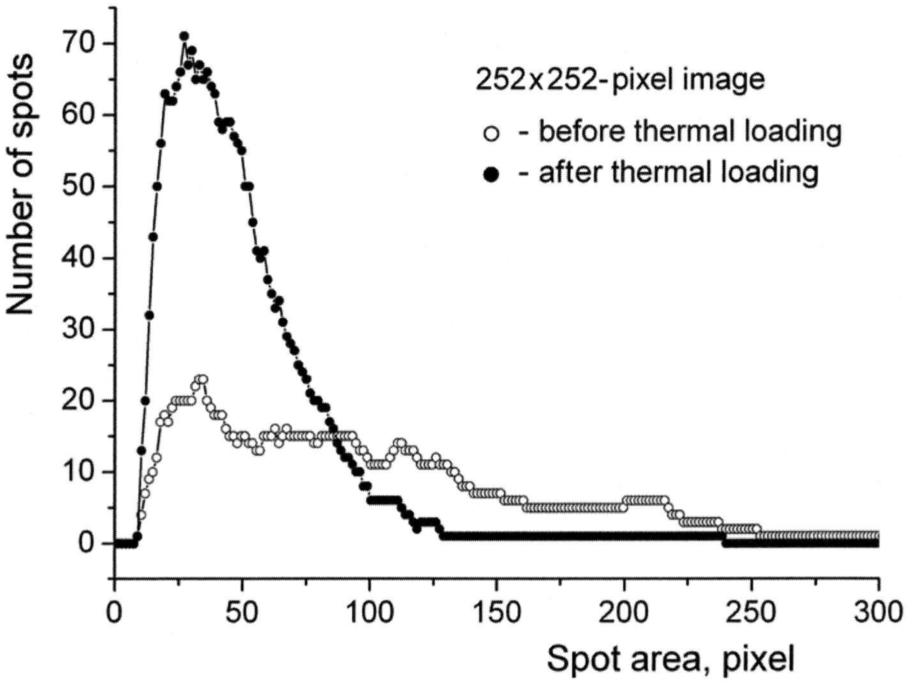

Figure 9. Spectral distributions of heat spots in terms of their areas: the results of computational analysis of thermograms shown in figure 8. The data are given for the artificially expanded (252×252) thermograms. Open and filled symbols represent the results obtained before and after whole-body thermal loading of the organism, respectively.

As the method in question is based on some properties of the normal distribution function used for evaluation of certain shape of each thermal focus, the last result may be of fundamental importance. It serves as an argument for the fact that the temperature distribution in cross sections of even asymmetric spots of hyperthermia on human skin predominantly obeys the normal distribution law.

10) *Temporal data.* Performance of the modern FPA-based IR systems that allows thermograms to be measured and saved in computer memory with a frame frequency of tens frames per second makes possible to study the real-time variations of skin thermal patterns. A great enough specific heat of superficial layers associated with a high content of water in living tissues and the limited speed of blood flow result, of course in delay of manifestation of internal processes on skin as compared to time of these processes true origin. At the same time, IR camera temperature sensitivity of the order of hundredth of a degree makes possible to catch the above-mentioned variations even at their very beginning. The study of temporal behavior of practically all characteristics mentioned above by items 1–9 gives a better insight into what happens in a human organism during different conditions.

It is possible to provoke (enhance) the nonstationarity of thermal manifestation of organs and tissues by excitation (stimulation) of living system to disturb it from equilibrium (normal) condition. It is evident that the peculiarities registered while the functioning system(s) of the organism is(are) reverting to the original state can disclose abnormalities in the organism functioning if any. To trigger the abnormal nonuniformity of surface thermal pattern in order to retrace temporal characteristics of the pattern, an intensity of blood perfusion and/or heat production in the interested part of body should be changed (increased or decreased). To realize it, different ways of internal and external influence on the total organism or on its separate parts are known. Among them are physical loads, heating or cooling of extremities, whole-body heating in sauna bath, etc. Among unusual tricks the attempt to cool ear lobe by wetting can be referred here [36].

Attractive opportunity to the tasks concerned physiology is that the correct processing of IRT data makes possible to estimate the amount of heat dissipated by a human body. Work on this problem carried out ever since temperature was evaluated from thermograms on the basis of shades-of-gray density measurement [37]. With the help of IR measurements and the followed computer processing, surface mean temperature was determined in such a manner, and the obtained results were used in subsequent calculations. At that, radiant, convective, and evaporative heat losses, as well as respiratory water lost were taken into account. By means of comparison between thermographic and calorimetric data, it was experimentally shown and concluded in [37] that IRT system permits a rapid, accurate determination of heat loss in humans.

It should be noted that the foregoing list of ten items is not exhaustive. Spatial and temporal characteristics of thermograms contain much more information about biological objects than it is possible to reveal today. Application and development of new methods of mathematical treatment and analysis of thermal images can enlarge substantially the informativity of IRT assisted investigations.

2.6. Computer Simulation of Human Skin Thermal Fields

Computer simulation can provide additional insights into dominating mechanisms of the formation of temperature fields on human skin. At present, more emphasis is placed on this approach to investigation of the heat exchange phenomena in humans [38, 39, etc.]. This method of attack is directly connected with the solving of reverse problem of IRT. The latter consists in identification of heat generating foci deepened in the body and in the revelation of their characteristics by means of computer analysis of thermograms.

Shown in figure 10 is the surface distribution of temperature numerically calculated for a specific case when a benign tumor composed of fatty tissue (lipoma) is situated under the skin, and a major blood vessel lies deeper in tissues (under the tumor). The really measured thermogram depicting this case and showing clearly the "shade" from the tumor on a skin surface is exhibited in the same figure on the right. Three-dimensional result represented in figure 10 was obtained by solving the heat conduction equation described in [40]. This equation indirectly takes into account isotropic convection associated with microcirculation and playing a principle role in the heat transport phenomena in living tissues [39].

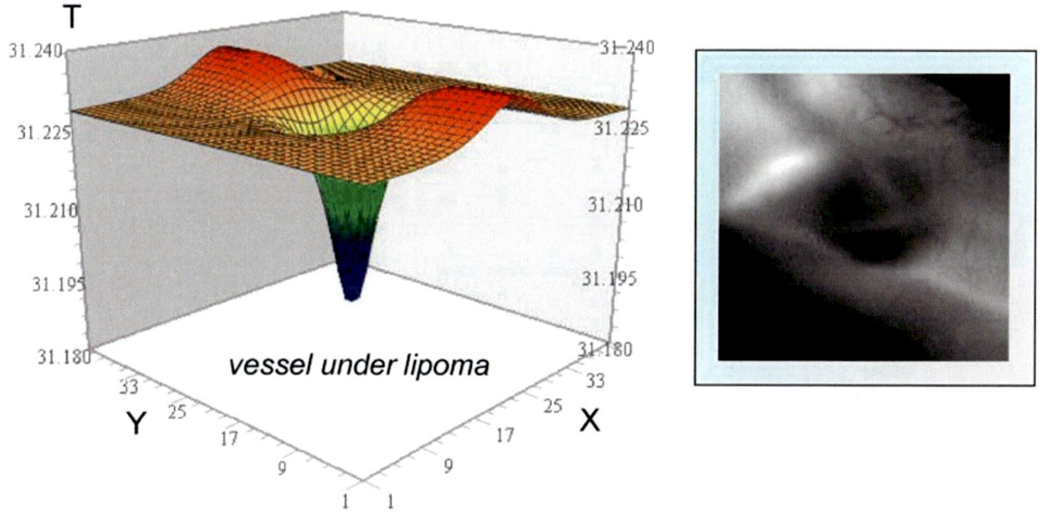

Figure 10. On the left: 3D-representation of the skin temperature spatial distribution calculated in the course of a computer simulation of human living tissues containing a blood vessel passing under the benign tumor composed of fatty tissue (lipoma). On the right: really measured thermogram representing lipoma (dark "shade") and blood vessels (light cords) crossing the tumor.

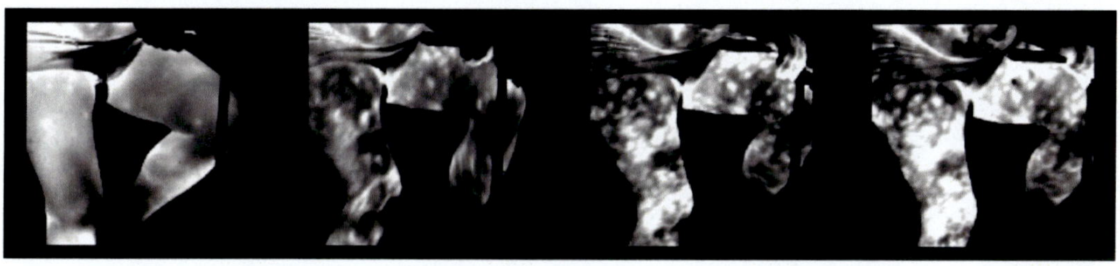

Figure 11. From left to right: evolution of thermal pattern of the thighs of healthy subject in the course of physical load (exercise bicycling).

3. IRT-RESULTS Associated with Human Physiology

3.1. Vascular and Hidropoetic Mechanisms of Thermoregulation

Sweating is an important executive mechanism of human body thermoregulation. Skin perspiration intensity is examined with the use of weight loss data [41], ninhydrin test [42–44], colorimetric tests [42, 45, 46], galvanic methods [47], etc. In physiology the ventilated-capsule method [4, 48], dew-point hygrometry [49, 50], as well as various modifications of these techniques [51] have been used. The sweating capability of individual sweat glands is determined using capillary tubes [42] and some other methods [52, 53]. The great diversity of scientific problems [54, etc.] and methods devoted to the study of hidropoiesis engendered the idea of consolidation them and creation a separate scientific area – sweatology [26–27].

Unfortunately, properties peculiar to the majority of the above-mentioned tools, namely, physical and chemical contacts with skin [55], as well as mechanical insulation of skin from natural environment, may disturb normal gland functioning. As it is confirmed by figures 3 and 4, IRT is capable of qualitative and quantitative analysis of separate sweat glands with high spatial resolution, and without any mechanical contact with skin at that [26–27].

The authors of [56] examined the distribution of sweat evaporation rate over the surface of human body and found that numerous body regions could be classed into groups according to their perspiration capability. While studying thermoregulation in humans, detailed infrared data obtained in a manner similar to those represented in figures 3 and 4 allow one to answer the question [5]: which of two factors, an increase in the density of activated sweat glands or an increase in the sweat output per gland dominates in the increased sweating?

Superficial vascular network is another important agent participating in liberation of extra-heat from body to environment. Light branches, nets, and spots caused by major surface vessels as well as dark zones frequently caused by intensive perspiration followed by evaporation create in the IR part of spectrum the motley heterogeneous patterns like those shown in figures 5 and 11. On basis of experimental fact that hyper- and hypothermic regions occupy quite comparable areas on human skin the next hypothesis was advanced in [26–27]. As a result of evolution, the human body has acquired a circulatory system originally “pre-programmed” to the existence of at least two alternative heat-removal mechanisms partly duplicating each other. In the case of exocrine organs pathology (anhidrosis, etc.), a vascular mechanism starts operating, whereas in the case of occlusion or at various forms of vasoconstriction a perspiratory mechanism comes into action. IR thermography uncovers for us the dominant mechanism of organism heat exchange at the moment.

The use of temperature histogram plotting technique for solving the problems of physiology makes possible to find out many nuances escaped investigator's attention at carrying him out the standard visual analysis of thermograms still popular in biomedical IRT. An example of statistical treatment of thermograms is given in figure 12. In this figure, it is shown how the temperature histograms corresponding to similar parts of body can differ from one another by their shapes even at identical load of the organism (thermograms of thighs of two women exposed to heating their body in sauna are processed). It is easily observable in figure 12 that in the case 1 the histogram center of gravity is shifted towards lower temperatures whereas in the case 2 it is shifted towards higher ones. The distinctiveness of each thermogram (and histogram) in figure 12 is based on the dominant mechanism of the investigated subject’s body heat exchange with environment. Case 1 denotes hidropoiesis as a key mechanism of heat dissipation, as opposed to vascular (and hence conductive-convective-irradiative) mechanism dominating in the case 2.

3.2. Heat Dissipation Conditioned by Breath

Human heat dissipation at breath is conducted by two basic mechanisms. On the one hand, it is the transition of water from liquid phase to gas that requires an additional heat for evaporation in much the same manner as at perspiration, for another, it is a heating of atmospheric air (in respiratory organs) which is normally cooler than the human core. The breath conditioned heat can be simply detected by IR camera that is demonstrated in figure 13. Here, in the left thermogram, the face of man breathed atmospheric air into his lungs is represented; in the right thermogram, the process of expiration through nose is depicted. The heating of moustache is clearly seen in the right thermogram. It should be noted that the increase moustache in temperature at expiration can be partly contributed by adsorption of exhalant water molecules on heir accompanied by heat evolution. Results of IRT investigation of water adsorption on surfaces of different matters including fleecy materials are published in [57].

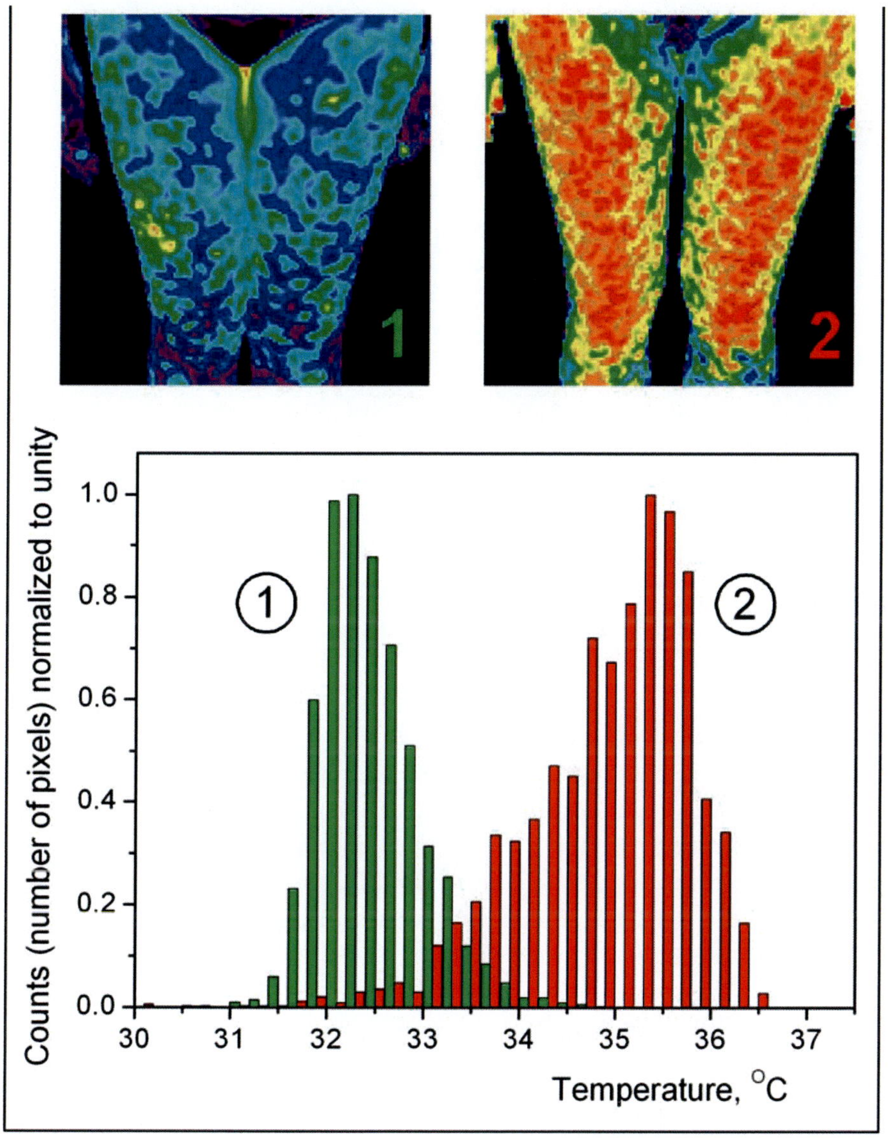

Figure 12. Qualitative (thermograms on top) and quantitative (temperature histograms in the bottom) IR-representations of thighs of two different women exposed by the similar whole-body thermal loadings. Dominating mechanism of heat dissipation: 1 – hidropoiesis; 2 – vasodilation.

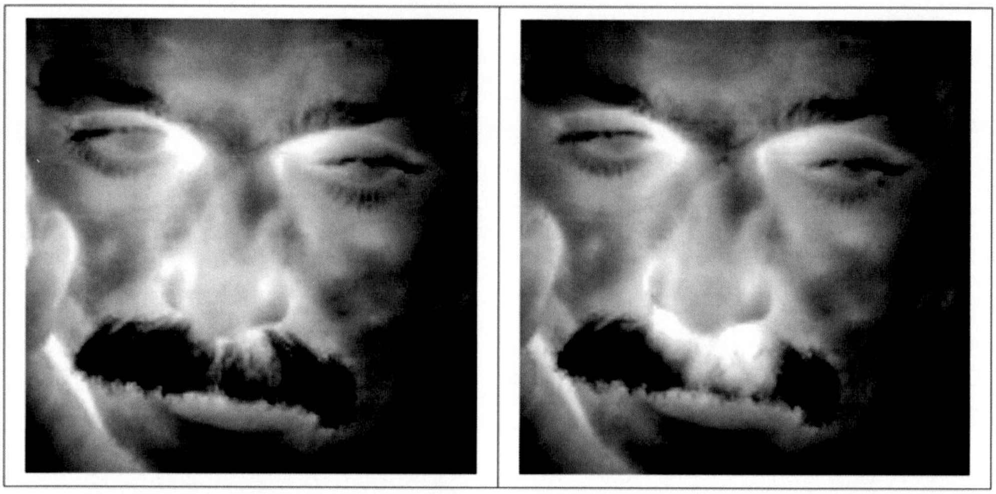

Figure 13. The breath conditioned heat. On the left – thermogram of the man breathed atmospheric air; on the right – the process of expiration through nose.

3.3. Autonomic Reactions Associated with Thermoregulation

Thermoregulatory mechanisms of a warm-blooded living organism are in uninterrupted "operational readiness". Thanks to this apt property, they respond to any deviation happened with internal (inside organism) or external conditions that could be resulted in overheating or overcooling of body. At the same time, it is well known that abrupt changes of scene such as sudden increase or decrease in environmental temperature, unexpected physical load, circumstances making the brain overtasked, heebie, irritation of gustatory receptors, fright, trauma, etc. are rather standard and even typical for human life. Hence it follows that drastic deviation of skin thermal pattern caused by the abrupt change of heat production normally accompanying the above-mentioned events is also the typical physiological phenomenon in humans. In this connection, of interest is the scientific study of dynamic thermoregulatory characteristics in humans, and physiology is in urgent need of the appropriate experimental tools capable of carrying out the objective quantitative control of temperature distribution on skin. It is easy to understand that IRT is best suited to this task.

The example demonstrating appropriateness of IRT for solving the problem in question is depicted in figure 14. This figure exhibits two states of the body of a volunteer treadling exercise bicycle. Time interval between the displayed snapshots is three minutes. With the help of thermograms presented in figure 14, it becomes possible to learn that the degree of heterogeneity of the volunteer's surface thermal pattern was dramatically increased during this short time interval.

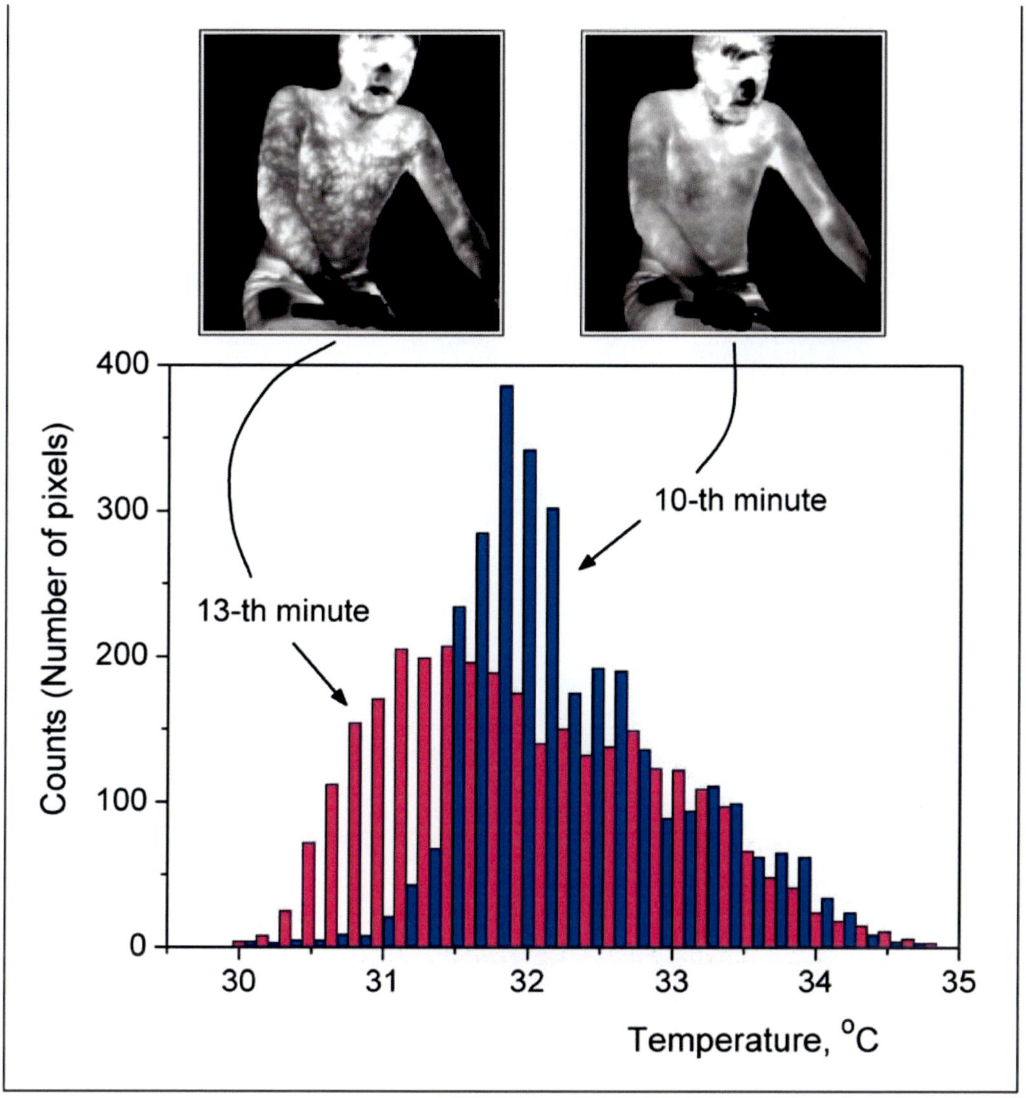

Figure 14. Thermograms and the corresponding hystograms disclosing abrupt increase of heat dissipation through hidropoiesis in the course of physical load (exercise bicycle). Hystogram discretization step ΔT = 0.15 °C.

In the wake of such a qualitative result, we can turn our attention to the quantitative processing of the presented data. Some approaches to quantitative analysis of heterogeneity were described in the section 2.5. However, temperature histograms alone represented in figure 14 unambiguously show that during three minutes the reinforcement of heat dissipation mechanism associated with perspiration took place, but the irradiative mechanism of heat dissipation kept unchanged. This conclusion is made on the basis of experimental fact showing that histogram's high-temperature "tail", the responsibility for which is on warm surface vessels, is remained practically unchanged, whereas the low-temperature edge and the center of gravity of the histogram is displaced towards lower temperatures.

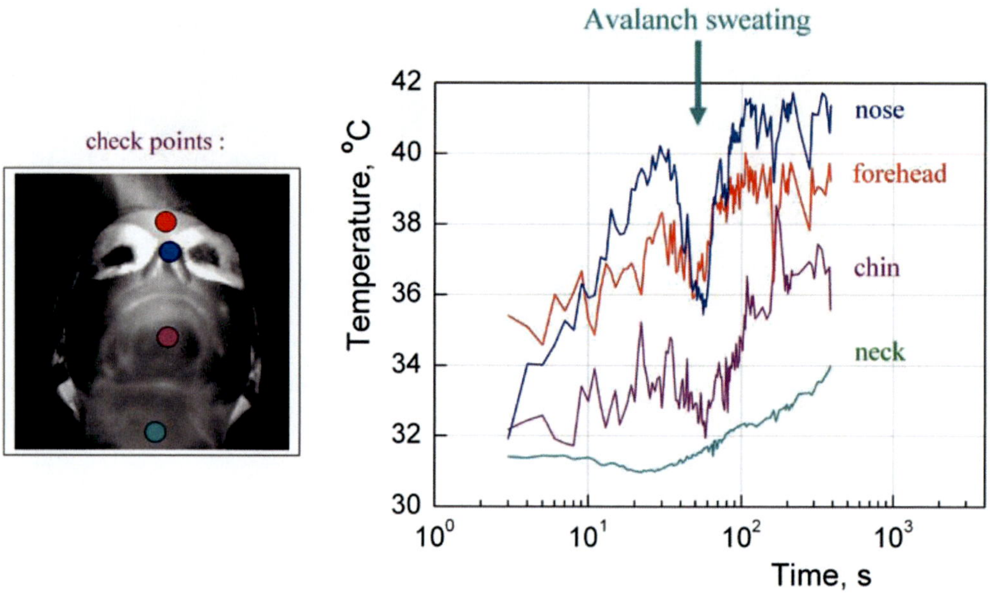

Figure 15. Face temperature instability detected by IRT system in the course of cosmetic steaming, and the manifestation of "avalanche perspiration" phenomenon during the steaming.

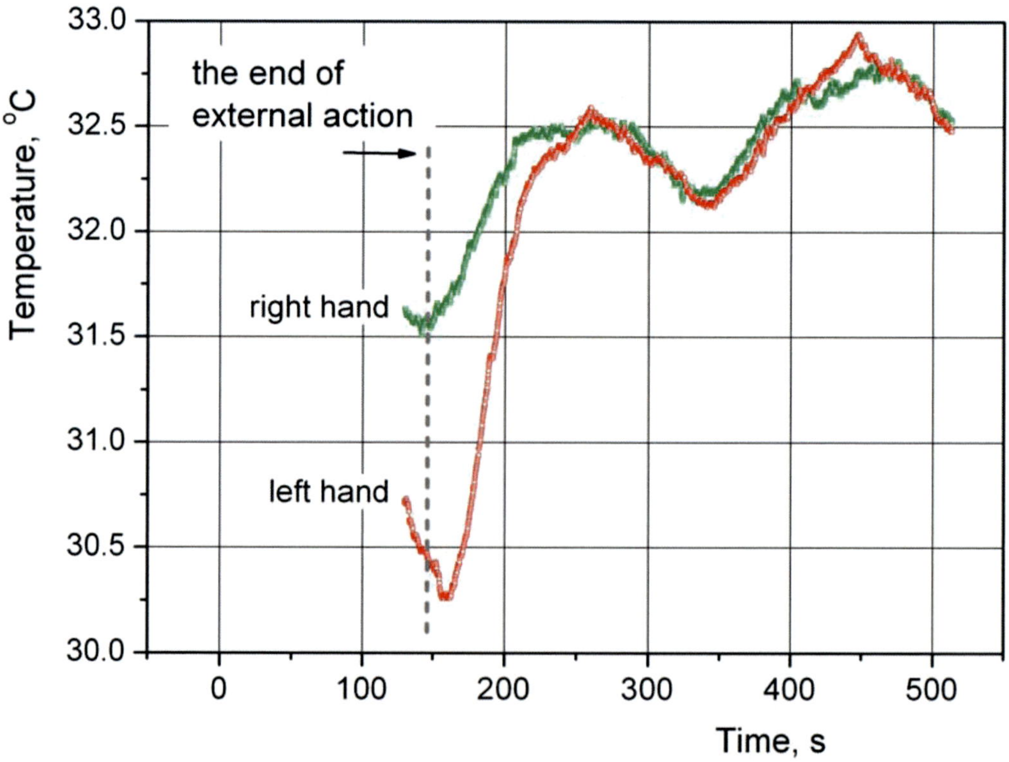

Figure 16. Synchronous wavy change in temperature of the third fingers of healthy subject revealed on completion of external mechanical action to one of his hands (left).

The threshold phenomenon manifested in figure 14 in the form of a profuse sweating abruptly appeared during the human strenuous activity (in more details, it is described in [26–27]) is akin to that we first registered at ozone-admixed steaming of face in cosmetic salon [24]. By the latter measurements, IR camera has given us the chance to monitor in details the manifestation of thermoregulation process which was called in [24] "avalanche perspiration" phenomenon because it had been developed just over a period of several (about 10) seconds. Thermal manifestation of avalanche sweating at face steaming is represented in figure 15. A noticeable downward excursion of temperature appeared in forty seconds after commencement of cosmetic procedure enables one to trap the moment of abrupt increase in activity of sweat glands. In such a manner, IRT gives investigator an insight into the processes usually hidden from his eyes.

With the help of IRT, one can register precisely with high sensitivity and temporal resolution autonomic thermoregulation-linked reactions triggered by psychogenic factors. We was favored to observe how the nose temperature of a human relaxed in armchair of cosmetic salon suddenly, during several minutes, increased by 9 (!) degrees centigrade, and this temperature increase took place before any contact with his face at that [31]. It should be noted that the temperature of the rest part of face was not changed appreciably during this record.

Another interesting phenomenon quantitatively and reliably monitored by IRT is demonstrated in figure 16. It bears on the autonomic work of neuroregulatory mechanisms controlling the blood flow and consequently influencing metabolic processes in tissues. In figure 16, shown is the temperature change of similar fingers of the subject's left and right hands measured after completion of external action on one (left) hand (mechanical occlusion of blood vessels) in one of our experiments devoted to investigation of vascular reactions in humans. Despite the fact that during a period depicted in figure 16 both hands exposed to IR control were at rest, and the external action was already finished, synchronous wavy change of blood perfusion intensity in the hands' surface vessels was found. The appearance of these synchronous waves, the form of which is individual for each subject, can be explained predominantly by psychogenic factor.

3.4. Regular Thermal Manifestations of the Organism at External Impacts

As it was proved in the section 2.2, IRT method enables a human to be examined over and over again without any detriment of his health. It offers unique advantages to IRT in using this method for scientific investigations requiring, in many cases, statistical approach. Psychoemotional manifestations are usually individual for each person, and some experimental results in this sphere can turn out to be irreproducible at cohort study. Contrastingly, external thermal, physical, mechanical, pharmaceutical, and some other actions usually require equal responses from the organism that is conditioned by the organism constitution and by the laws of human existence and survival. At such kind of impacts, acquisition of statistically corroborated data and disclosure of general statistical laws are quite possible. Two illustrative examples are given below which demonstrate IR representation of physiological manifestations regularly repeated in all examined healthy subjects.

3.4.1. Steaming in Cosmetic Salon

An example (avalanche perspiration) of IRT results obtained in cosmetic salon in the course of a face steaming was presented in the section 3.3. Cosmetic means and standard equipment of a modern ordinary cosmetic salon aimed at the care of healthy skin can be successfully used for some research directed to human physiology, in particular, to human body temperature regulation [31, 58]. These facilities can serve as the triggering agents starting up thermoregulatory and some other reflex conditioned processes in the organism. Favorable factors giving proof of such an idea are, for instance, the next. Equipment and other means of cosmetic salons are mostly harmless for human. Cosmetology belongs to one of the most speedy developing industries. Unlike conventional medical physiotherapeutic procedures, the predominating conception of cosmetology is not only safety, but also convenience and comfort that are not the least of the factors if humans are under investigation. Being free of lasting side effects, cosmetic means favor the development of one or another thermal reactions, IRT analysis of which can serve as a key to understanding of the organism functioning [24, 58, 59].

Thermograms represented in figure 17 demonstrate how the 5-minute long face steaming in cosmetic salon results in dilation of vessels (light zones in the middle and right thermograms) in submaxillary region and neck. It should be noted that the steam stream was directed not to the neck's part of body but to the region of forehead and nose. Figure 17 demonstrably shows how thermoregulatory dilation of vessels enables organism to dump the extra heat received from a hot steam.

3.4.2. Physical and Thermal Loadings of the Organism

Whole-body heating and intense physical loads are the experimentally available yet powerful enough expedients resulting in homeostasis shift. As compared with the stable conditions, in either case the body eventually receives extra heat, with the difference that at whole-body heating the heat comes from without, but at intensive work of muscles it comes from within. Detailed experimental investigations carried out in both this lines with the use of IRT are described in [26–27]. Homoeothermic organism is forced to pass (or return) the major portion of extra-heat to environment. The effective ways to do it are cooling by evaporation of sweat and exposure of heated parts of skin to the passing relatively cold atmospheric air. The latter is provided by admitting blood to dilated surface vessels.

To perform a physical load, exercise bicycles and bicycle ergometers are widely used in physiological tests [49, 60–62]. Two examples of skin thermal pattern transformation arisen from the organism physical load (using exercise bicycle) are presented in figures 11 and 14. In addition, figure 11 shows in consecutive order how the quite homogeneous thermal pattern becomes highly heterogeneous during volunteer's mechanical work during strenuous activity.

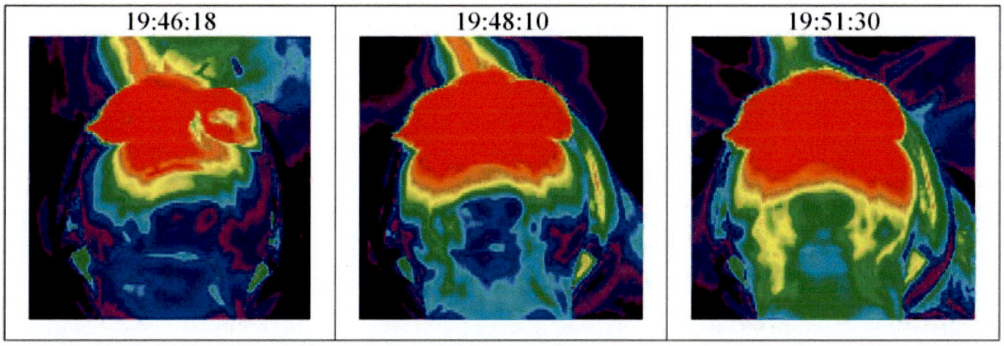

Figure 17. Vasodilation at submaxillary and neck regions in the course of face steaming in cosmetic salon. Above each thermogram the current time (hh:mm:ss) in the process of treatment is indicated.

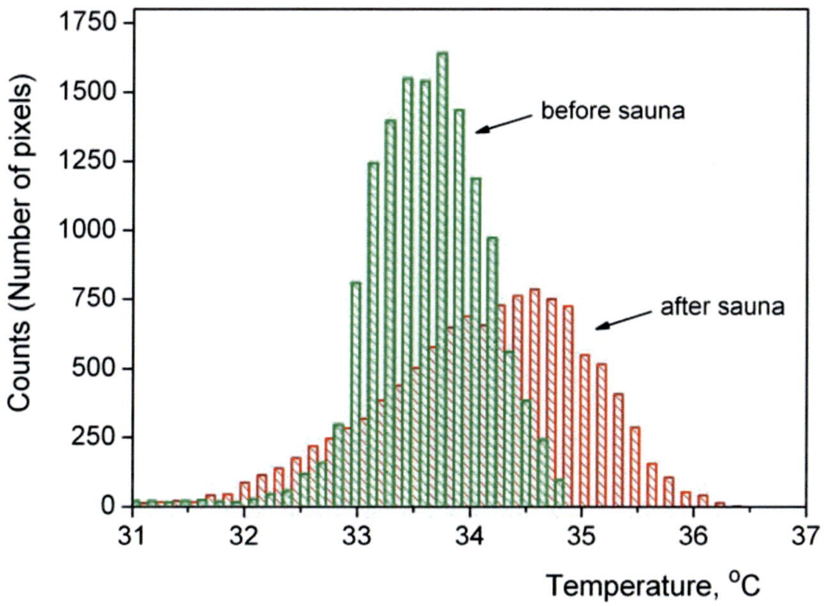

Figure 18. Temperature hystograms indicating reinforcement of vasculatory heat exchange mechanism after the body overheating in sauna. Hystograms are plotted from the thermograms of rear surface of woman thighs. Hystogram discretization step $\Delta T = 0.15$ °C.

To experience an outside thermal action, volunteers can be asked to take a sauna. The thermal action due to sauna makes the mean skin temperature rise, and this starts thermoregulatory mechanisms [63]. Figure 18 represents the initial and post-load temperature histograms plotted from the thermograms of rear surface of woman's thighs and demonstrating the organism overheating in sauna. It is seen that the hyperthermic impact shifts the center of gravity of the histogram to the right. This fact and the abrupt fall of the histogram's envelope line at high temperatures can be evidence of the predominance of vasculature mechanism in heat removal. To withstand the excess thermal stress, thermoregulation system "gave an order" for total intense dilation of thermoregulatory superficial vessels, that brought the organism near heat stroke [64] the initial signs of which were observed during the measurements.

4. Conclusion

IRT can be applied to experimental physiology as an independent convenient research tool intended for examination of human regulatory mechanisms, including the cases in which nervous regulation plays an important role. Here, the object under IR study, the human body surface, is treated as a complex non-stationary physical object whose temperature and peculiarities of thermal pattern are mainly determined by the processes developing inside the organism in the norm and at pathological deviations. It is questionable to expect success if laboratory animals are used as model objects in IRT investigations, the results of which are meant to be applied to humans. Present-day IR cameras using fast (tens of frames per second) and high-sensitive (hundredths of a degree) FPA-based IR detectors make it possible to reach a radically new level in the study of physical and physiological mechanisms responsible for the formation of a skin thermal pattern. Contact thermoelements traditionally used in physiology are incapable of attacking this problem.

The wide angle of view and high spatial resolution of IR cameras supporting the IRT method make it possible to investigate both the great parts of the human body and the excretory characteristics of separate sweat glands. It is obvious that the problem of body temperature regulation will keep pace with the new research field that is termed *sweatology*. IR imaging and IRT show considerable potential in quantitative examining of the perspiration, disposition on skin and functioning of active sweat glands, in monitoring the reflectory responses of these glands and superficial blood vessels to various external and psycho-emotional factors, in examining the neuroregulatory apparatus and some other systems of the organism.

To study the physiological response in answer to the whole-body thermal loads in humans, hyperthermic impacts provided by a sauna bath can be successfully employed. In turn, an exercise bicycle in the facilities can make it convenient to apply physical loads for the purpose of subsequent IRT-aided investigations. Hardware facilities and other means of cosmetic salons can also be considered as triggering agents in active IRT.

Quantitative analysis of digitized thermograms can be successfully performed using statistical methods of data processing, in particular, using a technique of temperature histogram plotting. Such a skin-temperature distribution function evaluated from IRT data is a new valuable characteristic in the study of body temperature regulation. Violation of the symmetry and variation of the Gaussian-like shape of this characteristic may serve as an indicator for homeostasis shift and can be used for evaluation of qualitative and quantitative criteria for the latter.

Analysis of heterogeneous surface thermal patterns displayed by organism under various impacts has allowed the advancement of the hypothesis that, as a result of human evolution, a special rarefied subnet of superficial veins (veins-collectors) had emerged in the superficial layer of a body, intended for efficient removal of excessive heat from organism into the environment. Computer analysis and quantification of sharply nonuniform thermograms as well as evaluation of the degree of heterogeneity of human IR images, combined with the possibility of thermograms quantitative comparison, may provide valuable information on the physiological status of the organism.

To gain a better understanding of the mechanisms of formation of thermal patterns on human skin, computer simulation should be pursued. The end purpose of this kind of scientific work is to find a solution of the inverse IRT problem consisting in the extraction of physical and anatomic-physiological parameters of subsurface heat sources, in particular, the sources whose behavior can bring additional information about functioning of the thermoregulation system.

As is evident from the foregoing, up-to-date IRT opens up new fields of use in the science concerned with the human being. Hopefully, this chapter has demonstrated, in particular, that it is impossible to obviate the need for the IRT when problem of body temperature regulation is under experimental study.

References

[1] Applications of Thermal Imaging. Editors SG, Burnay; TL, Williams; CH, Jones. Inst. of Physics Pub.; 1988.

[2] Quantitative InfraRed Thermography: Eurotherm Seminar 27, QIRT 92, Chatenay-Malabry, France, July 7–9, 1992, Proceedings. Editors D, Balageas; G, Busse; GM, Carlomagno. France: ONERA; 1992.

[3] 9-th International Conference on Quantitative InfraRed Thermography, QIRT2008, Proceedings, July 2–5, 2008, Krakow-Poland. Editor B, Wiecek. Poland: Technical University of Lodz, Institute of Electronics; 2008.

[4] Takano, S; Kondo, N; Shibasaki, M; Aoki, K; Inoue, Y; Iwata, A. The influence of work loads on regional differences in sweating rates. *Jpn. J. Physiol.*, 1996 46, 183–186.

[5] Kondo, N; Shibasaki, M; Aoki, K; Koga, S; Inoue, Y; Crandall, CG. Function of human eccrine sweat glands during dynamic exercise and passive heat stress. *J. Appl. Physiol.*, 2001 90, 1877–1881.

[6] Craig, FN; Cummings, EG. Dehydration and muscular work. *J. Appl. Physiol.*, 1966 21, 670–674.

[7] Michikami, D; Kamiya, A; Fu, Q; Iwase, S; Mano, T; Sunagawa, K. Attenuated thermoregulatory sweating and cutaneous vasodilation after 14-day bed rest in humans. *J. Appl. Physiol.*, 2004 96, 107–114.

[8] Stuttgen, G; Eilers, J. Reflex heating of the skin and telethermography. *Arch. Dermatol. Res.*, 1982 272, 301–310.

[9] Shuran, M; Nelson, RA. Quantitation of energy expenditure by infrared thermography. *Am. J. Clin. Nut.*, 1991 53, 1361–1367.

[10] Zontak, A; Sideman, S; Verbitsky, O; Beyar, R. Dynamic thermography: analysis of hand temperature during exercise. *Ann. Biomed. Eng.*, 1998 26, 988–993.

[11] Cassis, LA; Marshall, DE; Fettinger, MJ; Rosenbluth, B; Lodder, RA. Mechanisms contributing to angiotensin II regulation of body weight. *Am. J. Physiol.*, 1998 274, E867–E876.

[12] Adams, AK; Nelson, RA; Bell, EF; Egoavil, CA. Use of infrared thermographic calorimetry to determine energy expenditure in preterm infants. *Am. J. Clin. Nutr.*, 2000 71, 969–977.

[13] Klede, M; Handwerker, HO; Schmelz, M. Central origin of secondary mechanical hyperalgesia. *J. Neurophysiol.*, 2003 90, 353–359.

[14] Nagashima, K; Nakai, S; Matsue, K; Konishi, M; Tanaka, M; Kanosue, K. Effects of fasting on thermoregulatory processes and the daily oscillations in rats. *Am. J. Physiol. Regul. Integr. Comp. Physiol.*, 2003 284, R1486–R1493.

[15] Lloyd, JM. Thermal Imaging Systems, Optical Physics & Engineering Ser. Plenum Publishing Corporation; 1975.

[16] Cox, P; Driggers, RG; Edwards T. Introduction to Infrared & Electro-optical Systems. Artech House, Inc.; 1998.

[17] Caniou, J. Passive Infrared Detection: Theory and Applications. Kluwer Academic Publishers; 1999.

[18] Head, JF; Lipari, CA; Wang, F; Elliot, RL. Cancer risk assessment with a second generation infrared imaging system. *Proc. SPIE*, 1997 3061, 300–307.

[19] Vainer, BG. High-resolution medical infrared thermography: new abilities. *Vrach (Physician)*, 1999 2, 25–27. (in Russian)

[20] Vainer, BG. FPA-based infrared thermography systems in medicine. *Vrach (Physician)*, 1999 10, 30–31. (in Russian)

[21] Ecker, RD; Goerss, SJ; Meyer, FB; Cohen-Gadol, AA; Britton, JW; Levine, JA. Vision of the future: initial experience with intraoperative real-time high-resolution dynamic infrared imaging. Technical note. *J. Neurosurg.*, 2002 97, 1460–1471.

[22] Falk, V; Walther, T; Kitzinger, H; Rauch, T; Diegeler, A; Autschbach, R; Mohr, FW. An experimental approach to quantitative thermal coronary angiography. *Thorac. Cardiovasc. Surg.*, 1998 46, 25–27.

[23] Kurishev, GL; Kovchavtzev, AP; Vainer, BG; Guzev, AA; Bazovkin, VM; Stroganov, AS; Subbotin, IM; Zakharov, IM; Efimov, VM; Postnikov, KO; Lee, II; Valisheva, NA; Panova ZV. Medical infrared imaging system based on a 128 x 128 focal plane array for 2.8–3.05 □m spectral range. – *Optoel. Instrum. Data Process. (Autometria)*, 1998, no. 4, p. 5–10.

[24] Vainer, BG. Narrow spectral range infrared thermography in the vicinity of 3 μm operating wavelength. In: Editors D, Balageas; J-L, Beaudoin; G, Busse; GM, Carlomagno. *Quantitative InfraRed Thermography 5, Eurotherm Seminar 64, QIRT'2000, Reims, France, July 18–21, 2000. Proceedings*. France: UTAP URCA; 2000; 84–91.

[25] Vainer, BG. Limitary operation conditions affecting CID short-wave infrared detector performance. *Meas. Sci. Technol.*, 2004 15, 821–830.

[26] Vainer, BG. FPA-based infrared thermography in physiology: Investigation of vascular response, perspiration, and thermoregulation in humans. Novosibirsk: Publishing House SB RAS; 2004. (in Russian)

[27] Vainer, BG. FPA-based infrared thermography as applied to the study of cutaneous perspiration and stimulated vascular response in humans. *Phys. Med. Biol.*, 2005 50, R63–R94.

[28] Ivanov, KP. The principles of energetics in organism: Theoretical and practical aspects: Volume 1. The general energetics, heatexchange, thermoregulation. Leningrad: Nauka; 1990. (in Russian)

[29] Basics of Clinical Remote Thermodiagnostics. Editor LG, Rozenfel'd. Kiev: Zdorov'ya; 1988. (in Russian)

[30] Dragun, VL; Filatov, SA. Computational thermography. Application in medicine. Minsk: Navuka i Technika; 1992. (in Russian)

[31] Vainer, BG. Treated skin temperature regularities revealed by IR thermography. *Proc. SPIE*, 2001 4360, 470–481.

[32] Vainer, B; Moskalev, A. Heterogeneous thermograms: the methods of attack. In: Editor B, Wiecek. *QIRT2008, 9-th International Conference on Quantitative InfraRed Thermography, July 2-5, 2008, Krakow-Poland, Proceedings*. Poland: Technical University of Lodz, Institute of Electronics; 2008; 157–164.

[33] Tarkov, MS; Vainer, BG. Evaluation of a thermogram heterogeneity based on the wavelet Haar transform, In: *Proc. IEEE Int. Siber. Conf. on Control and Communications (SIBCON-2007)*. Tomsk: The Tomsk IEEE Chapter & Student Branch; 2007; 145–152.

[34] Vainer, BG; Moskalev, AS; Sapetina, AF. Computer-assisted analysis of skin thermal heterogeneity in humans. In: *The Sixth International Conference on Bioinformatics of Genome Regulation and Structure (BGRS'2008), Russia, Novosibirsk, June 22-28, 2008. Abstracts.* Novosibirsk: Institute of Cytology and Genetics; 245.

[35] Vainer, BG; Moskalev, AS. Heterogeneous thermograms: the methods of attack. *Acta Bio-Optica et Informatica Medica*, 2008 14, 143–144.

[36] Rustemeyer, J; Radtke, J; Bremerich, A. Thermography and thermoregulation of the face. *Head Face Med.*, 2007 3, 17 (8 pp)

[37] Shuran, M; Nelson, RA. Quantitation of energy expenditure by infrared thermography. *Am. J. Clin. Nutr.*, 1991 53, 1361–1367.

[38] Deng, Z-S; Liu, J. Mathematical modeling of temperature mapping over skin surface and its implementation in thermal disease diagnostics. *Comput. Biol. Med.*, 2004 34, 495–521.

[39] Vainer, B; Belozerov, P; Baranov, V. Use of IR thermography for bioheat transfer studies. In: *Proceedings of the 7th International Conference on Quantitative IR Thermography, July 5–8, 2004*. Belgium: von Karman Inst. for Fluid Dynamics; 2004; J.8.1–J.8.4.

[40] Pennes, HH. Analysis of tissue and arterial blood temperature in the resting human forearm, *J. Appl. Physiol.*, 1948 1, 93–102.

[41] Allen, JA; Grimley, JF; Roddie, IC. A body balance to measure sweat rates in man. *Biomed. Eng.*, 1971 6, 468–471.

[42] Big Medical Encyclopaedia in 30 volumes: Volume 20. 3-rd edition. Editor BV, Petrovskii. Moscow: Sovetskaya Entsiklopediya; 1983.

[43] Ponten, B. Quantative measurements of sweat gland activity using the ninhydrin method. *Acta Physiol. Scand.*, 1960 48, 20–28.

[44] Ahcan, U; Arnez, ZM; Bajrovic, F; Janko, M. Contribution of collateral sprouting to the sensory and sudomotor recovery in the human palm after peripheral nerve injury. *Br. J. Plast. Surg.*, 1998 51, 436–443.

[45] Rovensky, J; Toman, M. A colorimetric contact method for measurement of insensible skin perspiration. *Hautarzt.*, 1962 13, 464–467. (in German)

[46] De Marinis, M; Stocchi, F; Testa, SR; De Pandis, F; Agnoli, A. Alterations of thermoregulation in Parkinson's disease. *Funct Neurol.*, 1991 6, 279–283.

[47] Chateao-Thierry, M. Fruit acids. – *Les Nouvelles Esthetiques, Russian edition*, 1998 1, 43–44. (in Russian)

[48] Kenny, GP; Periard, J; Journeay, WS; Sigal, RJ; Reardon, FD. Effect of exercise intensity on the postexercise sweating threshold. *J. Appl. Physiol.*, 2003 95, 2355–2360.

[49] Kellogg, DL Jr; Johnson, JM; Kenney, WL; Pergola, PE; Kosiba, WA. Mechanisms of control of skin blood flow during prolonged exercise in humans. *Am. J. Physiol. Heart Circul. Physiol.*, 1993 265, H562–H568.

[50] Kingston, JK; Geor, RJ; McCutcheon, LJ. Use of dew-point hygrometry, direct sweat collection, and measurement of body water losses to determine sweating rates in exercising horses. *Am. J. Vet. Res.*, 1997 58, 175–181.

[51] Ariagno, RL; Glotzbach, SF; Baldwin, RB; Rector, DM; Bowley, SM; Moffat, RJ. Dew-point hygrometry system for measurement of evaporative water loss in infants. *J. Appl. Physiol.*, 1997 82, 1008–1017.

[52] Robinson, MP; Clegg, J; Stone, DA. A novel method of studying total body water content using a resonant cavity: experiments and numerical simulation. *Phys. Med. Biol.*, 2003 48, 113–125.

[53] Schmitt, JM. Array detection for speckle reduction in optical coherence microscopy. *Phys. Med. Biol.*, 1997 42, 1427–1439.

[54] Verde, T; Shephard, RJ; Corey, P; Moore, R. Sweat composition in exercise and in heat. *J. Appl. Physiol.*, 1982 53, 1540–1545.

[55] Ohhashi, T; Sakaguchi, M; Tsuda, T. Human perspiration measurement. *Physiol. Meas.*, 1998 19, 449–461.

[56] Park, SJ; Tamura, T. Distribution of evaporation rate on human body surface. *Ann. Physiol. Anthropol.*, 1992 11, 593–609.

[57] Vainer, BG. Focal plane array based infrared thermography in fine physical experiment. *Journal of Physics D: Applied Physics*, 2008 41, 065102 (12 pp).

[58] Vainer, BG. Infrared thermography monitoring in cosmetology. *Innovations*, 2005 7, 119–122. (in Russian)

[59] Vainer, BG. Short-wave FPA-based infrared thermographs – the advanced means for medical diagnostics and monitoring. *Bol'nichnyi List*, 2002 9, 14–21. (in Russian)

[60] Whipp, BJ; Wasserman, K. Efficiency of muscular work. *J. Appl. Physiol.*, 1969 26, 644–648.

[61] Smolander, J; Kolari, P. Laser-Doppler and plethysmographic skin blood flow during exercise and during acute heat stress in the sauna. *Eur. J. Appl. Physiol. Occup. Physiol.*, 1985 54, 371–377.

[62] Torii, M; Yamasaki, M; Sasaki, T; Nakayama, H. Fall in skin temperature of exercising man. *Br. J. Sports Med.*, 1992 26, 29–32.

[63] Leppaluoto, J. Human thermoregulation in sauna. *Ann. Clin. Res.*, 1988 20, 240–243.

[64] Knochel, JP. Heat stroke and related heat stress disorders. *Disease of the Month*, 1989 35, 306–377.

In: Body Temperature Regulation
Editor: A. B. Cisneros and B. L. Goins

ISBN: 1-60741-282-3

Chapter VI

Clinical Thermometry

Manuel Varela*[*]*, Maria Jose Fernandez – Cotarelo
Department of Internal Medicine, Hospital de Mostoles
Madrid, Spain

1. Historical Aspects

The earliest references to fever appear in an Akkadian cuneiform inscription dating from the sixth century BC(Mackowiak, 2000). In these times, fever was placed in a supernatural context, and dealt with magic.

In the 5th -4th century BC Hippocrates stated his four humour's theory and explained fever as an excess of animal heat. This was slightly modified by Galen of Pergamon in 2nd AD, who explained fever as an excess of heat due to the presence of alien factors. Fever was therefore considered a systemic disease *per se*. Because of their huge influence, the galenic theories persisted through several centuries.

During the 17th and 18th century, the classical theories faded away, and fever became more the consequence of the disturbance of certain systems (i.e. the central nervous system, in Boerhaave).

Finally, in the 19th century C. Bernard proposed the idea of homeostasis and the presence of a self-regulating "milieu interne". Fever would be a –hopefully temporary- alteration in the temperature of this internal environment.

Nevertheless, since Galen and well until the 19th century, fever remained a systemic disease. Only with the development of anatomopathology and the idea of lesion as a cause of disease would fever slowly change its character, and become a symptom of an underlying process, rather than a disease proper. However, reminiscences of the old paradigm persist, and, as will be discussed later, both physicians and patients have a compulsion to treat fever even though frequently this may be useless or even harmful. Furthermore, the ¿syndrome? of

[*] Contact author: Manuel Varela, mjvarelae@gmail.com / mvarela.hmtl@salud.madrid.org

Fever of Unknown Origin (FUO) is widely accepted, and in about a third of these patients, no definitive cause of fever is ever found.

Although the idea of fever has accompanied medical practice for many centuries, the difficulty of precisely measuring the temperature has hampered its use until recent times. Initially, fever was explored by hand, and probably the idea of "heat" was related but no identical to temperature. But during the XIVth – XVIIth centuries a "measuring obsession" spread through Europe, becoming one of the most prominent features of scientific thought (at least in our western world point of view)(Crosby, 1977). Thermometry was no exception. The first clinical thermometer we know was described by Santorio (Sanctorio Sanctorius) in the 17th century, although probably copying an air thermometer developed by Galileo Galilei. Santorio produced several designs (Figure 1) but they were cumbersome and required long time for each reading, and they did not gain wide acceptance. In the 18th century Boerhaave, van Swieten, de Haen and Martine started to use the thermometer at the bedside. Diurnal variations were described, as were the relation between fever and increased heart rate or shivering. However, thermometry remained a bothersome operation, and was not widely used. In the 19th century, several improvements (notably the use of mercury instead of alcohol, and the introduction of a narrower tube to prevent the drop back of the column) made thermometers much smaller and practical. Finally, Carl Rheinhold August Wunderlichs's seminal book *Das Verhalten der Eigenwärme in Krankenheten* (Leipzig 1868) established thermometry as an essential tool at the bedside, fixing what thereon would be considered the normal range of temperature. In his book, he reports the results of several million temperature readings on 25.000 patients. Although some of his results may be disputable (there were no formal statistics at that time, and the precision and calibration of his thermometers may be questionable)(Mackowiak et al., 1992a, Mackowiak and Worden, 1994), Wunderlich marks the starting point of temperature measuring as we still practice it.

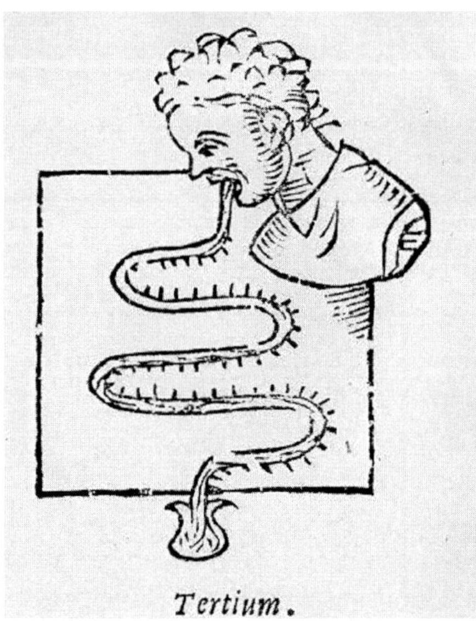

Figure 1. Santorio's thermometer (Clendening History of Medicine Library and Museum)

2. Measuring Body Temperature

Thermometers

Most of the thermometers used in clinic fall in the following groups:

- *Liquid* (usually mercury) *in glass*. This is the most classical thermometer, and the one Wunderlich used to establish the bases of modern thermometry. It is widely available, cheap, accurate and easy to use. Nevertheless, several problems have limited its use. Taking a temperature may be time consuming: it has been stated that the device should remain *in situ* for at least 3' for oral and rectal temperature, and not less than 5' for axillary measurement(Loudon, 1957). Some reports state that in elderly people, even after 10' in the axilla, the thermometer may have not reached a stable value. Furthermore, thermometers loose their precision with time(Abbey, 1978). Finally, certain issues on environmental risks with mercury have been raised(Canada, 2007) and the European Union is considering a ban on mercury thermometers(Commission, 2006)
- *Thermistor based devices*. The sensor tip has a semiconductor with an electrical conductance which varies depending on its temperature, thus allowing for its use as a thermometer. It is small, easy to use and has a quick equilibration time (aprox. 1'). Some models even have an anticipatory circuit based on the warm up rate, displaying the final equilibrium temperature even before it is reached. The drawbacks are its dependence on calibration and the possibility of dysfunction or calibration drift.
- *Infrared radiometers*. They are based on the "black box" principle, which states that for each wavelength, the maximum amount of radiation depends only on the temperature of the emitter and is a precise known value. Thus, the temperature of a surface may be calculated from the knowledge of its emissivity geometry and spectrum of radiation transmitted. Infrared radiometers operate in the range of 0.7 - 20 μm. Usually the tympanic membrane is used as the measuring target, although some thermometers may also assess skin temperature. Some models display the real (measured) temperature, while others make an estimated adjustment to the core temperature based on previous studies. These instruments are very comfortable, quick and precise, but have several limitations: it is quite operator-dependent, the results may be severely hampered by a bad technique (i.e. incorrectly facing the tympanic membrane) and agreement between tympanic and other central temperatures has been polemic (as discussed later)

Locations

Conventionally, body temperature is considered a composite measure of core and shell temperature. Although it may occasionally be helpful to measure shell (skin) temperature as an indirect assessment of peripheral haemodynamic status, it is usually the core temperature that is clinically relevant.

Core temperature can be measured in several sites, with differences in precision, invasiveness and convenience.

Wunderlich's classic studies employed axillary temperature, and this convenient location is still frequently used by many physicians. However, it demands for long equilibration times and may be affected by the haemodynamic status of the patient. From all the measuring locations it is probably the least "core-representing" one.

Oral temperature has been a standard of care in many settings. It is accessible, quick, precise and cheap. Its main drawbacks are that it can not be measured in several situations (very young children, agitated patients, patients at risk of convulsions). Furthermore, local factors (i.e. smoking, drinking, breathing through the mouth) may interfere with temperature readings.

Rectal temperature has also been widely used, and in certain settings it is considered standard of care. Usually rectal readings are higher than those from other sites, and it has been suggested that they may be influenced by metabolic activity of faecal bacteria or by nearby inflammatory processes. Furthermore, stool in the rectum may act as a heat sink, delaying or mitigating temperature changes.

A rather new but rapidly growing location to measure core temperature is the tympanic membrane, by means of an infrared radiometer. This is an extremely convenient and quick method, and one that is supposed to measure the temperature in a blood circuit in intimate connection with the thermoregulatory hypothalamic centres(Erickson and Kirklin, 1993, Edge and Morgan, 1993). Nevertheless, some authors have raised concern on its applicability in certain clinical settings(Rogers et al., 2007, Manian and Griesenauer, 1998, O'Brien et al., 2000, Hooker and Houston, 1996, Stavem et al., 1997, Betta et al., 1997).

Other more invasive locations are used only for monitorizing purposes. These include oesophagus (although this location may be interfered by food, if the patient is on an oral diet), bladder and pulmonary artery, which may be considered the gold standard.

There is a wide literature concerning the normal values and the relation between these different sites(Dinarello and Wolff, 1978, Sund-Levander et al., 2002, Rabinowitz et al., 1996, Varney et al., 2002, Smith, 2004). Table 1 displays the values suggested as normal in a systematic literature review involving 2614 subjects (Sund-Levander et al., 2002) .

Table 1. Normal temperatures (ºC)

	Men		Women	
	Mean	Range	Mean	Range
Oral	36.7	35.7-37.7	36.2	33.2-38.1
Rectal	37.0	36.7-37.5	37.0	36.8-37.1
Tympanic	36.5	35.5-37.5	36.6	35.7-37.5

3. Body Temperature Variability

Mainly based on Wunderlich's work, 37°C (98.6°F) degrees is considered the normal body temperature in healthy adults. There is a complex regulation of body temperature that involves a number of brain centers controlled in the preoptic area, in order to maintain it within the normal range under diverse external conditions. Body heat is mainly produced in abdominal and thoracic organs and in brain, and is transferred to distal parts of the body were vasodilatation and vasoconstriction mechanisms regulate heat loss. (Mackowiak, Wasserman et al. 1992; Mackowiak and Worden 1994; Krauchi 2002; Krauchi 2007).

On the other hand, body temperature varies due to many physiological conditions, which are reviewed below.

Circadian Variations

By the nineteenth century it was described that body temperature follows a 24-hour pattern. Nowadays it is known that body temperature has an endogenously generated circadian rhythm with a cycle that can be slightly modified by changes in external conditions such as light and dark schedule. As other circadian rhythms, body temperature can be described by its mesor (arithmetic mean), amplitude (half of the difference between the highest and the lowest value) and acrophase (time to reach the highest value from a reference point). Normal body temperature can fluctuate approximately 1°C (1.8°F) degree within the day, with the lowest value usually reached between 3:00 and 6:00 a.m., and the highest value between 4:00 and 9:00 p.m. in healthy individuals under stable conditions of lightning and social patterns. In free-running state (isolated from external conditions), it has a period of about 25 hours.

Body temperature is regulated by a feedback system, activating heat production mechanisms when temperature falls, and heat dissipation mechanisms when temperature rises. Under resting conditions, heat production mechanisms are activated to maintain body temperature. This occurs mainly in the morning, while heat dissipation mechanisms are activated in the evening. The pineal hormone melatonin plays an important role here, initiating distal vasodilatation. There also exists an influence of environmental factors; e.g. the exposure to bright light in the evening has shown to delay the circadian rhythm of body temperature.

Changes in the circadian rhythm of body temperature with a decrease in the amplitude have been described related to several diseases as well as with aging, thus suggesting that health means to some extent the maintaining of a stable and synchronized rhythm. Finally, many endogenous circadian rhythms (temperature, sleep-wake, activity...) interact, and it is not clear yet whether there is only one pacemaker to control all of them (Davy 1845; Mackowiak, Wasserman et al. 1992; Refinetti and Menaker 1992; Mackowiak and Worden 1994; Kelly 2006; Kelly 2007; Krauchi 2007; Weinert and Waterhouse 2007; Atkinson, Barr et al. 2008).

Menstrual Variations

In addition to the circadian rhythm of body temperature across the day, body temperature is known to vary across the menstrual cycle. In a healthy cycling woman, an elevation of body temperature around 0.5 °C degrees can be observed around ovulation. This change affects waking temperature and the mean day temperature, including a reduction in its nocturnal decline. However, the moment at which the highest temperature is reached can vary among individuals, and even among cycles in the same woman. This temperature increase in the luteal phase is mediated by progesterone, while estradiol and estrogens lower body temperature. The elevation of the progesterone/estradiol ratio has been associated to a higher mesor, a lower amplitude and a delayed acrophase of the temperature rhythm. Melatonin plays also a role in these changes, by its hypothermic effect. Increased activities of IL-1, IL-6 and TNF have been described after ovulation, but the inhibition of the prostaglandin synthesis has not proven to avoid the temperature increase in women in the luteal phase. The mechanism of temperature elevation in the luteal phase is therefore yet not clear, and prostaglandins don't seem to play a role in it. Elevated temperatures have been described in women with premenstrual syndrome, a number of physical and psychological symptoms that appear in the luteal phase (Refinetti and Menaker 1992; Cagnacci, Soldani et al. 1996; Coyne, Kesick et al. 2000; Cagnacci, Arangino et al. 2002; Kelly 2006; Baker and Driver 2007)

Body temperature is also influenced by exogenous factors. Oral contraceptives can increase body temperature approximately 0.6°C degrees, probably due to progestins. On the other hand, estrogen replacement therapy lowers mean body temperature (Baker and Driver 2007; Kelly 2007). The variability of body temperature across the menstrual cycle is one of the data classically used for family planning, although its real efficacy has not been proven in clinical trials to date (Grimes, Gallo et al. 2005).

Seasonal Variations

Some studies have reported changes in oral and rectal temperature along the year, with a decrease in the average temperature in winter. A delay in the time to reach the highest temperature of the day has also been described in spring and summertime. Other authors, however, found a circannual rhythm of body temperature in elderly individuals with the highest values in winter. It is difficult to draw conclusions from these contradictory results since the studies have been carried out in different populations and weather conditions (Kelly 2006). More recent researches have described seasonal changes in physiological variables in young healthy individuals. A higher mean tympanic temperature has been observed in winter, related to a higher body weight, body fat and an increased metabolic rate and adaptative preservation of heat in winter (Makinen, Paakkonen et al. 2004; van Ooijen, van Marken Lichtenbelt et al. 2004; Umemiya 2006).

Variations due to Aging

The circadian rhythm of body temperature is not fully developed in neonates. Mechanisms of thermoregulation are less effective, and the rhythm has a lower amplitude and a greater variability than in adults. Circadian rhythms help individuals maintain adaptability to the environment and performance, and in older ages these capacities tend to deteriorate. Mean body temperature decreases with aging, by means of lower physical activity and lower efficiency of heat-preserving mechanisms. With respect to the circadian rhythm, it seems to be phase-advanced in elderly people, but this may be influenced by different schedules and routine compared to young adults. It has been described a lower body temperature mesor and amplitude in the elderly (Kelly 2006; Weinert and Waterhouse 2007).

Gender Aspects

It is difficult to assess the normal body temperature in men and women, since the studies that deal with it show different results. Furthermore, many of these do not consider the influence of the menstrual cycle in women, and the different locations used for temperature measurement make them difficult to compare. There have been described higher tympanic temperatures in women, no significant differences in rectal temperature between sexes, and lower and higher oral temperatures in women compared with men. With regard to sleep influence, women seem to reach their nocturnal minimum temperature values earlier than men, while there is a deeper drop in nocturnal body temperature in men than in women (Baker, Waner et al. 2001; Sund-Levander, Forsberg et al. 2002).

Women and men show differences in thermoregulation due to many factors: different body composition, effect of sex hormones, and tolerance to exercise. Body composition differences include a smaller total mass, a higher body surface/mass ratio and a larger fat content in women. They have a higher sweating threshold than men, but it does not seem to affect significantly central or peripheral temperature when exposed to a hot ambient. The effect of sex hormones has been reviewed above. With respect to exercise, similar rectal temperature responses have been described in both sexes (Kaciuba-Uscilko and Grucza 2001; Deschenes, Hillard et al. 2006; Hazelhurst and Claassen 2006).

Variations due to the Sleep-wake Cycle

Temperature and sleep are closely interrelated. The activation of heat loss mechanisms in the evening is associated to the fall in physical activity and lightning. Sleep makes body temperature fall also by means of a decrease in the sympathetic tone and the metabolic rate, as well as lying down facilitates vasodilatation thus increasing skin temperature and heat loss. The nocturnal reduction in body temperature provides as a consequence some metabolic energy saving. However, we now know that sleep has little further influence in thermoregulation.

Similarly, the circadian changes in body temperature are involved in the regulation of the sleep-wake cycle, and skin temperature seems to play a more important role here than core temperature. Behavioural factors (bedding and clothing) and the increased skin blood flow interact to create a microclimate during sleep. It has been postulated that this favourable thermal condition is by some way reported to the sleep control centres to induce or maintain sleep. Sleep follows the heat loss onset facilitated by this skin blood flow increase, revealing the causal effect of temperature changes in sleep. The nocturnal melatonin secretion plays a role in both skin vasodilatation (and the resulting decrease in core body temperature) and the induction of sleepiness. Experimental cutaneous warming has also demonstrated to induce sleep. It is also known that the need to activate thermoregulatory mechanisms because of a high or low external temperature disrupts sleep. (Krauchi, Cajochen et al. 1997; Raymann, Swaab et al. 2005; Van Someren 2006; Baker and Driver 2007; Krauchi 2007; Weinert and Waterhouse 2007).

Changes in waking time lead to different body temperature recordings. Individuals usually alter their sleep-wake schedule from workdays to weekend, and this has shown to influence body temperature: body temperature measurements will show higher values the later an individual wakes-up, compared to his or her usual wake-up time. Based on the preferred waking time and activity during the daytime, three chronotypes can be defined: morning, evening and intermediate types. Differences in body temperature between these types have been described, with earlier minimum temperatures and a higher daily amplitude in the morning type (Waterhouse, Folkard et al. 2001; Baker and Driver 2007).

A higher decline in body temperature is associated to a better sleep quality. On the other hand, sleep deprivation has shown to influence body temperature by increasing the minimum values, thus increasing the mesor and decreasing the amplitude. The same effect has been described in travellers when arriving to a different time zone. Finally, shift workers undergo similar modifications in the circadian rhythm of body temperature, and this seems to be associated to a worse tolerance to schedule changes (Kelly 2007).

Variations due to Physical Activity

To dissipate the heat generated during exercise, mechanisms such as peripheral vasodilatation and sweating are activated, in order to avoid an extreme rise in body temperature. These responses are associated to cardiovascular changes, and are influenced by the phase of the temperature rhythm along the daytime. There is evidence that thermoregulatory mechanisms (cutaneous vasodilatation and sweating) act in a different way during the phase of rising temperature, so that physical activity produces a higher rise in temperature when it is performed in the "heat preserving" hours. Sports medicine explores strategies that can be used to reduce the risks associated to accumulation of heat in athletes (Kenny, Jay et al. 2007; Weinert and Waterhouse 2007; Wendt, van Loon et al. 2007). Some studies have shown that the circadian rhythm of body temperature is not a consequence of the circadian rhythm of activity (with the resulting heat gain generated), but the two rhythms are independent although closely related (Refinetti and Menaker 1992).

Some differences in body temperature have been found between physically active and inactive individuals. Active people tend to have a higher amplitude of body temperature during the day. Physical activity also helps to maintain the circadian rhythm of body temperature under changes in external conditions (Kelly 2006; Cain, Rimmer et al. 2007).

4. Fever

Fever is usually defined as a state of elevated core temperature which is often, but not necessarily, part of the defensive responses of organisms (hosts) to the invasion by live (microorganisms) or inanimate matter recognized as pathogenic or alien by the host(Commission and Sciences), 2001).

Physiopathology(Romanovsky et al., 2005, Moltz, 1993, Mackiowak, 1997b)

The most typical cause of fever is infection, with infecting agents producing several bioactive substances (i.e. bacterial lipopolysacharide, "exogenous pyrogen"). In response to these agents, phagocytic leukocytes release certain cytokines (mainly interleukin 1(IL-1), IL-6 and alpha-tumor necrosis factor (α-TNF)) that act as early mediators in the febrile response. Cytokines arrive to the preoptic anterior hypothalamus (POAH) (Figure 2) through four mechanisms:

- a transendothelial, saturable, transport system
- through the fenestrated capillaries of the *organum vasculosum* of the *lamina terminalis* (OVLT), a specialized structure lying in the midline of the medial POAH, in the anteroventral wall of the third ventricle
- the blood-brain barrier (BBB), which was previously considered a wall separating these structures but may rather be a transducer, able to detect and react to this family of cytokines.
- although of uncertain importance, endotoxin also may act on POAH through peripheral stimulation on vagal terminations

The next step on fever production is the synthesis of prostaglandin E_2 (PGE_2). This happens mainly *in situ* in the hypothalamic nuclei, although the influence of peripheral PGE_2 has not been ruled out.

PGE_2 plays a critical role in fever, and is central to our therapeutic armamentarium. It is synthesized through a cascade involving three enzymatic systems: phospholipase A_2, cyclooxygenase and terminal PGE synthase. Although there is a low-grade continuous PGE_2 synthesis, fever is produced through a brisk increase in PGE_2 production following cytokine release. The traditional theory stated that these increased PGE_2 levels would change the thermostat set-point and unleash heat conserving mechanisms (increased muscular tone, peripheral vasoconstriction, shivering). Once the set point temperature is achieved, heat conserving mechanisms would stop, giving way to heat diffusing processes (vasodilation, sweating).

Recently this "central thermostat" point of view has been challenged(Romanovsky, 2007), and the thermoregulatory system is increasingly described as a network of afferent and efferent circuits, responding to different ranges of temperature and generating different (and to a certain point, independent) physiologic reactions. This would help explaining the conspicuous finding of hypothermia in experimental work with endotoxin administration, not easily justified with the classical set point theory. This hypothermic phase is most dependent on external conditions, and thus should probably better be considered as sepsis related poikilothermia. With the new network approach, this phenomenon can be readily justified: considering separate (heat-preserving and heat-diffusing) circuits allows for a differential action on them, and thus permitting the appearance of a poikilothermic phase.

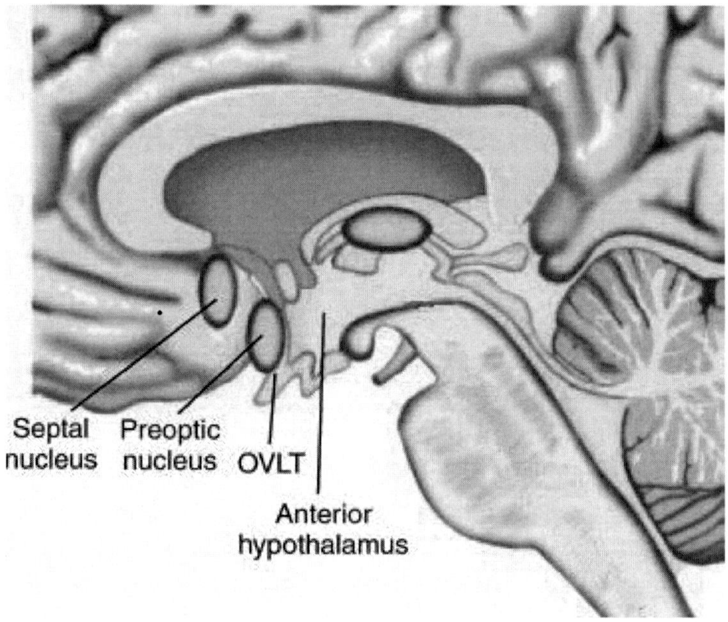

Figure 2. Thermoregulatory centres (from Mandell, Bennet and Dolin: Principles and Practice of Infectious Diseases, modified).

5. Clinical Significance of Fever

It is known that an increase in body temperature has the beneficial effect of helping the organism to fight against the noxious agent. Behavioural factors collaborate with the autonomous generation of fever, and the preference for warm ambient by the individuals in the rising phase of fever has been described even in ectothermic animals (Refinetti and Menaker, 1992).

It was thought that the magnitude of fever correlates with the severity of the underlying disease. Representing this thought, Wunderlich wrote: "fever can give more certainly than anything else information as to the grade of the disease. If the departure of the temperature from the normal range is wide, the case is severe, and viceversa" (Mackowiak and Worden, 1994). Now we know that this is not true, and furthermore, we have to keep in mind that hypothermia is one of the manifestations of the development of a sepsis. It is also known that old people don't develop fever as easily as young people, so we will often find elderly patients suffering from a severe infection, but without a high fever. Newborns, patients taking drugs such as corticosteroids and patients with chronic renal failure are other cases in which fever may not be present. However, sometimes certain characteristics of fever can help clinicians to identify the underlying disease, such as the febrile pattern of malaria and, to a lesser extent, Hodgkin lymphoma. In the case of typhoid fever, brucellosis and leptospirosis, there typically exists a relative bradycardia. Patients with a factitious fever will also lack other manifestations of the febrile state (Porat, 2006).

Causes

Most fevers are due to mild, self-limited infections, usually of viral causes. However, there is a never-ending list of processes that can cause fever. The causes of fever can be classified into various groups, beginning with the *infectious diseases* -bacterial, mycobacterial, fungal, viral and parasitic-, whose review is beyond the scope of this chapter. The second broad group are the non-infectious *inflammatory diseases*, also referred to as systemic, autoimmune or rheumatic diseases and collagen vascular disorders. It includes connective tissue diseases (systemic lupus erythematosus, rheumatoid arthritis,...), vasculitis syndromes and granulomatous disorders. Except for Still's disease, the fever is usually moderate, which may help in distinguishing them from infectious etiologies. However, fever in these cases can be due either to the disease itself or to infections or other complications that frequently develop in these patients. *Neoplasms* are other potential cause of fever, mediated by endogenous pyrogens release from tumoral cells. Hematological cancers and renal cell carcinoma often produce fever; osteosarcoma, Ewing´s sarcoma, atrial myxoma, feocromocitoma, hypothalamic tumors and liver metastases are other possible causes. These patients tend to develop mild fever, and commonly respond to NSAIDs.

Drug fever is an exclusion diagnosis that requires a high index of suspicion. Many drugs and mechanisms can be implicated. The most frequent cause is an immune-mediated or hypersensitivity response, that may produce also other symptoms like rash or urticaria. Exogenous thyroid hormone, sympathomimetic or anticholinergic agents can produce fever by altering thermoregulatory mechanisms. Antimicrobial and chemotherapy agents can cause fever as a part of their pharmacologic effect, because of the release of pyrogens from the dying cells and microorganisms. Malignant hyperthermia and neuroleptic malignant syndrome are examples of idiosyncratic reactions to some drugs. Finally, fever can appear during a drug administration because of contaminated solutions, intrinsic pyrogen properties, or inflammatory response at the site of administration (Johnson and Cunha, 1996).

Factitious fever should be considered in unexplained fever, specially in absence of other associated symptoms and signs. This psychiatric disorder is more common in women and health professionals.

Fever of unknown origin (FUO) is defined by a temperature higher than 38.3°C (101°F) without a diagnosis within 3 weeks, including three days of hospital investigation and microbiological cultures. Infections represent almost half of the cases of the "classical" FUO, where tuberculosis, endocarditis, abscesses, osteomyelitis and urinary tract infections have to be considered. Connective tissue diseases include Still's disease, systemic lupus erythematous and temporal arteritis. The relative frequency of neoplasms as a cause of FUO has decreased in recent years. Finally, there is a miscellanea of potential causes of FUO, and a number of cases that remain undiagnosed. More recent series have found a higher percentage of cases due to inflammatory diseases and undiagnosed cases, and a decrease in the number of infectious and tumoral diseases responsible for a prolonged fever. The future definition of FUO will probably consider the performance of certain diagnostic tests instead of a time-related criterion trying to identify the source of fever.

Besides the "classical FUO", there exist three other categories of FUO: nosocomial, neutropenic, and FUO in HIV-infected patients. Causes of *nosocomial FUO* include nosocomial infections (catheter related infections, *Clostridium difficile*-associated diarrhea...), pulmonary embolism, drug fever, and transfusion reactions. *Neutropenic FUO* is an increasing and potentially life-threatening condition commonly due to bacterial and fungal infections. These patients usually don't develop other signs and symptoms of infection. In *HIV-infected FUO* many infectious causes should be considered, including mycobacterial, leishmaniasis, Non-infectious diseases such as lymphoma account for a small number of cases of FUO in these patients (Durack, 1998, Vanderschueren et al., 2003, Zenone, 2007, Bleeker-Rovers et al., 2007, Cunha, 2007a, Dinarello, 2008, Mourad et al., 2003).

Ventilator-associated pneumonia, intravascular catheter-related infections and urinary tract infections are the most frequent infectious causes of fever in patients at the intensive care unit. Surgical site infection should be considered in post-operative patients. Non-infectious causes of fever include thromboembolic disease, transfusion reactions and drug fever (MacLaren et al., 2007).

There is finally a large list of miscellaneous causes of fever, including hereditary syndromes, inflammatory processes and disorders of poorly understood pathophysiology, most of which are very uncommon.

Fever in children is commonly due to viral infections, many of them affecting the ENT area and upper airway. With respect to bacteriemic infections, the spread use of pneumococcal vaccine has changed the spectrum of etiologies, making urinary tract infections the first cause in children up to 3 years. The most common causes of FUO in children are infectious diseases, followed by autoimmune disorders and malignancy. Kawasaki disease is also worth considering in these cases. As in adults, a percentage of cases of FUO in children remain undiagnosed (Antonyrajah and Mukundan, 2008).

Clinical Workup

In the initial approach to a febrile patient it is important to plan the diagnostic strategy according to the patient's characteristics. The clinical investigation and even the timing of the diagnostic procedures will be different in adults and children, immunocompetent and immunocompromised hosts, stable and unstable patients, outpatients and hospitalized patients. The first step in the evaluation of a febrile patient includes a complete anamnesis and physical examination. General aspects and travel history, risk contacts, drug administration, surgery or invasive procedures should be investigated. It is important to know the chronology of fever, and associated symptoms that lead to an specific organ or system for further investigation.

Physical examination must include evaluation of skin, lymph nodes, cardiovascular system, thorax, abdomen and pelvis (with genital and rectal examination), musculoskeletal and nervous system. Attention must be paid to some physical clues such as lymphadenopathy, hepatomegaly, splenomegaly, skin lesions, eye involvement, joint pain, heart murmurs, and local tenderness (Tolia and Smith, 2007).

Laboratory tests include a complete blood cell count, in which neutrophilia and neutrophilic band forms suggest a bacterial infection. Leucopenia with lymphocytosis or monocytosis can be seen in viral, mycobacterial and atypical germ infections; leucopenia can be also present in autoimmune disorders and severe sepsis. Eosinophilia suggests parasitic infection, but can also be found in vasculitis and certain tumours. Laboratory biochemistry blood analysis including liver function test and urinanalysis help to evaluate general and organ involvement. Ferritin, protein electrophoresis, antinuclear antibodies (ANA) and rheumatoid factor should also be performed in the diagnostic evaluation. C-reactive protein and more specifically procalcitonin are serum markers that can help distinguish bacterial infections from other infectious and non-infectious causes of fever. Erythrocyte sedimentation rate and other acute phase reactants will usually be elevated.

Blood and urine cultures, chest radiography and abdomen-pelvis ultrasonography are useful diagnostic tools in the initial approach to the febrile patient. Serologic tests have a limited value in patients without data pointing to a specific disease, although it is advisable to test for HIV antibodies. Imaging techniques and other diagnostic procedures will be performed based on the clinical suspicion. Bronchoscopy, gastroscopy, colonoscopy, leg doppler imaging, FDG-PET and echocardiography should be considered in the diagnostic strategy of a prolonged fever. Histologic studies (liver biopsy and temporal artery biopsy) play also a role in the investigation of certain cases of FUO (Dinarello, 2008, Antonyrajah and Mukundan, 2008, Zenone, 2007, Mourad et al., 2003, Cunha, 2007b).

6. *To Treat or not to Treat...*: The therapeutic Dilemma

Fever is a widespread mechanism of defence to infection, present in Chordata, Arthropoda and Annelida. This is an expensive tool, in terms of metabolic requirements, and it must have some definite advantages to explain its wide diffusion.

There are many studies proving the adaptive value of fever(Kluger et al., 1997). In a classic experiment, dessert iguanas (*Dipsosaurus dorsalis*) were kept in boxes with a gradient of temperature, ranging from 30° to 50°C. In this setting, body temperature ranged from 37.5°C to 39°C, with an average of 38°C. When infected with *Aeromonas hydrophila*, the body temperature raised to 41°C as a result of the lizard moving to the hot area of the box.

Furthermore, if the infected animal was maintained at an ambient temperature of 38ºC, the mortality was 50%. Allowing the animals to migrate to an environment that permitted an increase of their temperature to 40ºC resulted in a mortality of 15%. If the temperature was allowed to rise to 41ºC, all animals survived. Similar results have been obtained with several other species.

In clinical grounds, there are also data suggesting a protective role of fever. There is a correlation between "moderate" fever and survival in patients with bacterial peritonitis(Weinstein et al., 1978, Hoefs et al., 1980) and bacteraemia(Bryant et al., 1971, Mackowiak et al., 1980)

Fever may be protective through several ways:

- increasing neutrophil migration
- stimulating the production of antibacterial substances by neutrophils (i.e. superoxide anion)
- raising the synthesis of interferon
- increasing the antiviral and antitumor activity of interferon
- promoting T-cell proliferation
- decreasing microorganism growth in iron-poor environment.

On the other hand, fever is traditionally viewed not just as a symptom of disease, but as a therapeutic target worth treating. The relation between fever and malnutrition in admitted patients is well established, fever may produce convulsions in infants, and several studies show a correlation between high temperature and mortality. Furthermore, some of the agents producing the febrile response (IL-1, IL-6, TNF) are known to have deleterious effects by their own, can increase mortality when administered with LPS (bacterial lipopolysccharides)(Johnson et al., 1991, Heinzel, 1990), and blocking their effects may improve outcome(Abraham et al., 1995, Fisher et al., 1994). Fever is common in several neurological disorders, and has been associated with unfavourable outcomes (Cremer and Kalman, 2007, Kaamersgaard et al., 2002, Stochetti et al., 2002). Fever also increases oxygen consumption and therefore may hinder cardiopulmonary function. Often, the development of fever prevents the weaning from a respirator or triggers the destabilization of a septic patient.

Finally, one should keep in mind that fever is uncomfortable, and frequently patients feel better after antipyretics

It has been suggested(Mackiowak, 1997a) that this apparent contradiction between favourable and unfavourable effects of fever may be due to the point of view of the observer. Rather than favouring the individual, fever may act as a "time accelerator", hastening the course of the disease and thus protecting the group. If the individual is a healthy member of the community, it will rush his cure. If he is old or debilitated (and thus poorly fit to survive) it will shorten the time to death, optimizing the group's resources and minimizing the risks of spreading the disease.

So, when to treat? Although no clear guidelines can be offered, it should be kept in mind that fever is a time honoured defence mechanism, and it should not be systematically fought without considering its context.

In a frail, debilitated or malnourished patient, or a patient haemodynamically unstable(Aronoff and Neilson, 2001, Diringer et al., 2004), fever will probably increase the burden of disease and diminish his chances of survival. But in a young, stable patient running an infectious fever, maybe treating the disease, instead of focusing on the temperature, can be a better option.

7. Limitations and New Approaches

Limitations

Fever is a central concept in our clinical practice and will probably remain so for some time. Nevertheless, it has several important limitations:

- Although seemingly a quantitative value, fever is uniformly used as a qualitative, dichotomic variable (febrile / afebrile), with a rather arbitrary "red line". This is obviously an oversimplification. The separating line –if ever it could be drawn- could not be the same for both genders, at any moment of the daytime, the menstrual cycle or at any age.
- Although admittedly a flowing variable, temperature it is uniformly used as a "snapshot" measure, supposedly representing the whole time series. It is debatable to what extent do this punctual readings represent the whole time series.
- We seem to believe there is a unique core temperature. Taking a temperature is just trying to get as near as possible to this "real" temperature. Thermoregulatory centres would measure this "real" core temperature, check it against a set point and consequently start the appropriate heat-preserving or heat-dissipating mechanisms. Recent findings do not support these ideas(Kobayashi et al., 2006, Romanovsky, 2007).
- Apart from certain special circumstances (i.e. strenuous exercise, certain drugs), fever is a rather specific symptom (i.e. a patient running a fever is probably ill), but it has a discouragingly low sensitivity: not running a fever does certainly not mean that a patient is healthy. Furthermore, once proven that the patient is febrile, classic thermometry offers very little, if any, information on the origin of the fever(Musher et al., 1979).
- Although admittedly a symptom, physicians and nurses (and patients and relatives!) have a tendency to focus fever as a therapeutic target by itself. It is as if one would treat pallor on an anaemic patient. This is probably a relic of the old paradigm that conceived fever as a disease *per se*.

New Approaches

Recently, other approaches are emerging.

Certain methods derived from nonlinear dynamics and chaos theory are useful in analyzing time-series complexity. Given the fact that often the first sign of a failing complex system is the "decomplexification" of its output(Goldberger et al., 2002), this technique may be useful to explore the functionality of the thermoregulatory system. Specifically, this approach may overcome the traditional febrile/afebrile dichotomy, allowing for a truly quantitative analysis and giving an image of a whole time period, and not just a snapshot. Furthermore, this technique would give information on the physiological status of patients not running a fever, from whom classical thermometry could say nothing. This approach has been used in critical care patients and has shown to correlate with the patient's physiological status(Varela et al., 2003) and final outcome(Varela et al., 2006).

Furthermore, as an independent variable, complexity may differ between diverse disorders, and thus be useful in the differential diagnosis of febrile patients.

Monitoring simultaneously central and peripheral temperature may provide a real-time picture of the haemodynamic status of unstable patients(Lima and Bakker, 2005) and may display "pre-febrile" patterns allowing for earlier (and arguably more efficacious) blood cultures and therapeutic interventions.

Body temperature follows a well established circadian rhythm, and is often used as it's most conspicuous and easily monitored output. Therefore, it has been used to analyse circadian rhythm disorders, both due to disease or external disruption (jet lag, shift work)(Kelly, 2007a, Weinert and Waterhouse, 2007b, Volicer et al., 2001).

In conclusion, temperature regulation is a crucial part of our homeostasis. Changes in temperature regulation has been used for centuries as one of the most important signs of disease. Although thermometry will probably remain an invaluable tool in clinical practice, it has severe limitations as now applied. Hopefully, new approaches may boost its use and allow its application in new fields.

References

Abbey, J. (1978)·How long is that thermometer accurate? *Am J Nurs,* 78, 1375-6.

Abraham, E., Wunderink, R., Silverman, H., Perl, T. M., Nasraway, S., Levy, H., Bone, R., Wenzel, R. P., Balk, R., Allred, R. and et al. (1995)·Efficacy and safety of monoclonal antibody to human tumor necrosis factor alpha in patients with sepsis syndrome. A randomized, controlled, double-blind, multicenter clinical trial. TNF-alpha MAb Sepsis Study Group *Jama,* 273, 934-41.

Antonyrajah, B. and Mukundan, D. (2008)·Fever without apparent source on clinical examination *Curr Opin Pediatr,* 20, 96-102.

Aronoff, D. M. and Neilson, E. G. (2001)·Antipyretics: mechanisms of action and clinical use in fever suppression *Am J Med,* 111, 304-15.

Atkinson, G., Barr, D., Chester, N., Drust, B., Gregson, W., Reilly, T. and Waterhouse, J. (2008)·Bright light and thermoregulatory responses to exercise *Int J Sports Med,* 29, 188-93.

Baker, F. C. and Driver, H. S. (2007)·Circadian rhythms, sleep, and the menstrual cycle *Sleep Med,* 8, 613-22.

Baker, F. C., Waner, J. I., Vieira, E. F., Taylor, S. R., Driver, H. S. and Mitchell, D. (2001)·Sleep and 24 hour body temperatures: a comparison in young men, naturally cycling women and women taking hormonal contraceptives *J Physiol,* 530, 565-74.

Betta, V., Cascetta, F. and Sepe, D. (1997)·An assessment of infrared tympanic thermometers for body temperature measurement *Physiol Meas,* 18, 215-25.

Bleeker-Rovers, C. P., Vos, F. J., de Kleijn, E. M., Mudde, A. H., Dofferhoff, T. S., Richter, C., Smilde, T. J., Krabbe, P. F., Oyen, W. J. and van der Meer, J. W. (2007)·A prospective multicenter study on fever of unknown origin: the yield of a structured diagnostic protocol *Medicine (Baltimore),* 86, 26-38.

Bryant, R. E., Hood, A. F., Hood, C. E. and Koenig, M. G. (1971)·Factors affecting mortality of gram-negative rod bacteremia *Arch Intern Med,* 127, 120-8.

Cagnacci, A., Arangino, S., Tuveri, F., Paoletti, A. M. and Volpe, A. (2002)·Regulation of the 24h body temperature rhythm of women in luteal phase: role of gonadal steroids and prostaglandins *Chronobiol Int,* 19, 721-30.

Cagnacci, A., Soldani, R., Laughlin, G. A. and Yen, S. S. (1996)·Modification of circadian body temperature rhythm during the luteal menstrual phase: role of melatonin *J Appl Physiol,* 80, 25-9.

Cain, S. W., Rimmer, D. W., Duffy, J. F. and Czeisler, C. A. (2007)·Exercise distributed across day and night does not alter circadian period in humans *J Biol Rhythms,* 22, 534-41.

Canada, E. (2007), Vol. 2008.

Commission, E. (2006), Vol. 2008, pp. Press Release.

Commission, I. T. and Sciences), T. C. f. T. P. o. t. I. U. o. P. (2001)·Glossary of terms for thermal physiology *Jpn J Physiol,* 51, 245-280.

Coyne, M. D., Kesick, C. M., Doherty, T. J., Kolka, M. A. and Stephenson, L. A. (2000)·Circadian rhythm changes in core temperature over the menstrual cycle: method for noninvasive monitoring *Am J Physiol Regul Integr Comp Physiol,* 279, R1316-20.

Cremer, O. and Kalman, C. (2007)·Cerebral pathophysiology and clinical neurology of hyperthermia in humans *Progress in Brain Research,* 162, 153-169.

Crosby, A. (1977) *The measure of reality. Quantification and western society, 1250-1600,* Cambridge University Press, New York.

Cunha, B. A. (2007a)·Fever of unknown origin: clinical overview of classic and current concepts *Infect Dis Clin North Am,* 21, 867-915, vii.

Cunha, B. A. (2007b)·Fever of unknown origin: focused diagnostic approach based on clinical clues from the history, physical examination, and laboratory tests *Infect Dis Clin North Am,* 21, 1137-87, xi.

Davy, J. (1845)·On the temperature of man *Phylos Trans R Soc Lond A,* 135, 319-333.

Dinarello, C., Porat R (2008) In *Harrison's Principles of Internal Medicine*(Ed, Anthony S. Fauci, E. B., Dennis L. Kasper, Stephen L. Hauser, Dan L. Longo, J. Larry Jameson, and Joseph Loscalzo, Eds.) McGraw-Hill.

Dinarello, C. A. and Wolff, S. M. (1978)·Pathogenesis of fever in man *N Engl J Med,* 298, 607-12.

Diringer, M. N., Reaven, N. L., Funk, S. E. and Uman, G. C. (2004)·Elevated body temperature independently contributes to increased length of stay in neurologic intensive care unit patients *Crit Care Med,* 32, 1489-95.

Durack, D. T. (1998) In *Fever. Basic mechanisms and management*(Ed, Mackowiak, P. A.) Lippincot-Raven, pp. 237-249.

Edge, G. and Morgan, M. (1993)·The genius infrared tympanic thermometer. An evaluation for clinical use *Anaesthesia,* 48, 604-7.

Erickson, R. S. and Kirklin, S. K. (1993)·Comparison of ear-based, bladder, oral, and axillary methods for core temperature measurement *Crit Care Med,* 21, 1528-34.

Fisher, C. J., Jr., Slotman, G. J., Opal, S. M., Pribble, J. P., Bone, R. C., Emmanuel, G., Ng, D., Bloedow, D. C. and Catalano, M. A. (1994)·Initial evaluation of human recombinant interleukin-1 receptor antagonist in the treatment of sepsis syndrome: a randomized, open-label, placebo-controlled multicenter trial *Crit Care Med,* 22, 12-21.

Gisolfi, C. and Mora, F. (2000) In *The hot brain: survival, temperature and the human body*Massachusetts Institute of Technology, pp. 96.

Goldberger, A. L., Amaral, L. A., Hausdorff, J. M., Ivanov, P., Peng, C. K. and Stanley, H. E. (2002)·Fractal dynamics in physiology: alterations with disease and aging *Proc Natl Acad Sci U S A,* 99 Suppl 1, 2466-72.

Grimes, D. A., Gallo, M. F., Grigorieva, V., Nanda, K. and Schulz, K. F. (2005)·Fertility awareness-based methods for contraception: systematic review of randomized controlled trials *Contraception,* 72, 85-90.

Heinzel, F. P. (1990)·The role of IFN-gamma in the pathology of experimental endotoxemia *J Immunol,* 145, 2920-4.

Hoefs, j., Sapico, F., Canawati, H. and Montgomerie, J. (1980)·The relationship of white bood cell (WBC) and pyrogenetic response to survival in spontaneous bacterial peritonitis *Gastroenterology,* 78, 1308.

Hooker, E. A. and Houston, H. (1996)·Screening for fever in an adult emergency department: oral vs tympanic thermometry *South Med J,* 89, 230-4.

Johnson, D. H. and Cunha, B. A. (1996)·Drug fever *Infect Dis Clin North Am,* 10, 85-91.

Johnson, J., Brigham, K. L., Jesmok, G. and Meyrick, B. (1991)·Morphologic changes in lungs of anesthetized sheep following intravenous infusion of recombinant tumor necrosis factor alpha *Am Rev Respir Dis,* 144, 179-86.

Kaamersgaard, L., Jorgensen, H. and Rungby, J. (2002)·Admission body temperature predicts long-term mortality after acute stroke: The Copenhaguen stroke study *Stroke,* 33, 1759-1762.

Kaciuba-Uscilko, H. and Grucza, R. (2001)·Gender differences in thermoregulation *Curr Opin Clin Nutr Metab Care,* 4, 533-6.

Kelly, G. (2006)·Body temperature variability (Part 1): a review of the history of body temperature and its variability due to site selection, biological rhythms, fitness, and aging *Altern Med Rev,* 11, 278-93.

Kelly, G. (2007a)·Body temperature variability (Part 2): masking influences of body temperature variability and a review of body temperature variability in disease *Altern Med Rev,* 12, 49-62.

Kelly, G. S. (2007b)·Body temperature variability (Part 2): masking influences of body temperature variability and a review of body temperature variability in disease *Altern Med Rev,* 12, 49-62.

Kenny, G. P., Jay, O. and Journeay, W. S. (2007)·Disturbance of thermal homeostasis following dynamic exercise *Appl Physiol Nutr Metab,* 32, 818-31.

Kluger, M., Kozak, W., CAConn, Leon, L. and Soszynski, D. (1997) In *Fever: Basic Mechanisms and Management*(Ed, Mackiowak, P.) Lippincott-Raven, Philadelphia, pp. 255-266.

Kobayashi, S., Okazawa, M., Hori, A., Matsumura, K. and Hosokawa, H. (2006)·Paradigm shift in sensory system - Animals do not have sensors *J Thermal Bio,* 31, 19-23.

Krauchi, K. (2002)·How is the circadian rhythm of core body temperature regulated? *Clin Auton Res,* 12, 147-9.

Lima, A. and Bakker, J. (2005)·Noninvasive monitoring of peripheral perfusion *Intensive Care Med,* 31, 1316-1326.

Loudon, I. S. (1957)·On taking the temperature in the mouth and the axilla *Lancet,* 273, 233-5.

Mackiowak, P. (1997a) In *Fever. Basic Mechanisms and Management*(Ed, Mackiowak, P.) Lippincott-Raven, Philadelphia, pp. 279-286.

Mackiowak, P. (1997b) *Fever. Basic Mechanisms and Management,* Lippincott-Raven, Philadelphia.

Mackowiak, P. (2000) In *Mandell, Douglas and Bennett's Principles and Practice of Infectious Diseases*, Vol. 1 (Ed, Mandell GL, B. J., Dolin,R) Churchill Livingston, Philadelphia, pp. 604 -622.

Mackowiak, P., Wasserman, S. and Levine, M. (1992a)·A Critical Appraisal of 98.6°F, the Upper Limit of the Normal Body Temperature, and Other Legacies of Carl Reinhold August Wunderlich *JAMA,* 268, 1578-1580.

Mackowiak, P. A., Browne, R. H., Southern, P. M., Jr. and Smith, J. W. (1980)·Polymicrobial sepsis: an analysis of 184 cases using log linear models *Am J Med Sci,* 280, 73-80.

Mackowiak, P. A., Wasserman, S. S. and Levine, M. M. (1992b)·A critical appraisal of 98.6 degrees F, the upper limit of the normal body temperature, and other legacies of Carl Reinhold August Wunderlich *Jama,* 268, 1578-80.

Mackowiak, P. A. and Worden, G. (1994)·Carl Reinhold August Wunderlich and the evolution of clinical thermometry *Clin Infect Dis,* 18, 458-67.

MacLaren, G., Manaker, S., Wilson, K. and Spelman, D. (2007) In *UpToDate 16.1.*

Makinen, T. M., Paakkonen, T., Palinkas, L. A., Rintamaki, H., Leppaluoto, J. and Hassi, J. (2004)·Seasonal changes in thermal responses of urban residents to cold exposure *Comp Biochem Physiol A Mol Integr Physiol,* 139, 229-38.

Manian, F. A. and Griesenauer, S. (1998)·Lack of agreement between tympanic and oral temperature measurements in adult hospitalized patients *Am J Infect Control,* 26, 428-30.

Moltz, H. (1993)·Fever: causes and consequences *Neurosci Biobehav Rev,* 17, 237-69.

Mourad, O., Palda, V. and Detsky, A. S. (2003)·A comprehensive evidence-based approach to fever of unknown origin *Arch Intern Med,* 163, 545-51.

Musher, D., Fainstein, V., Young, E. and Pruett, T. (1979)·Fever patterns. Their lack of significance *Arch Intern Med,* 139, 1225-1228.

O'Brien, D. L., Rogers, I. R., Holden, W., Jacobs, I., Mellett, S., Wall, E. J. and Davies, D. (2000)·The accuracy of oral predictive and infrared emission detection tympanic thermometers in an emergency department setting *Acad Emerg Med,* 7, 1061-4.

Porat, R., Weller, PF, Thorner AR, Dinarello CA (2006) In *UpToDate 16.1.*

Rabinowitz, R. P., Cookson, S. T., Wasserman, S. S. and Mackowiak, P. A. (1996)·Effects of anatomic site, oral stimulation, and body position on estimates of body temperature *Arch Intern Med,* 156, 777-80.

Refinetti, R. and Menaker, M. (1992)·The circadian rhythm of body temperature *Physiol Behav,* 51, 613-37.

Rogers, I. R., Brannigan, D., Montgomery, A., Khangure, N., Williams, A. and Jacobs, I. (2007)·Tympanic thermometry is unsuitable as a screening tool for hypothermia after open water swimming *Wilderness Environ Med,* 18, 218-21.

Romanovsky, A. A. (2007)·Thermoregulation: some concepts have changed. Functional architecture of the thermoregulatory system *Am J Physiol Regul Integr Comp Physiol,* 292, R37-46.

Romanovsky, A. A., Almeida, M. C., Aronoff, D. M., Ivanov, A. I., Konsman, J. P., Steiner, A. A. and Turek, V. F. (2005)·Fever and hypothermia in systemic inflammation: recent discoveries and revisions *Front Biosci,* 10, 2193-216.

Salkeld, B. D., MacAulay, J. C., Ball, R. W. and Cannon, J. G. (2001)·Modulation of body temperature, interleukin-6 and leptin by oral contraceptive use *Neuroimmunomodulation,* 9, 319-25.

Smith, L. S. (2004)·Temperature measurement in critical care adults: a comparison of thermometry and measurement routes *Biol Res Nurs,* 6, 117-25.

Stavem, K., Saxholm, H. and Smith-Erichsen, N. (1997)·Accuracy of infrared ear thermometry in adult patients *Intensive Care Med,* 23, 100-5.

Stochetti, N., Rossi, S. and Zanier, E. (2002)·Pyrexia in head-injured patients admitted to intensive care *Intensive Care Med,* 28.

Sund-Levander, M., Forsberg, C. and Wahren, L. K. (2002)·Normal oral, rectal, tympanic and axillary body temperature in adult men and women: a systematic literature review *Scand J Caring Sci,* 16, 122-8.

Tolia, J. and Smith, L. G. (2007)·Fever of unknown origin: historical and physical clues to making the diagnosis *Infect Dis Clin North Am,* 21, 917-36, viii.

Umemiya, N. (2006)·Seasonal variations of physiological characteristics and thermal sensation under identical thermal conditions *J Physiol Anthropol,* 25, 29-39.

van Ooijen, A. M., van Marken Lichtenbelt, W. D., van Steenhoven, A. A. and Westerterp, K. R. (2004)·Seasonal changes in metabolic and temperature responses to cold air in humans *Physiol Behav,* 82, 545-53.

Vanderschueren, S., Knockaert, D., Adriaenssens, T., Demey, W., Durnez, A., Blockmans, D. and Bobbaers, H. (2003)·From prolonged febrile illness to fever of unknown origin: the challenge continues *Arch Intern Med,* 163, 1033-41.

Varela, M., Churruca, J., Gonzalez, A., Martin, A., Ode, J. and Galdos, P. (2006)·Temperature curve complexity predicts survival in critically ill patients *Am J Respir Crit Care Med,* 174, 290-8.

Varela, M., Jimenez, L. and Farina, R. (2003)·Complexity analysis of the temperature curve: new information from body temperature *Eur J Appl Physiol,* 89, 230-7.

Varney, S. M., Manthey, D. E., Culpepper, V. E. and Creedon, J. F., Jr. (2002)·A comparison of oral, tympanic, and rectal temperature measurement in the elderly *J Emerg Med,* 22, 153-7.

Volicer, L., Harper, D., Manning, B., Goldstein, R. and Satlin, A. (2001)·Sundowning and circadian rhythms in Alzheimer's disease *Am J Psichiatry,* 158, 704-711.

Waterhouse, J., Folkard, S., Van Dongen, H., Minors, D., Owens, D., Kerkhof, G., Weinert, D., Nevill, A., Macdonald, I., Sytnik, N. and Tucker, P. (2001)·Temperature profiles, and the effect of sleep on them, in relation to morningness-eveningness in healthy female subjects *Chronobiol Int,* 18, 227-47.

Weinert, D. and Waterhouse, J. (2007a)·The circadian rhythm of core temperature: effects of physical activity and aging *Physiol Behav,* 90, 246-56.

Weinert, D. and Waterhouse, J. (2007b)·The circadian rhythm of core temperature: effects of physical activity and aging *Physiol Behav,* 90, 26-256.

Weinstein, M. P., Iannini, P. B., Stratton, C. W. and Eickhoff, T. C. (1978)·Spontaneous bacterial peritonitis. A review of 28 cases with emphasis on improved survival and factors influencing prognosis *Am J Med,* 64, 592-8.

Wendt, D., van Loon, L. J. and Lichtenbelt, W. D. (2007)·Thermoregulation during exercise in the heat: strategies for maintaining health and performance *Sports Med,* 37, 669-82.

Zenone, T. (2007)·Fever of unknown origin in rheumatic diseases *Infect Dis Clin North Am,* 21, 1115-35, x-xi.

In: Body Temperature Regulation
Editor: A. B. Cisneros and B. L. Goins

ISBN: 1-60741-282-3

Chapter VII

Body Temperature Rise Caused by Diagnostic Imaging: Literature Review for Safer Use

Kiyotaka Nakagawa
Department of Anesthesiology, & Medical Engeneering laboratory, Jikei University School of Medicine, Tokyo, Japan
Department of Anesthesia, Kawaguchi Municipal Medical Center, Saitama, Japan
Koichi Yuki, Keiichiro Maniwa
Department of Anesthesiology, Itabashi chuo Medical Center, Tokyo, Japan

Abstract

In general, diagnostic imaging equipment creates images by exposing the living body to physical energy. In the living body, this applied energy can be transformed into heat. The resultant heating may have adverse effects on body temperature, which is otherwise strictly maintained within a narrow range by the thermo-regulatory system.

Physicians tend to have a general impression that ultrasonic imaging and magnetic resonance imaging (MRI) are safe diagnostic methods without radioactive exposure. However, the rise of body temperature due to these equipment has been reported as a side effect. In ultrasonic imaging, tissue temperature is increased by absorption of ultrasonic energy by the tissues. In MRI, applied radiofrequency and gradient magnetic field generate electric currents and ohmic heating in the body. The heating effect during ultrasonic imaging has been well studied, and many investigators have reported that local body temperature rise due to ultrasonic equipment may have potential adverse effects. Needless to say, the specification of ultrasonic diagnostic equipment, such as output power, is strictly regulated. Nevertheless, some use conditions, such as operating time, that still depend on the examiner. Therefore, safer use of ultrasonic diagnostic equipment cannot be accomplished unless the users of these equipment are well aware of the safety standards and potential adverse effects. This also applies to MRI. In MRI, especially conductive foreign bodies implanted inside the body and applied on the body surface can create problems of heating. Even without these objects, the human body itself also functions as a conductive object. In order to avoid thermal problems involving MRI, the examiner must pay careful attention to conductive objects including human body position.

The primary purpose of this paper is to draw attention to body temperature rise due to diagnostic imaging equipment and review the problems associated with the heating effect.

Introduction

In 2004, the Food and Drug Administration (FDA) cautioned against ultrasound 'Keepsake' images that give parents a prenatal portrait [1, 2]. Actually, this image uses the latest ultrasound technology to produce high-resolution three-dimensional and four-dimensional (moving) images showing the surface anatomy of a fetus developing in the uterus. Notably, this entertainment modality does not have any medical reason and is often performed by unskilled technicians without the supervision of a qualified physician. Under common and wrong impression, ultrasound imaging device is considered as a quite safe method because there is no exposure to radiation. What is the problem in this entertainment use of ultrasound imaging equipment?

Generally, diagnostic devices produce images by applying physical energy to the living body. This extrinsic energy can potentially harm the living body. DNA damage caused by ionizing radiation, such as X-ray, is a typical example [3, 4]. Remarkably, acoustic energy administration and non-ionizing radiation are not exceptional. The energy administered by these modalities is known to be transformed into heat. To inhibit body temperature elevation, the living body protects itself from this heat by transportation and diffusion to surrounding tissues via blood circulation and so on. Despite these protective mechanisms by the body, temperature can still increase in some situations [5-7]. These heating phenomena have been well described in ultrasound and magnetic resonance imaging (MRI) by many researchers. This heating can potentially cause some harmful effects to the living body. In this review, we primarily focus on the temperature elevation by diagnostic imaging devices, and discuss the safe usage of these modalities based on literatures.

Ultrasonic Diagnostic Device

The living body is composed of numerous tissues and materials. The ease of ultrasonic penetration depends on the properties of individual materials (acoustic impedance). Once the ultrasound is applied to the living body, the diagnostic device detects reflected waves produced at the junction of tissues with different acoustic impedance and creates images based on its analysis [8]. It can also measure the blood flow velocity by applying ultrasound to red blood cells and measuring the change in frequencies of reflection waves (ultrasonic

Doppler method) [9]. Ultrasonic imaging has obvious advantages over other diagnostic methods; (1) examiners can obtain real-time diagnostic image easily at bedside and (2) it involves no exposure to ionizing radiation. Based on these advantages, it has become one of the prerequisite diagnostic image modalities in current medical practice, especially in the obstetric field based on the assumption that it would not cause adverse effects to the fetus. However, the progress of ultrasonic diagnostic device led into the development of new high output devices, which requires reconsideration about their clinical application and user conditions.

The ultrasound is transmitted through the body as compressional waves and vibrates molecules in tissues. This vibration causes translational and rotational movement of molecules, resulting in heat production at the irradiation site. This process illustrates a mechanism of absorption of ultrasonic energy. Additionally, ultrasonic transducer itself in the probe can also be heated. This transducer self-heating is secondary to the energy that is lost while electric power is transformed into acoustic power. The heat is conducted to the irradiated site and will cause further local temperature elevation. This transducer self-heating is a significant component in considering tissue-heating effects of ultrasonic devices [10, 11].

To obtain clear images, it would be necessary to increase the ultrasound output power [12]. As a consequence, diagnostic devices with such a strong output power that could potentially produce biological effects have been developed [13, 14]. Following this trend, Center for Devices and Radiological Health in the FDA relaxed the regulation of ultrasonic output power while mandating manufactures to display the index of biological effects in a real-time fashion and present more detailed information on the machine (FDA 510(k) track 3) [15]. These indices include thermal index (TI) - the index of biological effect caused by heat - and mechanical index (MI) - the index of non-thermal mechanical effect (Please refer to [16] for MI for details). Both of them are advocated by the American Institute of Ultrasound in Medicine (AIUM) and the National Electrical Manufacturers Association (NEMA) [17, 18]. On this FDA's decision, output power limit had been increased in fetal diagnosis by a factor of 8. Furthermore, ultrasound exposure conditions such as output power, and continuous use time were largely left at the discretion of each user in exchange of displaying these indices.

TI is defined by the ratio of W_0 (total acoustic power at target site) and W_{deg} (the acoustic power required to rise the temperature at target site by 1°C) as follows;

$$TI = W_0 / W_{deg} \text{ (formula 1)}$$

For example, TI = 2 means that the device can potentially increase local tissues' temperature by 2°C. Devices are standardized to display this predicted values on the screen (Figure 1). Furthermore, TI is subdivided into three categories as TI of soft tissue (TIS), TI of cranium (TIC), and TI of bone (TIB) because acoustic properties of each tissue are different. Since TI displays the predicted value of temperature elevation at target sites on the screen in a real-time fashion, it can be a very useful tool to determine the irradiated conditions. However it is important to keep in mind that underestimation of actual temperature elevation might occur because this is only a predicted value of temperature elevation primarily dedicated to risk stratification [19, 20]. It is mainly because this calculation is based on the assumption that ultrasound as the modeled beam propagates through the homogenous tissues

[21]. However, actual living tissues are heterogenous. Furthermore, the conduction of transducer self-heating will further increase the temperature at irradiated sites. Therefore, it may not be prudent to think that TI value is a simple representation of actual elevation of temperature in tissues.

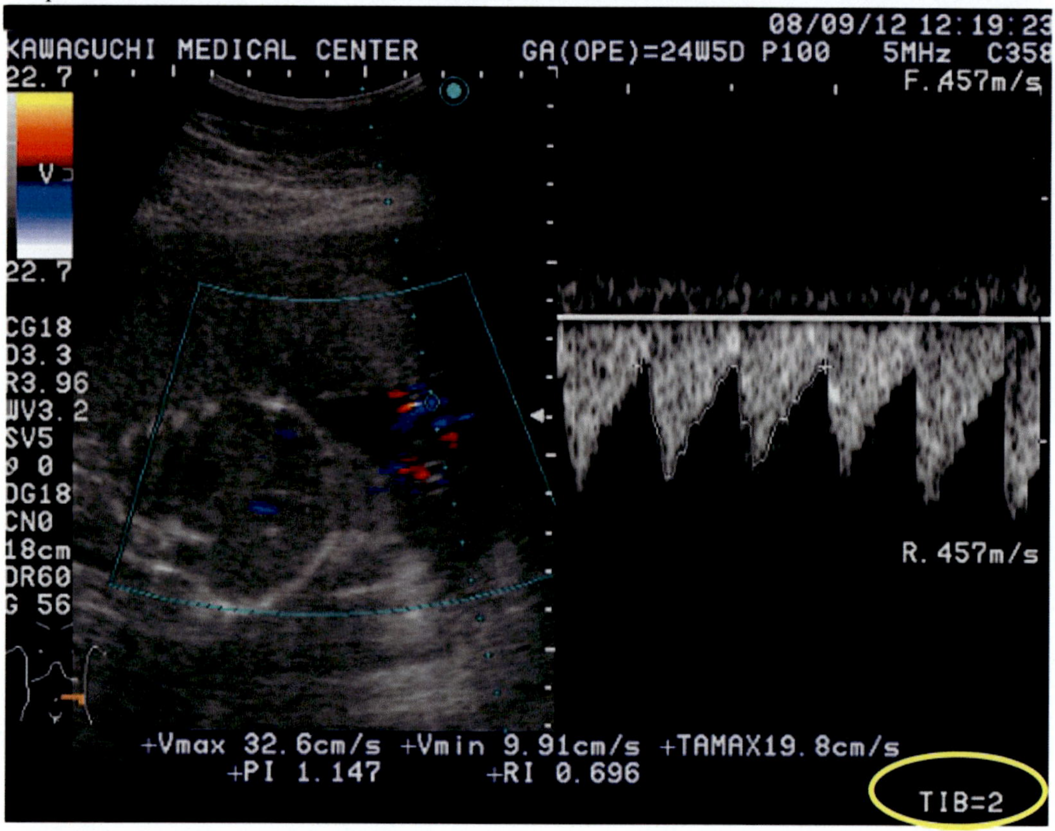

Figure 1. Real time display of thermal index.
Thermal index of bone (TIB circled in yellow) is displayed on the monitor of an ultrasonic imaging device (LOGIQ™ 500 version 5.0: GE medical Systems, Milwaukee, Wisconsin, USA) simultaneously with measured flow velocity of the umbilical artery.

In clinical obstetric practice, no clear causal relationship has been established between temperature elevation caused by ultrasonic diagnostic devices and their proposed harmful effects [21]. This topic has been thoroughly reviewed by Ziskin MC, and Abramowicz JS et al. [16, 22]. In the adult population, few reports described their significantly harmful effects, too. This is supposedly due to the fact that it will be difficult to perform strictly controlled studies from an ethical standpoint. Also it would be hard to control exposure time since it greatly depends on examiners' skill. Although there is no clear evidence to show that ultrasound will adversely affect health, it does not necessarily mean that it has no biological effects on the human body [23]. Rather, it should be interpreted that its biological effects have not been detected yet. In fact, thermal damage increases as the time at an elevated temperature increases. In vitro thermal effects are generally noted at 39 to 43°C; and at higher temperatures (44 to 46°C) coagulation of proteins occurs [24]. It implies that heat can be an assault on the human body even if there is a slight rise in temepature. Additionally, hyperthermia is a recognized teratogen in mammalian laboratory animals and is a suspected teratogen for humans [25].

Then, how much temperature elevation and continuous use will the body allow? Some of currently available diagnostic devices approved by FDA can operate with a displayed TI of 6.0 or higher, though depending on irradiation conditions [15]. The current worldwide consensus is represented by the recommendation of World Federation for Ultrasound in Medicine and Biology (WFUMB) [14, 23]. According to the recommendation, we are allowed to use devices with no limitations, if in the range of the temperature elevation < 1.5°C above normal physiological levels (37°C). In addition, it recommends us not to use more than 5 minutes when temperature elevates more than 4°C above normal temperature (37°C) in the field of obstetrics.

The following formula derived by Miller MW and Ziskin MC can be helpful in deciding the duration for continuous use [24].

$t = 4^{(43-T)}$ (formula 2)

In this formula, t [min] is the threshold time of continuous heat exposure causing fetal injury, and T [°C] is the actual temperature. This is a quantitative and simple formula. [TI + 37°C] can be used as [T] to estimate the threshold period of the allowable continuous use (Figure 2). However, when an elevation of 4°C occurs, allowable continuous use duration calculated using this formula is 16 minutes, which is three times longer than the duration recommended by the WFUMB guideline [23]. Furthermore, when T = 1.5 °C is used, t is calculated as 512 minutes, which is extremely long in real obstetric practice. This result corresponds to WFUMB recommendation stating that examiners can use the device unlimitedly at this level of temperature elevation. In any case, we need to keep in mind that this formula is derived from animal data and not from human ultrasonic experiments. Therefore, this formula is no more than an estimation of actual practice.

Irradiation mode and target location are also important in determining the allowable temperature elevation and duration of continuous use. Especially, the pulse Doppler, which measures blood flow, irradiates target sites with higher intensity than other modes [26, 27]. As a result, the probability to overheat tissues is the highest. Different from other modes (such as B- mode), pulse Doppler is applied to a pinpoint spot continuously to measure blood flow [21]. On the other hand, bone particularly absorbs ultrasound among living tissues, resulting in strong heating at the bone-tissue interface [27]. Since it is important to evaluate temperature elevation at tissues around the bone, TIB and TIC are defined. In the context of temperature elevation at the tissues surrounding bones, special attentions should be paid to fetal ultrasound following the progress of bone development with gestational age. Tissues such as the embryo and eyes may show more temperature elevation than that is expected since they have limited circulation and therefore do not have heat dissipated by blood flow [23].

Our study can be referred to as an example of using high output pulse Doppler around bone tissues, a worst case scenario from the heat exposure standpoint [28]. We measured intracranial temperature changes when transcranial Doppler (TCD) was applied continuously at the maximum output in animals. Approximately 3.5°C of temperature elevation was measured at the skull-brain interface, and we also theoretically proved that this result could be extrapolated to human adults. Based on the findings and from formula 2, we advocated that it would be necessary to let TCD stop for about 10 minutes following each 30 minutes of continuous usage.

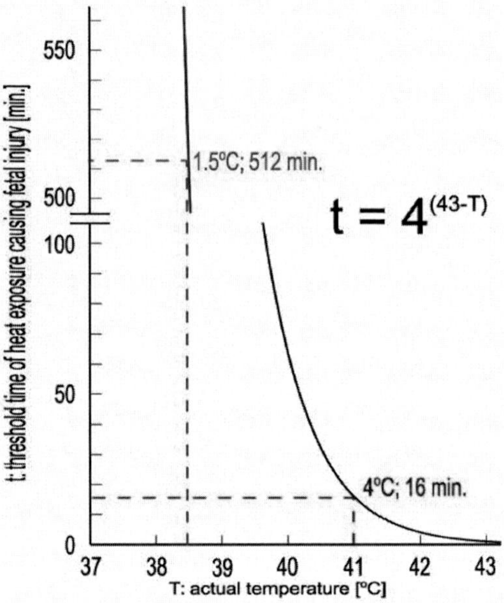

Figure 2. Safe threshold time of heat exposure.
The curve is obtained by formula 2 [24]. The right side of curve indicates the probability of tissue damage caused by heat. By this equation, threshold time at 38.5°C (37 + 1.5°C) and 41°C (37 + 4°C) can be calculated. Results are shown by blue and red broken line, respectively.

We reviewed that using ultrasound involves energy administration and therefore elevates local temperature in the living body. In addition to having the above-mentioned thermal effect, ultrasound has non-thermal mechanical effect that causes radiation force, acoustic streaming, and cavitation [16]. Examiners have to perform "an effective test with as low output and short duration as possible" understanding the meaning of TI displayed on the screen. This idea is phrased as ALARA (as low as reasonably achievable), and should be the principle of tests in various ultrasonic test guidelines.

Magnetic Resonance Imaging (MRI)

MRI has become one of the major diagnostic tools since it has no involvement in ionizing radiation just like ultrasound and it provides more detailed imaging information than computed tomography (CT). Presently, MRI with as strong magnetic field as much as 3 Tesla (T) is becoming popular rather than standard 1.5 T MRI [29]. (Generally, as the static magnetic field becomes stronger, you can obtain higher spatial resolution and / or shorter scan time [30]. Therefore, the capacity of MRI can be represented by the strength of static magnetic field (T)). To obtain more detailed imaging information, MRI devices with stronger magnetic fields have been constructed. FDA also relaxed the regulation of output conditions gradually in this trend [31]. In fact, the allowable maximal strength of static magnetic field was 2 T in 1982, whereas it was up to 8 T by 2003. In addition, 9.4 T MRI, which is far above the upper limit of 8 T, was developed and the result of human study for its safety was reported in 2007 [32].

In MRI, testees are placed within two powerful magnetic fields composed of static magnetic field (constant magnetic field) and gradient magnetic field (varying magnetic field), where radiofrequency (RF) of specific frequency to static magnetic fields strength (Lamor frequency) is applied. Protons of hydrogen atoms in the body become resonant with RF. Once the RF is stopped, the absorbed energy is emitted. MRI receives, analyzes, and visualizes the signal from this emission energy [33, 34]. MRI, therefore, has three energies of static magnetic field, gradient magnetic field, and RF applied to the body (Figure 3).

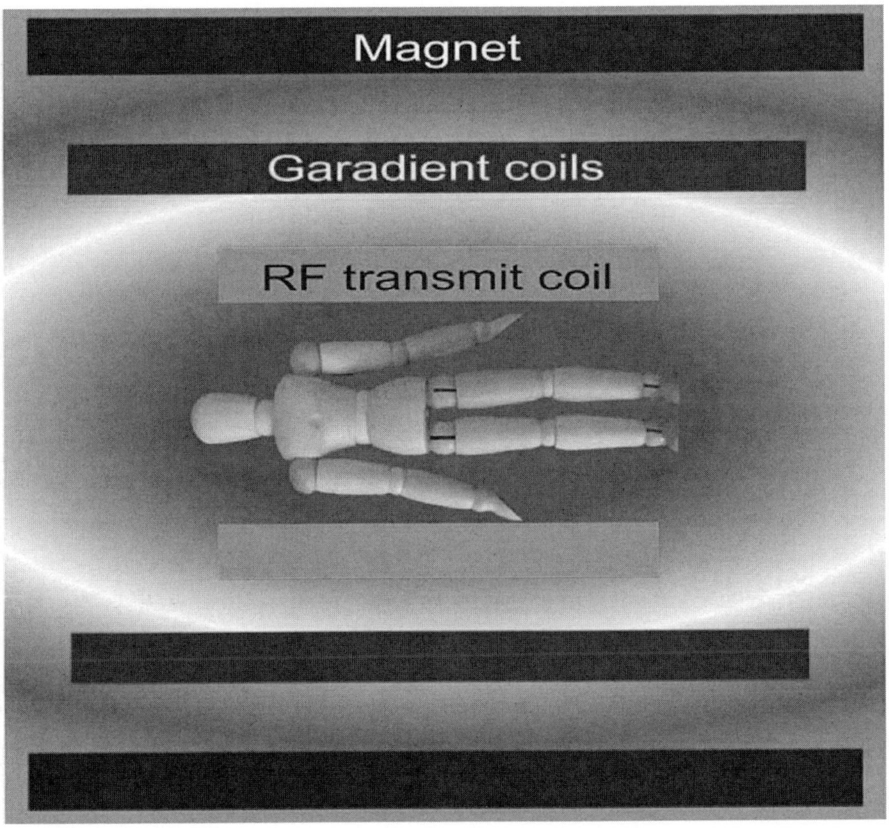

Figure 3: Schematic of magnetic resonance (MR) procedure.
A magnet applies a static magnetic field to the testee. Gradient coils generate time-varying magnetic fields. Additionally, radiofrequency (RF) coil sends RF to an object in order to resonate. Therefore, three types of energy are released in MR procedure.

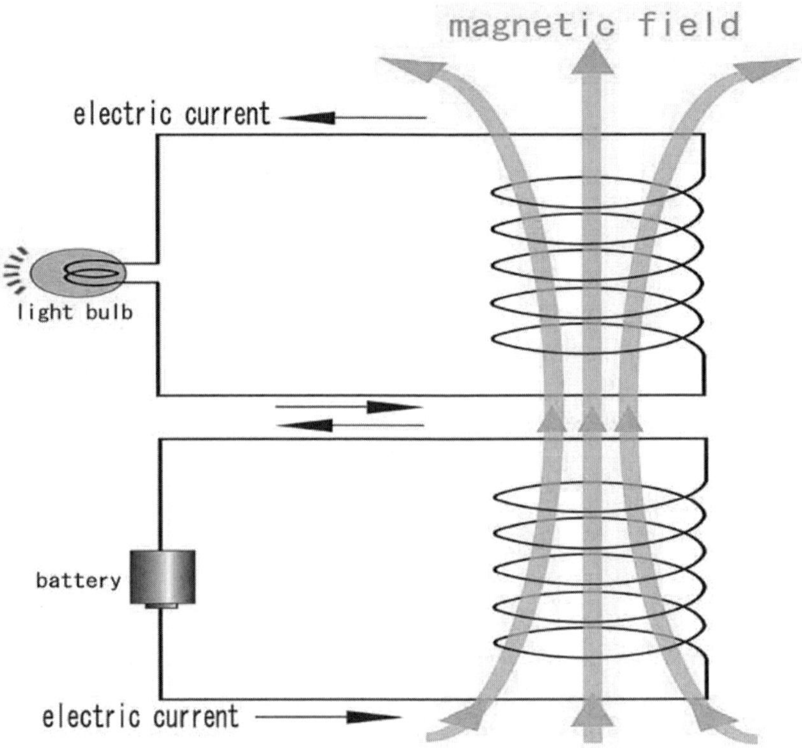

Figure 4. Mutual induction.
By sending an electric current through a coil, a magnetic field is generated and the magnetic field induces an electromotive force in an adjacent another coil [35]. This phenomenon of mutual induction is a good illustration of the linked relationship between magnetic field and electric field.

Varying magnetic field, such as gradient magnetic field produces electromotive force in a closed circuit (electromagnetic induction) [35]. Similarly, changing electric field produces a magnetic field. Both of phenomena are linked to each other (Figure 4). Moreover, these phenomena repeatedly occur when electrical and magnetic wave are alternating. As a result of these inductions, an eddy current is generated and produces ohmic heating due to resistance of conductive objects. In magnetic resonance (MR) procedures, gradient magnetic field has a probability of affecting human temperature regulation under the above-mentioned mechanism [36]. Additionally, since RF has both oscillating current and magnetic field components at the same time, RF also induces ohmic heating (Figure 5) [35]. RF gets absorbed in the body because conductive objects function as an antenna [36]. Consequently, currents are induced and transformed to ohmic heat in the process of resistance loss [37]. Tissue heating that results from the RF radiation used for MR procedures is primarily caused by magnetic induction, with a negligible contribution from the current field [38]. Moreover, the current induced by RF is 1000 times stronger than the current induced by gradient magnetic field [34, 39, 40]. Therefore, we need to take precautions against the heating by RF rather than gradient magnetic fields. Conductive foreign bodies implanted inside the body and applied on the body surface (eg. electrocardiogram cable) can especially create problems.

Even without these objects, heating phenomenon based on the above-mentioned mechanisms can be seen since the human body itself is conductive.

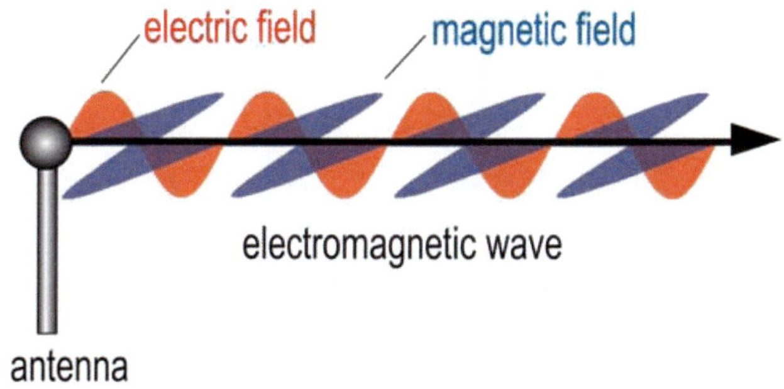

Figure 5. Electromagnetic wave.
Electromagnetic waves including radiofrequency (RF) created by electric field (red sine curve) and magnetic field (blue sine curve) [35]. Two components propagate varying with time. Generally, the two components invariably coexist. Oscillation of two fields is perpendicular to the direction of propagation. Both fields are perpendicular to each other.

The Presence of Conductive Foreign Bodies

When conductive foreign bodies including RF coil cable are present within the body or on the surface of the body, excessive local temperature elevation occurs through the electromagnetic induction of gradient magnetic fields and RF. If conductive objects are thin and long, it is especially problematic [41]. As practical examples, many burn injuries have been reported related to electrocardiogram monitor leads, pulse oximetries and pacemaker leads. This topic has been thoroughly reviewed by Dempsy MF et al. and Shellock FG at al. [36, 41]. When the cable of conductive objects makes a loop, gradient magnetic fields are especially prone to cause electromagnetic induction. Also loop cable is able to easily receive RF signal, functioning as an antenna. Even without the formation of a loop, conductive objects are resonant with RF, leading to current induction. If the length of conductive objects is one half of RF wavelength, it may possibly cause strong energy absorption and conductive currents with resonance (approximate wavelength is 4.7 m in 1.5 T MRI and 2.3 m in 3 T MRI) [38]. It is thought that this is the same principle with half-wave dipole antenna used in wireless transmission (Figure 6) [42]. In animal and phantom experiments, temperature elevation was recorded as high as 12°C in electrocardiogram electrodes, 14°C in spinal fusion stimulator leads, 20°C in pacemaker leads, and 48°C (from initial temperature of 26°C) in intravascular guide wires [43-46]. Metal components in tattoo are reported as causes of heating [47-49]. Needless to say, the thorough safety testing of these conductive devices are a must for MRI testing. The examiner must keep RF coil cables parallel to the main magnetic field and place sufficient insulation between RF coil cables and the patient's skin [50, 51]. Furthermore, it is important to exclude risks by screening implants within the body based on the suggestion of the Institute for Magnetic Resonance Safety, Education and Research (IMRSER), and so on [52]. When it cannot be avoided to use conductive bodies such as electrocardiogram monitors in MR environment, careful attention needs to be paid, for example, not to form unnecessary loops (circular, U shape, and S shape) or crossing of cables including RF coil cables [41].

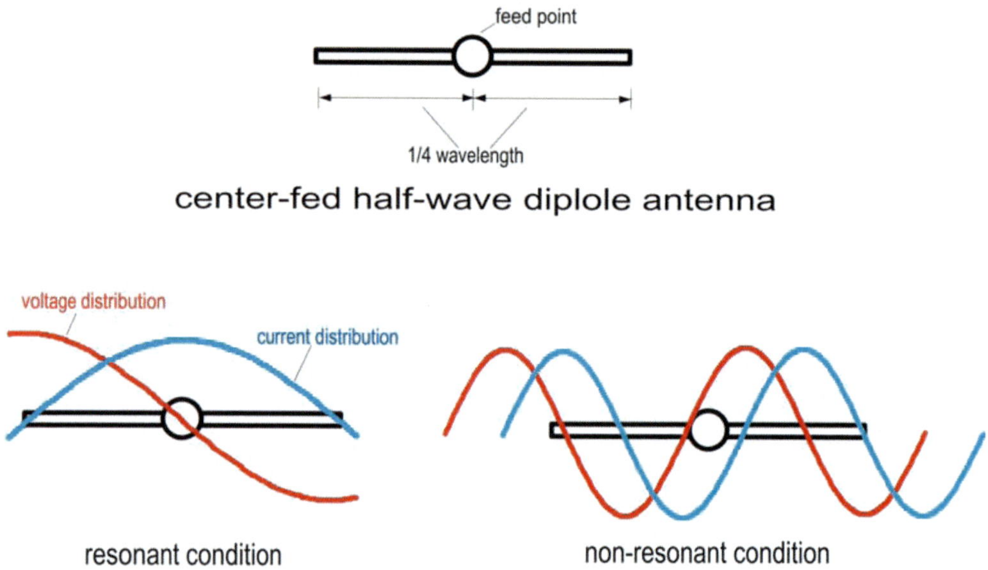

Figure 6. Half-wave dipole antenna.
This type of antenna has two quarter-wavelength elements and a feed point that is transmitting or receiving radiofrequency (RF) energy [42]. When RF is one half (or integer multiples of one half), voltage is at the maximum the bilateral ends of the antenna. Therefore, maximum electric current is generated in the antenna with this wavelength condition.

The Absence of Conductive Foreign Bodies

Since the living body is conductive, the heating phenomenon can be observed with gradient magnetic fields and RF in the absence of conductive foreign bodies [53]. We emphasize again that the heating by RF especially needs precaution. The index of energy absorption by RF is defined by the specific absorption rate (SAR) [34]. SAR is the energy absorbed by the living tissues per unit weight and time. Its unit is W/kg. Experiments have shown that temperature elevates less than 1°C when a person in a normal condition is exposed to electromagnetic field with SAR 1-4 W/kg for about 30 minutes [54]. The current FDA guideline advocates the usage of < 15 minutes at SAR 4 W/kg for the whole body, < 10 minutes at 3 W/kg for the head, < 5 minutes at 8 W/kg for the head or the torso, < 5 minutes at 12 W/kg for the extremities [31]. Additionally, FDA guidelines suggest that the prediction of the value of SAR that will be used during the scan should be displayed on the operator's console [55]. Under this regulation, human studies demonstrated that 0.3°C temperature elevation was recorded with MRI of systemic SAR at 4.0 W/kg and 0.2°C with MRI of SAR 0.42-1.2 W/kg [56, 57]. Furthermore, the exposure to systemic SAR 6.0 W/kg, which is above the FDA recommendation, was reported. Despite several degrees of temperature elevations observed on the skin, core temperature had an average elevation of 0.4°C (maximal 0.5°C) with a maximum temperature of 37.9°C [58]. Based on the result, it may be reasonable to consider that the heat accompanying MR usage was acceptable since normal body temperature regulation will compensate if MR is used within the SAR range of FDA recommendations. However, allowable temperature elevation can change depending on environmental temperature, humidity and air current when MRI is performed. Therefore, the National Radiological Protection Board (NRPB) in the UK and the International Electrotechnical Commission (IEC) regulates the upper limit of SAR on environmental temperature and humidity during the test to avoid excessive temperature elevation [50, 59]. We also need to pay attention to the fact that the ability of temperature regulation is diminished by aging, health state decline, and medications [42, 50]. Moreover, the examiner should schedule patient cool-down breaks between high SAR scans.

Notably, the above-mentioned report demonstrated that temperature had elevated by several degrees in the skin despite minimal core temperature change. It is considered that this mechanism is a local deposition of RF energy. Similar to this case was a report where third degree of burn developed on the skin in the absence of monitoring cables [60]. In this report, third degree burn was observed on the inner surface of both calves despite MRI being obtained by following FDA standards for 1.5 T MRI with systemic average SAR < 1.0. This burn resulted from the induction current of the conductive loop within the living body when skin contacted each other. The patient had the unusual anatomy of much thinner subcutaneous fat of the thigh after resection and radiation therapy for liposarcoma. The authors also suggested that a single small focal contact zone between patient's thighs may have potentially caused the burn. Another report described second degree burn of both elbows in a patient with both arm crossed touching the gantry's wall during MRI following FDA standard [61]. (Of course, it is recommended that the patient's skin does not touch the bore of the magnet [51].) The authors re-examined this result using living phantom. They demonstrated approximately 12°C temperature elevation in the closed part of loop phantom. Since they measured the temperature elevation during RF radiation, not during gradient magnetic fields, they speculated that this might be due to local deposition of RF energy. Even though several unforeseen event occured at the same time in these cases, precautions should be taken because of the possibility of serious health injuries accompanying temperature elevation if unintended loops are created by living body parts despite SAR limits of FDA standards being followed (Figure 7).

When the intensity of magnetic field doubles (eg. 1.5 T vs 3.0 T), the redistribution of RF power becomes four-fold by MR pulse sequence [42]. Therefore, the examination of temperature safety in high field strength MR system of 8 T needs more caution than the one in low field strength MR such as 1.5T. Currently, the examination has not yet been fully performed, although some are reported by Kangarlu et al. and so on [62]. The developer claims that 9.4 T ultra high field strength MR system above the FDA standard is safe on the observation that no significant vital sign change has been demonstrated in the examinee [32]. This result is currently under discussion [63]. Moreover, we should point out the fact that they evaluated the temperature only on the skin; no significant change was demonstrated on the skin using 9.4 T MRI, but core temperature change is unknown. Frankly speaking, the effect of ultra high field strength MR system on temperature regulation is not demonstrated at all. The demand for high field strength MRI would be inevitable worldwide in the future. The research on the effect of this high field strength MR on temperature regulation is not adequate at this point and remains to be pursued in the future.

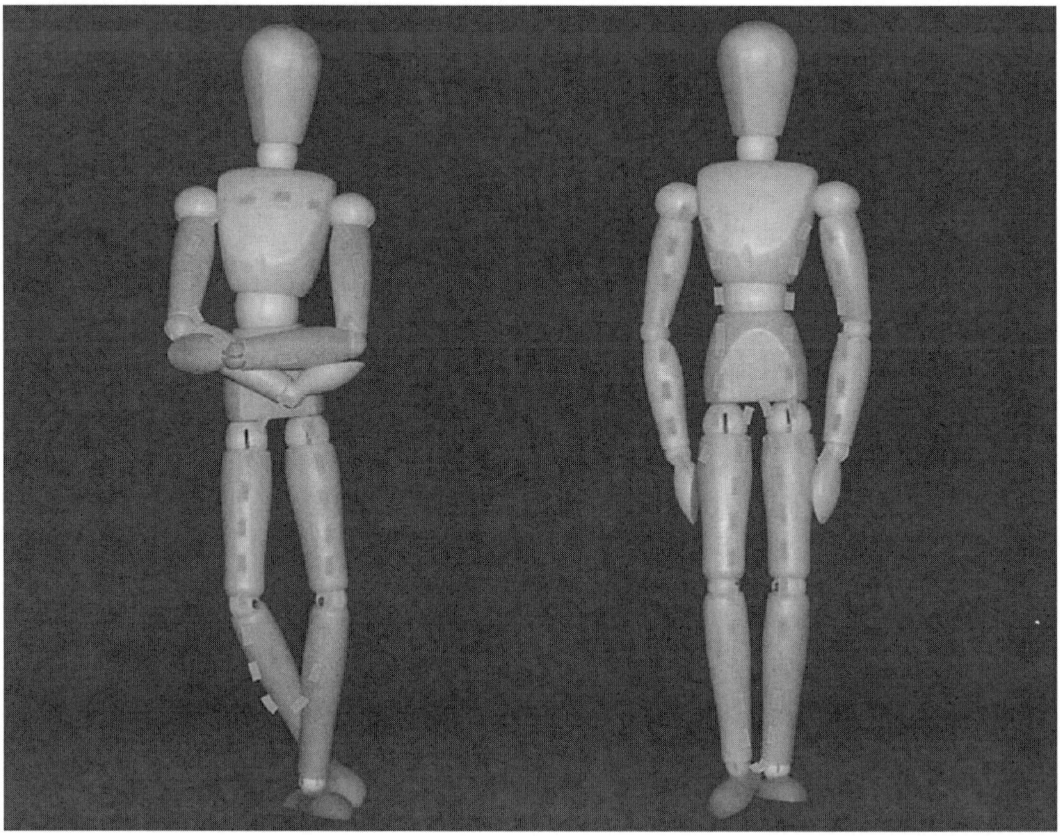

Figure 7. Unintended loop by body parts.
Left figure shows loops by crossing arms or legs. Without such positions, loops can be unintentionally formed as shown in the right figure. Electrical conductivity of skin-to-skin is especially increased by sweat.

Conclusions

Here we reviewed temperature elevation accompanying the usage of ultrasonic diagnostic devices and MRI, explaining its mechanism and relevant problems. Either device has been a widely used diagnostic imaging technique for the last 30 years. To obtain more detailed information of the living body, technical innovation increased at an extraordinary speed. Consequently, tissue perfusion can be visualized with ultrasound and MRI [65, 64]. Furthermore, even sodium in the living body can be visualized using MRI, although it is still in the stages of trial [32, 66]. This is remarkable progress. However, the examination of safety has not caught up with its technical progress. Further examination of safety is necessary as we pointed out in the review. We must be aware that the issue of safety on these devices has not met the progress yet, and therefore must choose its appropriate application with limited use.

References

[1] U. S. Food and Drug Administration (FDA) Consumer magazine. (2004). FDA Cautions Against Ultrasound 'Keepsake' Images. http://www.fda.gov/fdac/features/2004/104_images.html

[2] Tanne, J. H. (2004). FDA warns against commercial prenatal ultrasound videos. *BMJ, 328*, 853

[3] Yamaguchi, H., Ohara, H., & Waker, A. J. (2006). A model for the induction of DNA damages and their evolution into cell clonogenic inactivation. *J Radiat Res (Tokyo), 47*, 197-211.

[4] Wrixon, A. D. (2008). New recommendations from the International Commission on Radiological Protection--a review. *Phys Med Biol, 53,* R41-60.

[5] Duggan, P. M., Liggins, G. C., & Barnett, S. B. (1995). Ultrasonic heating of the brain of the fetal sheep in utero. *Ultrasound Med Biol, 21,* 553-60.

[6] Horder, M. M., Barnett, S. B., Vella, G. J., Edwards, M. J., & Wood, A. K. (1998). In vivo heating of the guinea-pig fetal brain by pulsed ultrasound and estimates of thermal index. *Ultrasound Med Biol, 24,* 1467-74.

[7] Horder, M. M., Barnett, S. B., Vella, G. J., Edwards, M. J., & Wood, A. K. (1998). Ultrasound-induced temperature increase in guinea-pig fetal brain in utero: third-trimester gestation. *Ultrasound Med Biol, 24,* 1501-10.

[8] Kelsey, C. A. (1993). Introduction. In Juhl, J. H., & Crummy, A. B. (Ed.), *Paul and Juhl's Essentials of Radiologic Imaging Sixth Edition* (pp. 1-18). Philadelphia: J.B. Lippencott Company.

[9] Mathews, L., & Singh, R. K. (2008). Cardiac output monitoring. *Ann Card Anaesth, 11,* 56-68.

[10] Wu, J., Cubberley, F., Gormley, G., & Szabo, T. L. (1995). Temperature rise generated by diagnostic ultrasound in a transcranial phantom. *Ultrasound Med Biol, 21,* 561-8.

[11] Killingback, A. L., Newey, V. R., El-Brawany, M. A., & Nassiri, D. K. (2008). Development of A Thermal Test Object For The Measurement Of Ultrasound Intracavity Transducer Self-Heating. *Ultrasound Med Biol,* (*in press*).

[12] Harris, G. R., Stewart, H. F., Leo, F. P., & Sanders, R. C. (1989). Relationship between image quality and ultrasound exposure level in diagnostic US devices. *Radiology, 173,* 313-7.

[13] World Federation for Ultrasound in Medicine and Biology (WFUMB). (1992). WFUMB Symposium on Safety and Standardisation in Medical Ultrasound. Issues and Recommendations Regarding Thermal Mechanisms for Biological Effects of Ultrasound. Hornbaek, Denmark, 30 August-1 September 1991. *Ultrasound Med Biol, 18,* 731-810.

[14] World Federation for Ultrasound in Medicine and Biology (WFUMB). (1998). Ultrasound. Update on thermal bioeffects issues. *Ultrasound Med Biol, 24,* Suppl 1, S1-10.

[15] U. S. Food and Drug Administration (FDA). (2008). Guidance for Industry and FDA Staff - Information for Manufacturers Seeking Marketing Clearance of Diagnostic Ultrasound Systems and Transducers. http://www.fda.gov/cdrh/ode/guidance/560.html

[16] Ziskin, M. C. (1999). Intrauterine effects of ultrasound: human epidemiology. *Teratology, 59,* 252-60.

[17] American Institute of Ultrasound in Medicine (AIUM)., & National Electrical Manufacturers Association (NEMA). (2004). Standard For Real-Time Display of Thermal and Mechanical Acoustic Output Indices On Diagnostic Ultrasound Equipment, Revision 2. http://www.aium.org/.

[18] Hedrick, W. R., & Hykes, D. L. (1993). An overview of thermal and mechanical acoustic output indices. *Journal of Diagnostic Medical Sonography, 9,* 228-236.

[19] Jago, J. R., Henderson, J., Whittingham, T. A., & Mitchell G. (1999). A comparison of AIUM/NEMA thermal indices with calculated temperature rises for a simple third-trimester pregnancy tissue model. *Ultrasound Med Biol, 25,* 623-8.

[20] Barnett, S. B. (2001). Intracranial temperature elevation from diagnostic ultrasound. Ultrasound Med Biol, 27, 883-8.

[21] Abramowicz, J. S., Barnett, S. B., Duck, F. A., Edmonds, P. D., Hynynen K. H., & Ziskin, M. C. (2008). Fetal thermal effects of diagnostic ultrasound. *J Ultrasound Med, 27,* 541-59.

[22] Abramowicz, J. S., Fowlkes, J. B., Skelly, A. C., Stratmeyer, M. E., & Ziskin, M. C. (2008). Conclusions regarding epidemiology for obstetric ultrasound. *J Ultrasound Med, 27,* 637-44.

[23] Barnett, S. B., Ter Haar, G. R., Ziskin, M. C., Rott, H. D., Duck, F. A., & Maeda, K. (2000). International recommendations and guidelines for the safe use of diagnostic ultrasound in medicine. *Ultrasound Med Biol, 26,* 355-66.

[24] Miller, M. W., & Ziskin, M. C. (1989). Biological consequences of hyperthermia. *Ultrasound Med Biol, 15,* 707-22.

[25] Miller, M. W., Nyborg, W. L., Dewey, W. C., Edwards, M. J., Abramowicz, J. S., & Brayman, A. A. (2002). Hyperthermic teratogenicity, thermal dose and diagnostic ultrasound during pregnancy: implications of new standards on tissue heating. *Int J Hyperthermia, 18,* 361-84.

[26] European Committee for Medical Ultrasound Safety (ECMUS) Safety Committee. (1995). Guidelines for the safe use of Doppler ultrasound for clinical applications. http://www.efsumb.org/guidelines/safe-use-doppler-clinical.pdf

[27] European Committee of Medical Ultrasound Safety (ECMUS). (2008). Clinical Safety Statement for Diagnostic Ultrasound – 2008 http://www.efsumb.org/guidelines/2008safstat.pdf

[28] Nakagawa, K., Ishibashi, T., Matsushima, M., Tanifuji, Y., Amaki, Y., & Furuhata, H. (2007) Does long-term continuous transcranial Doppler monitoring require a pause for safer use? *Cerebrovasc Dis, 24,* 27-34.

[29] Heidemann, R. M., Seiberlich, N., Griswold, M. A., Wohlfarth, K., Krueger, G., & Jakob, P. M. (2006). Perspectives and limitations of parallel MR imaging at high field strengths. *Neuroimaging Clin N Am, 16,* 311-20.

[30] Takahashi, M., Uematsu, H., & Hatabu, H. (2003). MR imaging at high magnetic fields. *Eur J Radiol, 46,* 45-52.

[31] U. S. Food and Drug Administration (FDA). (2003). Guidance for Industry and FDA Staff. Criteria for Significant Risk. Investigations of Magnetic Resonance Diagnostic Devices http://www.fda.gov/cdrh/ode/guidance/793.pdf

[32] Atkinson, I. C., Renteria, L., Burd, H., Pliskin, N. H., & Thulborn, K. R. (2007). Safety of human MRI at static fields above the FDA 8 T guideline: sodium imaging at 9.4 T does not affect vital signs or cognitive ability. *J Magn Reson Imaging, 26,* 1222-7.

[33] Berger, A. (2002). Magnetic resonance imaging. *BMJ, 324,* 35.

[34] U. S. Food and Drug Administration (FDA). (1997). A Primer on Medical Device Interactions with Magnetic Resonance Imaging Systems http://www.fda.gov/cdrh/ode/primerf6.html

[35] Halpern, A. M., & Erlbach, E. (1998). *Theory and Problems of Beginning Physics II: Waves, Electromagnetism, Optics, and Modern Physics (Schaum's Outlines).* New York, USA: Mcgraw-Hill

[36] Dempsey, M. F., & Condon, B. (2001). Thermal injuries associated with MRI. *Clin Radiol, 56,* 457-65.

[37] Kanal, E., Shellock, F. G., & Talagala, L. (1990). Safety considerations in MR imaging. *Radiology, 176,* 593-606.

[38] Shellock, F. G. (2000). Radiofrequency energy-induced heating during MR procedures: a review. J Magn Reson Imaging, 12, 30-6.

[39] Thermal injuries and patient monitoring during MRI studies. (1991). *Health Devices, 20,* 362-3.

[40] Schaefer, D. J., Bourland, J. D., & Nyenhuis, J. A. (2000). Review of patient safety in time-varying gradient fields. *J Magn Reson Imaging, 12,* 20-9

[41] Shellock, F. G., & Crues, J. V. (2004). MR procedures: biologic effects, safety, and patient care. *Radiology, 232,* 635-52.

[42] Straw, R. D. (Eds.), (2007). *The ARRL Antenna Book 21 ST Edition: The Ultimate Reference for Amateur Radio Antennas, Transmission Lines And Propagation. Newington, USA: ARRL the national association for amateur radio.*

[43] Felmlee, J. P., Rochester, M. N., Hokanson, D. A., Zink, F. E., & Parkins, W. J. (1995). Real-time temperature measurement of heating under electrocardiographic electrodes during MR imaging at 1.5 T. *Radiology, 197(suppl.),* 423, Abstract.

[44] Chou, C. K., McDougall, J. A., & Chan, K. W. (1997). RF heating of implanted spinal fusion stimulator during magnetic resonance imaging. *IEEE Trans Biomed Eng, 44,* 367-73.

[45] Luechinger, R., Zeijlemaker, V. A., Pedersen, E. M., Mortensen, P., Falk, E., Duru, F., Candinas, R., & Boesiger, P. (2005). In vivo heating of pacemaker leads during magnetic resonance imaging. *Eur Heart J, 26,* 376-83.

[46] Konings, M. K., Bartels, L. W., Smits, H. F., & Bakker, C. J. (2000). Heating around intravascular guidewires by resonating RF waves. *J Magn Reson Imaging, 12,* 79-85.

[47] Tope, W. D., & Shellock, F. G. (2002). Magnetic resonance imaging and permanent cosmetics (tattoos): survey of complications and adverse events. *J Magn Reson Imaging, 15,* 180-4.

[48] Wagle, W. A., & Smith, M. (2000). Tattoo-induced skin burn during MR imaging. *AJR Am J Roentgenol, 174,* 1795.

[49] Kreidstein, M. L., Giguere, D., & Freiberg, A. (1997). MRI interaction with tattoo pigments: case report, pathophysiology, and management. *Plast Reconstr Surg, 99,* 1717-20.

[50] Medicines and Healthcare products Regulatory Agency (MHRA). (2007). Safety Guidelines for Magnetic Resonance Imaging Equipment in Clinical Use. http://www.mhra.gov.uk/Publications/Safetyguidance/DeviceBulletins/CON2033018

[51] Emergency Care Research Institution (ECRI). (2001). Safety concerns in the MR environment: unique environment, unique risks. *Health Devices, 30,* 422-444.

[52] Shellock, F. G., & Spinazzi, A. (2008). MRI safety update 2008: part 2, screening patients for MRI. *AJR Am J Roentgenol, 191,* 1140-9.

[53] Kanal, E., Barkovich, A. J., Bell, C., Borgstede, J. P., Bradley, W. G. Jr., Froelich, J. W., Gilk, T., Gimbel, J. R., Gosbee, J., Kuhni-Kaminski, E., Lester, J. W. Jr., Nyenhuis, J., Parag, Y., Schaefer, D. J., Sebek-Scoumis, E. A., Weinreb, J., Zaremba, L. A., Wilcox, P., Lucey, L., & Sass, N.; ACR Blue Ribbon Panel on MR Safety. (2007). ACR guidance document for safe MR practices: 2007. *AJR Am J Roentgenol, 188,* 1447-74.

[54] International Commission on Non-Ionizing Radiation Protection (ICNIRP). (1998). Guidelines for limiting exposure to time-varying electric, magnetic, and electromagnetic fields (up to 300 GHz). *Health Phys*, 74, 494-522.

[55] U. S. Food and Drug Administration (FDA). (1998). Guidance for Industry. Guidance for the Submission Of Premarket Notifications for Magnetic Resonance Diagnostic Devices. http://www.fda.gov/cdrh/ode/mri340.pdf

[56] Shellock, F. G., & Crues, J. V. (1987). Temperature, heart rate, and blood pressure changes associated with clinical MR imaging at 1.5 T. *Radiology, 163,* 259-62.

[57] Schaefer, D. J., Barber, B. J., Gordon, C. J., Zielonka, J., & Hecker, J. (1985). Thermal effects of magnetic resonance imaging (abstract). In: Book of abstracts: *Society of Magnetic Resonance in Medicine. Vol.2.* (pp 925) Berkeley, USA: Society of Magnetic Resonance in Medicine

[58] Shellock, F. G, Schaefer, D. J, & Kanal, E. (1994). Physiologic responses to an MR imaging procedure performed at a specific absorption rate of 6.0 W/kg. *Radiology, 192,* 865-8.

[59] International Electrotechnical Commission (IEC). (1995 revised 2002). Medical electrical requirements – particular requirements for the safety of magnetic resonance equipment for medical diagnosis. IEC 60601-2-33.

[60] Knopp, M. V., Essig, M., Debus, J., Zabel, H. J., & van Kaick, G. (1996). Unusual burns of the lower extremities caused by a closed conducting loop in a patient at MR imaging. *Radiology, 200,* 572-5.

[61] Yamazaki, M., Yamada, E., Kudou, S., & Higashida, M. (2005). Study of temperature rise in RF irradiation during MR imaging: measurement of local temperature using a loop phantom. (in Japanese). *Nippon Hoshasen Gijutsu Gakkai Zasshi, 61,* 1125-32.

[62] Kangarlu, A., Shellock, F. G., & Chakeres, D. W. (2003). 8.0-Tesla human MR system: temperature changes associated with radiofrequency-induced heating of a head phantom. *J Magn Reson Imaging, 17*, 220-6.

[63] de Vocht, F., & Kromhout, H. (2008). Human MRI above the FDA 8 T guideline: can we conclude that it is safe? *J Magn Reson Imaging, 27*, 938-9.

[64] Meairs, S. (2008). Contrast-enhanced ultrasound perfusion imaging in acute stroke patients. *Eur Neurol, 59 Suppl 1*, 17-26.

[65] van Laar, P. J., van der Grond, J., & Hendrikse, J. (2008). Brain perfusion territory imaging: methods and clinical applications of selective arterial spin-labeling MR imaging. *Radiology, 246,* 354-64.

[66] Ouwerkerk, R. J. (2007). Sodium magnetic resonance imaging: from research to clinical use. *Am Coll Radiol, 4,* 739-41.

In: Body Temperature Regulation
Editor: A. B. Cisneros and B. L. Goins

ISBN: 1-60741-282-3

Chapter VIII

Energy Utilization in the Homeotherm Animal: The Effects of Environmental Temperature and Metabolic Behaviour Charactehristics

***M. R. Sanz Sampelayo*[1] *and I. Prieto Gómez*[2]**

[1]Instituto de Nutrición Animal, Estación Experimental del Zaidín (CSIC). Profesor Albareda, 1. 18008 Granada, Spain
[2]Departamento de Ciencias de la Salud, Universidad de Jaén. Campus Las Lagunillas. 23071 Jaén, Spain

Abstract

The purpose of this study was to examine the metabolic behaviour of pre-ruminant kid goats and lambs at different environmental temperatures (12, 24 and 30°C). The energy and protein retained were determined by means of a feeding experiment on different feeding levels and by the comparative slaughter method. The relationships between the total energy retained or those retained as protein or fat, and the metabolizable energy intake, were established by means of regression analyses. Energy requirements for maintenance were lower, and more closely grouped, among the kid goats than among the lambs. On the contrary, the efficiencies of energy utilization for energy retention were higher among the lambs. Under fasting conditions, the necessary thermogenesis among the kid goats was associated with the catabolism of fat. However, among the lambs, at 12°C there was found to be a high utilization of both protein and fat, while at the other two temperatures tested the corresponding utilization was mainly of fat. At satiety, the kid goats were able to regulate their intake, such that independently of the environmental temperature, their energy balance, and consequently the body composition, remained constant. On the contrary, under thermal stress induced by cold, the lambs were not capable of counteracting the greater thermal demands made by means of increasing their food intake, and thus fat was catabolized and their body composition became more lean. The temperature of 12°C was cold for both animals, located on the left hand side of their thermoneutrality zone, with the lambs being more sensitive to this cold than were the kid goats. For the goats, temperatures of 24°C and 30°C were identified as belonging to their thermoneutrality zone. However, for the lambs, 24°C could still be considered a cold temperature, while 30°C could be located within their thermoneutrality zone.

Introduction

Homeothermia is one of the many intriguing homeostatic functions of higher organisms. A living being cannot be considered in isolation from its environment (Bligh, 1973), any more than a candle flame can be contemplated separately from the atmosphere in which it burns. The structures and functions of every species have been shaped by natural selection, in accordance with the progressive changes and mutations that have granted each one certain advantages in its struggle for survival. Thus we see that it is the influence of the environment as a whole that has given rise to the prodigious variety presented by living beings. Each species, hence, has achieved adaptation to a particular environment, and the different forms and functions we now observe relate the historical environment to that of the present day.

For a homeotherm animal to maintain a stable core temperature, its bodily heat production must equal its heat loss. A basic concept to the consideration of energy exchange is that of thermal neutrality, i.e. the zone of environmental temperature within which the animal's heat expenditure is maintained constant and at a minimum, thus producing the maximum retention of dietary energy. The lower end of this zone is called the lower critical temperature, or simply the critical temperature, and the higher end is thus the higher critical temperature. The position of this zone of thermal neutrality within a temperature range is determined by nutritional and environmental factors. Within and above this zone, the animal has no need of heat for thermal regulation and so the heat associated with productive processes must be dissipated. The higher the level of nutrition, the higher the rate of heat production; as the level of food intake rises, there is an increased range within which the demand for thermoregulated heat can be satisfied, and in consequence the lower limit of the thermal neutrality zone falls. On the other hand, at levels above the thermal neutrality zone, high levels of food intake give rise to greater demands for heat loss, and hence the upper limit of the thermal neutrality zone is reached sooner. Accordingly, the response of an animal to its thermal environment depends on its food intake; for a given intake, the environmental temperature determines the rate of heat loss. These considerations, discussed in a review (Close, 1980) of the significance of the thermal environment with respect to energy utilization by pigs, highlight the importance of the environmental temperature in the utilization of metabolizable energy consumed.

When an animal is situated outside its thermal neutrality zone, this creates a situation of stress to which it reacts in order to maintain its thermal and energetic homeostasis, both of which depend on the environmental temperature, the nutritional status of the animal and its particular metabolic behaviour characteristics (Mount, 1979). From a productive point of view, this could be of considerable importance. A low environmental temperature always creates greater energy requirements for maintenance, which under *ad libitum* feeding conditions will generate a greater consumption of food. This is turn could generate more or equal growth depending on the nature of the food and the animal's intake capacity. On the other hand, a high environmental temperature will probably reduce growth, because of the fall in the intake that usually occurs under these circumstances, as the animal's response to avoid the production of heat (Close, 1980).

When an animal's food intake exceeds its requirements, it gains weight, which is associated with the depositing of protein and, above all, of fat. However, Miller et al. (1979) reported that various studies have long shown that rats, pigs and even human beings, under such conditions of over-feeding, are capable of maintaining an unchanged weight, and in some cases, body composition. Without reaching these extremes, it is well known that some domestic animals accumulate body fat very readily, while others are resistant to such accumulation. It has been pointed out that this type of classification can be used to explain more than the metabolic and physiological basis of body type. Different specific mechanisms might separately be employed by different genotypes to obtain their particular phenotypic characters. Miller et al. (1979) reported that thermogenesis, induced by the diet or by the physical environment, appears to be fundamentally based on protein utilization, with leaner animals being more resistant, in the sense that they are capable of maintaining their energy balance within a wider range of food intake and/or temperatures.

The corporal development of goat and ovine species is different, and this is manifest from early life. The most typical characteristic of goats' development is their resistance to accumulating fat (Sanz Sampelayo et al., 1995). Taking this into account, together with the fact that both kid goats and lambs are sensitive to low temperatures (Raggi et al., 1985), we carried out a series of trials with the young of both species in order to determine their response, in terms of energy metabolism, when subjected to each of three different temperatures (12°C, 24°C and 30°C).

Material and Methods

Animals and Experimental Design

Thirty three intact male kid goats of the Granadina breed and 33 intact male lambs of the Segureña breed were used. They were removed from their dams 1 or 2 days after birth, when six animals of each type were slaughtered. The remaining 27 kids and 27 lambs were fed individually with a milk replacer at three different feeding levels, ranging from near maintenance to satiety, and the animals were kept at three different environmental temperatures, 12°C, 24°C and 30°C. Three animals of each species were allocated at random to each treatment. All animals were slaughtered on day 61.

Experimental Procedure

The animals were individually housed in metabolism cages at the indicated temperatures and at a relative humidity of 60±5%. Until the end of the experiment, all the animals were fed a milk replacer containing 30% of protein and fat. The fat incorporated was obtained from pork lard. The protein was obtained 50% from skim milk powder, 25% from a soya concentrate specially produced for the milk replacer and 25% from casein. The energy maintenance requirements were assumed to be those previously estimated for milk-replacer-fed kid goats (Sanz Sampelayo et al., 1988) and for milk replacer-fed lambs (Sanz Sampelayo et al., 1995). The replacer was prepared at 170 g fresh powder per kg and was placed over a device designed to maintain the replacer temperature at 39°C and to keep it satisfactorily mixed. The animals had access to the feed every day from 09.00 h until 18.00 h. Daily food intake was determined by subtracting the weight of the container at the end of the period of feeding from that immediately prior to the feeding period. All animals were weighed twice weekly and their food intake determined in accordance with the weight gain. Two balance trials of 7 days duration were performed, starting when the animals were aged 23 and 57 days. During these trials in addition to the daily measurement of feed intake, quantitative tests of feces and urine were made; all samples taken were stored at -20°C until needed for analysis. The metabolizable energy values of the milk replacer, according to animal type and environmental temperature, were calculated from the gross energy content and that of the feces and urine. At the end of the experiment, all animals were slaughtered by carotid section following anaesthetization by intramuscular injection of xilazine (Rompum; Bayer. Barcelona, Spain). After slaughter, the skin, feet, head and all internal organs were removed and the stomach compartments emptied of digesta. The blood, skin, carcass and all internal organs were considered as different body components. All management and experimental procedures conducted in this study were performed in strict accordance with the requirements of the European Union and Spanish rules and guidelines regarding the ethical treatment of experimental animals (Presidencia del Gobierno, 2007).

Measurements and Analyses

A sample of each body component was minced independently and then analysed for dry matter (DM), nitrogen (N) and energy content. Energy retained (ER) and weight of protein deposited were measured by the comparative slaughter method, i.e. body composition of animals slaughtered shortly after birth and at 61 days of age. Energy incorporated as protein (ERp) was calculated by multiplying the weight of protein by 23.8 (kilojoules per gram of protein) (Brouwer, 1965; Assendelf et al. 1973). Heat loss (HL) was calculated as the difference between metabolizable energy intake (MEI) and ER (HL = MEI – ER). As ER is primarily constituted of protein and fat, the energy retained as fat (ERf) was calculated as the difference between ER and ERp (ERf = ER – ERp). The values for each individual animal were calculated from its mean body weight over the whole experimental period. Finally, for animals fed at satiety, heat production from protein oxidation was calculated as the difference between the corresponding MEI and ER. In the same way, heat production from fat oxidation was calculated under the assumption that fat deposits were entirely of dietary origin (Astrup et al., 2000).

The DM and N contents were determined using fresh samples and all other analyses were carried out using dry samples. The DM of the milk replacer was calculated by heating it on a stove at 100±2°C and that of the feces, urine and body component mass, by lyophilization. The N content was determined using Kjeldahl's method (AOAC, 2005). The fat content of the milk replacer was determined by Gerber's method (AOAC, 2005). The energy content of the samples was determined by adiabatic bomb calorimeter.

Statistical Analysis

Energy utilization was analysed according to the Agricultural Research Council (ARC, 1980) metabolizable energy (ME) system. The relationships between the rates of ER and those of MEI (kJ/kg$^{0.75}$.day) and between the rates of ERp or ERf and those of MEI (kJ/kg$^{0.75}$.day) were established using standard regression techniques. Statistical analyses were performed using the Statgraphics package (Statgraphics, 2001).

Results and Discussion

Analysis of Energy Utilization at Metabolic Level. Effects of Temperature and Animal Type

The utilization of a given energy intake for maintenance and growth was analyzed using the factorial method, examining the relationships, under different levels of intake, between the amounts of metabolizable energy consumed and that retained (ARC, 1980).

Energy Requirements for Maintenance and Efficiencies of Energy Utilization for Energy Retention

The relationship between the values of metabolizable energy consumed and those of effective energy retention, assuming this may be established using a general linear model, provides a means of estimating both the maintenance intake requirements (the intake producing null retention) and the efficiency of energy utilization for energy retention. The first of these values corresponds to the independent term of the relationship between metabolizable energy intake (the y value in the expression) and the energy retention attained (the x value in the expression), assuming the latter quantity to be zero. When the values for energy retained are used as the value of the dependent variable (y) and those of the energy intake as the value of the independent variable (x), the corresponding coefficient of regression provides an estimate of the efficiency of the energy utilization over maintenance for energy retention. Tables 1 and 2 show the values for the corresponding independent terms, regression coefficients, correlation coefficients and the residual standard deviations of these expressions.

Table 1. Regression coefficients for relationships between total energy retained (ER) and metabolizable energy intake (MEI) in kid goats and lambs kept at different environmental temperatures+

f(x)	Species	Temperature	Mean	SE	Mean	SE	r	Residual s d (%)
(1)	Kid goats	12°C	-208.8	20.4	0.436	0.043	0.981	7.61
(2)	Kid goats	24°C	-214.4	1.8	0.508	0.004	0.998	0.64
(3)	Kid goats	30°C	-182.2	14.0	0.481	0.044	0.983	4.64
(4)	Lambs	12°C	-772.8	50.3	0.999	0.170	0.959	26.33
(5)	Lambs	24°C	-301.4	23.9	0.646	0.074	0.981	7.22
(6)	Lambs	30°C	-381.2	10.6	0.808	0.037	0.996	3.35

+Mean values with their standard errors according to species and environmental temperature: Regression equation: $y = a + b\,x$, where y is ER (kJ/kg$^{0.75}$.day) and x is MEI (kJ/kg$^{0.75}$.day).

Table 2. Regression coefficients for relationships between metabolizable energy intake (MEI) and total energy retained (ER) in kid goats and lambs kept at different environmental temperatures+

f(x)	Species	Temperature	Mean	SE	Mean	SE	r	Residual s d (%)
(7)	Kid goats	12°C	496.2	46.1	2.21	0.22	0.981	3.66
(8)	Kid goats	24°C	422.4	3.5	1.97	0.02	0.998	0.30
(9)	Kid goats	30°C	393.7	28.8	2.01	0.19	0.983	2.50
(10)	Lambs	12°C	783.3	46.1	0.92	0.16	0.959	3.55
(11)	Lambs	24°C	479.6	35.8	1.49	0.17	0.981	3.01
(12)	Lambs	30°C	474.4	12.8	1.23	0.06	0.996	1.26

+Mean values with their standard errors according to species and environmental temperature: Regression equation: $y = a + b\,x$, where y is MEI (kJ/kg$^{0.75}$.day) and x is ER (kJ/kg$^{0.75}$.day).

The results obtained show that the kid goats' maintenance requirements, under each of the environmental temperatures used, were lower ($P<0.05$) than those of the lambs, as were the corresponding values for the efficiencies ($P<0.05$). Within each species, the highest value for maintenance was obtained at 12°C ($P<0.05$). These values fell with increasing temperature, and in this respect there were no differences ($P>0.05$) between those for 24°C and for 30°C. The values for efficiency, at 12°C and at 24°C, varied among the goats ($P<0.05$), with the lower such values being recorded at 12°C and the higher at 24°C. Among the lambs, however, all three such values varied ($P<0.05$), with the highest being obtained at

12°C and the lowest at 24°C. Together with these findings, and from the second of the relationships established, it was also possible to estimate heat loss during fasting, i.e. the energy retention when energy intake is zero - logically, these quantities were found to be negative -. Among the kid goats, these values were very similar for all three temperatures tested ($P>0.05$) and in every case lower ($P<0.05$) than those corresponding to the lambs. Among the latter, this value was very high at an environmental temperature of 12°C, and different ($P<0.05$) from those found for temperatures of 24°C and 30°C; of these, a higher value was recorded at 30°C ($P<0.05$) than at 24°C.

With respect to the information currently available on the effects of temperature on the parameters analyzed, a considerable amount has been published concerning growing pigs (including suckling pigs), but virtually none on pre-ruminant animals. In a review of the data available for pigs, Close (1980) reported that the thermal environment for any animal is capable of determining the proportion of a given level of energy intake that is utilized for maintenance (including here, necessities for thermoregulation) and for energy retention or growth. According to this author, for a specific level of energy intake, a fall in the environmental temperature to below the level of thermal neutrality produces an increase in maintenance energy requirements, with less energy being available for production (Close and Maunt, 1978). As regards the values for the efficiency of the energy consumed for retention, Close (1980) comments that these may be similarly affected; as the environmental temperature falls below the zone of thermal neutrality, efficiency may increase. In Close's opinion, this fact is due to the fact that, under these circumstances, part of the heat associated with the synthesis of protein and fat might compensate for the greater thermal demands made for thermoregulation, and would thus be employed for this purpose rather than being lost. This finding of high net efficiencies under relatively cold conditions, it should be noted, does not seem to be an undisputed fact for all classes of animals, but tends to be limited to those which, as a consequence of their particular metabolic behaviour, have a lower thermogenic capacity associated with the mobilization of their reserves. In such cases, there would arise a diet-induced thermoregulation-thermogenesis interaction, meaning that together with higher maintenance requirements, there would be a lower production of heat from the intake of energy. This conclusion was already confirmed by Rubner in 1903.

From the above remarks, we see that the results presented in this paper, concerning the pre-ruminant animal, and as regards its maintenance requirements in response to diverse temperatures, are in accordance with those reported by Close (1980). However, our findings on the efficiency of the energy utilization for energy retention vary widely depending on the type of animal in question; thus, the behaviour pattern of lambs appears to be closer to that of pigs and, in general, to that shown by animals more prone to fat gain and, hence, with a lower capacity for lipid mobilization. These findings echo those reported in earlier studies (Sanz Sampelayo et al., 1988; 1991). Moreover, and again in agreement with Close's findings (1980), the values of maintenance energy requirements clearly show that for both types of animals, 12°C represents a cold temperature, and therefore one that is located on the left hand side of the thermal neutrality zone. Among lambs, this temperature is probably further to the left, highlighting the fact that these animals are quite a lot more sensitive to low temperatures than are kid goats. In addition, in view of the fact that the maintenance requirements for the two species do not present significant differences between 24°C and 30°C, it can be concluded that both temperatures lie within the thermal neutrality zone. Nevertheless, this inference cannot be made as concerns the values for heat losses during fasting, which do vary between 24°C and 30°C. Thus, the role played by these two temperatures, in this respect, remains until this point, an open question.

Protein and fat Retention according to the Metabolizable Energy Intake

In order to examine diverse effects of environmental temperature and animal type on the energy retained as protein or as fat depending on the metabolizable energy intake, the simple relationships between these values were established. These results are shown in Tables 3 and 4.

Fasting Conditions

From the relationships established above, we may conclude that among kid goats, assuming a zero intake of metabolizable energy (i.e., a fasting situation, MEI=0), at temperatures of 12°C, 24°C and 30°C the energy utilization proceeds basically from fat reserves; this value is somewhat lower at 30°C ($P<0.05$). On the contrary, among lambs, at 12°C a large amount of both protein and fat is catabolized, while at 24°C and 30°C, in general, only fat is utilized. These values, in every case, are higher than those recorded for kid goats ($P<0.05$).

Energy homeostasis enables an animal to enjoy the continuous provision of energy needed, independently of its exogenous supply. The end products of the digestion of different nutrients reach the bloodstream, from where they are distributed to diverse parts of the body to be used or stored. Under fasting conditions, the mobilization of reserves provides the animal with the energy it needs. The peculiarities and intensity of these mechanisms determines the particular energy metabolism and body composition of the animal in question (Lara, 1991). Under conditions of cold, the heat needed for thermoregulation seems to originate from the catabolism of both protein and fat (Fuller and Boyne, 1972; Ames and Brink, 1977) or otherwise only from that of protein (Miller et al., 1979). As the animal approaches its thermal neutrality zone, protein utilization tends to decrease, while the catabolism of fat continues or begins (Fuller and Boyne, 1972; Ames and Brink, 1977; Geraert et al., 1996). In other cases, there is a reduction in the catabolism of both protein and fat (Yunianto et al., 1997). Under fasting conditions, kid goats seem to be moving from 12°C to 30°C to increasingly warm zones. On the contrary, for lambs (taking into account the catabolism of their body fat), it can be said that 12°C would correspond to a distinctly cold temperature, while one of 30°C would perhaps be a hot temperature, with 24°C lying within the zone of thermal neutrality. In addition to these findings, and with respect to the reactions of the two animal species, kid goats seem to present minimal variations with respect to changes in temperature, and to have a special ability to maintain their energy balance – which does not occur in the case of lambs.

Table 3. Regression coefficients for relationships between energy retained as protein (ERp) and metabolizable energy intake (MEI) in kid goats and lambs kept at different environmental temperatures⁺

f(x)	Species	Temperature	Mean	SE	Mean	SE	r	Residual s d (%)
(13)	Kid goats	12°C	-31.68	8.54	0.126	0.020	0.962	7.39
(14)	Kid goats	24°C	-10.82	9.44	0.117	0.023	0.946	7.83
(15)	Kid goats	30°C	-36.12	10.82	0.147	0.034	0.907	8.12
(16)	Lambs	12°C	-320.30	45.54	0.439	0.154	0.855	17.66
(17)	Lambs	24°C	-35.29	2.50	0.158	0.008	0.996	1.79
(18)	Lambs	30°C	1.26	14.62	0.116	0.050	0.756	11.91

⁺Mean values with their standard errors according to species and environmental temperature: Regression equation: $y = a + b\,x$, where y is ERp (kJ/kg$^{0.75}$.day) and x is MEI (kJ/kg$^{0.75}$.day).

Table 4. Regression coefficients for relationships between energy retained as fat (ERf) and metabolizable energy intake (MEI) in kid goats and lambs kept at different environmental temperatures⁺

f(x)	Species	Temperature	Mean	SE	Mean	SE	r	Residual s d (%)
(19)	Kid goats	12°C	-177.09	23.97	0.310	0.050	0.951	15.68
(20)	Kid goats	24°C	-203.59	10.18	0.391	0.025	0.994	6.53
(21)	Kid goats	30°C	-146.12	10.30	0.334	0.033	0.981	5.62
(22)	Lambs	12°C	-452.54	5.12	0.561	0.017	0.999	6.52
(23)	Lambs	24°C	-266.10	25.89	0.488	0.080	0.962	13.47
(24)	Lambs	30°C	-382.44	17.57	0.691	0.061	0.985	9.03

⁺Mean values with their standard errors according to species and environmental temperature: Regression equation: $y = a + b\,x$, where y is ERf (kJ/kg$^{0.75}$.day) and x is MEI (kJ/kg$^{0.75}$.day).

Intake at Satiety. Changes Associated with Reduced Environmental Temperature

The aspect analyzed above for fasting conditions can also be considered for situations in which food is consumed. In this case, it is especially interesting to observe what occurs when the animal is at satiety. The amounts of energy retained as protein and fat were determined from the relationships found between these retentions and the intake of metabolizable energy. The corresponding MEI, total energy retentions and total heat losses were also considered. Heat losses associated with the utilization of either protein or fat were calculated as described in the Material and Methods section. Perhaps the significance of each of the temperatures used, and that of the responses presented by each type of animal when food intake to satiety occurs, can be determined more easily by analyzing what happens when the environmental temperature falls. Tables 5, 6 and 7 present the corresponding values, showing the changes that take place when the environmental temperature falls from 30°C to 24°C (Table 5), from 24°C to 12°C (Table 6) and from 30°C to 12°C (Table 7). In the first of these cases, among the kid goats none of the values in question presented any change at all, once again highlighting the characteristics of this animal and the possible location of these two temperatures within its thermal neutrality zone. On the contrary, among the lambs there was a small increase in food intake and a loss of heat; this loss was associated only with the utilization of fat – there was apparently no oxidation of protein. It is concluded, therefore, that for lambs 24°C (in comparison with 30°C) can be considered a colder temperature, and that when this temperature decrease occurs the animal presents a thermoregulatory reaction based on the oxidation of fat.

When the environmental temperature falls from 24°C to 12°C, there is a higher level of energy consumption among the kid goats, which is entirely dissipated in the form of heat; there is a much higher involvement of fat than of protein in this heat loss. The retention of protein and fat remains constant, and again we see the animal's ability to maintain its energy balance and thus its body composition, independently of changes in environmental temperature. It is also concluded, from the higher level of energy intake and from the resulting heat loss, that the temperature fall in question moves the animal to a distinctly colder location with respect to its thermal neutrality zone. Under the same conditions, the lambs also consumed more food, but this did not compensate for the heat lost, which shows, in the first place, that this animal cannot counteract such a temperature fall by increasing its energy intake. In this case, the greater amount of heat lost is associated with the utilization exclusively of fat. Therefore, we conclude that the temperature change from 24°C to 12°C represents a move towards a colder environment, and also that, as in the previous case but much more clearly, this animal not only does not maintain its energy balance, but undergoes a change in body composition, becoming leaner.

Examination of the consequences of the more drastic fall in temperature (from 30°C to 12°C) shows that among the kid goats the values are identical to those found in the previous case, in both quantitative and qualitative terms. Similar results were found for the lambs, but because of the higher levels of food intake, there was a greater total heat loss, which as before, was only in the form of fat loss. Logically, 12°C was found to be a colder temperature with respect to 30°C than to 24°C, which again leads us to consider for this animal, the latter two temperatures to be different.

Table 5. Energy intake, energy retention and heat loss changes in kid goats and lambs maintained at satiety, when environmental temperature falls from 30°C to 24°C

	Kid goats							Lambs						
	MEI	ERp	ERf	ER	HP	HPp	HPf	MEI	ERp	ERf	ER	HP	HPp	HPf
30°C	909.6	97.6	157.7	255.3	654.3	219.2	435.1	835.6	98.2	195.0	293.2	542.4	200.7	341.7
24°C	912.3	95.9	153.1	249.0	663.3	214.2	449.1	888.5	105.1	167.5	272.6	615.9	205.1	410.8
Change	≈ 0	≈ 0	≈ 0	≈ 0	≈ 0	≈ 0	≈ 0	52.9	≈ 0	-27.5	-20.6	73.5	≈ 0	69.1

MEI = Metabolizable energy intake (kJ/kg$^{0.75}$.day), ERp = Energy retained as protein (kJ/kg$^{0.75}$.day), ERf = Energy retained as fat (kJ/kg$^{0.75}$.day), ER = Total energy retained (kJ/kg$^{0.75}$.day), HP = Total heat loss (kJ/kg$^{0.75}$.day), HPp = Heat loss associated with protein oxidation (kJ/kg$^{0.75}$.day), HPf = Heat loss associated with fat oxidation (kJ/kg$^{0.75}$.day).

Table 6. Energy intake, energy retention and heat loss changes in kid goats and lambs maintained at satiety, when environmental temperature falls from 24°C to 12°C

	Kid goats							Lambs						
	MEI	ERp	ERf	ER	HP	HPp	HPf	MEI	ERp	ERf	ER	HP	HPp	HPf
24°C	912.3	95.9	153.1	249.0	663.3	214.2	449.1	888.5	105.1	167.5	272.6	615.9	205.1	410.8
12°C	1054.9	101.2	149.9	251.1	803.8	249.2	554.6	960.6	101.4	86.4	187.8	772.8	199.4	573.4
Change	142.6	≈ 0	≈ 0	≈ 0	140.5	35.0	105.5	72.1	≈ 0	-81.1	-84.8	156.9	≈ 0	162.6

MEI = Metabolizable energy intake (kJ/kg$^{0.75}$.day), ERp = Energy retained as protein (kJ/kg$^{0.75}$.day), ERf = Energy retained as fat (kJ/kg$^{0.75}$.day), ER = Total energy retained (kJ/kg$^{0.75}$.day), HP = Total heat loss (kJ/kg$^{0.75}$.day), HPp = Heat loss associated with protein oxidation (kJ/kg$^{0.75}$.day), HPf = Heat loss associated with fat oxidation (kJ/kg$^{0.75}$.day).

Table 7. Energy intake, energy retention and heat loss changes in kid goats and lambs maintained at satiety, when environmental temperature falls from 30°C to 12°C

	Kid goats							Lambs						
	MEI	ERp	ERf	ER	HP	HPp	HPf	MEI	ERp	ERf	ER	HP	HPp	HPf
30°C	909.6	97.6	157.7	255.3	654.3	219.2	435.1	835.6	98.2	195.0	293.2	542.4	200.7	341.7
12°C	1054.9	101.2	149.9	251.1	803.8	249.2	554.6	960.6	101.4	86.4	187.8	772.8	199.4	573.4
Change	145.3	≈ 0	≈ 0	≈ 0	149.5	30.0	119.5	125.0	≈ 0	-108.6	-105.4	230.4	≈ 0	231.7

MEI = Metabolizable energy intake ($kJ/kg^{0.75}$.day), ERp = Energy retained as protein ($kJ/kg^{0.75}$.day), ERf = Energy retained as fat ($kJ/kg^{0.75}$.day), ER = Total energy retained ($kJ/kg^{0.75}$.day), HP = Total heat loss ($kJ/kg^{0.75}$.day), HPp = Heat loss associated with protein oxidation ($kJ/kg^{0.75}$.day), HPf = Heat loss associated with fat oxidation ($kJ/kg^{0.75}$.day).

Conclusions

With regard to the corresponding thermoneutrality zones, the significance of the three temperatures tested was found to vary between the animal species. In accordance with its own metabolic response, the kid goats were capable of regulating their voluntary intake of feed according to the thermal demands imposed by the environment, maintaining their energy balance, and thus body composition, independently of the environmental temperature. The lambs, on the other hand, were more sensitive to the cold and under thermal stress were not capable of maintaining their energy balance by means of regulating their food intake; thus, their body composition was dependent on the environmental temperature.

References

Agricultural Research Council (ARC). 1980. Requirements for energy. In: T*he Nutrient Requirements of Ruminant Livestock.* Commonwealth Agricultural Bureaux. London.

Ames, D.R., Brink, D.R. 1977. Effect of temperature on lamb performance and protein efficiency ratio. *J. Anim. Sci.* 44: 136-140.

Assendelf, O.W., van Mook, G.A., Zilstra, G.M. 1973. International systems of units in physiology. *Pfluegers Arch. Eur. J. Physiol.* 339: 265-272.

Association of Official Analytical Chemists (AOAC). 2005. *Official Methods of Analysis.* 18th ed. Gaithersburg, M.D.

Astrup, A., Buemann, B., Flint, A., Raben, A. 2002. Low-fat diets and energy balance. How does the evidence stand in 2002? *Proc. Nutr. Soc.* 61: 299-309.

Bligh, J. 1973. *Temperature regulation in mammals and other vertebrates.* Neuwerger, A. and Tatum. E.L. eds. Elsevier Publishing Company. Inc. New York.

Brower, E. 1965. *Report of Sub-committee on Constants. In: Energy Metabolism of Farm Animals.* Blaxter, K.L. ed. EAAP 11. Academic Press. London, pp. 441.

Close, W.H. 1980. The significance of the environment for energy utilization in the pig. *Proc. Nutr. Soc.* 39: 169-175.

Close, W.H., Mount, L.E. 1978. The effects of plane of nutrition and environmental temperature on the energy metabolism of the growing pig. I. Heat loss and critical temperature. *Br. J. Nutr.* 40: 413-421.

Fuller, M.F., Boyne, A.W. 1972. The effects of environmental temperature on the growth and metabolism of pigs given different amounts of food. 2. Energy metabolism. *Br. J. Nutr.* 28: 373-384.

Geraert, P.A., Padilha, J.C.F., Guillaumin, S. 1996. Metabolic and endocrine changes induced by chronic heat exposure in broiler chickens: growth performance body composition and energy retention. *Br. J. Nutr.* 75: 195-204.

Lara, L. 1991. Factores nutritivos y metabólicos que determinan el crecimiento y desarrollo del ganado caprino y ovino prerrumiante. *Lactancia artificial.* PhD. University of Granada. Spain.

Millar, B.G., Otto, W.R., Grimble, R.F., Cork, D.A., Taylor, T.G. 1979. The relationship between protein turnover and energy balance in lean and genetically obese (ob/ob) mice. *Br. J. Nutr.* 42: 185-199.

Mount, L.E. 1979. Adaptation to thermal environment. In: *Man and his productive animals.* Edward Arnold Limited, London.

Presidencia del Gobierno. 2007. Ley 32/7 de 7 de noviembre, para el cuidado de los animales en explotación, transporte, experimentación y sacrificio. *BOE Nº* 268. pp. 45914-45919.

Raggi, L., Sanz, R., Guerrero, E., Boza, J. 1985. Fisiología ambiental de la cabra. *Monografía Med. Vet.* 7: 27-32.

Rubner, M. 1903. Handbuch der Ernährungsterapie. Vol. I. Deutiche, Leipzig. Germany. Cited by: Prieto Gómez, I. 1993. *Comportamiento nutritivo y metabólico de cabritos y corderos prerrumiantes sometidos a diferentes temperaturas.* PhD. University of Granada. Spain.

Sanz Sampelayo, MR., Muñoz, F.J., Guerrero, J.E., Gil Extremera, F., Boza, J. 1988. Energy metabolism of the Granadina breed goat kid. Use of goat milk and a milk replacer. *J. Anim. Physiol. Anim. Nutr.* 59: 1-9.

Sanz Sampelayo, M.R., Ruiz, I., Valverde, A., Lupiani, M.J. 1991. Comparative study of the energy utilization among kid goats and lambs. Components of the energy balance and plasma levels of the energetically important substrates. In: Energy Metabolism of Farm Animals. Wenk, C. and Boessinger M. eds. *Schriftenreiche aus dem Institut für Nutztierwissenchafdtem.* Gruppe Ernährung. ETH-Zürich. Switzerland.

Sanz Sampelayo, M.R., Lara, L., Gil Extremera, F., Boza, J. 1995. Energy utilization for maintenance and growth in prerrumiant kid goats and lambs. *Small Rumin. Res.* 17: 25-30.

Statgraphics. 2001. *User manual: Statistical Graphics System by Statistical Graphics Corporation.* Statgraphics, Herndon. VA.

Yunianto, V.D., Hayashit, K., Kaiwda, S., Ohtsuka, A., Tomita, Y. 1997. Effect of environmental temperature on muscle protein turnover and heat production in tube-fed broiler chickens. *Br. J. Nutr.* 77: 897-909.

In: Body Temperature Regulation
Editor: A. B. Cisneros and B. L. Goins ISBN: 1-60741-282-3

Chapter IX

To Warm up or to Pre-cool? The Paradox of Optimal Strategies to Undertake Prior to Exercise in the Heat

Rob Duffield[1,*] ***and Ric Lovell***[2,†]

[1] School of Human Movement Studies, Charles Sturt University, Panorama Ave, Bathurst, NSW, Australia, 2795

[2] Department of Sport, Health and Exercise Sciences, University of Hull, Hull, HU6 7RX, UK

Abstract

Exercise in hot conditions alters the physiological response to the ensuing exercise bout and may hasten the onset of fatigue. Regardless of the environment, often pre-exercise procedures are employed in order to ensure the commencement of exercise in an optimally prepared state to perform. Two such pre-exercise procedures often proposed as being of benefit involve warming up and pre-cooling, respectively. While both have been shown to have ergogenic benefits, they are somewhat contradictory in nature. It is well documented that optimal muscular function of the contractile fibres occurs with an increased muscle temperature. Accordingly, the practice of a warm up prior to exercise commencement is endemic to most sports and athletes and is often based on the premise of increasing muscle temperature. In contrast, it is equally well documented that reducing body temperature by cooling the periphery of the body, including the musculature, is also ergogenic for exercise performance. Accordingly, the practice of pre-cooling is regularly used by athletes from a range of sports and environments. This contradiction between increasing and decreasing respective body temperatures to improve exercise performance raises interesting questions regarding the mechanisms behind the regulation of exercise

* Contact details: Rob Duffield PhD. 61 2 6338 4939 (Tel); 61 2 6338 4065 (Fax); rduffield@csu.edu.au

in the heat. Consequently, the underlying physiological mechanisms to both warm up and pre-cooling procedures are seemingly related to thermoregulatory control. As such, this chapter will review the respective literature on both warm up and pre-cooling in relation to exercise in the heat. Included in this synthesis of relevant literature will be the physiological and performance responses to these respective pre-exercise interventions in the heat. Additionally, a comparison of the respective roles and interaction of both procedures will be discussed and finally, recommendations for the integration of both practices will be provided.

Introduction

Core and muscle temperature have been respectively shown to affect exercise performance in both a positive and negative manner, depending on the type of exercise bout performed (Nybo and Nielsen 2001; Mohr et al., 2006). It is well documented that force production is enhanced in muscle fibres as muscle temperature is elevated above resting values (Brinkhorst et al., 1977; Bergh and Ekblom 1979). Conversely, it is also well documented that elevated core temperatures can be detrimental to performance during prolonged exercise (Gonzalez-Alonso et al., 1999; Drust et al., 2005; Uckert and Joch 2007). Accordingly, this presents a well documented dichotomy in body temperature regulation during exercise; where the respective maintenance of elevated muscle temperatures and suppression of core temperature are required for optimal, prolonged muscle force production (Bishop 2003a; Mohr et al., 2006; Uckert and Joch 2007). When the surrounding environmental temperature or humidity are elevated, these conditions compound the maintenance of elevated muscle and reduced core temperatures (Reilly et al., 2006). Regardless, prior to commencement of most competitive or training exercise sessions, it is common for athletes to perform pre-exercise strategies to optimise the ensuing exercise bout. Two common pre-exercise strategies that are often utilised involve warm-up and pre-cooling (Bishop et al., 2003a; Duffield et al., 2003), for which the respective benefits are seemingly opposed. Given the almost uniform use of warm-up procedures by athletes in all conditions and common implementation of pre-cooling procedures in warm to hot conditions, these two opposing pre-exercise strategies are common in the athletic environment (Marino 2002; Bishop 2003a). Accordingly, this chapter will explore this paradox of thermoregulatory function and the methods used by athletes to optimise ensuing exercise performance in hot environmental conditions.

The need for warm-up activities prior to exercise is heavily engrained in the psyche and attitudes of coaches and athletes, and the engagement in these procedures are almost universal to most athletes. Warm-up procedures often involve the use of passive or low to moderate exercise intensities and are often developed specifically for the requirements of the respective situation. To date, there are equivocal findings on the use of warm-up practices to improve sports-based exercise performance (Bishop 2003b); however, the general consensus is that the physiological responses to a warm-up benefit the ensuing exercise bout (Bishop

† 447 0148246 5569 (Tel); R.J.Lovell@hull.ac.uk

2003b). These physiological responses are generally related to the increase in temperature associated with either the passive or active procedures used (Morrison et al., 2004; Mohr et al., 2004). It is well documented that increased muscle temperatures result in an optimised contractile force production, and hence the primary benefit of a warm-up is to increase muscle temperature (Mohr et al., 2004). In addition, other non-temperature dependent mechanisms may exist, including an increased VO_2, increased post-activation potentiation, improved nerve conduction or altered biochemical environments (Bishop 2003a). During exercise in the heat, while the elevation of muscle temperature to improve contractile function is of benefit, the resulting increase in core temperature prior to exercise commencement can be detrimental during the ensuing bout (Gonzalez-Alonso et al., 1999; Drust et al., 2005; Uckert and Joch 2007). Despite this limitation, the benefits to other non-temperature related physiological mechanisms and psychological preparation, as well as muscle temperature, often results in the use of warm-up procedures regardless of environmental conditions.

The process of pre-cooling involves the application of cold micro-environments to reduce skin, muscle or core temperature (Drust et al., 2000; Castle et al., 2006; Duffield and Marino 2007). The procedures of pre-cooling for exercise bouts in warm to hot environments has grown in popularity in recent times, with procedures including whole-body procedures such as cold water immersion, and cold showers or part-body procedures such as a variety of cooling garments. To date, the laboratory evidence supports the use of pre-cooling to improve exercise performance and blunt the rise in core temperature during prolonged continuous and intermittent performance in the heat (Arngrímsson et al., 2004; Castle et al., 2006). However, it is apparent that a dose-response relationship may exist, in that larger interventions (whole-body) may result in a more effective blunting in the rise in core temperature and improved performance (Duffield and Marino 2007). Despite these results, pre-cooling in field-based environments where the use of larger, full-body cooling procedures is difficult is yet to be fully substantiated (Reilly et al., 2006). However, the current evidence indicates that the process of pre-cooling athletes prior to exercise in the heat is effective in blunting the rise in core temperature and improving performance (Reilly et al., 2006, Quod et al., 2006). Conversely, the process of cooling can also reduce muscle temperature (Castle et al., 2006), which as stated previously, reduces the ability of muscle fibres to maximise force output (Brinkhorst et al., 1977). Hence regardless of the potential improvements in thermoregulatory function, pre-cooling prior to exercise may act to blunt the ability to maximise power output and reduce ensuing exercise performance (Crowley et al., 1991).

Therefore, both warm-up and pre-cooling procedures are common practices for athletes in a range of environmental temperatures, including warm conditions. Both procedures have been shown to have benefits for ensuing exercise performance, yet the proposed mechanisms seem to be in opposition. While the benefits of a warm-up are proposed to stem from the increase in muscle temperature, the benefits of pre-cooling are conversely proposed to be based on the suppression of skin, muscle and core temperature (Bishop 2003a; Mohr et al., 2006). As such, the paradox of whether to engage in either warm-up or pre-cooling procedures prior to exercise in the heat seems to be related to thermoregulatory function. Accordingly, this chapter will review the respective literature on both procedures in relation to exercise in the heat. A discussion of the physiological responses and effects on exercise performance of the respective procedures and a comparison of the respective roles and interaction of both procedures will be covered. Finally, recommendations for the integration of both practices and the role for a combined intervention will be provided.

Warm-up Prior to Exercise in the Heat

Whilst there is some disparity in the research literature, it is generally agreed that undertaking an appropriately structured warm up will enhance the ensuing exercise performance (Bishop 2003a). The term 'warm-up' is used almost universally to describe preparative strategies prior to exercise, since temperature related mechanisms are generally considered to be the main source of this ergogenic activity (for a review see Bishop 2003a). The temperature effects associated with a warm-up include a catalysing of neural functioning (Davies and Young 1983; Gray et al., 2006), metabolic reactions (Febbraio et al., 1996; Gray et al., 2006), and oxygen delivery to the muscle (Burnley et al., 2000; Gray and Nimmo 2001), with a decreased viscous resistance of muscles and joints (Proske and Morgan 1999; Wiktorsson-Moller et al., 2003); the latter of which has also been purported to reduce the risk of injury (Woods et al., 2007). However there are also a number of non-temperature related benefits of warm-ups, and as such the term 'acid up' has also been used to describe the change in biochemical properties of the muscle fibre following warm up (Gerbino et al., 1996). The warm-up associated increase in baseline VO_2 would presumably suppress the requirement for glycolytic activity in the initial stages (Gray and Nimmo 2001; Robergs et al., 1991), sparing the anaerobic capacity for later in the bout. Other purported non-temperature related benefits of a warm up include post activation potentiation (Gossen and Sale 2000; Young et al., 1998) and an increased psychological preparedness.

Warm-ups typically involve active moderate-intensity continuous exercise of the major muscle groups. This metabolic conversion of chemical to mechanical energy required for muscle actions and limb movement operates at an efficiency of 25% at best and consequently, a large portion of energy produced is released as heat (Lindinger 1999). Exercising muscles generate heat at a rate directly proportional to the intensity of the work performed (Saltin et al., 1968), and at moderate work-rates muscle temperature begins to rise within 5 minutes, continuing to increase at a rate between 0.1 - 0.4 $°C \bullet min^{-1}$ (Gray and Nimmo 2001; Gregson et al., 2002). This heat moves from the contracting skeletal muscle to surrounding tissues by conduction and convective flow of lymph and venous blood. Accordingly, there is an ensuing accumulation of heat in the core (Gleeson 1998), occurring at a lower rate of 0.02 - 0.05 $°C \bullet min^{-1}$ (Bishop and Maxwell 2008; Gregson et al., 2002).

Alternatively, this increase in body and muscle temperature can also be achieved passively, with the added advantage that there is no depletion of energy substrates. Pre-warming via passive procedures can be broadly classified into two categories: wet-heating and dry-heating. Wet-heating techniques such as hot water immersion via a shower, bath, or sauna are the quickest way of raising body temperature because of the high temperature conductance of liquids. Hot water immersion (42 - 44 °C) can result in an elevation of core and muscle temperature by 0.03 - 0.06 $°C \bullet min^{-1}$ (Brown et al., 2008; Gregson et al., 2002, 2005) and 0.08 - 0.13 $°C \bullet min^{-1}$ (Gray et al., 2006; Gregson et al., 2002, 2005), respectively.

Alternatively, dry-heating methods such as hot-pads, diathermy, massage and high ambient temperatures have been administered with varied success. For example, Drust et al. (2003) observed comparatively low heat transfer to muscle (0.033 $°C•min^{-1}$) using massage, with no temperature change in muscle treated with ultra-sound. Whereas exposure to a very hot and humid environment (45 °C, 70% relative humidity) has increased core (0.033 $°C•min^{-1}$) and muscle (0.083 $°C•min^{-1}$) temperature at comparable rates to hot water immersion (Gray and Nimmo 2001).

The complex myriad of physiological responses to the varied warm up techniques means there are many factors to consider when designing an appropriate warm-up, such as the event characteristics, facilities, pre-event protocols and the recovery period between the warm up and the beginning of the event. The complication is compounded when the event is to be undertaken in hot environmental conditions. However, despite the physiological responses to warm-up procedures, the key requirement of these activities is to improve ensuring exercise performance.

Performance Benefits of a Warm-up

Power Performance

The performance enhancement of power following warm up activities are well established. Both active (Mohr et al., 2004; Racinais et al., 2005) and passive (Gray et al., 2006; Racinais et al., 2007) means of raising muscular temperature have been reported to increase power output. The mechanisms behind this phenomenon are still debated, possible temperature-related causes include the decreased resistance of the muscles and joints and the increase in nerve conduction rate. This would suggest a passive elevation of the local temperature would also provide the same benefits (Racinais et al., 2007). Therefore if power-based exercise bouts are to be undertaken in the heat, the ambient conditions alone might improve performance (Falk et al., 1998; Linnane et al., 2004). However combining the effects of a passive warm up with exposure to hot conditions is not additive, and points to a 'ceiling' effect, above which an increase in temperature does not further improve muscular performance (Racinais et al., 2004), potentially due to their similar effect on neuromuscular efficiency (Racinais et al., 2005). Alternatively, an active warm-up promotes non-temperature related benefits (e.g. post-activation potentiation), thus provided an appropriate active warm up is administered and the recovery period from this activity is sufficient to enable the resynthesis of high energy phosphates, this method may be of further benefit to power-related performance (O'Brien et al., 1997). Whilst a very high intra-muscular temperature can cause fatigue (Febbraio et al., 1996), it is very unlikely that an active warm up in hot conditions will negatively influence power output. Therefore since cooling of the muscle is detrimental to this type of performance (Cheung and Sleivert, 2004; Racinais et al., 2007), it is suggested that active warm ups are undertaken prior to power performance in hot environmental conditions.

High-intensity Performance

It is generally considered that high-intensity exercise of a short duration (< 10 min) is improved after a warm-up (Brown et al., 2008; Hajoglou et al., 2005). Although, to date there is limited data available on exercise of this nature performed in warm to hot conditions. O'Brien et al. (1997) observed an increase in a 60s supra-maximal bout in hot conditions, where in contrast, 4-km time-trial performance (Altareki et al., 2008) and high-intensity intermittent exercise to exhaustion (Maxwell et al., 1996) have been reported to be reduced in the heat. Whilst the heat generated from muscle metabolism during high-intensity exercise is considerable, and is compounded in intermittent exercise scenarios (Drust et al., 2000; Ekblom et al., 1971), the degree of heat stress attained in such short-duration activities would not reach the typical range in which hyperthermia related fatigue is evident. It has been speculated that the body regulates heat content (Webb, 1995) and that a given increase in heat content (McLellan et al., 2000), or a high rate of heat storage (Altareki et al., 2008) rather than the attainment of an absolute core temperature, might be responsible for the reduced performance in this exercise domain. Therefore, whilst raising the body temperature in a warm up as a consequence of either active or passive heating is recommended given its more comprehensive evidence base, pre-event core temperature should not exceed 38 °C, and the intensity and duration of the warm up should be manipulated to reduce the rate of heat storage. This might be achieved passively using hot water immersion, which has been shown to be successful for this type of exercise (Brown et al., 2008). Although since the temperature of the muscle is more important for producing muscular power than central body temperature (Falk et al., 1998; Stewart et al., 2003), adopting an active warm up in these circumstances is more appropriate, especially since non-temperature related effects might further enhance the performance potential.

Endurance Performance

During prolonged exercise bouts there is some evidence that an active warm up does not impact upon prolonged repeated sprint performance in either moderate (Bishop and Claudius, 2005) or hot environmental conditions (Bishop and Maxwell, 2008). Alternatively, there appears to be a consensus from prolonged steady-state (Gregson et al., 2002), incremental (Ukert and Jock 2007) and intermittent (Gregson et al., 2005) fixed-intensity exercise models that warm-up is detrimental to this type of performance. Furthermore, this decrement in prolonged exercise performance is apparent subsequent to both active and passive warm-ups (Gregson et al., 2002, 2005). Consequently it is thought that the increased body temperature generated by *a priori* warm-up, rather than alterations in metabolism, is deleterious to exercise performance. The resultant decreased capacity for heat storage and increased thermoregulatory strain leads to an earlier attainment of a high critical core temperature that has been shown to result in fatigue (Galloway and Maughan, 1997; Gonzalez-Alonso et al., 1999), even in moderate ambient temperatures (Gregson et al., 2002, 2005). In addition, the development of a high muscle temperature (> 40 °C) may result in dysfunction in intra-muscular metabolic processes and also contribute to fatigue (Febbraio et al., 1996). The evidence from fixed-intensity, time-to-exhaustion exercise models suggests that a warm up preceding prolonged activity is ergolytic, although when performance is self-paced, such as prolonged intermittent sprint work, there are no detrimental effects (Bishop and Claudius,

2005; Bishop and Maxwell, 2008), as long as the warm up intensity is not too high (Datson et al., 2007). The disparity between these aforementioned findings is most likely explained by the different exercise models. Fixed intensity protocols do not enable researchers to test the efficacy of warm up procedures on initial performance, which may have consequences for the ensuing pacing strategies adopted. Therefore if athletes abandon warm up strategies to alleviate the potential thermoregulatory strain, initial muscular performance may be inhibited (Hajoglou et al., 2005), not to mention the negative connotations for mental preparation and injury epidemiology. Thus there is a thermoregulatory paradox, between preparing the muscle optimally for work, whilst reducing the thermal strain on the muscle and whole body to alleviate thermal related fatigue.

One solution might be to reduce the overall intensity and/or duration of the active warm up in high ambient temperatures, as this has been shown to sustain repeated sprint performance during soccer-specific intermittent exercise (Datson et al., 2007). Alternatively, engaging in active exercise before a performance bout, which would elevate muscle temperature while concurrently operating strategies to decrease the thermal load via conductive, convective and evaporative heat transfer might be beneficial. The vaporisation of sweat is the major source of heat dissipation even in humid environments. Therefore wearing appropriate 'breathable' clothing and maintaining hydration status will enhance evaporative cooling without developing circulatory and thermal strain. Furthermore, increasing the convective air-flow using fans or preparing without stationary ergometers may also be efficacious. Finally, decreasing the peripheral temperature to increase the core-to-skin temperature gradient and facilitate conductive heat transfer to the shell will also enhance heat loss. Methods such as pre-cooling, and cooling of the body during the warm-up with ice-jackets, application of ice-packs to the trunk, neck and head have been shown to be successful (Marino 2002, Quod et al., 2006) and may allow the benefits of warm up procedures without associated negative responses for exercise in the heat.

Pre-cooling prior to Exercise in the Heat

Pre-cooling involves the reduction or suppression of either skin, muscle or core temperature by the application of cold micro-environments to the periphery of the body (Marino 2002). Common methods in laboratory studies involve cooling by cold-water immersion, cold showers or cold rooms (Olschewski and Brück 1988; Kay et al., 1999; Wilson et al., 2002), while more practical, part-body methods include cooling vests, cold water sprays and cold packs (Duffield et al., 2003; Arngrímsson et al., 2003; Castle et al., 2006). From a collection of research, it is evident that the greater the duration and extent of cooling stimulus applied, the greater the reduction in muscle and core temperature (Yanagisawa et al., 2007). Moreover, while a longer or more involved cooling stimulus may result in greater performance improvement (Duffield and Marino 2007) this is not always the case (Kay et al., 1999). However, a growing collection of research literature indicates that in warm environmental conditions, when an effective cooling procedure is implemented (greater than 15 min and to a reasonable proportion of the body), exercise performance benefits are likely (Marino 2002; Quod et al., 2006).

Physiological Responses to Pre-cooling for Exercise in the Heat

By providing a sufficient cooling stimulus, regardless of the method used, a range of systemic, contractile and neural physiological responses occur, which in turn may be of benefit for exercise in the heat (Marino 2002; Quod et al., 2006; Wendt et al., 2007). On a whole-body or systemic scale, pre-cooling results in significant local and whole-body reductions in skin temperature (Hessemer et al., 1984; Duffield and Marino 2007), and given the stimulus is applied for a sufficient duration, can reduce muscle temperature (Beelen and Sargeant 1991; Castle et al., 2006) and reduce the rise in core temperature (Olschewski and Brück 1988; Wilson et al., 2002). In turn the reduction in skin and muscle temperature increases the heat gradient between the internal and peripheral regions, providing what has been termed a 'heat sink' to improve heat storage (Marino 2002; Reilly et al., 2006). Additionally, given the increased heat gradient between areas of heat production (muscle) and heat release (cutaneous), it is proposed pre-cooling provides an improved heat transport system (Booth et al., 1997). Due to the improved efficiency of convective and conductive mechanisms, the reliance on evaporation is reduced, in turn preserving blood volume (Lee and Haymes 1995). Additionally, pre-cooling can either reduce core temperature prior to exercise commencement or alternatively reduce the rate of rise during exercise (Olschewski and Brück 1988; Wilson et al., 2002). It is proposed that the improved ability to store or dissipate heat via improved convective and conductive mechanisms results in a slower rise in core temperature for the same workload and muscle heat production (Olschewski and Brück 1988; Lee and Haymes 1995; Marino 2002). Therefore, an important aspect of the pre-cooling process is to improve the capacity of the body to store, transport and release heat generated from the exercising musculature.

The reduction in skin, muscle and/or core temperature prior to or during exercise in the heat, results in a reduced physiological load on various other systems. It is well established that exercising in cooler environments or following pre-cooling in the heat reduces cardiovascular strain (Hessemer et al., 1984; Olschewski and Brück 1988; Lee and Haymes 1995). It is proposed that the reduced heart rate values noted during exercise following pre-cooling in the heat results from the reduced need for peripheral vasoconstriction to facilitate heat loss (Lee and Haymes 1995). Additionally, pre-cooling is proposed to improve central blood volume, due to reduced peripheral (epidermal) vasoconstriction, thus resulting in an increased availability of blood flow to supply exercising musculature (Olschewski and Brück 1988). In turn the improved availability of muscle blood flow in an environment where demands on circulation are increased, results in improvements in O_2 pulse and arterial-venous O_2 difference (Hessemer et al., 1984; Olschewski and Brück 1988). However, despite reports of improved muscle O_2 uptake, there is equivocal evidence to suggest improvement in VO_2 following pre-cooling for exercise in the heat (Hessemer et al., 1984; Beelen and Sargeant 1991).

While pre-cooling prior to exercise in the heat improves heat storage and transport and reduces the cardiovascular load encountered during the bout, the effect of cooling the contractile element of the muscle does not optimally prepare the muscle for the ensuing exercise bout (Holewijn and Heus 1992). During exercise the afferent feedback from the contracting skeletal musculature overrides the effect of cooling on muscle blood flow;

however, at rest local vasoconstriction from cooling procedures results in a reduction in muscle blood flow (Taber et al., 1992). Moreover, reducing muscle temperature reduces cellular metabolism via both a reduction in the haemoglobin/myoglobin concentration and also the activity of rate limiting glycolytic enzymes (Yanagisawa et al., 2007; Febbraio et al., 1996). Additionally, reduced muscle temperatures may result in minor negative changes to the muscle and tendon stiffness and passive resistance encountered during contraction (Bishop 2003a). Finally, cooling a resting muscle has also been shown to slow nerve conduction velocity (Karvonen 1992). Thus while, pre-cooling may provide a reduced systemic load during exercise, the process of lowering muscle temperature can diminish contractile function (Holewijn and Heus 1992). Accordingly, early research literature highlights that the force-velocity contraction profile of the contractile unit is slowed at lower temperatures (Brinkhorst et al., 1977; Bergh and Ekblom 1979).

Thus we have a contradictory state following pre-cooling prior to exercise in the heat, with a reduction in the load placed on many of the physiological systems, yet muscle fibres that are not optimally prepared for the ensuing bout. However, overarching both these respective physiological structures are the effects on central nervous system (CNS) function. Reductions in core, and to a lesser extent muscle, temperature have been shown to prevent the reduction in voluntary activation that is present during exercise in the heat (Tucker et al., 2004). Passive heating experiments have highlighted that increases in core temperature suppress the activation of the exercising muscle (Morrison et al., 2004). It is thought this suppression of voluntary force production results in response or anticipation of the growing thermoregulatory load, and acts to protect the body from excess thermal stress (Marino 2002). Accordingly, during exercise in the heat, muscle fibre contraction and recruitment are proposed to be reduced by inhibitory neural signals from the CNS in response or anticipation of the growing thermal load. Hence, this reduced neural recruitment results in a reduced exercise performance to maintain the tolerable endogenous heat load (Marino 2002). Reductions and recovery of maximal voluntary force have been shown with manipulations of core temperature, independent of cardiovascular or exercise based responses (Morrison et al., 2004; Tucker et al., 2004). Thus, while yet to be shown in pre-cooling studies; it is thought that the reduction or suppression of core temperature reduces the inhibitory signals that normally result in a reduction in muscle recruitment during exercise in the heat (Marino 2002).

In summary, the prolonged application (>15-min) of cool micro-environments to the body can result in a reduction in the load on the thermoregulatory system, principally through the reductions in cardiovascular load. Paradoxically, the application of these cold micro-environments to the muscle can reduce the potential readiness for optimal force production during muscle contraction. However, for prolonged exercise in the heat, the prevention in the decline in voluntary force by the CNS may provide the stimulus for improved exercise performance noted with pre-cooling in the heat.

Performance Responses to Pre-cooling for Exercise in the Heat

The effects of pre-cooling on exercise performance, to some extent depend on the type and duration of exercise performed. In general, exercise of brief durations and maximal intensity are not improved by pre-cooling, while pre-cooling for prolonged continuous and intermittent exercise tends to be ergogenic. Again, given that a sufficient cooling stimulus is delivered, the physiological responses detailed earlier tend to be present for 20 – 30 min (Marino 2002) and accordingly, research involving prolonged exercise longer than 8-min in duration in warm conditions predominately reports performance benefits (Marino 2002; Quod et al., 2006). Alternatively, performance of short duration (less than 5-min), maximal efforts are often not improved, or may even be reduced by pre-cooling, regardless of environmental conditions (Crowley et al., 1991; Sleivert et al., 2001). As outlined previously, pre-cooling may slow nerve conduction, slow the function of glycolytic enzymes and increase muscle-tendon resistance (Bishop 2003a). Accordingly, given the primary mechanisms purported to be responsible for pre-cooling benefits relate to improved thermoregulatory function, reduced cardiovascular load and reduced CNS inhibition, it is not surprising the effects of pre-cooling on maximal intensity efforts are negligible.

Traditionally, the bulk of pre-cooling research literature has focussed on the effects of pre-cooling on prolonged, continuous, endurance exercise (Quod et al., 2006). Whether it be constant-intensity or free-paced (Lee and Haymes 1995; Kay et al., 1999), time-to-exhaustion, set distance or distance covered (Booth et al., 1997; Arngrímsson et al., 2003) the act of pre-cooling has generally been shown to be ergogenic in the heat. Early studies involving pre-cooling methods often used fixed-intensity protocols for finite durations or until volitional exhaustion, reporting either suppressed physiological responses to matched intensities for at least 20-min (Hessemer et al., 1984) or longer times to exhaustion (Lee and Haymes 1995). The extent of the increase in time to exhaustion following pre-cooling seems dependent on the intensity of exercise and type of intervention used, however, improvements range between 1 -15% (Olschewski and Brück 1988; Quod et al., 2006). More recent research has incorporated free or self-paced exercise protocols, also demonstrating improved performances in the heat following pre-cooling (Booth et al., 1997; Kay et al., 1999; Arngrímsson et al., 2003). While a wide selection of free-paced exercise protocols exists in the research literature, in general performance benefits are reported to range between 1 - 6 % (Kay et al., 1999; Quod et al., 2006). Traditionally, these improvements have been linked to the reduced cardiovascular load and improved O_2 delivery (Hessemer et al., 1984; Olschewski and Brück 1988). However, more recent research highlights the prevention of reduced muscle recruitment and selection of faster pacing strategies a mechanism of performance improvement (Kay et al., 1999; Marino 2002).

Initial results of research using prolonged intermittent-sprint protocols did not report performance improvements following pre-cooling (Drust et al., 2000; Duffield et al., 2003; Cheung and Robinson 2004). However, more recent evidence indicates pre-cooling may also benefit team sport type exercise (Castle et al., 2006; Duffield and Marino 2007). These differences may result in the use of fixed versus free paced intensities and the amount and duration of repeated sprint efforts. Castle et al. (2006) have recently reported a better maintenance of repeated 5-sec sprint efforts with various cooling methods while the control condition exhibited a greater decline over a 40-min protocol. Moreover, Duffield and Marino (2007) demonstrated improved sub-maximal, free-paced running between sprint efforts following full-body cooling, similar to previous studies using continuous endurance protocols. Given the similarity in recent intermittent-sprint data with that of previous continuous protocols, it is reasonable to assume similar mechanisms resulting in the performance improvements. Again, hypotheses of performance improvement relate to the prevention of reduced muscle recruitment in the heat due to the improved heat transport and storage properties following pre-cooling (Marino 2002; Nybo 2008). Therefore, while the body of research literature for simulated team-sport exercise is much smaller than that reporting continuous exercise, recent evidence indicates similarities in performance benefits following pre-cooling in the heat.

In summary, the implementation of pre-cooling exposures prior to performing prolonged duration exercise in the heat results in a reduced thermoregulatory and cardiovascular load. While the process of cooling may act to reduce the functionality of the muscle fibre, the resulting potential reduction in CNS inhibition and reduced physiological stress is potentially the cause of the ergogenic benefits noted during exercise performance in the heat. Accordingly, a predominance of research literature incorporating pre-cooling manoeuvres prior to endurance exercise in the heat show this intervention to be ergogenic in nature.

A Combined role For warm-up and Pre-cooling in the Heat

In regard to warm-up prior to exercise in the heat, evidence indicates that the increase in thermoregulatory strain may have negative consequences for ensuing prolonged exercise performance (Uckert and Joch 2007; Bishop and Maxwell 2008). However, the benefits for contractile function, VO_2, psychological arousal and athlete readiness cannot be ignored. Conversely, pre-cooling may have benefits for prolonged exercise in the heat, but commencing exercise with a cooler muscle temperature may have an initial ergolytic effect (Mohr et al., 2006). While both responses seem mutually exclusive, the practice of respective procedures can be mutually inclusive to allow optimal performance in the heat. Multiple studies have recently purported the benefits of engaging in pre-cooling, followed by warm-up activities where the pre-cooling stimulus is maintained during and after the warm-up (Arngrímsson et al., 2003; Castle et al., 2006; Duffield and Marino 2007). This combined process of warm-up with pre-cooling allows the active engagement in increasing muscle temperature, increasing VO_2 and improving muscle-tendon function, while providing a continued 'heat sink' to blunt the ensuing rise in core temperature. By allowing an increased muscle and suppressed core temperature respectively, contractile function is not inhibited yet the systemic physiological and thermoregulatory load is reduced and the associated reduction in voluntary force production evident when exercising in the heat may be delayed or prevented.

While these processes may seem in opposition, the benefits of both practices may be incorporated into pre-exercise strategies to limit the performance reduction in the heat. As highlighted in previous studies, cooling interventions may be used during warm-up procedures through the use of portable cooling devices including ice-vests and cold packs to the quadriceps during activities such as running and cycling (Arngrímsson et al., 2003; Castle et al., 2006; Duffield and Marino 2007). Cooling of the quadriceps seems in contrast to the aim of ensuring muscle temperature is optimal for force production. However, as long as a post-cooling warm up is employed to ensure muscle temperature is elevated for the start of exercise, these actions can have a powerful effect to act as a heat sink for the major heat producing muscles during exercise. Additionally, these procedures may be used to continue the cooling exposure during the warm-up, following prior cooling exposure (Duffield and Marino 2007). Given that the physiological responses to cooling last from 20 -40-min, the longer the cooling exposure can be maintained until exercise commencement, the more likely performance will be improved. Hence, in warm environments, by cooling the body prior to warm-up and then maintaining cooling procedures during or after warm-up, it permits an increase in muscle temperature, yet continues to blunt the rise in core temperature (Castle et al., 2006; Duffield and Marino 2007).

In a practical sense, this may have some logistical difficulties, however with practice; athletes can easily pre-cool prior to the warm-up with cold showers, ice-baths and cool-rooms, followed by warm-up activities with torso and neck cooling (ice-vests), before final game/training preparation. Previous research has used whole-body cooling (ice-bath) followed by warm-up with cooling vests (Duffield and Marino 2007), multiple cooling interventions pre-exercise (ice packs, cold baths, vests) followed by cold packs to the quadriceps during cycling warm-up (Castle et al., 2006) or cooling vests before and during a running warm-up (Arngrímsson et al., 2003). Regardless of the actual mode, ideally the aim is to limit the disruption to pre-exercise preparation and allow sufficient warm-up to be physiologically and psychologically prepared for exercise, yet start the exercise bout with a lower thermoregulatory strain via reduced cardiovascular and thermal stress and delay the possible protective CNS reduction of muscle recruitment.

Conclusion

Despite evidence to indicate that engagement in respective procedures of either warm-up or pre-cooling is a thermoregulatory paradox, both interventions can be used in a symbiotic fashion during pre-exercise routines. While the pre-exercise cooling of a muscle fibre may result in reduced power output and a warm-up induced increase in pre-exercise core temperature may reduce ensuing performance, if used properly, pre-cooling and warm-up procedures may provide optimal benefits for exercise in the heat. Accordingly, by engaging in pre-cooling procedures to reduce core body and muscle temperatures before using a warm-up, combined with pre-cooling procedures to increase muscle temperature, VO_2, neural drive and arousal while limiting the thermoregulatory strain, athletes may be optimally prepared to commence training or competition sessions in warm environmental conditions. In conclusion, it is recommended that athletes utilise both pre-cooling and warm-up procedures when exercising in warm environmental temperatures. While the best methods will depend on the individual circumstances, the use of portable and practical interventions such cooling garments to be worn before, during and after warm-up procedures is highly recommended.

References

Altareki, N; Drust, B; Atkinson, G; Cable, T; Gregson W. Effects of Environmental Heat Stress (35°C) with Simulated Air Movement on the Thermoregulatory Responses during a 4-km Cycling Time Trial. *International Journal of Sports Medicine*, 2008, In press.

Arngrímsson, SA; Petitt, DS; Stueck, MG; Jorgensen, DK; Cureton, KJ. Cooling vest worn during active warm-up improves 5-km run performance in the heat. *Journal of Applied Physiology*, 2004,96:1876–1874.

Beelen, A; Sargeant, AJ. Effect of lowered muscle temperature on the physiological response to exercise in men. *European Journal of Applied Physiology*, 1991,63,387-392.

Bergh, U; Ekblom, B. Physical performance and peak aerobic power at different temperatures. *Journal of Applied Physiology*, 1979,46,885-889.

Bishop, D. Warm Up I. Potential mechanisms and the effects of passive warm-up on exercise performance. *Sports Medicine*, 2003a,33,439-454.

Bishop, D. Warm Up II. Performance changes following active warm up and how to structure the warm up. *Sports Medicine*, 2003b,33,483-498.

Bishop, D; Claudius, B. The effects of warm up on simulated team-sport performance. *Acta Kinesiologiae Universitatis Tartuensis*, 2005, 9,7—21.

Bishop, D; Maxwell, N. Effects of active warm up on thermoregulation and intermittent-sprint performance in hot conditions. *Journal of Science and Medicine in Sport*, 2008, In press

Booth, J; Marino, F; Ward, JJ. Improved running performance in hot humid conditions following whole body precooling. *Medicine and Science in Sports and Execise*, 1997,29:943-949.

Brinkhorst, RA; Hoofd, L; Vissers, ACA. Temperature and force-velocity relationship of human muscles. *Journal of Applied Physiology*, 1977,42,471-475.

Brown, PI; Hughes, MG; Tong, RJ. The effect of warm-up on high-intensity, intermittent running using nonmotorized treadmill ergometry. *Journal of Strength and Conditioning Research*, 2008, 22, 801-808.

Burnley, M; Jones, AM; Carter, H; Doust, JH. Effects of prior heavy exercise on phase II pulmonary oxygen uptake kinetics during heavy exercise. *Journal of Applied Physiology*, 2000, 89, 1387–1396.

Castle, PC; Macdonald, AL; Philip, A; Webborn, A; Watt, PW; Maxwell, NS. Precooling leg muscle improves intermittent sprint exercise performance in hot, humid conditions. *Journal of Applied Physiology*, 2006,100,1377-1384.

Cheung, SS; Sleivert, GG. Lowering of skin temperature decreases isokinetic maximal force production independent of core temperature. *European Journal of Applied Physiology*, 2004, 91, 723-728.

Cheung, SS; Robinson, AM. The influence of upper-body pre-cooling on repeated sprint performance in moderate ambient temperatures. *Journal of Sports Science*, 2004,22,605-612.

Crowley, GC; Garg, A; Lohn, MS; Van Someren, S; Wade, AJ. Effects of cooling the legs on performance in a standard Wingate anaerobic power test. *British Journal of Sports Medicine*, 1991,25,200-203.

Datson, N; Drust, B; Atkinson, G; Gregson, W. Effects of pre-cooling and warm-up intensity on soccer-specific intermittent exercise performance in heat. *Journal of Sports Science and Medicine*, 2007, S10, 185.

Davies, CTM; Young, K. Effect of temperature on the contractile properties and muscle power of triceps surae in humans. *Journal of Applied Physiology*, 1983, 55, 191-195.

Duffield, R; Dawson, B; Bishop, D; Fitzsimons, M; Lawrence, S. Effect of wearing and ice cooling jacket on repeat sprint performance in warm/humid conditions. *British Journal of Sports Medicine*, 2003,37,164 – 169.

Duffield, R; Marino, FE. Effects of pre-cooling procedures on intermittent-sprint exercise performance in warm conditions. *European Journal of Applied Physiology*, 2007,100,727-735.

Drust, B; Cable, NT; Reilly, T. Investigation of the effects of the pre-cooling on the physiological responses to soccer-specific intermittent exercise. *European Journal of Applied Physiology*, 2000,81,11-17.

Drust, B; Atkinson, G; Gregson, W; French, D; Binningsley D. The effects of massage on intra muscular temperature in the vastus lateralis in humans. *International Journal of Sports Medicine*, 2003, 24, 395-399.

Drust, B; Rasmussen, P; Mohr, M; Nielsen, B; Nybo, L. Elevations in core and muscle temperature impairs repeated sprint performance. *Acta Physiologica Scandinavica,* 2005,183,181-190.

Ekblom, B; Greenleaf, CJ; Greenleaf, JE; Hermansen, L. Temperature Regulation during continuous and intermittent exercise in man. *Acta Physiologica Scandinavia*, 1971, 81, 1-10.

Falk, B; Radom-Isaac, S; Hoffmann, JR; Wang, Y; Yarom, Y; Magazanik, A; Weinstein, Y. The effect of heat exposure on performance of and recovery from high-intensity, intermittent exercise. *International Journal of Sports Medicine*, 1998, 19, 1–6.

Febbraio, MA; Carey, MF; Snow, RJ; Stathis, CG; Hargreaves, M. Influence of elevated muscle temperature on metabolism during intense, dynamic exercise. *American Journal of Physiology*, 1996,271,R1251-1255.

Galloway, SDR; Maughan, RJ. Effects of ambient temperature on the capacity to perform prolonged cycle exercise in man. *Medicine and Science in Sports and Exercise*, 1997, 29, 1240 – 1249.

Gerbino, A; Ward, S; Whipp, B. Effects of pulmonary gas-exchange kinetics during high intensity exercise in humans. *Journal of Applied Physiology*, 1996, 80, 99-107.

Gleeson, M. Temperature regulation during exercise. *International Journal of Sports Medicine*, 1998, 19, 96-99.

Gonzalez-Alonso, J; Teller, C; Anderson, SL; Jensen, FB; Hyldig, T; Nielsen, B. Influence of body temperature on the development of fatigue during prolonged exercise in the heat. *Journal of Applied Physiology*, 1999,86,1032-1038.

Gossen, ER; Sale, DG. Effect of postactivation potentiation on dynamic knee extension performance. *European Journal of Applied Physiology*, 2000, 83, 524-530.

Gray, SC; Nimmo, MA. Effects of active, passive or no warm-up on metabolism and performance during short-duration high-intensity exercise. *Journal of Sports Sciences*, 2001, 19, 693–700.

Gray, SR; De Vito, G; Nimmo, MA; Farina, D; Ferguson, RA. Skeletal muscle ATP turnover and muscle fiber conduction velocity are elevated at higher muscle temperatures during maximal power output development in humans. *American Journal of Physiology*, 2006, 290, R376–R382.

Gregson, WA; Batterham, A; Drust, B; Cable, NT. The influence of pre-warming on the physiological responses to prolonged intermittent exercise. *Journal of Sports Sciences*, 2005, 23,455—464.

Gregson, WA; Drust, D; Batterham, A; Cable, NT. The effects of pre-warming on the metabolic and thermo-regulatory responses to prolonged submaximal exercise in moderate ambient temperatures. *European Journal of Applied Physiology,* 2002, 86, 526 – 533.

Hajoglou, A; Foster, C; De Koning, JJ; Lucia, A; Kernozek, TW; Porcari, JP. Effect of Warm-Up on Cycle Time Trial Performance. *Medicine and Science in Sports and Exercise*, 2005, 37, 1608-1614.

Hessemer, V; Langusch, D; Bruck, K. Effect of slightly lowered temperatures on endurance performance in humans. *Journal of Applied Physiology*, 1984,57,1731-1737.

Holewijn, M; Heus, R. Effects of temperature on electromyogram and muscle function. *European Journal of Applied Physiology and Occupational Physiology*, 1992,65,541-545.

Karvonen, J. Importance of warm up and cool down on exercise performance. In Karvonen J, Lemon PWR, Iliev I editors. *Medicine and Sports Training and Coaching*. Basel Karger. 1992:190-213.

Kay, D; Taafe, DR; Marino, FE. Whole-body pre-cooling and heat storage during self paced cycling performance in warm humid conditions. *Journal of Sports Science*, 1999,17,937-944.

Lee, D; Haymes, E. Exercise duration and thermoregulatory responses after whole body precooling. *Journal of Applied Physiology,* 1995,79,1971-1976.

Lindinger, MI. Exercise in the heat: thermoregulatory limitations to performance in humans and horses. *Canadian Journal of Applied Physiology*, 1999, 24, 152-163.

Linnane, DM; Bracken, RM; Brooks, S; Cox VM; Ball, D. Effects of hyperthermia on the metabolic responses to repeated high-intensity exercise. *European Journal of Applied Physiology,* 2004, 93, 159–166.

Marino, FE. Methods, advantages and limitations of body cooling for exercise performance. *British Journal of Sports Medicine*, 2002,36,89-94.

Maxwell, NS; Aitchison, TC; Nimmo, MA. The effect of climatic heat stress on intermittent supramaximal running performance in humans. *Experimental Physiology*, 1996, 81, 833-45.

McLellan, TM; Smith, IF; Gannon, GA; Zamecnik, J. Melatonin has no effect on tolerance to uncompensable heat stress in man. *European Journal of Applied Physiology*, 2000, 83, 336–343.

Mohr, M; Rasmussen, P; Drust, B; Nielsen, B; Nybo, L. Environmental heat stress, hyperammonemia and nucleotide metabolism during intermittent exercise. *European Journal of Applied Physiology*, 2006,97,89-95.

Mohr, M; Krustrup, P; Nybo, L; Nielsen, JJ; Bangsbo, J. Muscle temperature and sprint performance during soccer matches-beneficial effects of re-warm up at half-time. *Scandinavian Journal of Medicine and Science in Sports*, 2004,14,156-162.

Morrison, S; Sleivert, GC; Cheung, SS. Passive hyperthermia reduces voluntary activation and isometric force production. *European Journal of Applied Physiology* 2004,91,729-736.

Nybo, L; Nielsen, B. Hyperthermia and central fatigue during prolonged exercise in humans. *Journal of Applied Physiology* 2001,91,1055-1066.

Nybo, L. Hyperthermia and fatigue. *Journal of Applied Physiology* 2008,104,871-878.

O'Brien, B; Payne, W; Gastin, P; Burge, C. A comparison of active and passive warm ups on energy system contribution and performance in moderate heat. *Australian Journal of Science and Medicine in Sports*, 1997, 29, 106–109.

Olschewski, H; Brück, K. Thermoregulatory, cardiovascular, and muscular factors related to exercise after precooling. *Journal of Applied Physiology,* 1988,64,803-811.

Proske, U; Morgan, DL. Do cross-bridges contribute to the tension during stretch of passive muscle? *Journal of Muscle Research Cell Motility*, 1999, 20, 433-442.

Quod, MJ; Martin, DT; Laursen, PB. Cooling athletes before competition in the heat. Comparison of techniques and practical considerations. *Sports Medicine* 2006,36,371-382.

Racinais, S; Blonc, S; Jonville, S; Hue, O. Time-of-day influences the environmental effects on muscle force and contractility. *Medicine and Science in Sports and Exercise,* 2005, 37, 256–261.

Racinais, S; Blonc, S; Oksa, J; Hue, O. Does the diurnal increase in central temperature interact with pre-cooling or passive warm-up of the leg? *Journal of Science and Medicine in Sport*, 2007,

Racinais, S; Hue, O; Blonc, S. Time-of-day effects on anaerobic muscular power in a moderately warm environment. *Chronobiology International*, 2004, 21, 485–495.

Reilly, T; Drust, B; Gregson, W.: Thermoregulation in elite athletes. *Current Opinion in Clinical and Nutritional Metabolic Care*, 2006,9,666–671.

Robergs, RA; Pascoe, DD; Costill, DL; Fink, WJ; Chwalbinska-Moneta, J; Davis, JA; Hickner, R. Effects of warm-up on muscle glycogenolysis during intense exercise. *Medicine and Science in Sports and Exercise*, 1991, 23, 37-43.

Saltin, B; Gagge, AP; Stolwijk, JAJ. Muscle temperature during submaximal exercise in man. *Journal of Applied Physiology*, 1968, 25, 679–688.

Sleivert, GG; Cotter, JD; Roberts, WS; Febbraio. MA. The influence of whole-body vs torso precooling on physiological strain and performance of high intensity exercise in the heat. *Comparative Biochemical Physiology* 2001,128,657-666.

Stewart, D; Macaluso, A; De Vito, G. The effect of an active warm-up on surface EMG and muscle performance in healthy humans. *European Journal of Applied Physiology*, 2003, 89, 509–513.

Taber, C; Contryman, K; Fahrenbruch, J; LaCount, K; Cornwall, MW. Measurement of reactive vasocdilation during cold gel pack application to nontraumatized ankles. *Physical Therapy*, 1992,72,294-299.

Tucker, R; Rauch, L; Harley, YX; Noakes, TD. Impaired exercise performance in the heat is associated with an anticipatory reduction in skeletal muscle activation. *Pflugers Archives,* 2004,448,422-430.

Uckert, S; Joch, W. Effects of warm-up and pre-cooling on endurance performance in the heat. *British Journal of Sports Medicine*, 2007,41,380-384.

Webb, P. The physiology of heat regulation. *American Journal of Physiology*, 1995, 268, R838-R850.

Wendt, D; van Loon, LJC; van Marken Lichtenbelt WD. Thermoregulation during exercise in the heat. *Sports Medicine*, 2007,37,669 – 682.

Wiktorsson-Moller, M; Oberg, B; Ekstrand, J; Gillquist, J. Effects of warming up, massage, and stretching on range of motion and muscle strength in the lower extremity. *American Journal of Sports Medicine*, 1983, 11, 249-252.

Wilson, TE; Johnson, SC; Petajan, JH; Davis, SL; Gappmaier, E; Luetkemeier, MJ; White, AT. Thermal regulatory responses to submaximal cycling following lower-body cooling in humans. *European Journal of Applied Physiology,* 2002,88,67-75.

Woods, K; Bishop, P; Jones, E. Warm-Up and stretching in the prevention of muscular injury. *Sports Medicine*, 2007, 37, 1089-1099.

Yanagisawa, O; Homma, T; Okuwaki, T; Shimao, D; Takahashi, H. Effects of cooling on human skin and skeletal muscle. *European Journal of Applied Physiology,* 2007,100,737-745.

Young, WB; Jenner, A; Griffiths, K. Acute enhancement of power performance from heavy load squats. *Journal of Strength & Conditioning Research*, 1998, 12, 82-84.

In: Body Temperature Regulation
Editor: A. B. Cisneros and B. L. Goins
ISBN: 1-60741-282-3

Chapter X

The Effect of Absolute Exercise Intensity on Core Temperature Responses of Athletes with a Spinal Cord Injury

Michael J. Price
Department of Biomolecular and Sports Sciences, Faculty of Health and Life Sciences, Coventry University, Priory Street, Coventry, UK

Introduction

Within studies of exercise physiology exercise intensity can be set at either an absolute workload or a relative workload. Absolute workloads, for example, involve participants exercising at the same power output or treadmill speed regardless of fitness. Relative exercise intensities are specific to the individuals and take into account the functional capacity or fitness level. Participants may therefore exercise at intensities relating to given percentages of maximal or peak oxygen uptake (Galloway and Maughan, 1997; Price and Campbell, 2002), peak power (Boot et al., 2006), heart rate (Price and Campell, 1997), ventilatory threshold (Green et al, 2004a, 2004b; Abiss et al., 2008), lactate threshold (Buresh et al., 2005) or given blood lactate levels (George and MacLaren, 1988).

When considering groups of differing fitness levels or those undergoing training physiological responses such as heart rate are generally related to relative exercise intensities (Bottoms, 2008; Gant et al., 2004). However, relatively few studies have specifically examined the roles of absolute and relative workloads on thermoregulatory responses to exercise. This chapter will examine the potential roles of absolute and relative exercise intensities during upper body exercise in able-bodied athletes and those with a spinal cord injury. The introduction will firstly provide a chronological review of studies which have examined absolute and relative workloads in relation to thermoregulatory responses and secondly provide a rationale for why these roles may be important for different populations regularly undertaking upper body exercise.

Saltin and Hermansen (1966) exercised a range of participants (6 male, 2 female) for 60 minutes of cycle ergometry at each of 25, 50 and 70% maximum oxygen uptake ($\dot{V}O_{2max}$). A further comparison between two of these participants with the same body surface area (1.90 and 1.86 m^2), and therefore the same SA:mass, but differing $\dot{V}O_{2max}$ values (5.35 and 3.54 $l.min^{-1}$) was also undertaken. The latter exercise trials were undertaken at 52% $\dot{V}O_{2max}$. In all participants there was an increase in rectal, oesphageal and muscle temperature with absolute exercise intensity. However, responses across the group were much more closely related to the relative exercise intensity (i.e. %$\dot{V}O_{2max}$).

A later study (Greenhaff, 1989) expanded upon these findings by examining a larger group of participants (n=9) with a range of $\dot{V}O_{2max}$ values (43 – 64 $ml.kg^{-1}.min^{-1}$). Participants exercised on a cycle ergometer for 60 min at a relative intensity of 70% $\dot{V}O_{2max}$ and an absolute exercise intensity of 140W. There was a more marked increase in rectal temperature during the relative intensity trial (39.0°C) when compared to the absolute trial (38.2°C). However, the absolute exercise intensity trial demonstrated rectal temperature to be inversely related to $\dot{V}O_{2max}$ whereas it was positively related during the relative exercise intensity. Thus, the data was in support of the previous results of Saltin and Hermansen, (1966).

Gant et al., (2004) expanded upon previous findings in this area by applying the same research question to treadmill running with a larger participant number and a cross sectional designed study. Two specific groups of athletes of either moderate (59 $ml.kg.min^{-1}$) (n=10) or high $\dot{V}O_{2max}$ (73 $ml.kg.min^{-1}$) (n=10) were examined. Both groups undertook one hour of exercise at 65% $\dot{V}O_{2max}$ and a common running speed of 10.5 $km.h^{-1}$. Coincidently, the body surface area of moderately trained (1.90 m^2) and highly trained (1.86 m^2) athletes was similar to those compared by Saltin and Hermansen (1966). The results demonstrated no differences between groups for rectal temperature after 60 min of running at 65% $\dot{V}O_{2max}$ (38.72 and 38.41°C for moderate and highly trained groups, respectively). Significant differences though were observed during the absolute trial, with higher rectal temperatures for the moderately trained (38.77°C) than highly trained group (38.14°C). Consequently, the previous results of Saltin and Hermansen (1966) and Greenhaff (1989) can be expanded to treadmill running.

From the above studies is clear in that during cycle ergometry and treadmill running, rectal temperature responses are related to the relative exercise intensity (%$\dot{V}O_{2max}$). A different relationship to those described above however, has been reported when the same individuals are compared between cycle and arm crank ergomtery (Sawka et al., 1984). Sawka et al. (1984) observed that when comparing 60 minutes of arm crank ergometry and cycle ergometry at 60% peak oxygen uptake ($\dot{V}O_{2peak}$) lower rectal temperatures were observed for arm cycling. When both exercise modes were compared at the same absolute metabolic rate (~1.6 $l.min^{-1}$) near identical rectal temperature responses were observed.

Consequently, the authors noted that core temperature responses were independent of the skeletal muscle mass and dependant on the absolute metabolic intensity. It would appear therefore that across individuals of differing $\dot{V}O_{2max}$ or $\dot{V}O_{2peak}$, but undertaking the same exercise mode, core temperature responses are related to the relative exercise intensity. Within individuals undertaking different exercise modes the absolute metabolic rate is of key importance.

To illustrate the above responses more clearly the mean rectal temperature values for each study in relation to relative exercise intensity are shown in Figure 1. Although the plot incorporates data for both cycle ergometry and treadmill running both modes demonstrated similar relationships and subsequently both included. When a linear regression line is fitted to the data a correlation of R = 0.908 (R^2 = 0.826) is observed. If the lowest relative exercise intensity data point is considered as an outlier (25% $\dot{V}O_{2max}$; Saltin and Hermansen, 1966) only a slightly improved relationship is found R = 0.917 (R^2 = 0.841). However, as may be expected from the study of Sawka et al., (1984), when the rectal temperature data for arm crank ergometry is included a much weaker relationship of R = 0.716 is evident (R^2 = 0.513).

The above studies have all compared the role of absolute and relative exercise intensities in able-bodied individuals. When comparing athletes with a disability to able-bodied athletes they can be matched by training status (i.e. training frequency, intensity, duration, competitive level) but may demonstrate different $\dot{V}O_{2peak}$ values. Although this scenario may not often present itself, an example is when it does is in examining the physiological responses of athletes with a spinal cord injury (SCI). As a result of SCI there is a reduction in both the amount of muscle mass which can be recruited during exercise and sympathetic nervous system activity (Schmid et al, 1998a, 1998b). Consequently, $\dot{V}O_{2peak}$ and peak aerobic power values are reduced when compared to athletes matched by training status (Price and Campbell, 1997). Furthermore, functional capacity, as evidenced by $\dot{V}O_{2peak}$ and peak power output, is generally proportional to the level of lesion (Price and Campbell, 2003). From a thermoregulatory perspective paraplegics, who have thoracic spinal cord injury or below, demonstrate sweating and fluid loss during exercise in proportion to the level of spinal cord lesion (Price and Campbell, 2003; Hopman et al., 1992). In other words, a paraplegic with a high level SCI will demonstrate a greater surface area for sweating than one with a lower level of injury. Individuals with tetraplegia generally have an absent or much reduced sweating response due to the level of injury (cervical spinal injury) being above the sympathetic nervous system (Guttman et al., 1958).

Previous studies examining thermoregulatory responses of athletes with SCI during exercise have shown similar increases in core temperature at the same relative exercise intensity (Price and Campbell, 1997, 1999; Dawson et al., 1994). However, due to the reduction in muscle mass, although participants can be matched for training status, the energy expenditure and subsequent heat production representing 60% $\dot{V}O_{2peak}$ is significantly lower than for a matched group of able-bodied athletes. This subsequently suggests that more heat is being stored in the body of athletes with SCI, presumably due to the reduced ability to sweat and dissipate heat below the level of lesion. However, the role of absolute and relative exercise intensities during exercise in athletes with SCI, and also within the able-bodied population during upper body exercise has yet to be reported. Therefore, the aim of this study was to examine the role of absolute and relative exercise intensities during arm crank ergometry in cool conditions in athletes with SCI in comparison to able-bodied upper body trained athletes. This was done from a retrospective approach by combining and re-evaluating two previous studies (data not previously presented) from our laboratory (Price and Campbell, 1997, 1999) in order to provide a range of metabolic rates and increases in core temperature. The hypothesis that athletes with a SCI will demonstrate greater increases in core temperature for a given metabolic rate was tested.

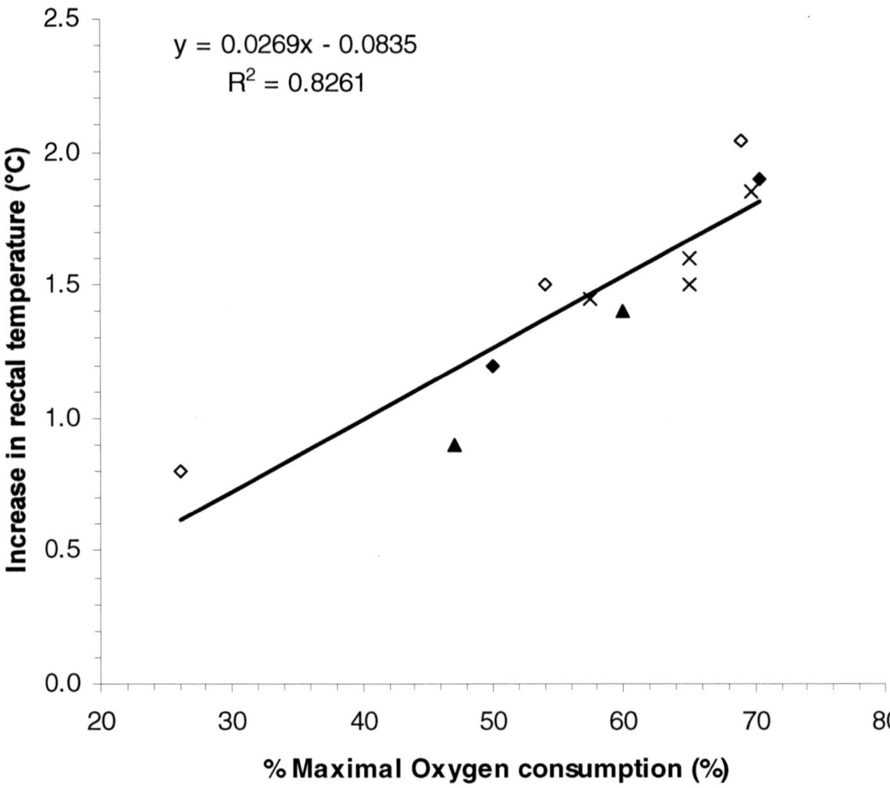

Figure 1. Increase in rectal temperature against relative exercise intensity ($\dot{V}O_{2max}$) during cycle ergometry and treadmill running. Data plotted from Saltin and Hermansen (1966) (empty diamond), Greenhaff (1989) (filled diamond), Sawka et al (1984) (filled triangle) and Gant et al., 2004) (cross).

Methods

Data from 14 able-bodied (AB), upper body trained athletes (21 trials) and 14 athletes with SCI (16 trials) who had previously undertaken a range of prolonged exercise trials in our laboratory (Price and Campbell, 1997, 1999) was analysed. Athletes with SCI had spinal cord lesions levels between T3/4 and L1. Exercise trials involved prolonged arm crank ergometry at either 80% peak heart rate for 90 minutes (17 trials) or 60% $\dot{V}O_{2peak}$ for 60 minutes (20 trials) at 21.5 ±0.9°C and 43.9 ±10.1% relative humidity. Only data up to 60 min of exercise was analysed in the present study. Combining the data sets provided a broad range of absolute metabolic rates from 1.24 – 2.84 $l.min^{-1}$ for AB athletes and 1.01 – 1.99 $l.min^{-1}$ for athletes with SCI. The range of increases in aural temperature for both groups was between 0.1 – 1.4°C and 0.3 – 1.0°C, respectively.

Preliminary Tests

Prior to all trials each athlete undertook a preliminary incremental arm crank ergometry test for $\dot{V}O_{2peak}$ on an adapted Monark cycle ergometer (Monark 814E, Vaarberg, Sweden). In the first instance (Price and Campbell, 1997) the protocol consisted of arm crank ergometry an initial power output of 35W followed by further increases of 35W every two minutes at a cadence of 70 $rev.min^{-1}$ until volitional exhaustion. In the second instance (Price and Campbell, 1999) the protocol involved 4 minute stages at 35, 70, 105, and 140 W separated by 2 minutes recovery. After the fourth rest period participants then exercised until volitional exhaustion at the next power output as in instance one. Expired gas was collected during the final minute of each exercise stage via the Douglas bag technique. Samples were analysed for fractions of expired oxygen and carbon dioxide (Servomex Analyser Series 1400, Crowborough, England) and evacuated (Harvard Dry Gas Meter, Harvard Apparatus Ltd, Kent, England) to determine ventilation rate ($\dot{V}_E$). Values for oxygen consumption ($\dot{V}O_2$), carbon dioxide production ($\dot{V}CO_2$) and respiratory exchange ratio (RER) were subsequently calculated. Both analysers were calibrated before each series of measurements with nitrogen, a calibration gas and room air. Heart Rate (HR) was continually monitored (HR; Polar Sports Tester PE4000, Kempele, Finland). In the 24 hours prior to all tests subjects were requested to maintain normal exercise and dietary habits. Subjects were also asked to drink plenty of water in order to maintain adequate hydration.

Prolonged Exercise Tests

On subsequent visits to the laboratory subjects performed either 60 min of arm-crank exercise at a pre-determined workload established from the preliminary tests, set to elicit 80% peak heart rate or 60% $\dot{V}O_{2peak}$. On arrival at the laboratory for the prolonged tests body mass was recorded (Seca 710, seated scales, Hamburg, Germany) after individuals had evacuated their bladder. Subjects then rested quietly for 15 min during which thermistors and a heart rate monitor were attached.

Aural temperature was measured by an aural thermistor (Edale Instruments, Cambridge, England) inserted into the subjects auditory canal (Benzinger and Taylor, 1963) and securely plugged and taped in position. The external ear was insulated with cotton wool. Values were recorded via a data logger (Squirrel 1250 Series, Grant Instruments, Cambridge, England). Aural temperature was employed in the present study as subject discomfort is often reported from the use of rectal and oesophageal probes (Armstrong et al. 1994; Sato et al. 1996).

Furthermore, the use of rectal temperature has been considered inappropriate for use in this population (Gass et al., 1988).

After the aural thermistor was secured participants then undertook a standardised 5 min warm-up at a work rate of 25-35 W on the arm crank ergometer. Once the warm-up was completed, subjects were allowed to undergo their usual stretching routine and then exercised at the appropriate exercise intensity for 60 min. Subjects were able to drink plain tap water *ad libitum*. Once exercise had been completed subjects were re-weighed. Expired gas samples were collected over a one minute period via the Douglas bag technique at 15, 30, 45 and 60 min along with ratings of perceived exertion (RPE; Borg Scale) (Borg, 1973). Aural temperature was recorded at rest, post warm-up and every 5 min during exercise. HR was continually monitored from rest, throughout exercise and recovery.

Statistical Analysis

All values are reported as means ± standard deviation. Although differences existed between the AB and SCI athlete groups for $\dot{V}O_2$ responses during exercise, no differences were observed between studies for each group. Therefore, data for each group was combined. All other parameters (HR, RER, RPE) were similar between the AB and SCI athletes across the two studies analysed and data also combined. Thermoregulatory and physiological data from prolonged exercise tests were compared using two-way Analysis of Variance with repeated measures on both factors (time × group). Significance was accepted at the $P<0.05$ level. Where significance was obtained Tukey post-hoc analysis was undertaken. Resting energy expenditure and resting $\dot{V}O_2$ between groups were compared by independent t-tests. Data for the increase in aural temperature from rest to the end of 60 min exercise (ΔT_{aur}) was calculated for each trial and plotted against both the absolute metabolic rate ($\dot{V}O_2$; $l.min^{-1}$). Regression equations for each plot were determined for each group of athletes. The ΔT_{aur} in relation to the metabolic rate ($\dot{V}O_2$) for each athlete was also calculated (i.e. $\Delta T_{aur} / \dot{V}O_2$; °C. $l.min^{-1}$) Differences between groups for this parameter were analysed via an independent t-test.

Results

There was no significant interaction between time and group for aural temperature ($P>0.05$). Resting aural temperature was similar for both groups (36.25 ±0.39 and 36.32 ±0.28°C for the AB athletes and athletes with SCI, respectively; $P>0.05$) increasing by 0.63 ±0.34°C for the AB athletes and by 0.56 ±0.20°C for the athletes with SCI over the 60 min exercise period ($P<0.05$).

Resting RER was similar between groups (0.94 ±0.20 and 0.95 ±0.21, for the AB and SCI, respectively, $P>0.05$). However, resting $\dot{V}O_2$ was lower for the athletes with SCI than that of the AB group (0.32 ±0.10 and 0.22 ±0.08 $l.min^{-1}$, $P<0.05$). Consequently, resting energy expenditure was lower for the athletes with SCI when compared to the AB athletes (1.07 ± 0.41 and 1.53 ±0.49, respectively, P<0.05). Differences were observed for $\dot{V}O_2$ during exercise between AB and SCI athletes at all exercise time points (Figure 2; P<0.05). The mean $\dot{V}O_2$ for the SCI and AB groups at 60 min were 1.27 ±0.34 and 2.17 ±0.46 $l.min^{-1}$, respectively (P<0.05) representing relative exercise intensities of 65.4 ±10.4 and 65.6 ±9.3%, for (P>0.05).

Energy expenditure during exercise was greater for the AB athletes at both 30 min (10.3 ±1.7 $Kcal.min^{-1}$) and 60 min (10.8 ±2.3 $Kcal.min^{-1}$) when compared to the athletes with SCI (5.3 ±2.7 and 6.4 ±1.7 $Kcal.min^{-1}$, respectively, P<0.05). No differences were observed between groups for heart rate with values at 60 minutes of exercise of 154 ±17 and 152 ±18 $beats.min^{-1}$ for the SCI and AB groups, respectively; P>0.05). The ΔT_{aur} in relation to metabolic rate at 60 min was greater for the athletes with SCI (0.41 ±0.17°$C.l.min^{-1}$) when compared to the AB athletes (0.29 ±0.16°$C.l.min^{-1}$; P<0.05).

Regression analysis of the increase in aural temperature against absolute metabolic rate ($\dot{V}O_{2;}$ x) is shown in Figure 3. Regression equations for absolute exercise intensity of y = 0.2627x + 0.0618 for the AB athletes and y = 0.2428x + 0.213 for the athletes with SCI were generated. Low correlations (R = 0.361 and R = 0.381, respectively; P>0.05) and subsequent R^2 values (0.131 and 0.145, respectively) were evident for both groups. Although similar gradients were observed for both groups, different intercepts were noted. Consequently, for a given metabolic rate, increases in aural temperature were approximately 0.15°C greater for the athletes with SCI than for the AB athletes. As little relationship exists between the parameters plotted this should be considered when discussing these specific results. However, a discussion of potential responses based on the data is provided in order to theorise the roles of absolute and relative exercise intensities.

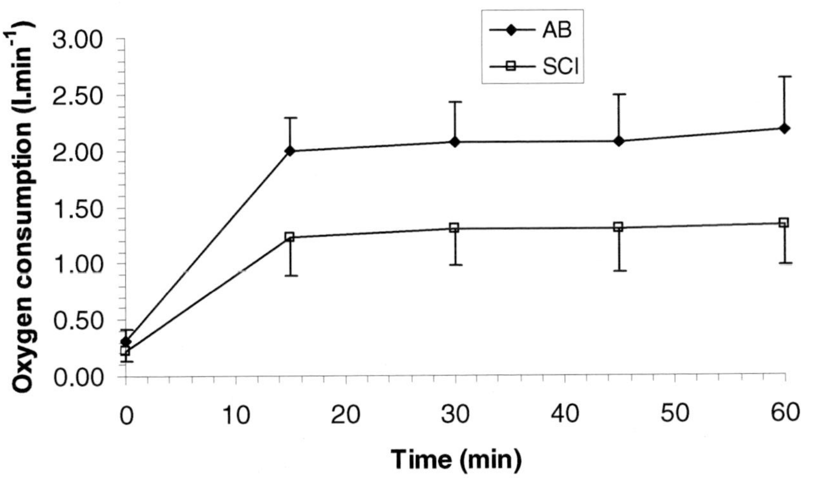

Figure 2. Oxygen consumption at rest and during 60 min of arm crank ergometry in able-bodied athletes (AB) and athletes with a spinal cord injury (SCI). A significant main effect for group was observed (P<0.05).

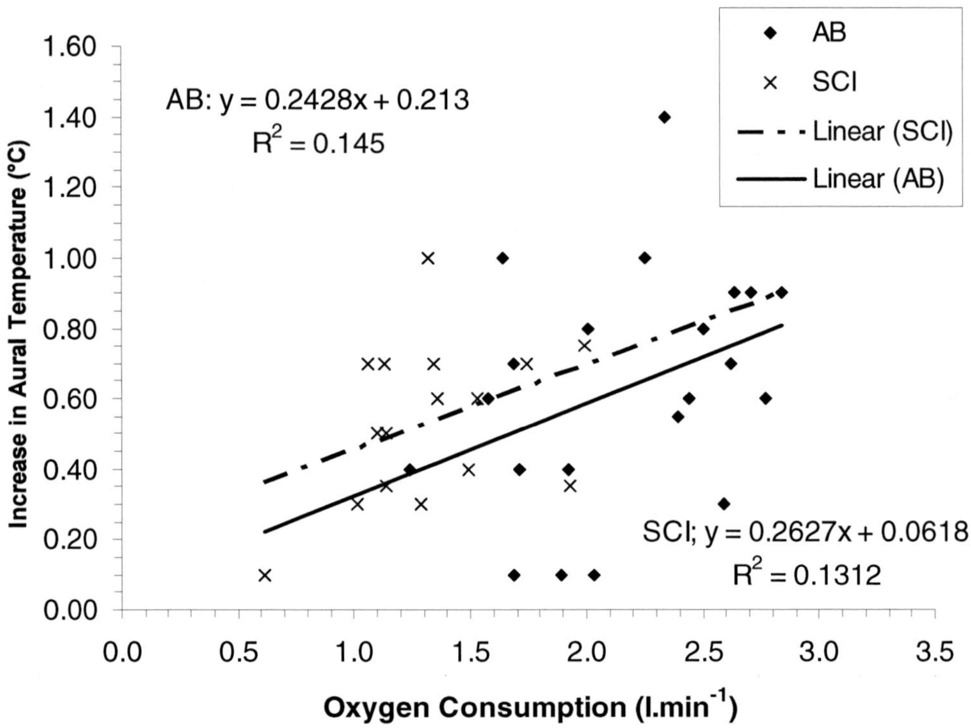

Figure 3. Relationship between the increase in aural temperature and absolute metabolic rate after 60 min exercise in able-bodied athletes (AB) and athletes with a spinal cord injury (SCI).

Discussion

The main finding of this preliminary and retrospective study was athletes with a spinal cord injury elicit a greater increase in aural temperature for every litre of oxygen consumed when compared to their able-bodies counterparts.

Resting Energy Expenditure

At rest oxygen consumption and energy expenditure were lower for the athletes with SCI. Lower resting metabolic rates have been observed previously for spinal cord injured individuals when compared to the able-bodied (Mollinger et al., 1984; Monroe et al., 1998) particularly on inactive days (Yamasaki et al, 1992). A recent review has shown that resting energy expenditure values are often between 5-32% lower than predicted for the able-bodied (Buchholz and Pencharz, 2004). This may be explained by the reduced active musculature and sympathetic nervous system observed in this population secondary to SCI and suggests that there may a difference in resting heat production. It is important to note that athletes in the present study demonstrated greater RER values than may be expected for a resting state. However, group means for RER were similar so any systematic error in subsequent calculations would most likely equal for both groups. For a lower, and more expected, RER value at rest specific energy expenditure values would be only slightly lower than those reported here.

Although resting energy expenditure, and therefore heat production, was lower in the athletes with SCI than the AB athletes, resting aural temperature was no different, a response commonly observed between such groups (Price and Campbell, 1997, 1999; Dawson et al, 1994) or between groups of athletes of different lesion levels (Price and Campbell, 2003). Although less heat may be produced at rest the reduced heat dissipation below the level of injury most likely results in maintenance of body heat levels. For example, Price and Goosey-Tolfrey (2008) recently demonstrated lower heat flow at the thigh (16 $W.m^2$) and calf (15 $W.m^2$) for athletes with paraplegia at rest in cool conditions when compared to able-bodied controls (33 and 39 $W.m^2$, respectively). This means that less heat would be lost from the lower limb in athletes with SCI in these conditions. Reduced heat flow from the insensate lower body may therefore contribute to maintenance of body temperature at rest in cool conditions.

Peak Physiological Responses

The peak physiological responses of the AB athletes and those with SCI have been described in the original publications. In summary, $\dot{V}O_{2peak}$ values of 3.19 ±0.38 and 3.45 ±0.40 $l.min^{-1}$ were observed for the AB athletes and values of 2.12 ±0.42 and 2.04 ±0.32 $l.min^{-1}$ for the athletes with SCI (Price and Campbell, 1997, 1999). The AB values were lower than for elite kayakers and canoeists (Vrijens et al., 1975; Tesch et al., 1976) but greater than for those who are non-specifically upper body trained (Davis et al., 1976; Toner et al., 1984).

Both data sets examined in the present study showed that $\dot{V}O_{2peak}$ values for the athletes with SCI were approximately 60-65% of those reported for the AB athletes. This no doubt reflects the loss of muscle mass and sympathetic nervous system which can be recruited as a result if SCI and noted earlier. However, $\dot{V}O_{2peak}$ values for the athletes with SCI are lower than for elite wheelchair athletes during wheelchair ergometry (Gass and Camp, 1984; Coutts, 1990, Campbell et al., 1997) but similar to active persons with a SCI during arm crank ergometry (Steinberg et al., 1998). Greater maximal exercise responses have been reported previously for homogenous groups of elite athletes when compared to heterogenous groups (Tesch et al., 1976; Seals and Mullin, 1982; Washburn and Seals, 1984). Furthermore, elite able-bodied athletes demonstrate similar maximal values during specific exercise modes to those achieved during running and cycling whereas recreational athletes have demonstrated lower values during specific exercise modes (Vrijens et al., 1975; Stamford et al., 1978, Bunc and Heller, 1991). The peak physiological responses are therefore in accordance with previous studies of able-bodied athletes and athletes with SCI.

Exercise Responses

Oxygen consumption during the 60 minute exercise bouts for athletes with SCI and the able-bodied group represented ~65% $\dot{V}O_{2peak}$ with similar changes in aural temperature (0.56 and 0.63°C). Therefore, core temperature responses may be considered to be related to relative workload as observed previously for cycle ergometry (Saltin and Hermansen, 1966; Greenhaff 1989) and treadmill running (Gant et al., 2004). However, as a result of SCI the reduced muscle mass and sympathetic innervation results in the absolute metabolic rate representing 65% $\dot{V}O_{2peak}$ being significantly lower than for the able-bodied (i.e. ~1.3 vs. ~2.2 $l.min^{-1}$, respectively). Therefore, similar core temperature values were observed in light of lower metabolic rates. As both groups were matched for training status similar core temperatures are most likely due to reduced heat dissipation. The greater increase in T_{aur} for a given metabolic rate in the present study would certainly be consistent with a reduced ability to dissipate heat from below the level of lesion in cool conditions noted earlier. As core temperature responses during exercise were similar, central heat storage would appear similar between athlete groups. However, differences in skin (peripheral) temperatures have been observed during exercise (Price and Campbell, 1997, 1999).

Previous studies have consistently shown differences in calf skin temperature responses between athletes with SCI and the able-bodied (Price and Campbell, 1997, 1999; Fitzgerald et al., 1990). For able-bodied individuals a decrease in calf skin temperature is generally seen during prolonged continuous upper body exercise in cool conditions (Price and Campbell, 1997, 2002) whereas and increase is observed for athletes with SCI (Price and Campbell, 1997, 1999; Fitzgerald, 1990). The former has been considered to be due to the redistribution of blood away from the relatively inactive lower body to the active upper body. However, although this has yet to be specifically proven studies measuring calf volume via strain gauge plethysmography during prolonged arm crank exercise have consistently shown a decrease in calf volume in able-bodied individuals (Hopman et al, 1993b, Bottoms, 2008) whereas this is not observed for individuals with SCI. Instead, an increase in calf skin temperature is observed with no change in calf volume, which most likely illustrates a reduced sympathetically induced vasoconstriction (Hopman et al., 1993b). In the absence of sweating for athletes with SCI any heat delivered to this area by the blood cannot be lost by evaporation of sweat. Skin temperature consequently rises resulting in peripheral heat storage. Heat gain by the lower body of athletes with SCI during resting heat exposure has also been observed suggesting that external heat gains in this part of the body are greater than those from metabolic sources in this population (Price and Tolfrey, 2008).

Absolute vs. Relative Workloads

As noted earlier similar increases in aural temperature were observed at similar relative workloads of ~65% $\dot{V}O_{2peak}$.When the change in aural temperature was plotted against the absolute metabolic rate similar regression equations were observed in that the aural temperature increased as metabolic rate increased for both groups. However, the two lines plotted demonstrated different intercepts and suggest that for a given absolute metabolic rate the athletes with SCI elicit a greater change in aural temperature when compared to the able-bodied of ~0.15°C. This would indeed be consistent with a reduced heat loss capacity but is within the repeatability of most aural and skin thermistors (~0.2°C) and most likely not to be of any physiological importance. Further examination using a specifically designed study would be of use to determine this response more specifically.

The data may also be viewed by determining the metabolic rate required to elicit a given change in aural temperature. In this instance, a gain in aural temperature of 0.5° occurred with a metabolic rate of ~1.15 $l.min^{-1}$ for the SCI and 1.65 $l.min^{-1}$ for the AB. When considering the gradients of the regression line for the relationship between ΔT_{aur} and $\dot{V}O_2$, the athletes with SCI would demonstrate an increase from 0.46 to 0.70°C (an increase of 0.24°C) between 1.0 and 2.0 $l.min^{-1}$ $\dot{V}O_2$ whereas the able-bodied would elicit aural temperatures of between 0.32 to 0.58°C (increase of 0.26°C). If a more realistic range of oxygen consumption values for the athletes with SCI is used (e.g. 0.5 – 1.5 $l.min^{-1}$) a similar increase in aural temperature to that of the able-bodied is predicted (0.24°C). Therefore, the rate of increase in T_{aur} in relation to increasing exercise intensity appears similar between groups, but the increase at a given exercise metabolism differs.

Limitations, Considerations and Future Work

When the change in aural temperature was plotted against absolute metabolic rate low correlation coefficients and R^2 values were noted and as noted, warrants a cautious interpretation. Low R^2 values would generally suggest a wide range of values or large individual variation was present. The large inter-individual variation also suggests that metabolic rate is certainly not the only contributing factor to increases in core temperature. Differences in dry heat exchange, individual rates of sweating, the surface area available for sweating and the range of spinal cord lesions examined are most likely other important contributors. The variation in physiological responses between persons with SCI has been observed previously (Normell, 1974) and is an important consideration here also. The fact that the study was retrospective may also have an effect on the data as it was not specifically designed to examine the role of absolute and relative exercise intensities. However, an insight into potential roles of absolute and metabolic rate has been presented.

Saltin and Hermansen (1966) evaluated core temperature responses during one hour of exercise at 25% $\dot{V}O_{2max}$ followed by, in the same exercise session, one hour of exercise at 70% $\dot{V}O_{2max}$. The intensity of 50% $\dot{V}O_{2max}$ was examined over one hour on a separate occasion. A study designed to assess a number of sub-maximal work intensities over 30 min would most likely be of great enough duration to allow steady state core temperature responses to be determined. Individual regression equations for the increase in core temperature in relation to metabolic rate, as well as for the group as a whole, could then be determined. This may provide a much more precise assessment of the role of absolute and relative exercise intensities in the SCI population.

Future work may also want to consider the affects of training on core temperature increases. Gant et al. (2004) examined this via a cross sectional design. Assessing changes in groups undertaking a training regime may improve our knowledge in this area. However, little data is available regarding thermoregulatory responses during upper body exercise training. Data from our laboratory (Bottoms, 2008) suggests that during and eight week training programme similar thermoregulatory adaptations for core temperature as observed for lower body exercise training do occur. Thermoregulatory responses in persons with SCI pre and post training have yet to be reported in the literature.

The current study also only considered the responses of athletes with paraplegia. Athletes with tetraplegia often have a reduced or absent sweating response and a faster rate of core temperature increase than athletes with paraplegia during steady state exercise in cool conditions (Price and Campbell, 1999) and warm conditions (Price and Campbell, 2003). A study requiring steady state core temperature values may be difficult in athletes with tetraplegia due to the continual nature of the increase in core temperature (Price and Campbell, 1999, 2003) but would be of value in determining the effects of lesion level and thermoregulatory adaptations in those with a reduced thermoregulatory capacity.

A final area of research which would be of interest for this population would be the combination of metabolic rate in relation to environmental temperature. Due to the reduced heat dissipation potential of persons with SCI gains in heat for the environment may be difficult to lose. This is particularly true if maximal whole body sweat rates are possibly reached at lower environmental temperatures than for the able-bodied (Price, 2006).

In summary, the results of the present study show that athletes with SCI demonstrate a greater increase in aural temperature per litre of oxygen consumption than for AB upper body trained athletes. Also, for a given absolute metabolic rate aural temperature of athletes with a spinal cord injury during exercise appears greater than for able-bodied athletes. To determine the specific roles of absolute and relative exercise intensities on thermoregulatory responses during exercise in athletes with SCI requires specifically designed studies to be undertaken. A range of future studies which may aid this process have been suggested.

References

Abbiss CR, Peiffer JJ, Peake JM, Nosaka K, Suzuki K, Martin DT, Laursen PB. (2008). Effect of carbohydrate ingestion and ambient temperature on muscle fatigue development in endurance-trained male cyclists. *J Appl Physiol.* 104(4):1021-8.

Armstrong, L.E., Maresh, C.M., Crago, A.E., Adams, R., Roberts, W.O., (1994). Interpretation of aural temperature during exercise and cooling therapy. *Medicine Exercise Nutrition and Health.* 3, 9-16.

Benzinger, T.H. and Taylor, G.W. (1963). Cranial measurements of internal temperature in Man. In Hardy, J.R. (Ed), *Temperature: Its Measurement and Control in Science and Industry. 3. Biology and Medicine*, New York, Reinhold, pp111-120.

Boot CR, Binkhorst RA, Hopman MT. (2006). Body temperature responses in spinal cord injured individuals during exercise in the cold and heat. *Int J Sports Med.* 27(8):599-604.

Borg GA. (1973). Perceived exertion: a note on "history" and methods. *Med Sci Sports.* 5(2):90-3.

Bottoms, L.M. (2008). Themoregulaotry responses during upper body exercise, thermal stress, training and heat acclimation. PhD thesis, Coventry University.

Bunc V, Heller J. (1991). Ventilatory threshold and work efficiency on a bicycle and paddling ergometer in top canoeists. *J Sports Med Phys Fitness.* 31(3):376-9.

Buchholz AC, Pencharz PB. (2004). Energy expenditure in chronic spinal cord injury. *Curr Opin Clin Nutr Metab Care.* 7(6):635-9.

Buresh R, Berg K, Noble J. (2005). Heat production and storage are positively correlated with measures of body size/composition and heart rate drift during vigorous running. *Res Q Exerc Sport.* 76(3):267-74.

Campbell IG, Williams C, Lakomy HK. (1997). Physiological responses of wheelchair athletes at percentages of top speed. *Br J Sports Med.* 31(1):36-40.

Coutts, K.D. (1990). Peak oxygen uptake of elite wheelchair athletes. *Adapted Physical Activity Quarterly,* 7, 62-66.

Davis JA, Vodak P, Wilmore JH, Vodak J, Kurtz P. (1976). Anaerobic threshold and maximal aerobic power for three modes of exercise. *J Appl Physiol.* 41(4):544-50.

Dawson B, Bridle J, Lockwood RJ. (1994). Thermoregulation of paraplegic and able bodied men during prolonged exercise in hot and cool climates. *Paraplegia* 32(12):860-70.

Fitzgerald PI, Sedlock DA, Knowlton RG. (1990). Circulatory and thermal adjustments to prolonged exercise in paraplegic women. *Med Sci Sports Exerc.* 22(5):629-35.

Galloway, S.D., Maughan, R.J. (1997). Effects of ambient temperature on the capacity to perform prolonged cycle exercise in man. *Med Sci Sports Exerc.* 29(9):1240-9.

Gant N, Williams C, King J, Hodge BJ. (2004). Thermoregulatory responses to exercise: relative versus absolute intensity. *J Sports Sci.* 22(11-12):1083-90.

Gass GC, Camp EM, Nadel ER, Gwinn TH, Engel P. (1988). Rectal and rectal vs. esophageal temperatures in paraplegic men during prolonged exercise. *J Appl Physiol.* 64(6):2265-71.

Gass GC, Camp EM. (1984). The maximum physiological responses during incremental wheelchair and arm cranking exercise in male paraplegics. *Med Sci Sports Exerc.* 16(4):355-9.

Green JM, Pritchett RC, Tucker DC, Crews TR, McLester JR. (2004a). Sweat lactate response during cycling at 30 degrees C and 18 degrees C WBGT. *J Sports Sci.* 22(4):321-7.

Green JM, Pritchett RC, Crews TR, McLester JR Jr, Tucker DC. (2004b) Sweat lactate response between males with high and low aerobic fitness. *Eur J Appl Physiol.* 91(1):1-6.

Greenhaff PL. (1989). Cardiovascular fitness and thermoregulation during prolonged exercise in man. *Br J Sports Med.* 23(2):109-14.

George KP, MacLaren DP. (1988). The effect of induced alkalosis and acidosis on endurance running at an intensity corresponding to 4 mM blood lactate. *Ergonomics.* 31(11):1639-45.

Guttman, L., Silver, J. an dWyndham, C.H. (1958). Thermoregulation inspinal man. *Journal of Physiology,* 142, 406-419.

Hopman MT, Oeseburg B, Binkhorst RA. (1993a). Cardiovascular responses in persons with paraplegia to prolonged arm exercise and thermal stress. *Med Sci Sports Exerc. 25(5):577-83.*

Hopman MT, Verheijen PH, Binkhorst RA. (1993b). Volume changes in the legs of paraplegic subjects during arm exercise. *J Appl Physiol.* 75(5):2079-83.

Mollinger LA, Spurr GB, el Ghatit AZ, Barboriak JJ, Rooney CB, Davidoff DD, Bongard RD. (1985). Daily energy expenditure and basal metabolic rates of patients with spinal cord injury. *Arch Phys Med Rehabil.* 66(7):420-6.

Monroe MB, Tataranni PA, Pratley R, Manore MM, Skinner JS, Ravussin E. (1998). Lower daily energy expenditure as measured by a respiratory chamber in subjects with spinal cord injury compared with control subjects. *Am J Clin Nutr.* 68(6):1223-7.

Normell LA. (1974). Distribution of impaired cutaneous vasomotor and sudomotor function in paraplegic man. *Scand J Clin Lab Invest.* 138:25-41.

Price, M.J. (2006). Thermoregulation during exercise in individuals with spinal cord injuries. *Sports Med.* 36(10):863-79.

Price MJ, Campbell IG. (1999). Thermoregulatory responses of spinal cord injured and able-bodied athletes to prolonged upper body exercise and recovery. *Spinal Cord.* 37(11):772-9.

Price, M. J. and Campbell, I. G. (1997)'Thermoregulatory Responses of Paraplegic and Able-Bodied Athletes at Rest and During Prolonged Upper Body Exercise and Passive Recovery. *Eur. J. Appl. Physiol Occup. Physiol* 76, (6) 552-560.

Price, M. J. and Campbell, I. G. (2002) 'Thermoregulatory responses during prolonged upper-body exercise in cool and warm conditions. *J. Sports Sci.* 20, (7) 519-527.

Price MJ, Campbell IG. (2003). Effects of spinal cord lesion level upon thermoregulation during exercise in the heat. *Med Sci Sports Exerc.* 35(7):1100-7.

Price, M.J. Goosey-Tolfrey, V.L. (2008) Heat Flow of the paraplegic and Able-bodied lower limb during resting heat exposure. *Journal of Thermal Biology,* 33; 255-260.

Saltin B, Hermansen L. (1966). Esophageal, rectal, and muscle temperature during exercise. *J Appl Physiol.* 21(6):1757-62.

Sato KT, Kane NL, Soos G, Gisolfi CV, Kondo N, Sato K. (1996). Reexamination of tympanic membrane temperature as a core temperature. *J Appl Physiol.* 80(4):1233-9.

Sawka MN, Pimental NA, Pandolf KB. (1984). Thermoregulatory responses to upper body exercise. *Eur J Appl Physiol Occup Physiol.* 52(2):230-4.

Schmid A, Huonker M, Barturen JM, Stahl F, Schmidt-Trucksäss A, König D, Grathwohl D, Lehmann M, Keul J. (1998a). Catecholamines, heart rate, and oxygen uptake during exercise in persons with spinal cord injury. *J Appl Physiol.* 85(2):635-41.

Schmid A, Huonker M, Stahl F, Barturen JM, König D, Heim M, Lehmann M, Keul J. (1998b). Free plasma catecholamines in spinal cord injured persons with different injury levels at rest and during exercise. *J Auton Nerv Syst.* 68(1-2):96-100.

Seals DR, Mullin JP. (1982). VO2 max in variable type exercise among well-trained upper body athletes. *Res Q Exerc Sport.* 53(1):58-63.

Stamford BA, Cuddihee RW, Moffatt RJ, Rowland R. (1978). Task specific changes in maximal oxygen uptake resulting from arm versus leg training. *Ergonomics.* 21(1):1-9.

Steinberg LL, Sposito MM, Lauro FA, Tufik S, Mello MT, Naffah-Mazzacoratti MG, Cavalheiro EA, Silva AC. (1998). Serum level of serotonin during rest and during exercise in paraplegic patients. *Spinal Cord.* 36(1):18-20.

Tesch P, Piehl K, Wilson G, Karlsson J. (1976). Physiological investigations of Swedish elite canoe competitors. *Med Sci Sports.* 8(4):214-8.

Toner MM, Sawka MN, Pandolf KB. (1984). Thermal responses during arm and leg and combined arm-leg exercise in water. *J Appl Physiol.* 56(5):1355-60.

Vrijens J, Hoekstra P, Bouckaert J, Van Uytvanck P. (1975). Effects of training on maximal working capacity and haemodynamic response during arm and leg-exercise in a group of paddlers. *Eur J Appl Physiol Occup Physiol.* 34(2):113-9.

Washburn RA, Seals DR. (1984). Peak oxygen uptake during arm cranking for men and women. *J Appl Physiol.* 56(4):954-7.

Yamasaki M, Irizawa M, Komura T, Kikuchi K, Sasaki H, Kai K, Ohdoko K. (1992). Daily energy expenditure in active and inactive persons with spinal cord injury. *J Hum Ergol* (Tokyo). 21(2):125-33.

In: Body Temperature Regulation
Editor: A. B. Cisneros and B. L. Goins

ISBN: 1-60741-282-3

Chapter XI

The Brain Dopaminergic System and Hyperthermia: The Neuroleptic Malignant Syndrome and The Parkinsonism-hyperpyrexia Syndrome

Nicola Tambasco, Annalisa Marchio, Vincenzo Belcastro, Chiara Menichetti, Laura Pierguidi, and Aroldo Rossi
Clinica Neurologica, Azienda Ospedaliera-Università di Perugia, Loc. S. Andrea delle Fratte, 06156 – Perugia, Italy

Abstract

The neuroleptic malignant syndrome (NMS) and the Parkinsonism-hyperpyrexia syndrome (PHS) are two clinically indistinguishable pathologies clinically characterized by hyperthermia. Their pathophysiologic mechanisms depend on decreased levels of dopamine that in NMS are due to dopamine receptor blockade by neuroleptic drugs, whereas in PHS they occur after withdrawal of dopaminergic drugs in the absence of neuroleptic medication.

This work has the aim to review the malignant syndromes depending on neuroleptic and parkinsonian drugs.

Introduction

Hyperthermia is a condition that occurs when the body balance between heat production and heat dissipation is altered. In the brain the heat production is due to neural metabolism, whereas the cerebral circulation provides the heat dissipation. When these mechanisms are altered, the local brain temperature increases modifying neural processes and functions.

There are different causes of brain hyperthermia which include, for example, the neuroleptic malignant syndrome, the Parkinsonism-hyperpyrexia syndrome, the serotoninergic syndrome, cerebral nervous system infections, malignant hyperthermia, heatstroke, and lethal catatonia.

In this chapter we discuss the neuroleptic malignant syndrome (NMS) and the Parkinsonism-hyperpyrexia syndrome (PHS). The pathologies of these two are clinically indistinguishable. Their pathophysiologic mechanisms depend on decreased levels of dopamine that in NMS are due to dopamine receptor blockade by neuroleptic drugs, whereas in PHS they occur after withdrawal of dopaminergic drugs in the absence of neuroleptic medication. Both NMS and PHS are rare but life-threatening syndromes, so timely treatment is very important for the prognosis.

Brain Dopaminergic System

Dopamine, discovered in 1952, is a member of the catecholamine family and a brain neurotransmitter. The biosynthesis of dopamine takes place first by the hydroxylation of the amino acid L-tyrosine to Levodopa (L-DOPA) via the tyrosine hydroxylase enzyme and then by the decarboxylation of L-DOPA by dopa decarboxylase enzyme (DDC). Dopamine is inactivated and degradated in homovanillic acid (HVA) by two different enzymatic pathways: catechol-O-methyltransferase (COMT) enzyme and monoamine oxidase (MAO) enzyme (Figure 1). In neurons, after synthesis, dopamine is packaged into vesicles and, in response to a presynaptic action potential, are released into the synapse, binding to specific dopaminergic receptors (Figure 2). Dopamine exerts its physiologic role on behaviour, cognition, motor activity, motivation, inhibition of prolactin production, sleep, mood, attention and learning.

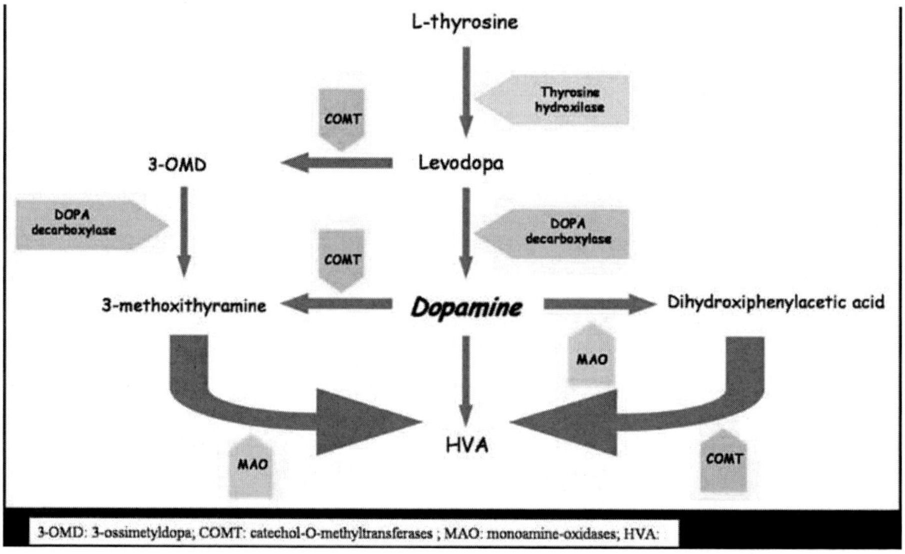

Figure 1. Metabolic pathway of levodopa.

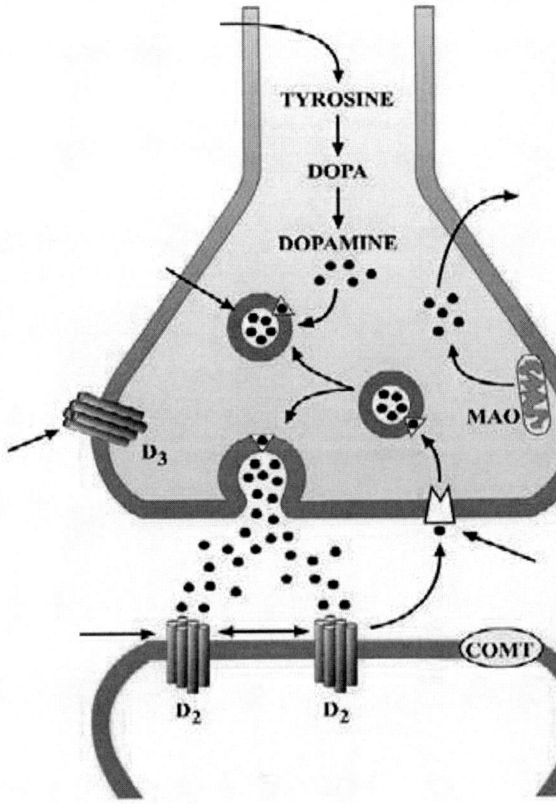

Figure 2. Dopaminergic synaptic complex.

Localization of Dopaminergic Neurons

Dopaminergic neurons were identified in the ventral midbrain and were classified into three nuclei: retrorubral area (A8), pars compacta of substantia nigra (A9) and ventral tegmental area (A10). Neurons of A8 and A9 areas contribute to the nigrostriatal system, whereas those of A10 area provide both the mesolimbic and the mesocortical system [1]. These are the major dopamine pathways in the brain.

The nigrostriatal pathway connects the pars compacta of substantia nigra to the corpus striatum into the system called basal ganglia [2]. These are a set of nuclei deep into the white matter of the cerebral cortex and they functionally include: corpus striatum (which includes caudate and putamen), globus pallidus, subthalamic nucleus, pars compacta (SNc) and pars reticulate of substantia nigra (SNr) .

The Substantia Nigra (SN) contains not only neurons projecting into the striatum, but also neurons that innervate the cortical and limbic areas. The organization of SN has been studied in detail and contains a slightly different parcellation of the nucleus based on a combination of calbindin and tyrosine hydroxylase immunocytochemistry, allowing a functional division of the SNc into two components: calbindin-rich areas (matrix) and calbindin-poor areas (nigrosomes).

The nigrostriatal pathway appears to play a crucial role in the control of movement and a loss of dopaminergic neurons in this pathway is the main pathological features of Parkinson's disease [2,3,4]. Moreover, it is indirectly implicated in tardive dyskinesias, one of the side effects of antipsychotics drugs.

Both mesocortical and mesolimbic pathways originate in the ventro tegmental area (VTA), which are located medially to the SN and consist of a few heterogeneous groups of cells grouped into nuclei [5]. Cytoarchitectonic data have long suggested four major nuclei within the VTA: paranigral nucleus (PN) and parabrachial pigmented area (PBP), that are dopaminergic cell-body rich zones, and parafasciculus retroflexus area and ventral tegmental tail (VTT), that are dopaminergic cell-body poor zones [6].

The mesocortical pathway originates in the dorsorostral part of the VTA and projects into the prefrontal cortex. It is essential for normal cognitive function; in fact, the reduction of dopaminergic activity in this pathway is involved in the motivation and emotional response in normal subjects and in the negative symptoms of schizophrenia.

The mesolimbic pathway projects from the ventrocaudal part of VTA to the nucleus accumbens, a part of the limbic system. Abnormal functioning in this area is linked to psychosis and the positive symptoms of schizophrenia.

Finally, besides these principal pathways, there is another one called the tuberoinfundibolar that projects from the hypothalamus to the hypophysis, and is concerned with neuronal control of the hypothalamic pituitary endocrine system.

Brain Control of Temperature

The brain represents about 2% of human body mass, but it consumes about 20% of its total oxygen. The most energy used for neuronal metabolism is spent on electrical neural activity, the remaining energy is used in other neural processes, especially for synthesis of macromolecules and transport of protons across mitochondrial membranes [7,8].

Since all energy used for neural metabolism is finally transformed into heat, the brain maintains a relatively stable temperature through heat dissipation from the brain to the body and then to the external environment via skin and lung surfaces. This process is warranted by cerebral circulation, particularly the arterial blood, which removes heat from the brain whereas venous blood is warmer. Therefore, increased brain temperature resulting from increased neural metabolism can increase local cerebral blood flow that provides oxygen and nutrients for the brain metabolism and also removes potentially dangerous metabolic heat from neural tissue contributing, thus, to brain temperature fluctuations occurring under behavioural conditions. When heat regulating mechanisms become unable to effectively remove the excess heat, it leads to brain hyperthermia.

The changes in brain temperature affect several neural processes and functions. Irreversible damage of neurons starts at about 40 °C and increases exponentially with slight increases above that level. Destruction of endothelial cells of the brain and leakage of serum proteins across the blood-brain barrier are important factors in determining brain oedema, the most dangerous acute complication of brain hyperthermia [9]. Brain damage induced by heat concerns not only the neurons, but also glial cells and cerebral microvessels. In addition hyperthermia-related leakage of the blood brain barrier results in alterations of the ionic environment leading some toxic substances into brain tissue.

Diminished heat outflow from the brain, accompanied by intra-brain heat production, appears to be the primary cause of intra-brain heat accumulation, but hyperthermia can also be caused by drugs, which increase brain metabolism and impair the heat dissipation mechanism.

Neuroleptic Malignant Syndrome

Neuroleptic malignant syndrome (NMS), described for the first time in the French literature by Delay and colleagues in 1960 [10], is a rare and life-threatening complication of treatment with neuroleptic or antipsychotic drugs. The incidence of NMS is low and probably falling, since there was, with time, a decreased use of older antipsychotics, preventive measures and an increase in awareness. Therefore, the exact incidence is difficult to calculate but is estimated about 0.5–1% among patients treated with neuroleptics.

Pathophysiology and Risk Factors

Although the pathophysiology mechanism of NMS is not completely understood it is thought to depend on decreased levels of dopamine [11,12,13,14] due to dopamine receptor blockade or a genetically reduced function of dopamine receptor D_2. This hypothesis is based on several lines of evidence: withdrawal of dopaminergic drugs can precipitate an NMS-like syndrome; all drugs associated with NMS produce dopamine receptor blockade; the risk of NMS appears to be correlated with the dopamine receptor binding affinity of drugs and dopaminergic drugs have been used in treatment of NMS symptoms.

The principal risk factors, besides exposure to neuroleptics, include psychomotor excitement, dehydration, hyponatremia, thyrotoxicosis, pre-existing abnormalities of cerebral nervous system dopamine activity or receptor function and concomitant use of antiparkinsonian drugs, prochlorperazine, anticholinergic drugs, metoclopramide or lithium carbonate [13,14,15,16].

Clinical and Laboratory Aspects

NMS is clinically characterized by hyperthermia, alterations in both autonomic nervous and extrapyramidal systems (muscular rigidity) and mental status changes (Table 1). The body temperature is above 38 °C, in same cases can be as high as 40 °C [17], and usually appears in the early NMS stage at the beginning or after the appearance of motor symptoms and muscle rigidity [15]. When hyperthermia appears early it is a centrally mediated event, whereas when it appears after rigidity, the high temperature is also due to the peripheral muscle pathology.

Muscle rigidity was reported in 90% of patients [17,18] and it is, frequently, lead-pipe type or plastic type whereas cogwheel type is rare. Dysphagia, dysarthria and sialorrhea are the consequence of the rigidity of the muscle of the oesophagus and pharynx; finally muscular rigidity is responsible of the onset of musculoskeletal complications that consist of muscle loss with secondary avulsions and orthopaedic deformities.

Autonomic instability affects, mainly, both cardiovascular and respiratory systems and may manifest with tachycardia, arrhythmia, fluctuating of blood pressure, tachypnea, dyspnoea, pulmonary oedema. Other symptoms are pallor, excessive sweating and salivation, dysuria and urinary incontinence [11,12,13,15,17,18]. Mental status change may vary: some patients are alerts and responsive, others are lethargic. In the advanced stages mental status results in depressed consciousness, stupor and coma. In addition to these principal features, there are several associated signs as akinesia and/or bradykinesia (frequently described in NMS), tremor (that occurs in 50% of the patients), dystonia (30% of the patients), occasionally it can be seen myoclonus and has several infrequent signs such as seizures, gaze paresis, reflex changes, Babinski signs, ataxia, and nystagmus.

Table 1. Clinical signs of Neuroleptic Malignant Syndrome

- **Hyperthermia and:**
  - *Muscle Rigidity:* lead pipe, plastic, cogwheel;
  - *Autonomic Instability:*
    tachycardia, arrhythmia, fluctuating of blood pressure, tachypnea, dyspnea, pallor, urinary incontinence, dysuria, excessive sweating/ salivation;
  - *Mental Status Change:*
    confusion, agitation, delirium, catatonia, muteness, lethargy, stupor, coma;
- **Associated signs:**
  - *Movement* Disorders:
    akinesia, bradykinesia, tremor, dystonias, myoclonus, chorea;
  - *Other* Neurological Signs:
    ataxia, nystagmus, ocular fluttering, seizures, gaze paresis, reflex changes, Babinski +.

Several laboratory findings have been linked to NMS (table 2), but they are non-specific and used as a support to the diagnosis. The main abnormality is the elevation of creatine phosphokinase (CPK), that is the consequence of muscle rigidity and temperature elevation. In a report by Rosebush and Stewart [18], on 24 cases of NMS was reported that 86% of patients had CPK elevations above 1000 IU/L and 33% had elevations of CPK above 10000 IU/L. Moreover, elevations of transaminases, aldolase and lactic acid dehydrogenase and very low levels of serum calcium have been reported as a result of rhabdomyolysis. The 95% of patients present iron deficiency [17,19], which can be related to muscle injury. Another laboratory feature is represented by polymorphonuclear leukocytosis that is present in 75% of cases. The white blood cell counts ranging from 10000 to 30000 per mm. Results of CSF analysis is usually normal as well as findings of neuroimaging studies, while electroencephalography can show diffuse slowing consistent with metabolic encephalopathy.

Table 2. Laboratory features of Neuroleptic Malignant Syndrome

Laboratory Features	
	INCREASED CREATINE PHOSPHOKINASE
	INCREASED TRANSAMINASE
	INCREASED ALDOLASE
	INCREASED LACTIC ACID DEHYDROGENASE
	HYPOCALCEMIA
	HYPOFERREMIA
	LEUKOCYTOSIS

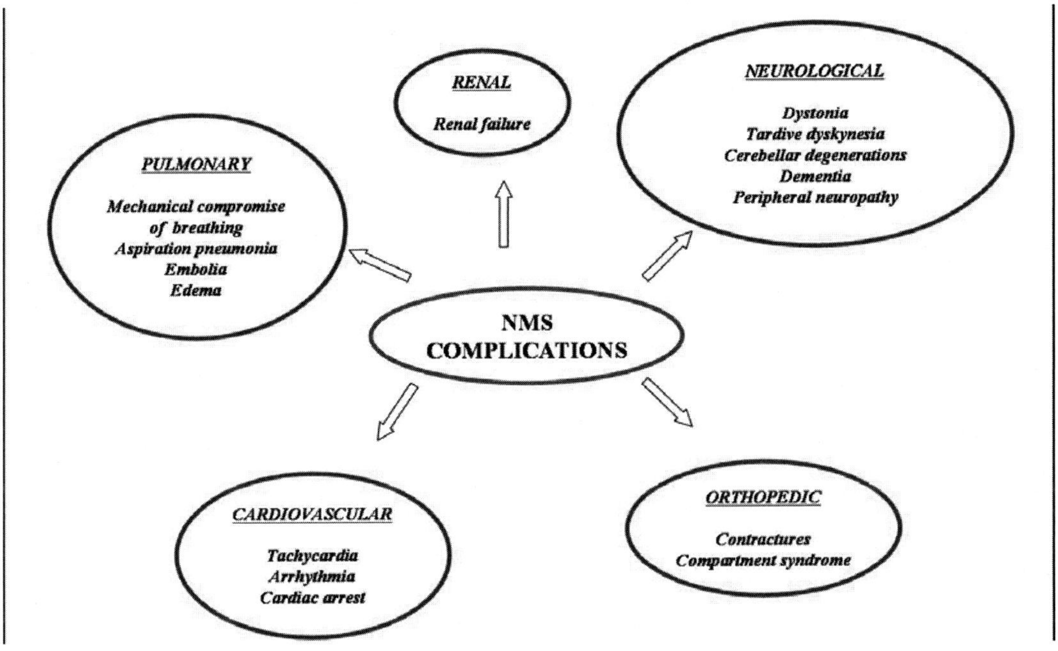

Figure 3. Complications of Neuroleptic Malignant Syndrome.

Diagnosis and Clinical Course

Although diagnostic criteria for NMS have been established and are widely accepted, the diagnosis of NMS is difficult and often a diagnosis of exclusion; in fact, similar symptoms are present also in the other disorders as, for example, the serotoninergic syndrome, malignant hyperthermia, heatstroke, lethal catatonia, cerebral nervous system infections (especially viral encephalitis) and associated to several classes of drugs. The DSM-IV criteria include for the diagnosis, in patients using neuroleptics from 1 to 4 weeks, muscle rigidity and hyperthermia associated to two of the following: diaphoresis, dysphagia, tremor, incontinence, changed mental status, mutism, tachycardia, fluctuating of blood pressure, leucocytosis and elevated CPK [20]. The temporal progress of signs and symptoms may provide important clues to diagnosis and severity of illness. The disorder typically develops within two weeks of the initial treatment with the drug, but may develop at any time the drug is being taken. In fact some cases of NMS develop within 24-72 hours after initiation of neuroleptic treatment, some within the first week, and occasional cases may have a fulminant onset within hours after drug administration [11]. These differences depend on long-acting depot of neuroleptics; in fact the average duration after oral administration lasts about two weeks, while depot preparations can present longer duration. Secondary complications are present in about 40% of patients with NMS and these may be cause of death (Figure 3).

Pulmonary alterations consists in mechanical compromise of breathing, aspiration pneumonia due to dysphagia and pulmonary embolism resulting from thrombophlebitis and oedema. Thromboemboli can result from immobilization, hypovolemia and hyperthermia. Additionally, cardiovascular complications as tachycardia, arrhythmia, cardiac arrest can lead to cardiopulmonary failure. Several cases, during the course, may require circulatory and ventilator support. Rhabdomyolysis provokes myoglobinuria which, associated to hemodynamic factors, causes acute renal failure. Finally, long-term consequences consist in neurological and orthopaedic complications. The first are represented by dystonia, tardive dyskynesia, cerebellar (degenerations due to hyperpyrexia), dementia and peripheral neuropathy. Orthopaedic complications consist in contractures and compartment syndrome secondary to rhabdomyolysis. In conclusion NMS is unpredictable and variable in both its presentations and its course with average mortality about 11-30% [12,14,17,21].

Treatment

NMS is a medical emergency requiring intensive and aggressive treatment which must begun in the early stage. In those patients, first of all, the offending drug must be suspended. The treatment includes two types of interventions: non-pharmacologic and pharmacologic.

Non-pharmacologic intervention is characterized by the administration of intravenous fluids for dehydration, mechanical cooling for hyperthermia, cardiopulmonary monitoring, intubation and ventilatory support if airway and breathing are compromises, preventative measures for thrombophlebitis as motor exercises, antiembolic stockings and kinetic nursing beds. In addition can be used massage to increase peripheral vasodilatation and heat dissipation.

Pharmacologic intervention consists in several different medications [11,22,23]. Bromocriptine, the dopamine direct receptor agonist, was showed an important reduction of extrapyramidal aspects [24]. It is administred orally with an initial dose of 2.5 mg incremented of 2.5 mg every 24 hours until a total of 45 mg [11]. Principal collateral effect in patient that use bromocriptine is represented by brief or worsening of psychosis. Furthemore, amantadine, identified in 1977 by Gelemberg and Mandel as the first drug in treatment for lethal catatonia, and levodopa can be used in treatment of NMS, as well as apomorphine, the parenteral form of dopaminergic drugs in patients presenting dysphagia.

Dantrolene sodium, a muscle relaxant which blocks intracellular calcium efflux, is supplied for oral or intravenous via. The oral dose is of 50 to 600 mg/day, whereas intravenous dose is of 10 mg/Kg/day. The effect of dantrolene is the diminution of CPK levels and hyperthermia, but use of higher dose may result in hepatotoxicity.

For their positive actions the association of bromocriptine and dantrolene can shorten the clinical course of NMS and decrease mortality.

Anticholinergic drugs can be administered in early stage to reduce extrapyramidal symptoms, if there are no other alterations and the diagnosis is uncertain.

Benzodiazepines have been utilized when the patients present catatonia. They act linking the gamma aminobutyric acid (GABA) receptors and increasing the activity of this neurotransmitter. It is necessary to monitor respiratory function using higher dose of benzodiazepines for their role on respiratory depression.

Finally, electroconvulsive therapy (ECT) was used in treatment of NMS when other drug have failed or when the patients have presented residual catatonia and parkinsonism after resolution of the acute metabolic symptoms of NMS [22,25]. Response to ETC is independent from age, sex or any particular feature of NMS and atypical ETC regimen includes until ten treatments with bilateral electrode placement.

Moreover, it is important individualize a therapy depending on the severity and the primary symptoms on NMS.

Parkinsonism-Hyperpyrexia Syndrome

Parkinson's disease (PD) is a slow but constant degenerative process in which a progressive neuronal loss in the substantia nigra and in other pigmented nuclei occurs [26]. Consequently, the dopaminergic nigrostriatal pathway is affected (Figure 4), as well as serotonergic and noradrenergic pathways [3,4,5].

The cardinal features of PD are represented by tremor at rest, rigidity, bradykinesia/akinesia and postural instability [27,28]. Motor symptoms can be preceded by non motor symptoms as disautonomic, olfactory, sleep and neuropsychiatric symptoms (Table 3). Clinical presentation is initially unilateral, but in the advanced stage became bilateral. The pharmacologic treatment includes levodopa that is metabolised to dopamine, and dopamine agonists that acting directly on dopamine receptors. Besides common collateral effects of this therapy, PD patients can develop, often after withdrawal of dopaminergic drugs, a life-threatening syndrome called Parkinson-hyperpyrexia syndrome (PHS) [29,30,31]. This is a rare, but very serious syndrome that is clinically indistinguishable from NMS, except that the symptoms occur in the absence of neuroleptic assumption. In same cases has been reported the PHS also after a reduction in non-dopaminergic drugs used to treat PD, such as amantadina or anticholinergics, and in the parkinsonism forms. About 30% of patients fail to recovery from it and about 4% die [31,32].

Table 3. Motor and non motor symptoms of Parkinson's disease

- **Motor symptoms**
  - Tremor: normally 4-6 Hz tremor, maximal when the limb is at rest and decreased with voluntary movement. It is typically unilateral at onset. In the 30% of patients there is little perceptible tremor which was classified as akinetic-rigid;
  - Rigidity: increased muscle tone and in combination with a resting tremor produces a cogwheel rigidity when the limb is passively moved;
  - Bradykinesia/Akinesia: respectively, slowness or absence of movement;
  - Postural instability: failure of postural reflexes.
- **Non motor symptoms**
  - Autonomy Disorders: urinary incontinence, constipation and gastric dysmotility, altered sexual function;
  - Sleep Alterations: alterations of physiological beat sleep and disturbances in REM sleep;
  - Iposmia/Anosmia
  - Neuropsychiatry: executive dysfunction, hallucinations, delusions and paranoia, dementia.

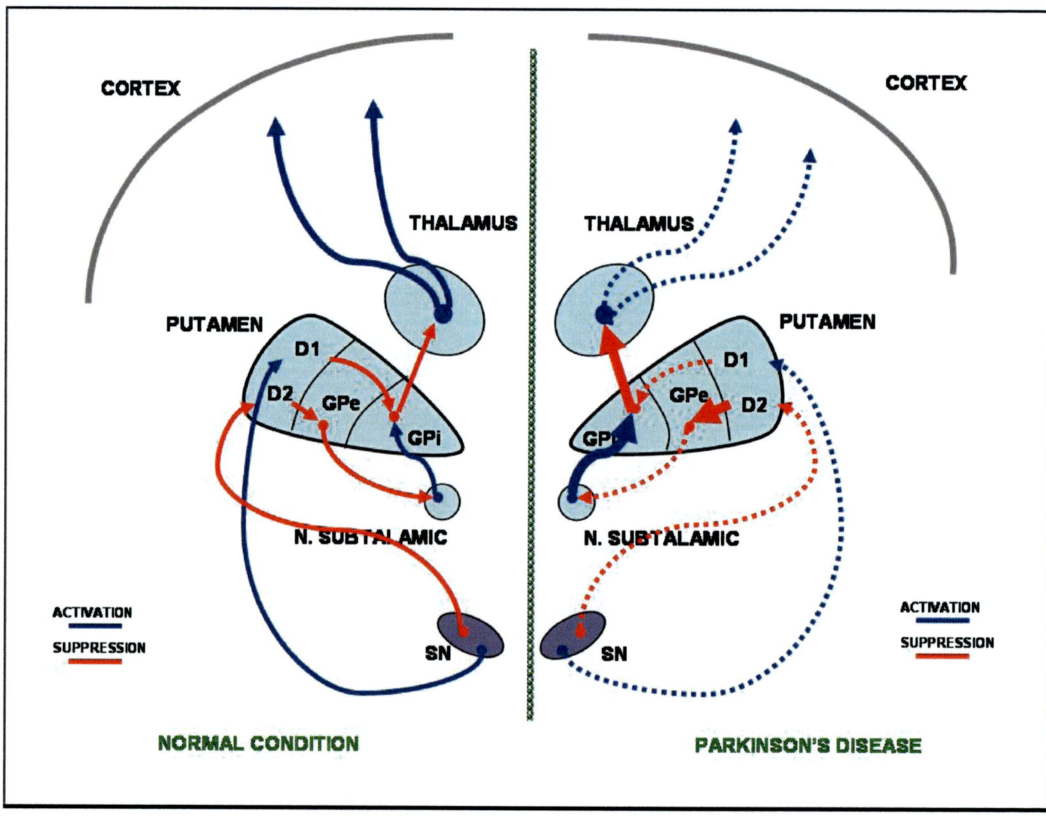

Figure 4. Dopaminergic pathways of the human brain. SN. (Substantia Nigra; D1 and D2: receptors of dopamine; GPe: external Globus Pallidus; GPi: internal Globus Pallidus).

According to the data reported in literature, all the dopaminergic systems, nigrostriatal, hypothalamic and mesolimbic, could be involved in the pathogenesis of PHS. The underlying disease is usually PD, but it may be other parkinsonisms [32].

Therefore withdrawal of dopaminergic drugs, abrupt discontinuation of medications or sudden alteration of drug regimen are the principal risk factors associated with the PHS development [30,33,34,35]. Other risk factors are represented by undercurrent infections, hot weather, dehydration, motor fluctuations and intestinal ileus.

Particularly hyperthermia is the most frequent symptom of levodopa withdrawal [36]. The body temperature can be as high as 40 °C and can be seen from the early stage of PHS when there is no clinical or laboratory evidence of infections. Clinical symptoms of PHS occur from 18 hours to 7 days after the interruption or withdrawal of dopaminergic drugs and they are very similar to NMS [31]. Important consequences are represented by deep venous thrombosis, pulmonary embolism, disseminated intravascular coagulation, rhabdomyolysis, and renal failure resulting from myoglobinuria. The risk to develop PHS is greater in patients with advanced PD because of the complicated therapy management.

Treatment consists of dopaminergic therapy replacement, administering orally or intravenously (when oral is contraindicated). The dose recommended is the same as those given before the PHS onset. Supportive care such as fluid replacement, dantrolene sodium, antipyretics and cooling measures, must be considered in patients at the onset of the symptoms [29].

References

[1] Björklund A, Dunnett SB. Dopamine neuron system in the brain: an update. *Trends in Neurosciences* 2007; 30:194-202.

[2] Moore RY. Organization of midbrain dopamine system and the pathophysiology of Parkinson's disease. *Parkinsonism Rel Disord* 2003; 9:65-71.

[3] Ehringer H, Hornykiewicz O. Distribution of noradrenalina and dopamine (3-hydroxytyramine) in the human brain and their behaviour in disease of the extrapyramidal system. *Parkinsonism Rel Disord* 1998; 4:53-57.

[4] Torta DME, Castelli L. Reward pathways in Parkinson's disease: Clinical and theoretical implications. *Psychiatry Clin Neurosci* 2008; 62:203-213.

[5] Oades RD, Halliday GM. Ventral segmental (V10) system: neurobiology. Anatomy and connectivity. *Brain Research Reviews* 1987; 12:117-165.

[6] Ikemoto S. Dopamine reward circuit: two projection systems from the ventral midbrain to the nucleus accumbens-olfactory tubercle complex. *Brain Res Rev* 2007; 56:27-78.

[7] Kiyatkin EA. Brain temperature fluctuations during physiological and pathological conditions. *Eur J Appl Physiol* 2007;101:3-17.

[8] Kiyatkin EA: Brain hyperthermia as physiological and pathological phenomena. *Brain Research Review* 2005;50:27-56.

[9] Sharma HS, Hoopes PJ. Hyperthermia-induced pathophysiology of the central nervous system. *Int J Hyperthermia* 2003;19:325-354.

[10] Delay J, Pichot P, Lemperiere T. et al. Un neuroleptique majeur non-phenotiazinique et non-reserpenique, l'haloperiol dans le traitment des psycosis. *Ann Med. Psychol* 1960;118:142-145.

[11] Susman VL. Clinical management of neuroleptic malignant syndrome. *Psy Quarterly* 2001;72:325-336.

[12] Caroff SN, Mann SC. Neuroleptic malignant syndrome. *Psychopharmacol Bull* 1988; 24:25-29.

[13] Strawn JR, Keck PE Jr, Caroff SN. Neuroleptic malignant syndrome. *Am J Psychiatry* 2007; 164:870-876.

[14] Buckley PF, Hutchinson M. Neuroleptic malignant syndrome. *JNNP* 1995;58:271-273.

[15] Scrinvasan AV, Murugappan M, Kristamurty SG, Sayeed ZA. Neuroleptic malignant syndrome. *JNNP* 1990; 53:514-516.

[16] Keck PE Jr, Pope HG Jr, Cohen BM et al. Risk factors for neuroleptic malignant syndrome. A case-control study. *Arch Gen Psychiatry* 1989; 46:914–918.

[17] Addonizio G, Sussman VL, Roth SD. Neuroleptic malignant syndrome: review and analysis of 115 cases. *Biol Psychiatry* 1987; 22:1004-1020.

[18] Rosebush P, Stewart T. A prospective analysis of 24 episodes of neuroleptic malignant syndrome. *Am J Psychiatry* 1989; 146:717-725.

[19] Rosebush P, Mazurek MF. Serum iron and neuroleptic malignant syndrome. *Lancet* 1991; 338:149-151.

[20] American Psychiatric Association: *Diagnostic and Statistical Manual of Mental Disorders*, 4th edition 1994.

[21] Shalev A, Hermesh H, Munitz H. Mortality from neuroleptic malignant syndrome. *J Clin Psychiatry 1989*; 50:18-25.

[22] Davis JM, Caroff SN, Mann SC. Treatment of neuroleptic malignant syndrome. *Psych Ann* 2000; 30:325-331.

[23] Sakkas P, Davis JM, Janicak PG, Wang ZY. Drug treatment of the neuroleptic malignant syndrome. *Psychopharmacol Bull* 1991; 27:381-384.

[24] Mueller PS, Vester JW, Fermaglich J. Neuroleptic malignant syndrome. Successful treatment with bromocriptine. *JAMA* 1983; 249:386-388.

[25] Nisijima K, Ishiguro T. Electroconvulsive therapy for the treatment of neuroleptic malignant syndrome with psychotic symptoms: A report of five cases. *J ECT* 1999; 15:158-163.

[26] Tambasco N, Pelliccioli GP, Chiarini P, Montanari GE, Leone F, Mancini ML, Paciaroni M, Gallai V. Magnetization transfer changes of grey and white matter in Parkinson's disease. *Neuroradiology.* 2003; 45:224-30.

[27] Marsden CD. Parkinson's disease. *JNNP* 1994; 57:672-681.

[28] Fox SH, Lang AE. Levodopa-related motor complications-phenomenology. *Mov* Disord 2008; 23:S509-14.

[29] Factor SA, Santiago A. Parkinsonism-hyperpyrexia syndrome in Parkinson's disease. In *Movement disorder emergencies: Diagnosis and treatment.* New York: Humana Press 2005; 29–41.

[30] Keyser D, Radnitzky R. Neuroleptic malignant syndrome in Parkinson's disease after withdrawal or alteration of dopamine therapy. *Arch Intern Med* 1991; 151:794-796.

[31] Newman EJ, Grosset DG, Kennedy PGE. The Parkinsonism-Hyperpyrexia Syndrome. *Neurocrit Care.* 2008; Aug (in press).

[32] Mizuno Y, Takubo H, Mizuta E, Kuno S. Malignant syndrome in Parkinson's disease: concept and review of the literature. *Parkinsonism Rel Disord* 2003; 9:S3-S9.

[33] Friedman JH, Feinberg SS, Feldman RG. A neuroleptic malignant-like syndrome due to levodopa therapy withdrawal. *JAMA* 1985; 254:2792-2795.

[34] Serrano-Dueñas M: Neuroleptic malignant syndrome-like, or dopaminergic malignant syndrome due to levodopa therapy withdrawal. Clinical features in 11 patients. Parkinsonism *Rel Disord* 2003; 9:175-178.

[35] Gordon PH, Frucht SJ. Neuroleptic malignant syndrome in advanced Parkinson's disease. *Mov Disord* 2001; 16:960-974.

[36] Sechi GP, Tanda F, Mutani R. Fatal Hyperpyrexia after withdrawal of levodopa. *Neurology* 1984; 34:249-250.

In: Body Temperature Regulation
Editor: A. B. Cisneros and B. L. Goins

ISBN: 1-60741-282-3

Chapter XII

Body Temperature and Eating Behavior

Marcellino Monda*, Giovanni Messina, Andrea Viggiano[1], Domenico Tafuri[1], Emanuela Viggiano, Sergio Chieffi, Vincenzo De Luca

Department of Experimental Medicine, Section of Human Physiology, and Clinical Dietetic Service, Second University of Naples, Via Costantinopoli 16, 80138 Naples, Italy
[1]Faculty of Motor Sciences, University of Naples "Parthenope", Naples, Italy

Abstract

Body temperature and eating behavior are two inter-dependent variables. Thermal set-point influences food intake, so that the introduction of energy can be considered as a system to reach a prefixed body temperature. Substances resetting the thermal set-point cause modifications of eating behavior. More precisely, pyrogens induce hyperphagia, while substances reducing the thermal set-point cause hypophagia. On the other hand, changes in body temperature (without changes in the thermal set-point) induce modifications of the ingestive behavior. Indeed, conditions inducing hyperthermia cause a reduction of food intake, while situations reducing body temperature are associated with an increase in food intake. The upon-exposed theory is validated by various evidences and it is in accord with the "thermoregulatory hypothesis of body weight". The present article reports our experimental observations, which corroborate this theory. The relationship between thermoregulation and body weight control are amply discussed.

* Corresponding author: Prof. Marcellino Monda, MD, Dipartimento di Medicina Sperimentale, Sezione di Fisiologia Umana, Seconda Università di Napoli, Via Costantinopoli 16, 80138 Napoli, Italy, Tel. +39 +81 566 5804, Fax +39 +81 5667500. E-mail: marcellino.monda@unina2.it

Introduction

The control of body temperature is strictly associated with the control of body weight, in accord with the "thermoregulatory hypothesis" of food intake. The autonomic nervous system is involved in the control of eating behavior through influences exerted on the production and loss of heat.

Since the incidence of body weight superior to normal values is increasing in the industrialized word, the World Health Organization considers obesity as a "pandemia" in rich countries. Investigation into the mechanisms that control body weight give growing relevance to the possibilities of new strategies to reduce the incidences of overweight and obesity, which are frequently associated with metabolic and cardiovascular diseases.

The aim of this article is to report our evidences which demonstrate that modifications in body temperature affect food intake, so that body weight can be controlled by changes in body temperature.

First Demonstration

This experiment tests the effect of intracerebroventricular injection of prostaglandin E_1 (PGE_1) on: 1) sympathetic activity and body temperature; 2) food intake. The firing rate of the sympathetic nerves to interscapular brown adipose tissue (IBAT), along with IBAT and colonic temperatures (T_{IBAT} and T_C) were monitored in male Sprague-Dawley rats before and for 90 min after food presentation. An intracerebroventricular injection of PGE_1 (500ng) or saline was made immediately before food presentation. The amount of food ingested was also measured.

Upper panel of figure 1 illustrates cumulative curves of food intake. PGE_1 reduces the ingestion of food. Middle-upper panel shows the percentage changes in firing rate of nerves to IBAT. The increase was higher in the animals with PGE_1 injection. Middle-lower and lower panels illustrate T_{IBAT} and T_C changes. The elevation of body temperature was higher in the animals with PGE_1 injection. Briefly, PGE_1 induces a higher elevation of body temperature and a reduction of food intake.

This experiment analyzes contemporaneously food intake and body temperature during the reactions due to an icv injection of PGE_1: body temperature elevation is inversely proportional to food intake [25].

In response to a pyrogen, such as PGE_1, animals reduce heat loss and increase heat production in order to raise body temperature to a new, elevated level [15, 19]. The findings of the present experiment suggest that PGE_1 controls both sympathetic activation (and related body temperature elevation) and food intake. Preoptic area/anterior hypothalamus (PO/AH) is a responsive structure to PGE_1, whereas this neural mediator slightly stimulates the other hypothalamic areas. Indeed, an injection of PGE_1 into the ventromedial hypothalamus (VMH) is much lower than that induced by a same dose of PGE_1 injected into the PO/AH [18, 36]. The present experiment suggests that PGE_1 injected into a cerebral ventricle acts on the PO/AH, which, in turn, could influence the VMH, a key structure in the control of sympathetic activity and food intake [14, 36]. This view is in agreement with our previous experiment showing that a VMH lesion reduces sympathetic and thermogenic reactions due PGE_1 [24].

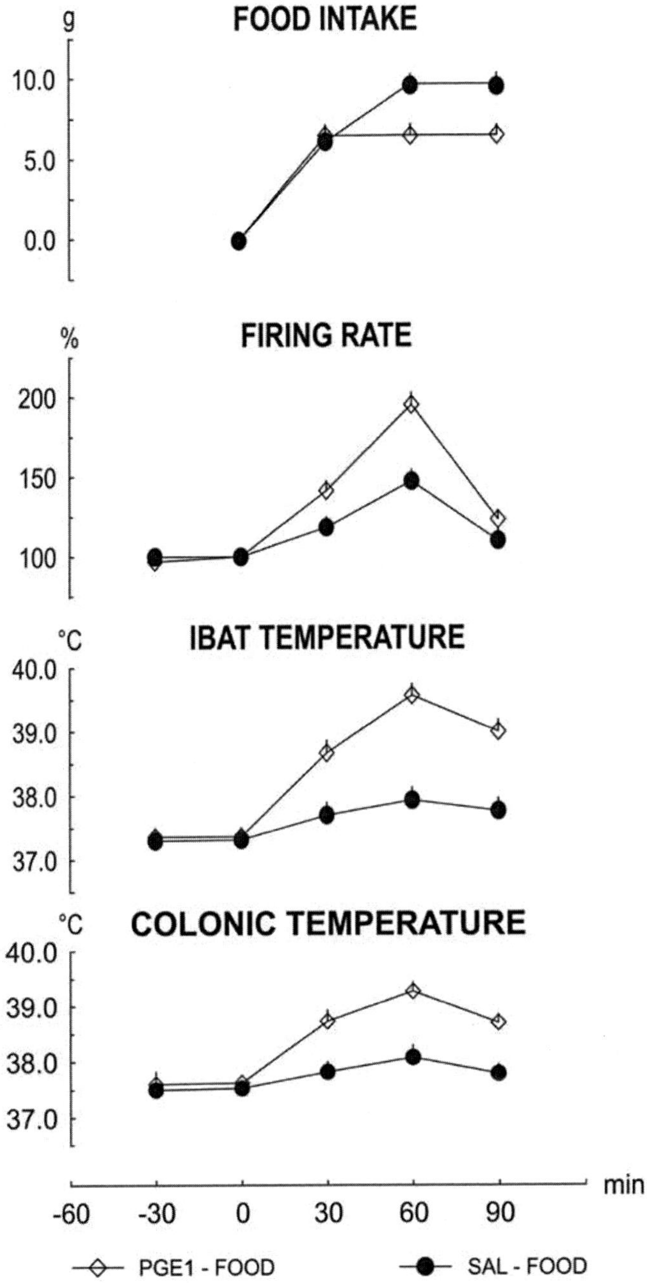

Figure 1. Cumulative values of food intake (upper panel), changes in firing rate (middle-upper panel), changes in temperature of interscapular brown adipose tissue (IBAT), changes in colonic temperature in the rats with prostaglandin E_1 (PGE_1) or saline (SAL) injection. Injection and food presentation at time 0.

Throughout our experiment, we report direct evidence of sympathetic discharge of nerves to IBAT after PGE_1 injection. IBAT is the organ responsible for evoking 35-65% of the total increase in metabolic heat production unrelated to shivering during various experimental manipulations in rodents [9, 13, 20].

The simultaneous measurements of the sympathetic firing rate and food intake is the nicest demonstration of the feed-back between the body temperature and food intake. The rise in sympathetic discharge and body temperature is a signal of satiety. Indeed, an activation of ß-3 adrenergic receptors reduces food intake [17]. In the present experiment, the increse in body temperature due to PGE_1 can be recognized as a signal of satiety, which reduces food intake.

Second Demonstration

This experiment tested the effect of the thermogenic activation induced by orexin A on eating behavior. Food intake, T_{IBAT} and abdominal temperatures (T_{ab}) were monitored in 24h-fasting male Sprague-Dawley rats for 12h after food presentation. Orexin A (1.5 nmol) was injected into the lateral cerebral ventricle 6h before food presentation, while T_{IBAT} and T_{ab} were also monitored. The same variables were controlled in rats receiving orexin A contemporaneously to food presentation.

Figure 2 illustrates cumulative food intake, T_{IBAT} and T_{ab}. Orexin elevated T_{IBAT} and T_{ab}. The rats receiving orexin 6h before food presentation showed a reduction in food intake with respect to other animals.

The effects due to orexin A on food intake depends on body temperature at time of food presentation. Then, body temperature influences eating behavior [26]. The name assigned to this peptide, called primarily hypocretin-1 [6], is due to the influence exerted by orexin A on eating behavior: induction of food intake [32, 35, 39]. The results of this experiment underline the importance of orexin A in the control of body temperature, which in turn affects eating behavior. In this experiment, an icv injection of orexin A induces an increase in the sympathetic activity and in the T_{IBAT} independently of food ingestion, that is reduced in the rats with a delayed presentation of food. Then, the effects on body temperature are predominant with respect to food intake. Orexin A can induce not only an increase, but also a decrease in food intake. This is in contradiction with the significance of name of this peptide, that could exert a primary hyperphagic effect

Primary hyperphagic substances, as neuropeptide Y or galanin, induce a reduction of the sympathetic discharge and a decrease in body temperature [2, 7, 29, 31]. Primary hypophagic substances, as leptin, cause an increase in the sympathetic activity and an increase in food intake [10, 12, 30, 32, 38]. For this reason, orexin A cannot be considered a substance with a primary hyperphagic effect. The orexin A can induce hypophagia, as in the present experiment, or hyperphagia, but it always induces an activation of thermogenesis. This peptide, called erroneously orexin A, has a primary thermogenic effect. We can suppose that this peptide elevates the thermal set-point, inducing the reactions to reach the new level of body temperature. The increase in food intake, obtained in the rats with a non-delayed presentation of food, could be a reaction aimed to reach an elevated body temperature.

Indeed, food ingestion induces a rise in body temperature due to post-prandial thermogenesis [13, 22]. The hyperphagic effect of orexin A disappears when the body temperature is already increased, so that a reduction in food intake can happen in this condition.

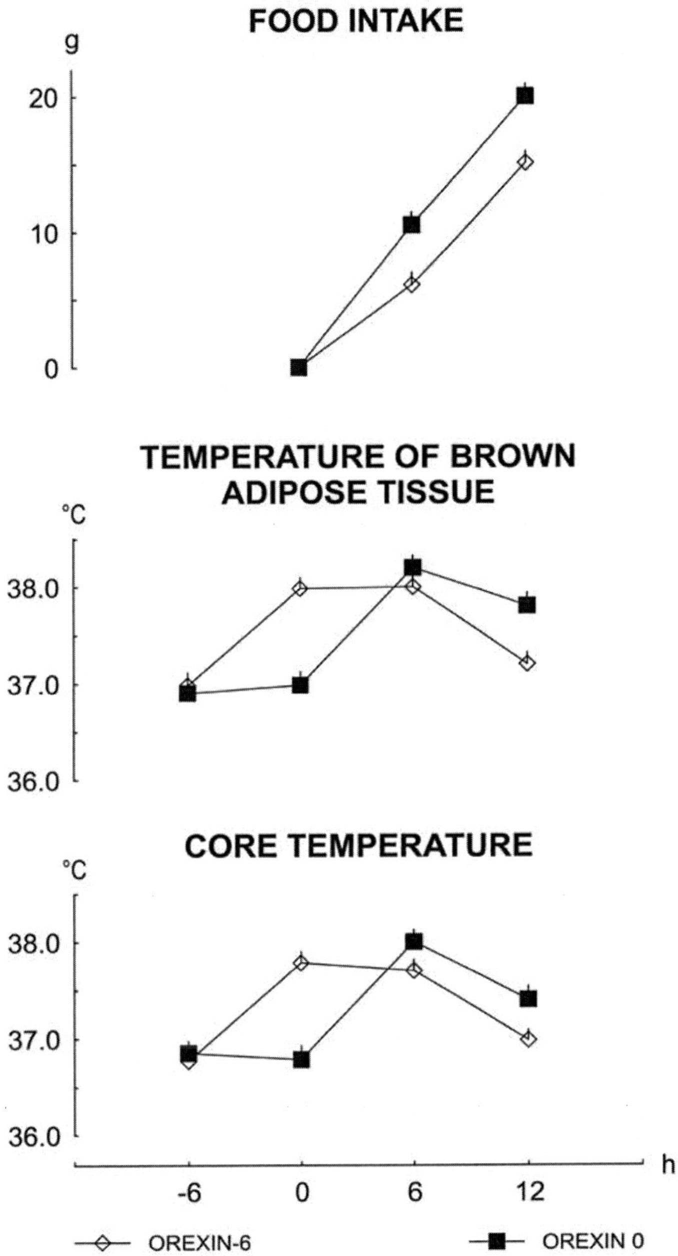

Figure 2. Cumulative food intake, interscapular brown adipose tissue (IBAT) and core temperatures. Food presentation at time 0. Orexin was injected 6 h before food presentation (-6) or contemporaneously to food presentation (0).

The orexin A affects the temperature of IBAT, which is the most important effector of non shivering thermogenesis in the rat [4], illustrating that the rise in heat production is also due to the activation of thermogenesis unrelated to muscle activity. This confirms the role of the sympathetic nervous system on IBAT activity.

The increase in abdominal temperature emphasizes the effect of orexin A on "core" temperature suggesting the inclusion of orexin A among the peptides controlling body temperature. The anterior hypothalamic/preoptic area is considered the most important region in the control of body temperature. Pyrogens, as prostaglandin of E series, act on this region to induce hyperthermia [21, 34] and inhibitors of synthesis of prostaglandins block the hyperthermia [5]. Recently, we demonstrated that lysine acetylsalicylate, an inhibitor of prostaglandins synthesis reduces hyperthermia induced by orexin A in anesthetized rats [27]. A possible explanation is that orexin A could induce a cerebral synthesis of prostaglandin E, which acts on the anterior hypothalamic/preoptic area [33], a very responsive structure to prostaglandin E.

The influence of orexin A in the control of eating behavior should be revised, assigning to this peptide a primary role in thermoregulation. Although this experiment could be repeated with an antagonist of the orexin receptors and/or a blocker of the sympathetic activity, the possibility that orexin A can induce hypophagia is well demonstrated by the present findings, so that the scientific community should use a different name for this peptide.

Third Demonstration

This experiment tests the effect of intraperitoneal (ip) injection of lysine acetylsalicylate on food intake and body temperature in rats with a lesion of the lateral hypothalamus (LH). Food intake, T_{IBAT} and T_C were monitored in male Sprague-Dawley rats lesioned in the LH. These variables were measured before and after ip injection of lysine acetylsalicylate. The same variables were also monitored in: a) lesioned rats with ip administration of saline; b) sham-lesioned animals with ip injection of lysine acetylsalicylate; c) sham-lesioned rats with ip injection of saline. The results show that lysine acetylsalicylate modifies the aphagia by increasing food intake and it also reduces the enhancements in T_{IBAT}, and T_C induced by LH lesion.

The results of the first experiment are represented in figure 3. Upper panel shows the changes in food intake. The LH lesion blocked food intake almost completely. The aphagia was modified by administration of L-ASA. In the sham-lesioned rats, L-ASA or saline injection did not exert any effect on food intake. T_{IBAT} changes are illustrated in middle panel. The LH lesion causes a rise in T_{IBAT}. This increase was reduced by L-ASA. In the sham-lesioned rats, L-ASA or saline did not cause any modification in T_{IBAT}. Lower panel shows T_C modifications. T_C rose in LH lesioned rats. This rise was reduced by L-ASA. In the sham-lesioned rats, L-ASA or saline did not cause any change in T_C.

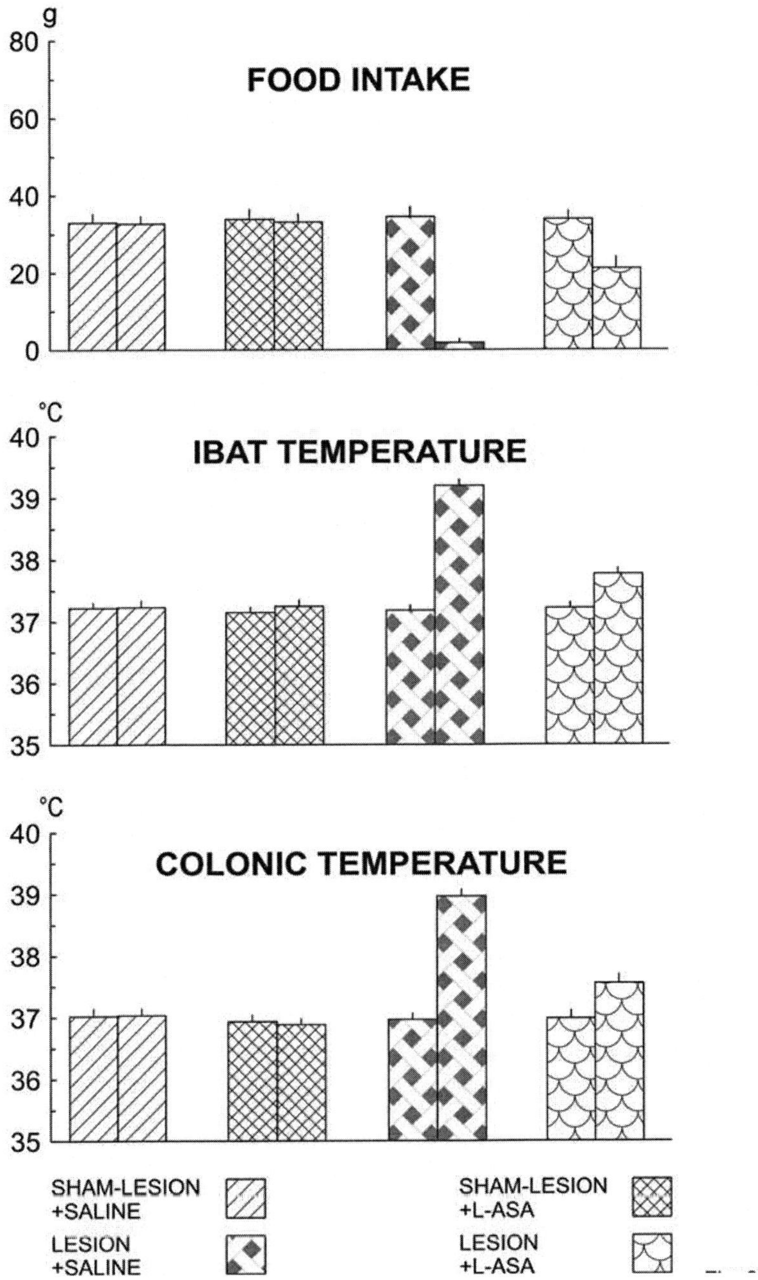

Figure 3. Changes in food intake (upper panel) in interscapular brown adipose tissue (IBAT) and colonic temperatures. The first bars indicate the amount of food eaten over a period of 24 hours before the LH lesion or the sham-lesion in rats injected with saline or lysine acetylsalycilate (L-ASA). The second bars indicate food intake over a period of 24 hours after the lesion or the sham-lesion.

The findings of this experiment demonstrates that the inhibition of PG synthesis is able: 1) to modify eating behavior in LH lesioned rats, and 2) to reduce the body temperature elevation [23]. The electrolytic lesion in the LH regulates body weight at a lower level. LH lesioned rats lose body weight at a faster rate than sham-lesioned controls subjected to the same degree of food deprivation [11]. The sympathetic activation and rise in body temperature are specific for LH lesion [16]. Indeed, lesions of other areas of hypothalamus, such as the ventromedial area [1] or paraventricular nucleus [37] fail to cause an increase in sympathetic discharge and a robust hyperthermia. It has been accepted that the mechanism for increasing heat production following LH lesion includes activation of non-shivering thermogenesis due to IBAT [13].

The present findings emphasize the role of PG synthesis during the hyperthermia induced by LH lesion. L-ASA reduces the hyperthermia induced by LH lesion. These results suggest that PGs are involved in the control of body temperature in this experimental model. The ability to inhibit PG biosynthesis by nonsteroidal antipyretic drugs is generally accepted as the mechanism by which these drugs reduce core temperature during hyperthermia [8]. These experiments confirm that an increase in body temperature reduces food intake, in accord with the findings of Bray [3]. On the other hand, our data show that an inhibitor of PG synthesis can modify the aphagia induced by LH lesion. This suggests that the prostaglandinergic system is strongly involved in the control of eating behavior. In the other words, PG biosynthesis may be activated by electrolytic lesion of the LH and the PGs, in turn, act on the ventromedial hypothalamus [28] and/or anterior hypothalamus/ preoptic area [34]. The aphagia and hyperthermia induced by LH lesion is strongly modified by L-ASA. The reduction of body temperature in hypertermic rats with LH lesion is associated to increase in food intake, confirming the strict relation between food intake and body temperature. This experimental model may be useful for a new approach to the investigation of the patho-physiologic mechanisms of eating disorders.

References

[1] Berthoud, HR; Jeanrenaud, B. Acute hyperinsulemia and its reversal by vagotomy after lesions of the ventromedial hypothalamus in anesthetized rats. *Endocrinology* 1979 105, 146-151.

[2] Bouali, SM; Fournier, A; St-Pierre, S: Jolicoeur, FB. Effects of NPY and NPY2-36 on body temperature and food intake following administration into hypothalamic nuclei. *Brain Res. Bull.* 1995 36, 131-135.

[3] Bray, GA. Reciprocal relation between the sympathetic nervous system and food intake. *Brain Res. Bull.* 1991 27, 517-520.

[4] Cannon, BJ; Houstek, J; Nedergaard, J. Brown adipose tissue. More than an effector of thermogenesis. *Ann. N. Y. Acad. Sci.* 1998 856, 171-187.

[5] Cooper, KE; Veale, WL; Kasting, NW: In:. Barnett, HJM; Hirsch, J; Mustard, JF. (Eds.), *Acetylsalicylic Acid: New Uses for an Old Drug.* Raven, New York, 1982, pp.153-165.

[6] De Lecea, L; Kilduff, TS; Peyron, C; Gao, XB; Foye, PE; Danielson, PE; Fukuhara, C; Battenberg, ELF; Gautvik, VT; Bartlett II, FS; Frankel, WN; Van Den Pol, AN; Bloom, FE; Gautvik, KM; Sutcliffe, JG. The hypocretins: two hypothalamic peptides with neuroexcitatory activity. *Proc. Natl. Acad. Sci.* USA. 1998 5, 322-327.

[7] Egawa, M; Yoshimatsu, H; Bray, GA. Neuropeptide Y suppresses sympathetic activity to interscapular brown adipose tissue in rats. *Am. J. Physiol.* 1991 260, R328-334.

[8] Flower, RJ; Vane, JR. Inhibition of prostaglandin synthetase in brain explains the anti-pyretic activity of paracetamol (4-acetamindophenol). *Nature London.* 1972 240, 410-411.

[9] Foster, DO. Quantitative contribution of brown adipose tissue to overall metabolism. *Can. J. Biochem. Cell. Biol.* 1984 62, 618-622.

[10] Haque, MS; Minokoshi, Y; Hamai, M; Iwai, M; Horiuchi, M; Shimazu, T. Role of the sympathetic nervous system and insulin in enhancing glucose uptake in peripheral tissues after intrahypothalamic injection of leptin in rats. *Diabetes.* 1999 48, 1706-1712.

[11] Harrell, LE; Decastro, JM; Balagura, S. A critical evaluation of body weight loss following lateral hypothalamic lesions. *Physiol. Behav.* 1975 15, 133-136.

[12] Haynes, WG; Morgan, DA; Djalali, A; Sivitz, WI; Mark, AL. Interactions between the melacortin system and leptin in control of sympathetic nerve traffic. *Hypertension.* 1999 33, 542-547.

[13] Himms-Hagen, J. Non-shivering thermogenesis. *Brain Res. Bull.* 1984 12, 151-160.

[14] Keesey, RE; Powley, TL. The regulation of body weight. *Ann. Rev. Psychol.* 1986 37,109-133.

[15] Kluger, MJ. The febrile response. In: Morimoto, RJ; Tissieres, A; Georgopulos, C. (Eds) *Stress Proteins in Biology and Medicine* (New York: Cold Spring Harbor) 1990 pp. 61-78.

[16] Yoshida, T; Kemnitz, JW; Bray, GA. Lateral hypothalamic lesions and norepinephrine turnover in rats. *J. Clin. Invest.* 1983 72, 919-927.

[17] Mantzoros, CS; Qu, D; Frederich, RC; Susulic, VS; Lowell, BB; Maratos Fier, E; Flier JS. Activation of beta-3 adrenergic receptors suppresses leptin expression and mediates a leptin-independent inhibition of food intake in mice. *Diabetes* 1996 45, 909-914.

[18] Monda, M; Amaro, S; De Luca, B. The influence of exercise on energy balance changes induced by ventromedial hypothalamic lesion in the rat. *Physiol. Behav.* 1993 54, 1057-1061.

[19] Monda, M; Amaro, S; Sullo, A; De Luca, B. Muscimol injection in the posterior hypothalamus reduces the hyperthermia induced by prostaglandin E1 in the rat. *Brain Res. Bull.* 1995 37, 575-580.

[20] Monda, M; Amaro, S; Sullo, A; De Luca, B. Lateral hypothalamic lesion induces sympathetic stimulation and hyperthermia by activating synthesis of cerebral prostaglandins. *Prostaglandins.* 1996 51, 169-178.

[21] Monda, M; Amaro, S; Sullo, A; De Luca, B. Injection of muscimol in the posterior hypothalamus reduces the PGE_1 -hyperthermia in the rat. *Brain Res. Bull.* 1995 6, 575-580.

[22] Monda, M; Sullo A; De Luca, B. Lesions of ventromedial hypothalamus reduce postingestional thermogenesis. *Physiol. Behav.* 1997 61, 687-691.

[23] Monda, M; Sullo, A; De Luca, E; Pellicano, MP. Lysine acetylsalicylate modifies aphagia and thermogenic changes induced by lateral hypothalamic lesion. *Am. J. Physiol.* 1996 271, R 1638-1642.

[24] Monda, M; Sullo, A; De Luca, V; Viggiano, A; Pellicano, MP. Acute lesions of the ventromedial hypothalamus reduce sympathetic activation and thermogenic changes induced by PGE_1. *J. Physiol.* (Paris) 1997 91, 285-290.

[25] Monda, M; Sullo, A; Viggiano, A; De Luca, V. Eating behavior is related to sympathetic activation induced by icv injection of prostaglandin E1 in the rat. *Nutr. Neurosci.* 1999 2, 69-74.

[26] Monda, M; Viggiano, A; De Luca, V. Paradoxical effect of orexin A: the hypophagia induced by hyperthermia. *Brain Res.* 2003 961, 220-228.

[27] Monda, M; Viggiano, A; Mondola, P; De Luca, V. Inhibition of prostaglandins synthesis reduces hyperthermic reactions induced by hypocretin-1/orexin A. *Brain Res.* 2001 909, 68-74.

[28] Morimoto, A; Murakami, N; Nakamori, T; Watanabe, T. Ventromedial hypothalamus is highly sensitive to prostaglandin E2 for producing fever in rats. *J. Physiol. London.* 1988 397, 259-268.

[29] Nagase, H; Bray, GA; York, DA. Effect of galanin and enterostatin on sympathetic nerve activity to interscapular brown adipose tissue. *Brain Res.* 1996 709, 44-50.

[30] Okamoto, S; Kimura, K; Saito, M. Anorectic effect of leptin is mediated by hypothalamic corticotropin-releasing hormone, but not by urocortin, in rats. *Neurosci. Lett.* 2001 307, 179-182.

[31] Patel, S; Hutson, PH. Effects of galanin on 8-OH-DPAT induced decrease in body temperature and brain 5-hydroxytryptamine metabolism in the mouse. *Eur. J. Pharmacol.* 1996 317, 197-204.

[32] Shiraishi, T; Oomura, Y; Sasaki, K; Wayner, MJ. Effects of leptin and orexin-A on food intake and feeding related hypothalamic neurons. *Physiol. Behav.* 2000 71, 251-261.

[33] Simpson, CW; Ruwe, WD; Myers, RD. Prostaglandins and hypothalamic neurotransmitter receptors involved in hyperthermia: a critical evaluation. *Neurosci. Biobehav. Rev.* 1994 18, 1-20.

[34] Stitt, JT. Prostaglandin E as the neural mediator of the febrile response. *Yale J. Biol. Med.* 1986 59, 137-149.

[35] Sweet, DC; Levine, AS; Billington, CJ; Kotz, CM. Feeding response to central orexins. *Brain Res.* 1999 821, 535-538.

[36] Thornhill, JA; Halvorson, I. Electrical stimulation of the posterior and ventromedial hypothalamic nuclei causes specific activation of shivering and nonshivering thermogenesis. *Can. J. Physiol. Pharmacol.* 1994 72, 89-96.

[37] Tokunaga, K; Fukushima, M; Kemnitz, JW; Bray, GA. Effect of vagotomy on serum insulin in rats with paraventricular or ventromedial hypothalamic lesion. *Endocrinology.* 1986 119, 1708-1711.

[38] Ukropec, J; Sebokova, E; Klimes, I. Nutrient sensing, leptin and insulin action. *Arch. Physiol. Biochem.* 2001 109 38-51.

[39] Wolf, G. Orexins: a newly discovered family of hypothalamic regulators of food intake. *Nutr. Rev.* 1998 56, 172-173.

In: Body Temperature Regulation
Editor: A. B. Cisneros and B. L. Goins
ISBN: 1-60741-282-3

Chapter XIII

Dietary-induced Thermogenesis and Perioperative Thermoregulation

Toshiki Mizobe[1,2,*], ***Yasufumi Nakajima***[1]

[1]Department of Anesthesiology, Kyoto Prefectural University of Medicine, Kyoto 602-8566, Japan

[2]The Outcomes Research Consortium[†], Cleveland, USA

Abstract

Perioperative hypothermia results from a combination of 1) anesthetic-induced impairment of thermoregulatory control, 2) decreased metabolism due to anesthesia, 3) cool operating room environment, and 4) factors unique to surgery that promote excessive heat loss. Perioperative hypothermia shows a characteristic three phase pattern: 1) redistribution phase, an initial rapid reduction in core temperature during the first hour resulted from an internal redistribution of body heat from the core to the peripheral shell caused by anesthetic-induced vasodilation, 2) linear phase, 2-3 hour of a slower and linear decrease at a rate determined by the difference between heat loss and heat production, 3) plateau phase shown after 2-4 hours of anesthesia, which develops due to delayed reemergence of thermoregulatory vasoconstriction restricting core to peripheral heat flow.

Perioperative Hypothermia is associated with several complications including morbid myocardial outcomes, surgical bleeding, wound infection, impaired immune function, delayeded postanesthetic recovery and prolonged hospitalization. All these data strongly suggest that we, anesthesiologists, have to take special thermal precautions to prevent perioperative hypothermia. Diverse devices have been used to prevent hypothermia such as circulating water mattress and forced air warming device.

[*] Address correspondence to Toshiki Mizobe, MD, PhD. Email: toshim@koto.kpu-m.ac.jp

[†] www.or.org

After the ingestion or infusion of nutrients, there is an increase in energy expenditure which has been referred to as dietary or nutrient-induced thermogenesis. This thermogenesis induced by protein or amino acids is well known to be largest and most prolonged. According to these physiological backgrounds, preoperative amino acid infusion was reported to prevent postoperative hypothermia during general anesthesia and spinal anesthesia. Also, perioperative amino acid infusion is reported to improve the outcome of the patients undergoing off-pump CABG. Fructose is also known to elicit the greatest increase in thermogenesis among various carbohydrates such as glucose. Fructose infusion prevents perioperative hypothermia during surgery. Both of amino acid infusion and fructose infusion were observed to shift upward the threshold core temperature for thermoregulatory vasoconstriction as well as to increase energy expenditure. The thermic effect of these nutrients might show the novel possibility of infusion therapy.

Mechanism of Perioperative Thermoregulation

Perioperative hypothermia results from a combination of anesthetic-induced impairment of thermoregulatory control, a cool operating room environment, and other factors promoting excessive heat loss such as cold fluid administration. Perioperative hypothermia in general anesthesia develops with a characteristicthree-phase pattern, the redistribution phase, linear phase, and plateau phase [1] (Figure 1).

1) Redistribution phase

   Core temperature decreases 1-1.5 °C during the first hour. This initial rapid reduction in core temperature after induction of anesthesia results from an internal redistribution of body heat. Anesthetic-induced vasodilation causes an internal redistribution of heat from the core to the peripheral shell.

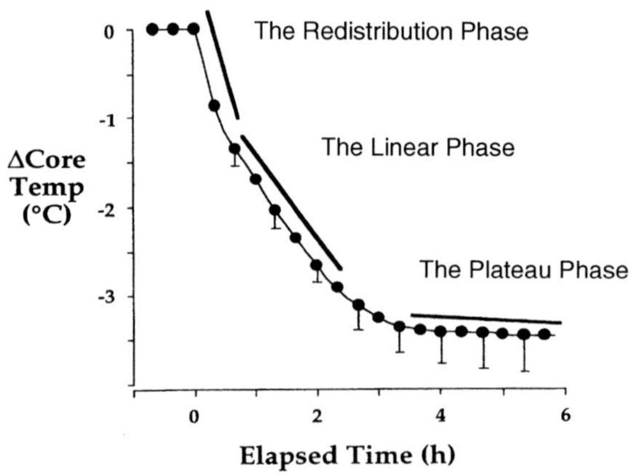

Figure 1. Three-phase pattern of hypothermia during general anesthesia.
During the first hour, core temperature usually decreases 1-1.5 °C. This is followed by a slower, nearly linear decrease in core temperature. Finally, core temperature reaches a plateau and does not decrease further. Reprinted with permission from ref 1, with some modifications.

2) Linear phase

The initial hypothermia is followed by 2 to 3 hours of a slower, linear, decrease in core temperature. Core temperature decreases linearly at a rate determined by the difference between heat loss and production. Energy expenditure is decreased by 25 % during general anesthesia.

3) Plateau Phase

The final phase of the hypothermia curve is a core temperature plateau that is usually shown after 2-4 hours of anesthesia. This plateau phase develops due to delayed reemergence of thermoregulatory vasoconstriction, which restricts core-to-peripheral heat flow by reestablishment of arteriovenous shunt tone.

Adverse Effects of Perioperative Hypothermia

Perioperative Hypothermia is associated with several complications including morbid myocardial outcomes, surgical bleeding, wound infection, impaired immune function, delayeded postanesthetic recovery and prolonged hospitalization [2-5]. All these data strongly suggest that we, anesthesiologists, have to take special thermal precautions to prevent perioperative hypothermia. (Tables 1-3)

Table 1. Postoperative wound infection (n=200)

	normothermia	hypothermia	P value
Temperature (°C)	36.6 ± 0.5	34.7 ± 0.6	<0.001
Infection (%)	6	19	<0.01
Hospitalization (days)	12.1 ± 4.4	14.7 ± 6.5	=0.01

Cited from ref 3 with some modifications.

Table 2. Postoperative coagulopathy (n=60, Total hip arthroplasty)

	normothermia	hypothermia	P value
Temperature (°C)	36.6 ± 0.4	35.0 ± 0.5	<0.001
Blood Loss (L)	1.7 ± 0.4	2.2 ± 0.6	<0.001
Allogeneic Blood (ml)	10 ± 55	80 ± 154	=0.02

Cited from ref 4 with some modifications.

Table 3. Postoperative myocardial outcomes (n=300)

	normothermia	hypothermia	P value
Temperature (°C)	36.7 ± 0.1	35.3 ± 0.1	<0.001
Morbid Cardiac Events (%)	2	7	=0.04
Ventricular Tachycardia (ml)	3	8	=0.03

Cited from ref 5 with some modifications.

How to Prevent Perioperative Hypothermia

The skin is the predominant source of heat loss during surgery. An ambient temperature above 25 °C would be able to maintain intraoperative normothermia in the patient, which might be uncomfortable for surgeons, unfortunately. Cutaneous heat loss can be decreased by covering the skin with surgical drapes or blankets. A single layer of an insulator reduces the heat loss by approximately 30 %, but adding additional layers does not proportionately increase this benefit [6]. Less than 10 percent of metabolic heat is reported to be lost through respiration, even if dry anesthetic gases are used during general anesthesia. Therefore, passive or active airway heating and humidification contribute little to perioperative thermoregulatory management [7]. Each liter of intravenous fluid on adult patients at ambient temperature, or each unit of blood at 4 °C, decreases the mean body temperature approximately 0.25 °C. Thus, heating fluids to near 37 °C is useful to prevent hypothermia and indispensable when large volumes are administered [8].

So far, forced-air warming is generally the most effective available method to maintain normal intraoperative core temperature [9] (Figure 2). Especially, the prewarming by the forced air warming devices (warming peripheral tissues before induction of general anesthesia) can minimize redistribution hypothermia [10]. (Figure 3)

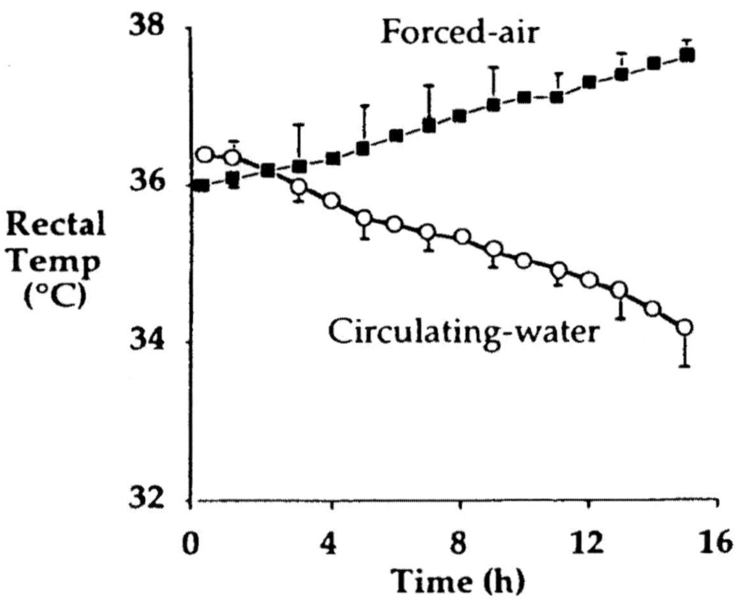

Figure 2. Forced-air warming is the most effective warming technique.
Core temperature in the patients given forced-air warming exceeded those in the circulating water group by 3.5 °C at the end of surgery in this study.
Reprinted with permission from ref 9, with some modifications.

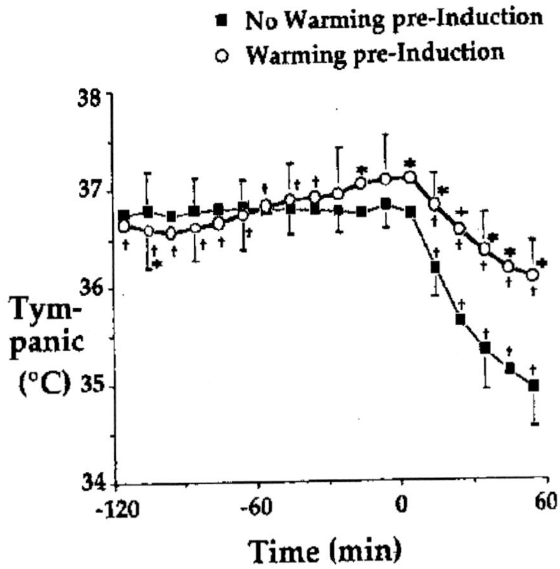

Figure 3. Prewarming minimizes redistribution hypothermia.
Active warming during the preinduction period for 120 min effectively prevented hypothermia, compared with no warming (passively cooling).
Reprinted with permission from ref 10, with some modifications.

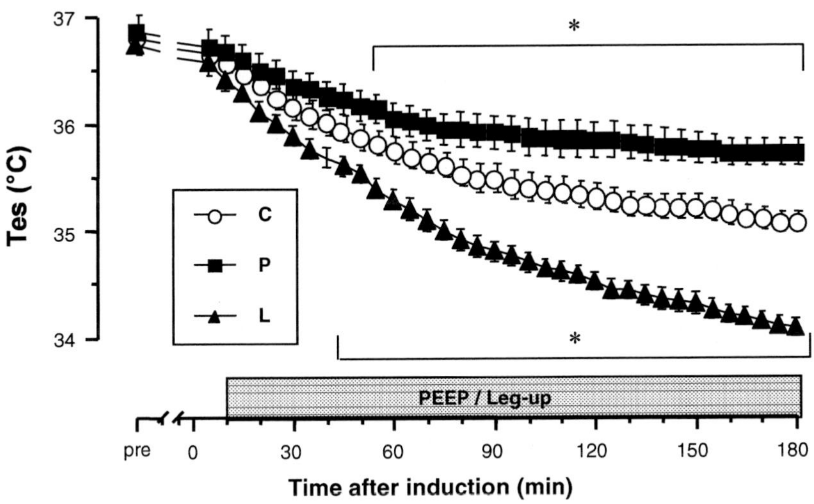

Figure 4. Excessive hypothermia in leg up position and the attenuated hypothermia in PEEP during general anesthesia.
Depending on the baroreceptor reflex loading or unloading, leg up position exaggerates hypothermia, and positive end-expiratory pressure (PEEP) ventilation attenuates hypothermia. The leg up position (L), or PEEP (P: 10 cmH_20) was applied 10 min after induction of anesthesia. The change in the distal esophageal temperature (Tes) in the P group became significantly less than that in the control (C) group starting 55 min after induction ($p<0.05$), whereas, the decrease in Tes in the L group became significantly greater than that in the C group starting 45 min after induction ($p<0.05$). Tes at 180 min after induction was 35.1 + 0.1 °C in C, 35.7 + 0.1 °C in P, and 34.1 + 0.1 °C in L, showing significant differences ($p<0.01$).
Reprinted with permission from ref 11, with some modifications.

Baroreceptor reflex modulation can also alter perioperative thermoregulation. Baroreceptor reflex is one of the reflexes which regulate blood pressure and blood volume. Through the central sympathetic nervous system and Renin – Angiotensin pathway, baroreceptor alters peripheral vascular resistance in response to BP and blood volume, which affects thermoregulatory vasoconstriction. Baroreceptor reflex loading by lithotomy position results in the exaggeration of perioperative hypothermia, and its unloading by positive end-expiratory pressure (PEEP) ventilation attenuates hypothermia [11]. This means PEEP can prevent perioperative hypothermia (Figure 4). Furthermore, clonidine premedication is reported to reduce dose-dependently this beneficial effect of PEEP on thermoregulation [12].

Dietary-induced Thermogenesis

Basic Backgrounds

The increase in energy expenditure after the ingestion has been referred to as dietary-induced thermogenesis, which was originally called specific dynamic action by Rubner [13,14]. This thermogenesis reaches a maximum after 1 h, and last for 4-6 h [15]. As is observed not only by oral intake of food, but also by intravenous administration of nutrients, this response is called thermic effect of nutrients, or nutrient-induced thermogenesis as well.

Nair et al. [16] measured the thermic response in volunteers by indirect calorimetry after a meal of protein, carbohydrate or fat, each of which provided 300 kcal (1.25 MJ). The thermic effect of protein was 30 %, to fat and to carbohydrate 13 %, respectively. The thermic response to protein and fat was still near the peak value 150 min after meal, whereas after carbohydrate meal the peak was reached at about 45 min.

When given intravenously, glucose induces an increase of energy expenditure of 6-8 %, whereas lipid infusion stimulates by 2-3 % [17]. The stimulation of energy expenditure due to amino acid infusion is 30-40 % of the energy infused as amino acids. This thermogenesis is reported to depend mainly on the nutrient composition, but is little affected by the administration route. The thermogenic responses induced by either parenteral (iv) or enteral (oral) administration of nutrients were comparable, because the energy cost of digesting and absorbing nutrients might be low. The rate of nutrient administration can influence the thermogenic response. Since the the thermic effect is mainly due to the energy cost of storage, one can understand that oral load of nutrients as a bolus induces a larger thermic effect than the continuous oral intake. It is concluded that the dietary-induced thermogenesi induced by protein or amino acids is largest and most prolonged. Nowadays, the thermic effect of protein or amino acid mixtures is known to be 4 – 6 times greater than that of carbohydrates, whereas fat seems to almost lack a thermic effect. But precise mechanisms whereby the nutrients stimulate energy expenditure are still poorly understood.

Amino acids-Induced Thermogenesis

Brundin et al measured energy expenditure and other physiological parameters such as cardiac output, arterial blood oxygen saturation, and arterial blood temperature, simultaneously after a meal in healthy subjects. They observed the increase in oxygen uptake by 20 % of the basal value after a mixed meal, with no change in arterial blood temperature [18].

The increase in blood temperature was reported in the dietary-induced thermogenesis after 900 kJ of fish protein meal in healthy volunteers [19]. The average increase in energy expenditure above basal was 13 % during 2 h study period along with the increase in arterial blood temperature by 0.22 °C at the end of study. Surprisingly, this increase in energy expenditure showed no tendency to level off at the end of the study period. This strongly suggests that protein differs from other nutrients not only by causing a quantitatively more marked thermogenesis, but also with regard to the time course and duration of the thermogenesis. Also, the increase in blood temperature showed no indication to level off and a further increase would probably have been observed. They speculated that protein induced increase in blood temperature represents an accumulation of heat in the body, and that protein ingestion elicits a pyrogen-like effect, causing the internal thermosensors to adjust the setpoint for blood temperature to a level above basal.

Intravenous administration of amino acids (600 kJ = 35 g) for 2.5 h (240 kJ/h) in healthy volunteers was first reported to be elicited blood temperature increase by 0.24 °C together with 19 % increase in energy expenditure [20] (Figure 5). Intravenous administration of amino acids stimulates the energy expenditure larger but more gradually compared with oral protein ingestion.

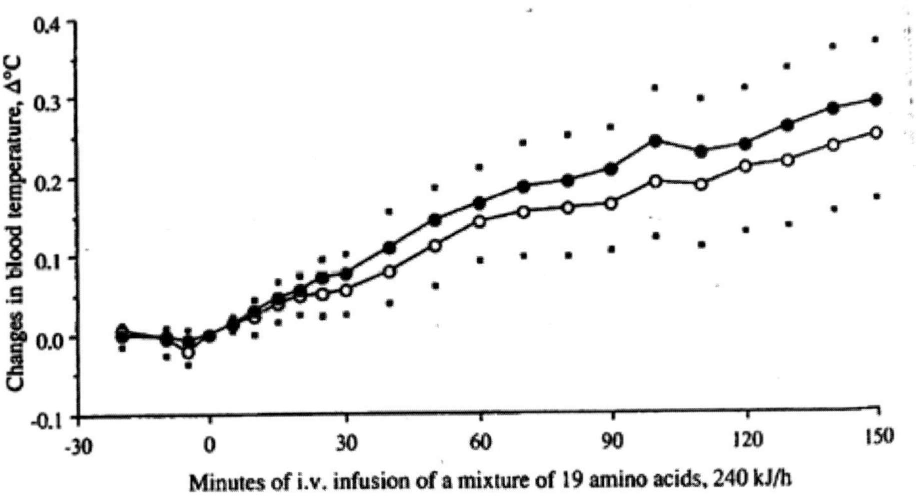

Figure 5. Changes in arterial (open circle), and hepatic venous (closed circle) blood temperatures during 2.5 h of intravenous infusion of amino acids, 240 kJ/h. Dots indicate ± SEM from healthy volunteers. Reprinted with permission from ref 20, with some modifications.

Intravenous administration of amino acids of 720 kJ for 3 h was reported to increase blood temperature by 0.34 °C along with energy expenditure increase by 24 % in healthy subjects [21]. The magnitude of thermogenesis was not directly related to the absolute blood concentration of amino acids, which indicates the thermogenesis induced by amino acids or protein is limited by other physiological regulatory mechanisms controlling heat generation.

The site and the regulation mechanism of dietary-induced thermogenesis remain undetermined. Whether brown adipose tissue or skeletal muscle is a site of increased metabolism after a meal in humans remains unknown. Yamaoka et al. have recently revealed the molecular mechanism of amino acids-induced thermogenesis [22]. The most important stage where nutritional changes play a major role in protein synthesis in skeletal muscle is translation initiation where, the translation initiation factors eukaryotic initiation factor (eIF), 4E-binding protein-1 (4E-BP1) and ribosomal protein S6 protein kinase (S6K1) play a central role through the mammalian target of rapamycin (mTOR) pathway. Enhancement of these elements of the translation initiation cascade by intravenous or oral amino acids administration is thought to require both stimulation by amino acids and insulin. They investigated the contribution of muscle protein synthesis to the prevention of hypothermia that results from amino acids administration in rats during anesthesia. They demonstrated that the stimulation of protein synthesis in skeletal muscle contributes in some way to the attenuation of anesthesia-induced hypothermia by amino acids administration, and that administration of both amino acids and the anesthetic exerts a synergic effect on the phosphorylation of S6K1 and 4E-BP1 via insulin-mTOR signaling, suggesting that an elevation of translation initiation may play a primary role in AA-induced stimulation of protein synthesis under anesthesia. These responses are likely to be triggered by the markedly elevated insulin levels that result from administering amino acids in rats under anesthesia compared with those in the conscious state. These finally facilitate thermal accumulation in the body.

Application to Clinical Session

Epoc-making study was reported by Sellden et al. that preoperative infusion of amino acids prevented perioperative hypothermia in general anesthesia [23]. They administered 480 kJ of amino acids mixture at an infusion rate of 240 kJ/h 1 – 2h before general anesthesia on patients undergoing hysterectomy, and observed surprisingly the increases in rectal temperature by 0.5-1.2 °C and energy expenditure by 40-50 % compared with those of saline infusion group at awaking after anesthesia (Figure 6). They demonstrated the thermic effect of amino acids is augmented during general anesthesia probably by increasing heat accumulation and causing delayed stimulation of heat production [24].

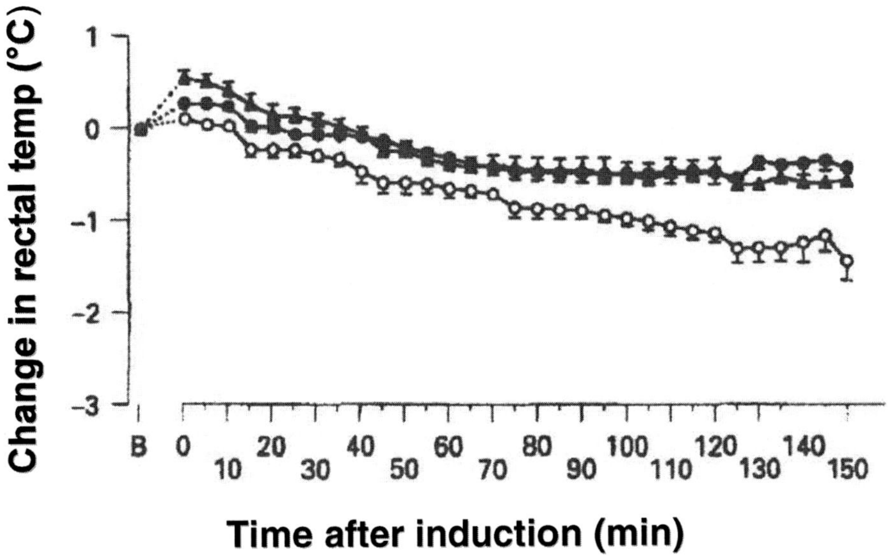

Figure 6. Changes in rectal temperature from baseline.
Amino acid mixture was infused at 240 kJ/h for 2 h starting from 1 h (closed circle), and 2 h (closed triangle) prior to induction. Amino acids infusion prevented perioperative hypothermia.
Reprinted with permission from ref 23, with some modifications.

The concept of two components is well known in the thermogenic response to glucose infusion; an obligatory thermogenesis which represents the theoretical metabolic costs for processing and storing ingested nutrients, and a facultative component which augments systemic energy expenditure and is attributable to stimulation of sympathetic activity. Intravenous administration of amino acids solution was reported to increase the metabolic rate in tetraplegic patients in whom the pathways between the brain and the peripheric sympathetic nerves were impaired [25]. Therefore, we examined the effect of intravenous amino acids infusion (4kJ $\cdot kg^{-1} \cdot h^{-1}$: starting 2 h before surgery) during spinal anesthesia, which partially blocks the sympathetic nervous activity [26]. Pre-operative amino acids infusion effectively prevents hypothermia by 0.8 °C of tympanic core temperature, due to maintain a higher metabolic rate by 18 % and to 0.3 °C of upward-shifting in the threshold for thermal vasoconstriction, compared with the saline infusion 90 min after the onset of spinal anesthesia. (Figure 7)

We also investigated the effect of amino acids infusion (4kJ $\cdot kg^{-1} \cdot h^{-1}$ for 2.5 h, mean=690 kJ) on the central thermoregulatory control in awake healthy volunteers under the both conditions of skin surface warming and cutaneous cooling [27]. Amino acid infusion significantly increased the thresholds for sweating (by 0.3 °C), and cutaneous vasodilation (by 0.2 °C). Also, amino acids significantly increased the thresholds for thermogenesis (by 0.4 °C), and cutaneous vasoconstriction (by 0.3 °C). In awake subjects (with intact thermoregulation), it is likely that amino acids induced hyperthermia results from a set point increase rather than an augmented metabolic rate per se. (Figure 8)

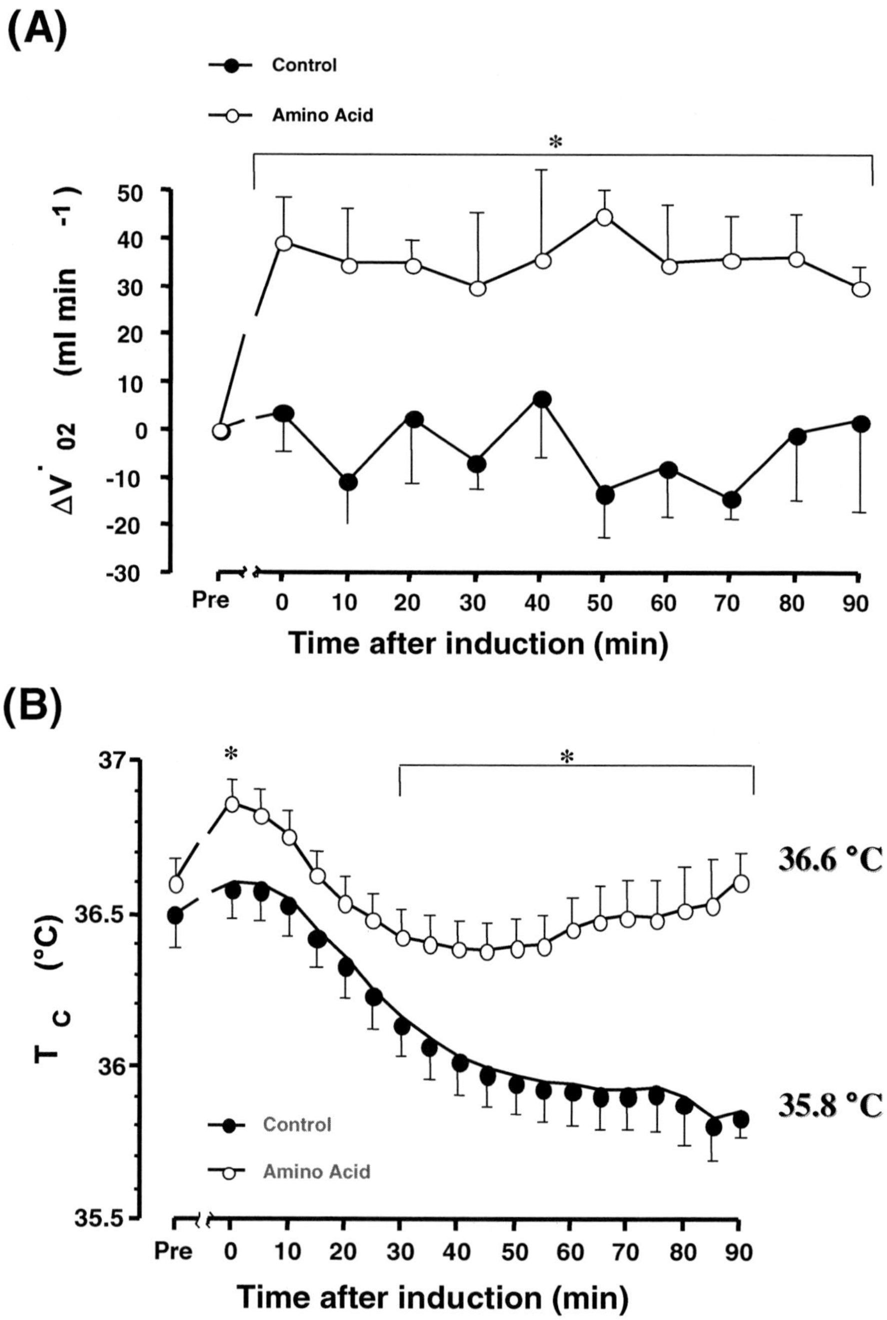

Figure 7. (A) Change in oxygen consumption during spinal anesthesia. All values after amino acid infusion in the amino acid group were significantly higher than in the saline group. (B) Change in tympanic membrane core temperature (Tc) during spinal anesthesia. Tc values at o min and after 30 min were significantly higher in the amino acid group than in the saline group. Pre indicates the period before amino acid infusion.

Reprinted with permission from ref 26, with some modifications.

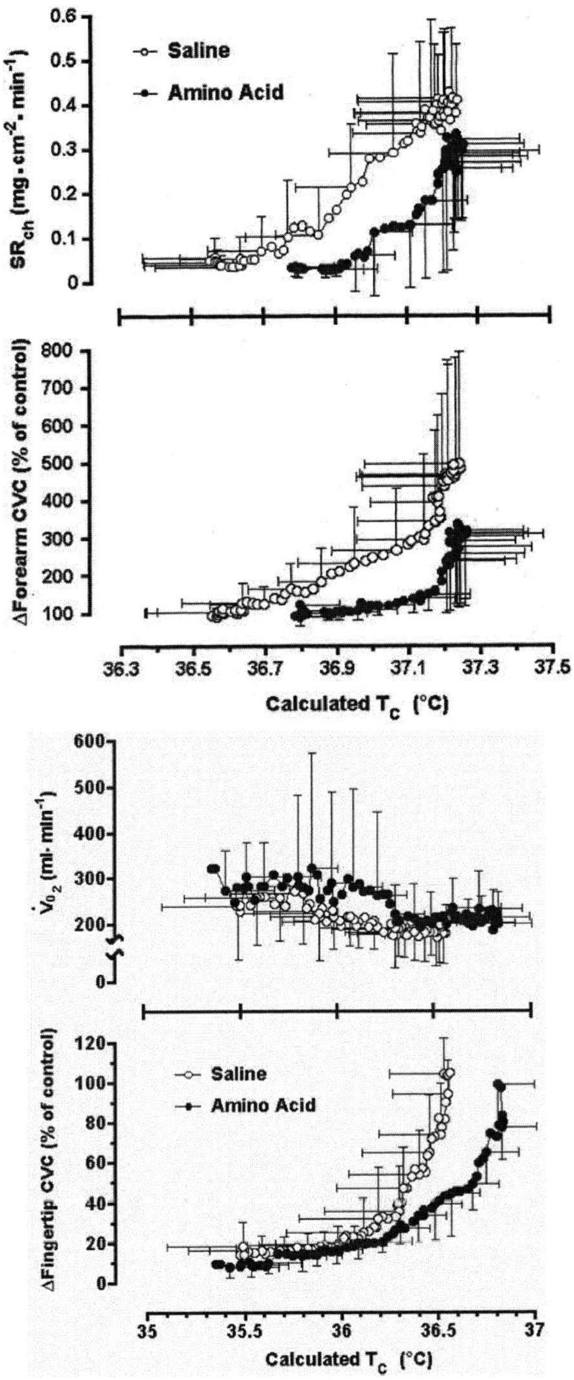

Figure 8. (A) Sweating rate on chest (SRch; top) or forearm cutaneous vascular conductance (CVC; bottom) as a function of calculated core temperature (Tc) during heating. Amino acid infusion significantly increased the threshold for sweating and cutaneous vasodilation in awake healthy volunteers. (B) Oxygen consumption (VO2; top) or fingertip cutaneous vascular conductance (CVC; bottom) as a function of calculated core temperature (Tc) during cooling. Amino acid infusion significantly increased the threshold for thermogenesis and cutaneous vasoconstriction in awake healthy volunteers.
Reprinted with permission from ref 27, with some modifications.

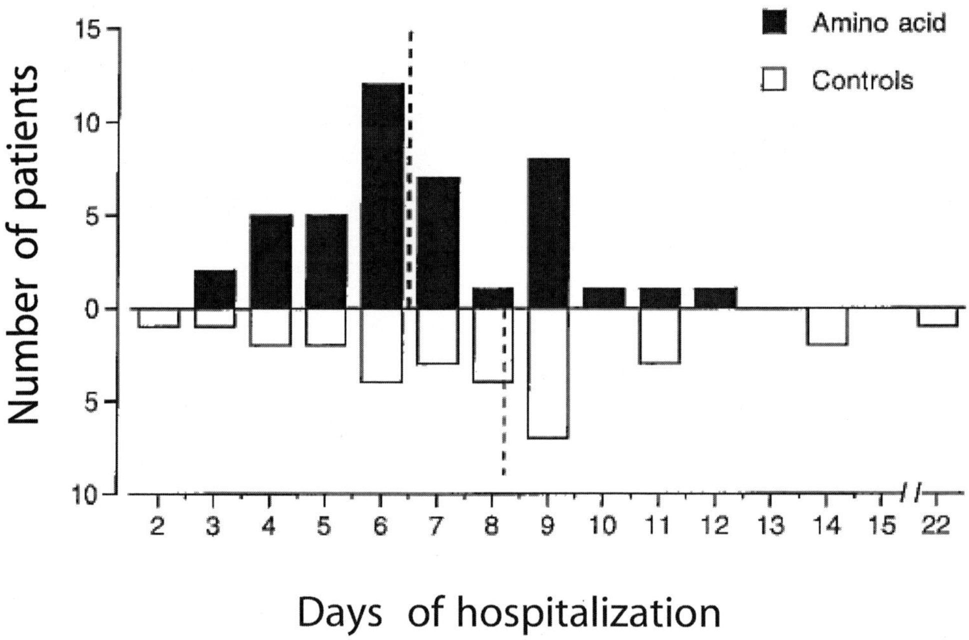

Figure 9. Days of hospitalization after abdominal surgery.
Perioperative amino acid administration at a rate of 240 kJ/h for 2-2.5 h shortened hospital stay. The estimated difference in hospital stay between the amino acid group and the control group was 2.7 days. Reprinted with permission from ref 28, with some modifications.

Sellden et al. first reported that amino acids administration improved the outcome of the patients in general anesthesia undergoing lower abdominal surgery [28]. Perioperative amino acids infusion (240 kJ/h for 2.0-2.5 h) shortened hospital stay by 2.7 days along with significantly higher core temperature and higher metabolic rate at awaking, compared with acetated Ringer's solution infusion (Figure 9). However, amino acids administration (960 kJ) for 4 h, starting 30 min before the end of a cardiopulmonary bypass in open heart surgery, did not improve the outcomes such as core temperature, extubation time, and blood loss, though oxygen uptake was significantly increased [29]. It was concluded that amino acids induced thermogenesis could not overcome the heat loss during a cardiopulmonary bypass.

Off-pump coronary artery bypass grafting (CABG) is one of the minimally invasive cardiac surgeries performed on a beating heart without using a cardiopulmonary bypass. However, off-pump techniques are often associated with core hypothermia because the heart is exposed to ambient temperature without compensatory heating from the bypass machine, and because there is little skin available for conventional surface warming during off-pump grafting. We thus investigated the effect of amino acid infusion (4 kJ kg^{-1} h^{-1} starting 2 h before surgery and lasting for 6 h) on esophageal core temperature and postoperative outcomes during off-pump CABG [30]. The esophageal core temperature at the end of surgery was significantly higer in the amino acid infusion group (36.1°C) than in the saline infusion group (35.6 °C) (Figure 10). Kaplan–Meier analysis demonstrated (Figure 11) that patients given amino acids required a significantly shorter duration of postoperative mechanical ventilation than patients given saline, [3.0 (2.5–3.9) vs 4.5 (3.8–5.8) h].

Furthermore, intensive care unit stay [20 (19.5–38.4) vs 44 (21–45) h], and days until fit for discharge from hospital [10 (9–11) vs 12 (11–13) days] were significantly shorter in patients given amino acid. Perioperative amino acid infusion in patients undergoing off-pump CABG effectively minimizes intraoperative hypothermia and improves postoperative recovery.

Fructose-induced Thermogenesis

Fructose, a natural carbohydrate found in fruit and honey is known to elicit the greatest increase in thermogenesis among various carbohydrates. Sharief et al. first observed the greater thermogenesis after 5 g/kg of sucrose ingestion than after glucose [31]. They also reported that both of sucrose ingestion, and glucose and fructose mixture ingestion showed a significant increase in metabolic rate, whereas neither of glucose, galactose, lactose, nor maltose showed any change [32] (Table 4). They concluded that it was fructose responsible for this greater increase because sucrose is metabolized to glucose and fructose.

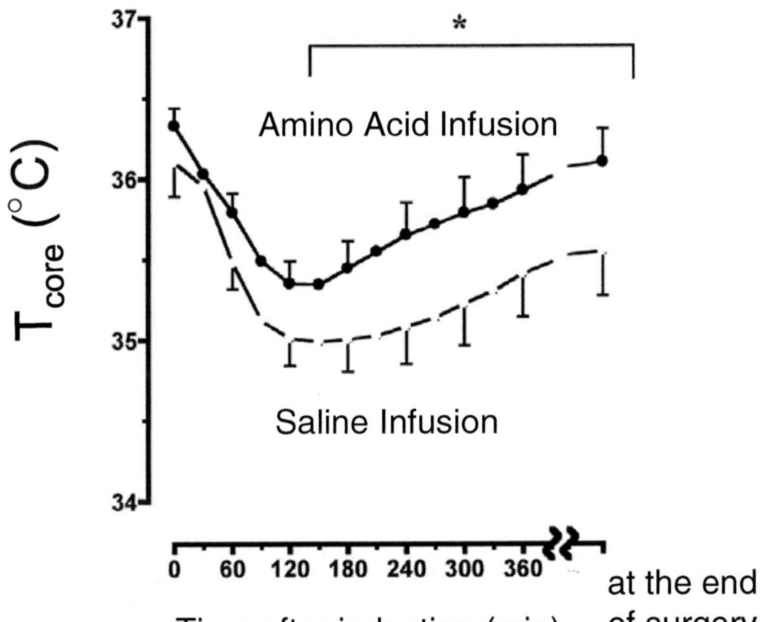

Figure 10. Distal esophageal core temperature (T_{Core}) during the study.
All temperatures after 150 min were significantly higher in the amino acid infusion group (closed circle), compared with those in the saline group (open circle). T_{core} at the end of surgery was significantly higer in the amino acid infusion group (36.1°C) than in the saline infusion group (35.6 °C).
Reprinted with permission from ref 30, with some modifications.

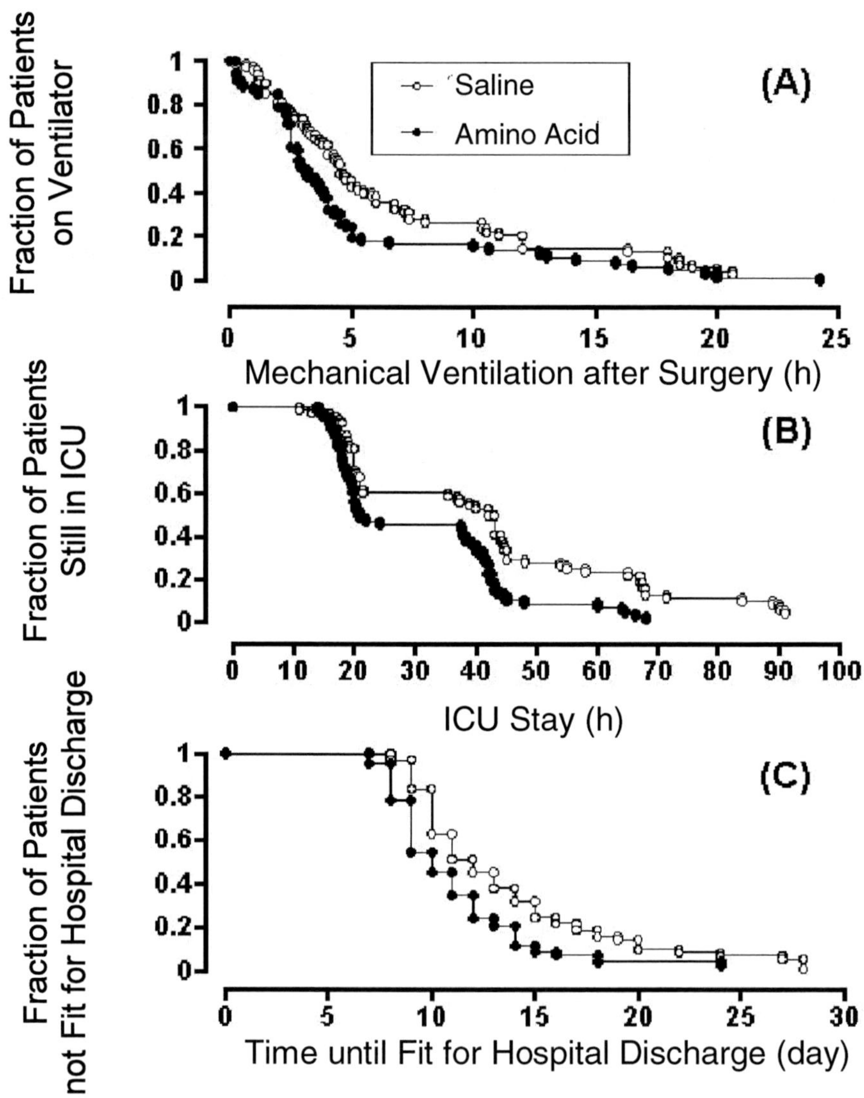

Figure 11. Kaplan–Meier curves of the probability of remaining on mechanical ventilation (A), intensive care unit stay (B), and hospitalization (C) over time comparing patients managed with amino acid infusion with those receiving saline infusion. Perioperative amino acid infusion in patients undergoing off-pump CABG improves postoperative recovery.
Reprinted with permission from ref 30, with some modifications.

A rapid increase in energy expenditure after 75 g fructose ingestion was 62 % greater than observed after a comparable dose of glucose, and 11 % greater than observed before ingestion [33]. The thermogenesis induced by fructose was accompanied by little change in the plasma insulin concentration, which suggests the increase in energy expenditure after fructose ingestion is related to the augmentation of cellular metabolism. Because both of increments in energy expenditure after both of glucose and fructose were partially inhibited with continuous infusion of propranolol, it is likely that β̃adrenergic activity (sympathetic nervous activity) somewhat contributes to the carbohydrate-induced thermogenesis.

Intravenous administration of fructose at a rate of 50 μmol/kg/min for 4 h elicited the increase by 22 % in the energy expenditure, compared with the basal value [34]. Also, oral fructose (75 g) increased the arterial blood temperature by 0.2 °C 120 min after ingestion, whereas a comparable dose of glucose showed no change [35] (Figure 12).

Table 4. The mean increase in metabolic rate (kJ/min) for men and women during 180 min after ingesting various carbohydrates (2 g/kg).

	Men	Women
Glucose	10.3 ± 2.33	11.8 ± 1.02
Galactose	12.3 ± 2.75	9.4 ± 2.18
Surcrose	17.7 ± 1.71	17.6 ± 1.92
Lactose	11.1 ± 3.52	9.2 ± 2.23
Galactose and Glucose	9.5 ± 0.70	12.3 ± 1.29
Fructose and Glucose	18.6 ± 1.53	--
Maltose	11.6 ± 1.87	--

Cited from ref 32 with some modifications.

Although fructose and glucose are both hexose monosaccharides, they are metabolized quite differently [36]. A large proportion of either intravenous or orally administered glucose escapes liver uptake and is disposed of by peripheral muscle tissue, while fructose is essentially metabolized in the liver. The enzyme fructokinase, which is responsible for the phosphorylation of fructose, is about 10 times more active than the combined activities of glucokinase and hexokinase, which are necessary for the phosphorylation of glucose. Because fructose avoids the rate-limiting enzyme of glycolysis, phosphofructokinase, it is metabolized more rapidly than glucose, resulting in the accumulation of three-carbon compounds, which are available for gluconeogenesis, and the production of glucose. Therefore, the large thermic response to fructose administration is probably due to stimulation of gluconeogenesis and glycogen synthesis in the liver [35]. The stimulatory effect of fructose on lipid synthesis may also contribute, in part, to its greater thermogenic effect when compared with glucose. Fructose has been shown to increase the rate of fatty acid synthesis [33].

Then, we undertook the study to test the hypothesis that fructose infusion would cause thermogenic stimulation in anesthetized humans during surgery [37]. Fructose infusion group was given fructose at a rate of 0.5 g/kg/h for 4 h (starting 3 h before the induction of anesthesia). Control group was administered the same amount of saline solution. Core temperature represented by esophageal temperature, forearm-minus-fingertip skin temperature gradient, and oxygen consumption were measured for 3h after the induction of anesthesia. Oxygen consumption increased by 33 ml/ min in the fructose group, and the significant increase was maintained for 135 min after the induction of anesthesia, compared with the saline group (Figure 13). Fructose group vasoconstricted at significantly higher core temperature than the control group (36.2 ± 0.3 °C vs 35.6 ± 0.3 °C) (Figure 14). Fructose infusion caused a significant increase in core temperature starting 20 min after induction of anesthesia, compared with saline infusion. Mean final core temperature 3 h after induction of anesthesia was significantly higher in the fructose group (35.7 ± 0.4 °C vs 35.1 ± 0.4 °C) (Figure 15) . Taken together with our previous studies of amino acids, fructose as well as amino acids prevent perioperative hypothermia as a result of both an increased level of energy expenditure and an upward shift in the threshold core temperature for thermoregulatory vasoconstriction (Figure 16).

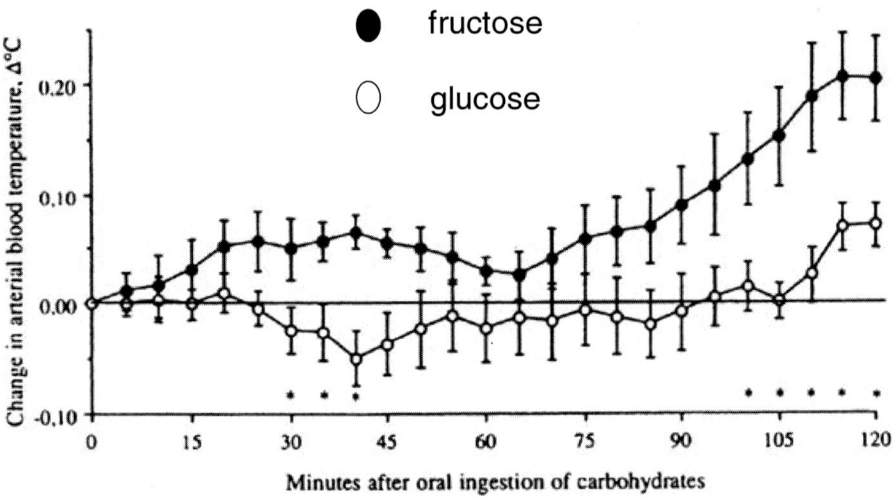

Figure 12. Change in arterial blood temperature after oral ingestion of 375 ml water solution of 75 g fructose (n=6: closed circle) or glucose (n=6: open circle).
Reprinted with permission from ref 35, with some modifications.

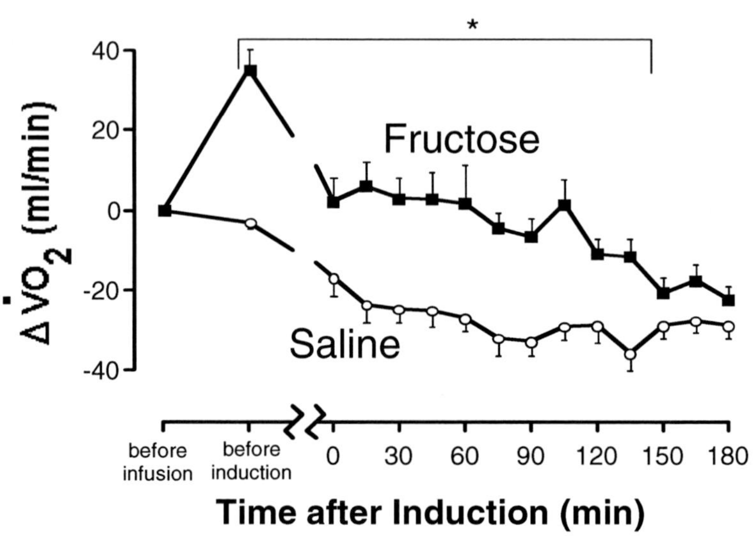

Figure 13. Change in oxygen consumption during the study. Oxygen consumption increased by 33 ± 16 ml/min in the fructose infusion group immediately before induction. It remained greater than that of the saline infusion group for 135 min after induction of anesthesia ($p<0.001$). Data are presented as mean ± SD for 10 patients.
Reprinted with permission from ref 37, with some modifications.

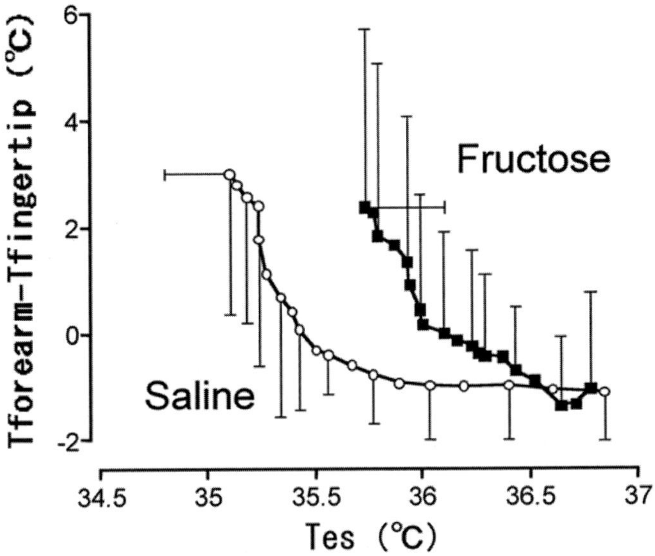

Figure 14. Relationship between the forearm minus fingertip temperature gradient (Tforearm – Tfingertip) and distal esophageal core temperature (Tes). Patients assigned to fructose infusion vasoconstricted at a significantly higher core temperature than did those receiving saline infusion (36.2 ± 0.3 °C vs 35.6 ± 0.3 °C) ($p<0.001$). Data are presented as mean ± SD for 10 patients.
Reprinted with permission from ref 37, with some modifications.

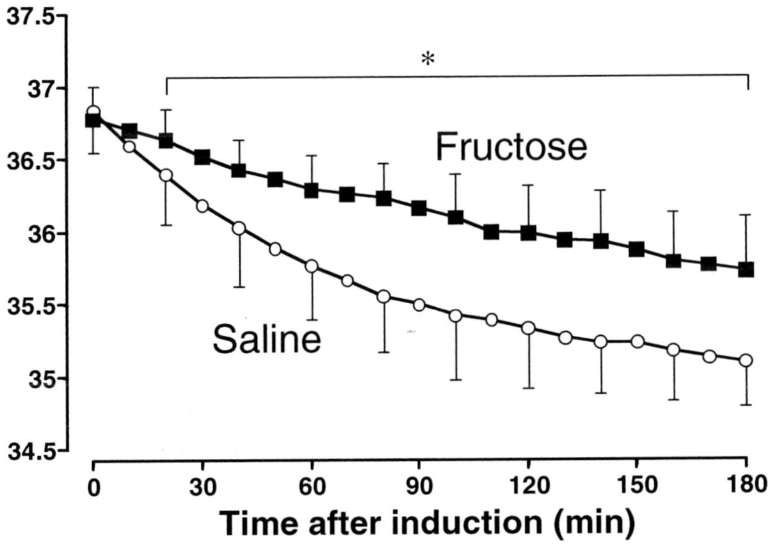

Figure 15. Core temperature measured at the distal esophagus (Tes) during surgery. Patients receiving the fructose infusion had significantly greater core temperatures than those receiving saline ($p<0.001$) starting 20 min after induction of anesthesia until the end of the monitoring period (180 min after induction). Mean final core temperature 3 h after induction of anesthesia was significantly higher in the fructose group (35.7 ± 0.4 °C vs 35.1 ± 0.4 °C) Data are presented as mean ± SD for ten patients in each group.
Reprinted with permission from ref 37, with some modifications.

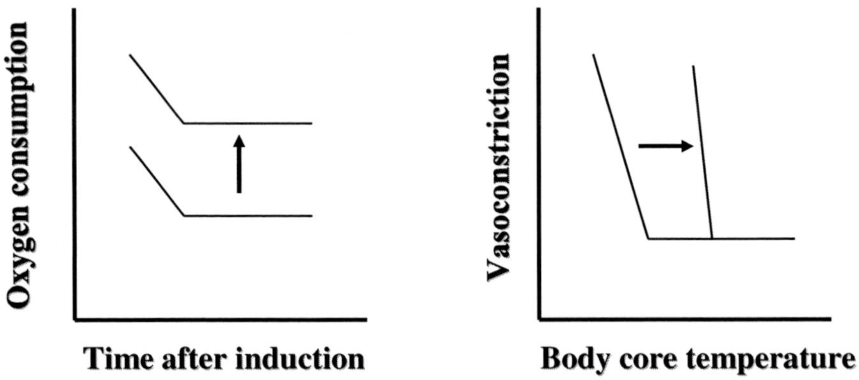

Figure 16. Mechanism of preventing perioperative hypothermia by dietary-induced thermogenesis. Fructose as well as amino acids prevent perioperative hypothermia as a result of both an increased level of energy expenditure and an upward shift in the threshold core temperature for thermoregulatory vasoconstriction.

It would be an effective method to improve temperature balance by using amino acids-induced and/or fructose-induced thermogenesis, which would be the furnace during anesthesia [38].

References

[1] Sessler DI. Perioperative heat balance. *Anesthesiology* 2000;92:578-96.

[2] Sessler DI: Complications and treatment of mild hypothermia. *Anesthesiology* 2001;95:531-543.

[3] Kurz A, Sessler DI, Lenhardt R. perioperative normothermia to reduce the incidence of surgical-wound infection and shorten hospitalization. *N Engl J Med* 1996; 334: 1209-15.

[4] Schmied H, Reiter A, Kurz A, Sessler DI, Kozek S. Mild hypothermia increases blood loss and transfusion requirements during total hip arthroplasty. *Lancet* 1996; 347: 289-92.

[5] Frank SM, Fleisher LA, Breslow MJ, Higgins MS, Olson KF, Kelly S, Beattie C. Perioperative maintenance of normothermia reduces the incidence of morbid cardiac events. A randomized clinical trial. *JAMA* 1997; 277: 1127-34.

[6] Sessler DI, Schroeder M. Heat loss in humans covered with cotton hospital blankets. *Anesth Analg* 1993;77:73-7.

[7] Deriaz H, Fiez N, Lienhart A. Influence d'un filtre hygrophobe oud'un humidificateur-reÅLchauffeur sur l'hypothermie peropeÅLratoire. *Ann Fr Anesth Reanim* 1992;11:145-9.

[8] Sessler DI. Mild perioperative hypothermia. *N Engl J Med* 1997;336:1730-37.

[9] Kurz A, Kurz M, Poeschl G, Faryniak B, Redl G, Hackl W. Forced-air warming maintains intraoperative normothermia better than circulatingwater mattresses. *Anesth Analg* 1993;77:89-95.

[10] Hyson JM, Sessler DI, Moayeri A, Mcguire J, Schroeder M. The effects of preinduction warming on temperature and blood pressure during propofol/nitrous oxide anesthesia. *Anesthesiology* 1993;79:219-28.

[11] Nakajima Y, Mizobe T, Takamata A, Tanaka Y: Baroreflex modulation of peripheral vasoconstriction during progressive hypothermia in anesthetized humans. *American Journal of Physiology* 2000;279:R1430-36.

[12] Mizobe T Nakajima Y, Sunaguchi M, Ueno H, Sessler DI. Clonidine produces a dose-dependent impairment of baroreflex-mediated thermoregulatory responses to positive end-expiratory pressure in anaesthetized humans. *British Journal of Anaesthesia* 2005; 94:536-41.

[13] Rubner, M. Die Quelle der thierischen Warme. *2. BioZ.* 1894;30: 73-142.

[14] Rubner, M. *Die Gesetze des Energiverbrauchs bei der Ernahrung,* 1902, Leipzig and Vienna.

[15] Pittet, P, Gygax PH, Jequier E. Thermic effect of glucose and amino acids in man studied by direct and indirect calorimetry. *Br. J. Nutr.* 1979;31: 343-349.

[16] Nair KS, Halliday D, Garrow JS. Thermic response to isoenergetic protein, carbohydrate or fat meals in lean and obese subjects. *Clinical Science* 1983; 65:307-12.

[17] Jequier E. The influence of nutrient administration on energy expenditure in man. *Clinical Nutrition* 1986;5:181-86.

[18] Brundin T, Wahren J. Influence of a mixed meal on splanchnic and interscapular energy expenditure in man. *Am J Physiol* 1991;260:E232-7.

[19] Brundin T, Warren. Influence of protein ingestion on human splanchnic and whole-body oxygen consumption, blood flow, and blood temperature. *Metabolism* 1994;43:626-32.

[20] Brundin T, Wahren J. Effects of iv amino acids on human splanchnic and whole body oxygen consumption, blood flow, and blood temperatures. *Am J Physiol* 1994;266:E396-402.

[21] Brundin T, Wahren J. Renal oxygen consumption, thermogenesis, and amino acid utilization during iv infusion of amino acids in man. *Am J Physiol* 1994; 267:E648-55.

[22] Yamaoka I, Doi M, Nakayama M, Ozeki A, Mochizuki S, Sugahara K, et al. Intravenous administration of amino acids during anesthesia stimulates muscle protein synthesis and heat accumulation in the body. *Am J Physiol* 2006;290:E882-8.

[23] Sellden E, Branstrom R, Brundin T. Preoperative infusion of amino acids prevents postoperative hypothermia. *Br J Anaesth* 1996; 76:227-34.

[24] Sellden E, Lindahl SG. Postoperative nitrogen excretion after amino acid-induced thermogenesis under anesthesia. *Anesth Analg* 1998;87: 641-6.

[25] Aksnes AK, Brundin T, Hjeltnes N, Wahren J. Metabolic, thermal, and circulatory effects of intravenous infusion of amino acids in tetraplegic patients. *Clin Physiol* 1995; 15:377-96 .

[26] Kasai T, Nakajima Y, Matsukawa T, Ueno H, Sunaguchi M, Mizobe T. Effect of preoperative amino acid infusion on thermoregulatory response during spinal anaesthesia. *Br J Anaesth* 2003; 90:58-61.

[27] Nakajima Y, Takamata A, Matsukawa T, Sessler DI, Ueno H, Mizobe T. Effect of amino acid infusion on central thermoregulatory control in humans. *Anesthesiology* 2004; 100:634-9.

[28] Sellden E, Lindahl SG. Amino acid-induced thermogenesis reduces hypothermia during anesthesia and shortens hospital stay. *Anesth Analg* 1999;89:1551-6.

[29] Sellden E, Rimeika D, Settergren G. Thermogenic effect of amino acids not demonstrated in heart surgery with cardiopulmonary bypass. *Acta Anaesthesiol Scand* 2005;49:35-40.

[30] Umenai T, Nakajima Y, Sessler DI, Taniguchi S, Yaku H, Mizobe T. Perioperative amino acid infusion improves recovery and shortens the duration of hospitalization after off-pump coronary artery bypass grafting. *Anesth & Analg* 2006;103:1386-93.

[31] Sharief NN, Macdonald. Differences in dietary-induced thermogenesis with various carbohydrates in normal and overweight men. *Am J Clin Nutr* 1982; 35: 267-72.

[32] Macdonald I. Differences in dietary-induced thermogenesis following the ingestion of various carbohydrates. *Ann Nutr Metab* 1984; 28: 226-30.

[33] Tappy L, Randin JP, Felber JP, Chiolero R, Simonson DC, Jequier E et al. Comparison of thermogenic effect of fructose and glucose in normal humans. *Am J Physiol* 1986; 250: E718-24.

[34] Schwarz JM, Acheson KJ, Tappy L, Piolino V, Muller MJ, Felber JP et al. Thermogenesis and fructose metabolism in humans. *Am J Physiol* 1992; 262: E591-8.

[35] Schwarz JM, Acheson KJ, Tappy L, Piolino V, Muller MJ, Felber JP et al. Thermogenesis and fructose metabolism in humans. *Am J Physiol* 1992; 262: E591-8.

[36] Brundin T, Wahren J. Whole body and splanchnic oxygen consumption and blood flow after oral ingestion of fructose or glucose. *Am J Physiol* 1993; 264: E504-E513.

[37] Mizobe T, Nakajima Y, Ueno H, Sessler DI. Fructose administration increases intraoperative core temperature by augmenting metabolic rate and the vasoconstriction threshold. *Anesthesiology* 2006; 104:1124-30.

[38] Lindahl S. :Setting the furnace during anesthesia. *Nesthesiology* 2006;104:1119-20.

In: Body Temperature Regulation
Editor: A. B. Cisneros and B. L. Goins
ISBN: 1-60741-282-3

Chapter IVX

Thermoregulation in a Hornet Nest, *Vespa Orientalis* (Hymenoptera: Vespinae): The Interaction between Workers and the Pupal Brood

Marian Plotkin[1,*], Stanislav Volynchik[1], David J. Bergman[2] and Jacob S. Ishay[1]

[1]Department of Physiology and Pharmacology, Sackler Faculty of Medicine, [2]Tel-Aviv University, Ramat-Aviv Israel 69978

[2]School of Physics and Astronomy, Beverly and Raymond Sackler Faculty of Exact Sciences, Tel-Aviv University, Ramat-Aviv Israel 69978

Abstract

Observations and experiments were carried out to assess continuous thermoregulation in the nest of the Oriental hornet *Vespa orientalis* (Hymenoptera, Vespinae). The temperature of a pupating larva inside a nest without adult hornets was found to be higher by 1°C from the nest temperature in the vicinity of the brood comb. However, the temperature of larvae subjected to diethyl ether anesthesia equaled the ambient temperature. Observations showed that adult worker hornets occasionally warm the pupae from outside their cells and this by blowing warm air through the silk cap, of each cell. Upon artificial warming of combs whose adult population was intentionally removed, the temperature around these combs (at the beginning of the warming process)

---

* Corresponding author: Marian Plotkin. E mail: physio7@post.tau.ac.il; Phone: 972-3-6409138; Fax: 972-3-6409113. Note: The Corresponding author of this manuscript (JSI) has passed away after this article was accepted for publication. For that reason, the corresponding author has now been changed to be MP.

was lower than that of the silk-encased pupae inside the comb cells. Beyond 31.1°C, the temperature exceeded that of the pupae. The findings suggest that of the normal nest temperature is of 29°C, the pupal temperature is 30°C, which is apparently optimal for pupal development. Results of the present study prompted construction of a simple model that describes the interaction between the adults and the pupal brood which enables to maintain the latter at a constant temperature.

Keywords: Thermoregulatory center, Hornet silk, Closed-loop control, Phase transition, Oriental hornet

Introduction

Temperature measurements in the nests of hornets have long been carried out, as witness Himmer (1927) and his measurements in the nests of *Paravespula vulgaris* and *Vespa crabro* and also those of Ishay and Ruttner (1971) on *V. orientalis*. In all these measurements the temperature was higher and steadier in the nest than outside it. In fact, anybody observing a hornet nest must consider and evaluate the conditions extant inside and outside the nest. First, the nest may house thousands of individual hornets, all of which breathe and maintain metabolic processes that require a continuous supply of oxygen throughout the day and night. Secondly, hornet activity in the nest produces a rise in the relative humidity, due to the breathing of the hornets, their excretions and metabolic processes, and also the building activities which release humidity as the moist building mash dries and hardens. Thirdly, the nest activities release quantities of CO_2 (metabolic and respiratory) which need to be removed from the nest. Additionally, one needs mention that the nest is situated in a 'closed pocket' and could therefore develop parasitic fungi as well as other parasites when it is dark and moist. All the above potential hazards demand that the nest be aerated and kept dry, a task that is achieved *via* active ventilation and is carried out throughout most of the warm hours of the day by a group of ventilating workers whose number befits the needs of the nest and reflect the total number of workers. In other words, the warmer the nest (especially in the afternoon) the greater the number of workers engaged in ventilation (Riabinin et al., 2004). Of course, the prevailing climatic conditions dictate the extent and nature of the ventilatory activity, so that for hornet and social wasps species living in a cold climate the need for ventilatory activity is low or even non-existent (Klingner et al., 2005).

One, of course, has to take into account the need for exchange of gases, i.e., the removal from the nest of CO_2 and the introduction of O_2. This, however, occurs spontaneously by simple diffusion, so that ventilation is not really essential for the gas exchange process in the case of hornets. Yet, in the nest of the Oriental hornet, we found that such gas exchange is probably improved by the presence of tiny holes (of about 50 μm) both in the 'button' at the center of each larval silk cap as well as in the envelope (up to 100μm) encasing the nest proper.

In hot climates where the temperature outside the nest is higher or equal to that inside it, there is a need both for ventilatory activity as well as the introduction into the nest of water droplets to cool it. The obvious interplay between all the afore-mentioned factors clearly demands more precise assessment of the factors that affect growth of the colony, and in particular, the behavior of the adult hornets and the way whereby they help the pupal brood to develop. It turns out that maintenance of a high and steady temperature in the nest is a hard task demanding individual attention and handling by the adult population, as will be detailed hereinafter.

Materials and Method

In the summer of 2007, combs of the Oriental hornet were collected from nests in the environs of Tel-Aviv by a method described previously (Ishay, 1964). In order to most closely simulate the conditions extant in nature for these hornets, the collected combs with contained workers and brood were transported to our laboratory and introduced into artificial breeding boxes (ABBs), one wall of which was made of transparent glass, and the other walls made of solid wood. Each ABB was supplied with water *via* a glass vial whose open end was sealed with cotton, which enabled the hornets to drink by licking the water off the water-soaked cotton. Additionally each ABB was supplied with food *ad libitum* in the form of sugar cubes, providing a source of carbohydrates and ready energy. This type of food is preferred by the adult hornets. However, food of a protein source, like eggs and tuna fish, was also provided, mostly for the feeding of the developing brood. Finally, each ABB was supplied with red loam soil for building purposes. Photographs of the behavior of hornets were taken *via* digital camera Dimage Z1 of Konica Minolta. Temperature measurements were obtained through use of CHY W506R thermocouples, with thermal recordings made with R 506T software. One thermocouple was inserted immediately upon completion of the pupal weave beneath the silk cap, while the other was affixed in the ABB outside the comb, at a distance of 2cm from the comb. Slow artificial warming of the sensor - equipped ABB was achieved *via* the pumping of hot air on the ABB from the outside, whereas a rapid rise in temperature was obtained by the introduction of hot air directly through one of the inlets into the ABB. The results of measurements and the subsequent statistical analyses (two-sample independent t test) were processed through use of Originpro 7.5 software.

Results

Is the temperature in the nest dependent primarily on the ambient temperature? We attempted to obtain at least a partial answer to this question, and this by maintaining the entire breeding facilities and their adjuncts within the laboratory room while enabling the hornets to obtain their least dietary needs *ad libitum* from outside the laboratory room, and also providing them with building materials (soil and paper).

Preferable layout of figures

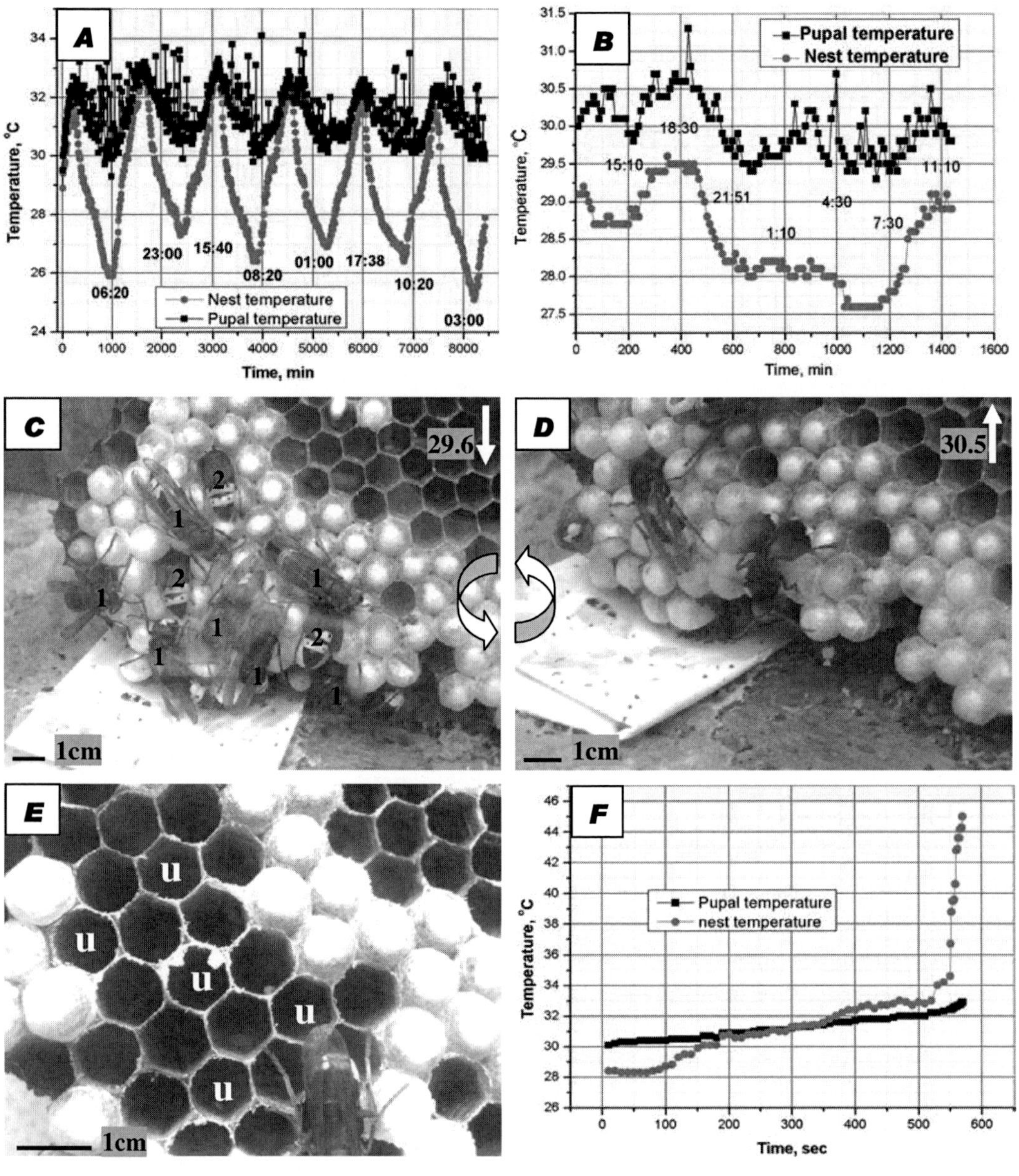

Figure 1. Depicts temperature changes in the nest of the Oriental hornet.
A. The temperature in a breeding box situated outside the laboratory with its exit opening out to the field. Temperature measurements were carried out over a period of 6 days. One can see that the temperature in the interior of the nest rises during the daytime and drops at night. The thermal difference between the pupa and its environment is significant $p<0.01$ (●=nest temperature, ■ = pupal temperature).
B. Temperature measured over 24 hours in a breeding box kept inside the laboratory. One can see that the temperature of the nest interior rises during the day and falls at night. The temperature of the pupa resembles only to a small extent the changeable temperature of the nest interior but rather remains relatively stable and does not exceed 30°C. The difference between the pupal and ambient temperature is significant $P<0.01$ (●=nest temperature, ■ = pupal temperature).
C. When the temperature drops below 29.6°C, the hornets aggregate over the comb surface (1), or else intrude head and thorax into a vacant nearby comb cell (2) and proceed to boost their respiration rate, thereby

warming the region. Bar=1cm.
D. Once the temperature rises and attains 30.5°C, the hornets reduce their respiration rate and leave the region. Upon cessation of the warming, the temperature continues to drop, and once it drops again to below 29.6°C, the hornets reassemble to warm the silk caps (Figure 1C). Bar=1cm.
E. Following several days of temperature regulation by the hornets, one can see that the area in the comb that was warmed by the hornets is now unoccupied (u) owing to eclosion of the pupae that had been warmed. Note also the silk caps covering pupae on both sides of the combs. The pupae have succumbed to accidental overdosing with anesthetic and have ultimately failed to eclode. Bar=1cm.
F. Graph depicting behavior of the temperature inside (black line) and outside (grey line) the silk cap when all the worker hornets had been removed from the ABB and the nest was warmed artificially. Upon slow warming of the ABB, the rate of warming of the pupa is slower than that of the ambience outside the comb cell. Temperature tends to equalize at 31.1°C following which, with continuation of the warming, the temperature outside the silk cap exceeds that of the pupa beneath the silk cap. Increase in the rate of warming of the ABB induces a sharp 14°C rise in the nest temperature whereas the pupal temperature increases by a mere 1°C (●=nest temperature, ■ = pupal temperature).

Figure 1A depicts temperature measurement in an active hornet nest located in an Artificial Breeding Box (ABB) that bears an outlet allowing the hornets to fly out freely to the field. Indeed after a while the hornets accustomed themselves to fly out and behave in the field as normal hornets do. In this particular nest, two thermocouples were affixed for continuous measurement of the temperature, namely, one underneath the silk cap of one of the pupae and the other in the space within the nest. The temperature measurements were carried out uninterruptedly for 6 days. The results revealed the presences of fluctuations associated with nycthemeral conditions, and also fluctuations depending on the site of measurement. Regarding the latter we found that measurements made in the space of the nest fluctuated between 32°C (in the warm hours of the afternoon) and 26°C (in the pre-dawn hours),with the disparity between the two thermal values amounting to about 6°C. On the other hand, with the thermocouples inserted beneath the silk cap, the hottest temperature was recorded in the warm hours of the day and ranged between 32-33°C, while the coldest temperature was about 30°C. Thus the disparity between low and high temperature in measurements pertaining to the pupa itself was 2-3°C. However, the thermal difference between the warmest point in the pupa and that in the nest is quite small and even overlaps occasionally. The disparity between the coldest temperature in both measurements undertaken was about 6°C. The difference between the pupal temperature and the ambient temperature proved significant at the $p<0.01$ level. If to recap, it was quite clear that the pupal temperature does not usually drop below 30°C.

How does the behavior of the worker hornets impact on the pupal temperature?

We attempted to obtain an answer by keeping the hornet nest inside the laboratory as before, except that now their ABB was sealed (i.e. the hornets were prevented from leaving the ABB), and they could only obtain their least dietary needs and building materials from us (i.e., they were provided inside the ABB).

Figure 1B shows thermal measurements attained by the use of two thermocouples, as in Figure 1A, but this time depicting 24-hour period selected from measurements run continuously over a fortnight. Regarding measurements undertaken on the pupa, we recorded fluctuation of the temperature around 30°C (29.5-30.5°C), as indicated by the upper line, whereas the thermal values recorded within the interior of the ABB ranged between 27.6°C – 29.5°C. The lowest temperature in the nest was recorded at dawn, between 6-7am. The differences here between the pupal and ambient temperatures were smaller, with the pupal temperature stable at about 30°C, while the ambient temperature was lower than the pupal one by up to 2.5°C. The thermal differences between the pupa and its ambience proved significant at the $p<0.01$ level. Changes in comb temperature elicited the following response in adult worker hornets near the comb cells, namely, they approached (by walking) the region housing the silk-caped pupae, bent over the comb, and where feasible and available intruded their head and thorax into vacant comb cells in the vicinity of or abutting the pupal cell. Next they increased their respiratory rate and thereby raised the temperature in the pupal cells to about 30°C. This attained, the workers proceeded to other parts of the comb that needed similar treatment. In this regard it is noteworthy that only pupae warmed to optimal temperature complete their development to the point of eclosion. Other pupae not receiving such treatment are likely to stop developing and to die off (Ishay & Ruttner, 1971). Following removal of the adult hornet population from the ABB , the temperature of the pupating larvae became higher by 1°C than the nest temperature in the vicinity of the brood comb. We have observed that worker hornets walking upon the pupal silk caps, seemingly engage in examining which sites need warming. Presumably, to ascertain the need for warming they intermittently touch the substrate with the tip of one of their antennae, the substrate in this case being the stretch of pupal silk caps, and finding the silk caps to have cooled, they remain to warm them. An illustration of a warming cycle with pauses between warming stages is given in Figure 1C, D. As shown in Figure 1C, when the temperature dropped to below 29.6°C, there was an immediate recruitment of workers from nearby areas to come and warm the brood. Once the warming phase was tackled successfully (within a matter of minutes), the temperature rose to 30.5°C (Figure 1D) and the 'warming' workers left the site and sought other regions needing similar treatment. Occasionally a few workers remained behind, but these temporarily stopped their warming activity. This process was self-repeating over time, and by the end of our fortnightly observation period, the cells that had received worker visitation and warming were vacated by ecloding pupae (Figure 1E). In contrast, we observed 'stretches' of comb with cells containing pupae that had failed to complete their development and died before ecloding. Such regions, in fact, did not receive warming treatment by the workers, possibly because they were earlier over-exposed to the anesthetic diethyl ether.

Figure 1F assesses the contribution of the silk weave produced by the pupating larvae to the overall comb temperature. To elucidate this contribution we compared the rate of warming of the nest interior with that of a comb cell housing a silk – capped pupa. What we actually did here was to remove all the workers from the nest, then subject the entire nest to warming by hot air. Initially the nest temperature was 28.3°C, while the pupal temperature was 30°C. The warming by hot air was carried out at first at a slow rate (see two lines at bottom), and the rate of warming of the pupa was slower than that of the ambience outside the pupal cell, but, the temperature of both evened out at 31.1°C. Subsequently, however, with continuation of the warming phase, the temperature outside the pupal silk weave exceeded that inside it. In all, the warming phase lasted close to 10 minutes and the overall increase in temperature of the nest interior was a mere 3°C (from 28°C to 31°C). At this stage we boosted the rate of warming, and in about a minute, the temperature of the nest interior attained 45°C while the pupal temperature remained at 33°C.

Discussion

The present study deals with thermoregulation of the brood by the Oriental hornet *Vespa orientalis* L., specifically insofar as the pupal stages of this social insect are concerned. The way whereby the worker hornets attain changes in the temperature of the brood, especially of the pupae encased in a silk weave, certainly earns them the status of being endothermic insects (Heinrich, 1993, 1996), much the same as the honeybee is (Grodzicki & Caputa, 2005), yet in complete contrast to the American cockroach *Pariplaneta americana,* which behaves as a poikilothermic insect, as do also the dung beetles (Verdú et al., 2006).

In a previous study, we showed that hornets can regulate their body temperature by means of a thermogenic or thermoregulatory center located in their prothorax (Plotkin et al., 2005a). This center was subsequently found to be rich in tracheae and mitochondria which supply the considerable energy needed for the thermoregulatory process (Ishay et al., 2006). Possibly there is in the Oriental hornet a local proliferation of mitochondria, as the occasion demands and in accordance with the need to supply heat (Sogl et al., 2000). By now, a thermogenic center or organ has also been encountered in social and solitary bees (Volynchik et al., 2006). In bee hives in the winter there is also heat production, despite the fact that in the winter bees neither fly nor have any brood, yet the production of heat enables them to survive the winter (Stabentheiner et al., 2003). In hornets, the mentioned thermogenic organ is undeveloped in the pupal stage but develops during the first few days after eclosion, and once this organ is activated, the nascent adults join in the thermoregulation of themselves as well as of the pupae. We need reiterate that no thermoregulatory center is detected in the pupal stage and even though the nest temperature is close to optimal also for the pupae, the pupal temperature needs to be precisely geared to optimum to ensure survival of the pupa. In this respect, the situation is similar to that extant in the hive of the honeybee (Kleinhenz et al., 2003). The temperature of the hornet pupa ranges around 30°C (29.99±0.38). We have previously reported (Ishay and Ruttner, 1971) that if hornet pupae develop under unsuitable thermal conditions, they subsequently eclode with numerous malformations. In the past, temperature measurements were carried out in a number of social hornets and wasps prevalent in West and North Europe, which showed the mean nest temperature to be 29°C (Spradbery, 1973; Edwards, 1980). In the case of the Oriental hornet, we established that the nest temperature is indeed 29°C, but that of the pupa beneath its silk cap is 30°C. This slight thermal disparity is normally retained owing to metabolic processes of the pupa and the humidity level beneath the silk cap as compared to that in the nest proper, and both the temperature and humidity in the nest and beneath the silk cap vary slightly from the ambient temperature and humidity (Klingner et al., 2005). Apparently the detected discrepancies are valid and essential, something that we have been able to confirm in two different ways: first, by observing the behavior of workers as they patrol the surfaces containing the silk caps and commence warming activities wherever the temperature drops below a certain threshold, and secondly by thermal measurements confirming that the temperature of the pupae is invariably close to 30°C while that of the comb is always lower. In this regard we need point out that dead pupae do not receive any heat increment from the patrolling worker (Figure 1E), and also that it is very easy to rapidly warm the interior of the nest, but not at all easy to raise the temperature of the pupae (Figure 1F). The explanation for the latter observation may reside in the pupal silk which, to begin with, seals the outlet of the pupal cell and thereby provides additional isolation of the internal cell environment (beyond the isolation of the nest proper by layers of soil), and secondly, contributes to sealing of the envelopes normally built around the combs. In the case of hornets, we need point out that we failed to detect any specific inclination of any group of them to engage in warming activity. In a recent work (Plotkin et al., 2007) we studied warming of the hornet comb when filled with or devoid of brood. We found that the warming of vacant comb cells or ones with open brood resembles warming of the ambient air in both rate and extent, whereas combs with encased brood, i.e., pupae under silk cap, are slow to warm up and the maximal temperature attained is considerably lower. Combs containing silk only (i.e., combs from which the pupae had been extracted) behaved similarly, in that the silk here warmed to a much lower extent (amounting to tens of degrees Celsius) than the comb itself. Interestingly the silk cap was the coldest part in the comb and the 'button' at the center of the cap – even colder. The ability of hornet silk to withstand heat may be attributed to a phase transition of the first order, in the course of which the excess heat is absorbed by the structural change undergone by the silk as a result of the thermal changes (Xu, 1991; Brokate et al., 1993; Anderson, 1997; Drummy et al., 2005). In testing the electric resistance, we found that at a certain point in the course of warming the silk, there is a rise in the resistance by 3 orders of magnitude without any further change in the temperature, indicative of an isothermic phenomenon (Plotkin et al., 2005b). It is at this point that the silk undergoes probably morphological transformation, and the event occurs at about 31°C indicating a phase transition of the first order.

In cases of overheating, how does the hornet colony handle the situation or in other words, how do the hornets cool the nest? Well, primarily through ventilation. To this end, we find hornets standing at the entrance to the nest and beating their wings in cadence, thereby effecting air exchange, drop of humidity in the nest, and consequently more effective evaporation and gas exchange. Secondly, at high nest temperatures, the hornets introduce water droplets which they 'hang' upon the silk caps. The Oriental hornet in Israel does not usually build an envelope around their nest, but do so in deserts or when the nest is exposed to light. The envelope around the nest is capable of absorbing water to the extent of up to 50% of its own weight (Golodnitsky et al., 2003), and it is the water that evaporates off the envelope which enables cooling of the nest ambience. In sum, the cooling process (Figure 2) is secondary to the primary system of heating because it is activated only in the warm hours of the day. Returning now to the sensory arsenal of the worker hornets, we point out again that the hornets assess the temperature of the silk by touching the silk stretches with the tips of their antennae. Previously we found that the surface of the silk is amenable to change upon exposure to an electric voltage and that changes in the voltage are dependent on the ambient temperature (Benshalom-Shimony & Ishay, 1990). Based on the accumulated data, both observational and measuremental, we can now offer a summarizing theoretical model of behavioral temperature control in the nest of the Oriental hornet (see Figure 3). The model explains the primary system of temperature control in the nest which is activated when the ambient temperature outside the nest is relatively low (throughout most hours of the day and night). To understand the model, we need clarify a primary definition already provided in Barak and Waks (1992), namely, that by regulation is meant a process entailing regulation of a physical dimension which is the controlled variable. The controlled variable is a physical entity that can be measured and quantified, and in the present case is the temperature of the pupae. The main problem in a control system is to prevent deviation of the controlled variable owing to external influence stemming in our case from the temperature in the nest interior and therefore necessitating a constant measuring system of the controlled dimension and immediate correction of any deviations. To this end, use is made of a control system with feedback (feedback control) or control in a closed system, else known as closed-loop control.

In a control system with feedback, three basic actions are attained:

1. Measurement of the controlled variable.
2. Comparison between the measurement value and the desired value, with identification of any error in order to negate or at least limit the error.
3. Correction of the control mechanism in case of any error in order to negate or at least limit the error.

The diagram provided in Figure 3 exemplifies the way the control system works. In this diagram: C (controlled) = the controlled variable; R (reference) = the affinity letter representing the desired value to be reached by the control variable C; B (back) = the value C after measurement; E (error) = the error letter representing the discrepancy between the desired and measured value E=R-B; M (manipulated variable) = the correction activity.

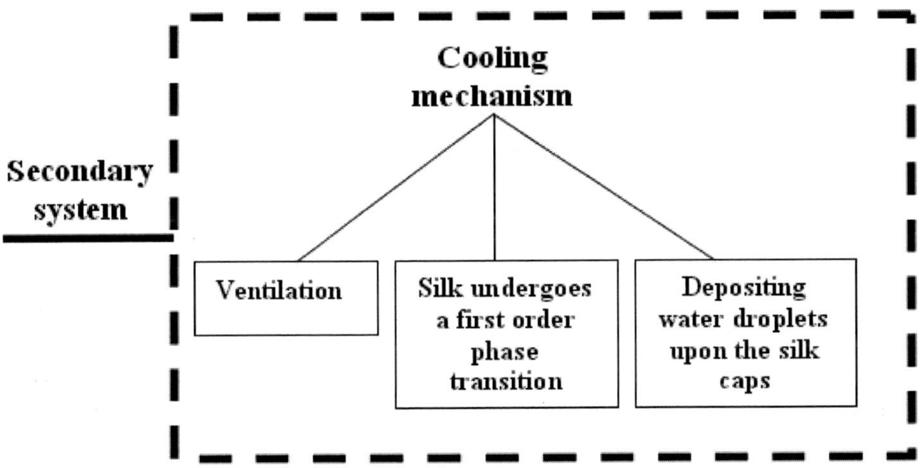

Figure 2. Secondary Control System is activated during the warm hours of the day, when the ambient temperature is relatively high, i.e., in excess of 26-27°C. Thereupon the hornets commence ventilatory activity at the entrance to the nest, such as enables exchange of water – saturated air between the interior and exterior of the nest and thereby enhances the rate of evaporation and cooling of the pupae. When the temperature is very high – in excess of 32-33°C – the hornets carry water droplets from the field and deposit them upon the silk caps. As for the silk itself, it undergoes a first – order phase transition process (at temperature above 31°C) and consequently the latent heat is absorbed, thereby helping to stabilize the pupal temperature.

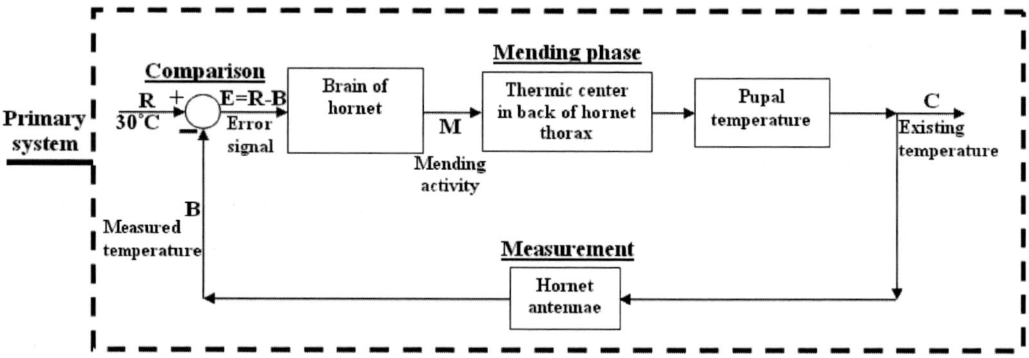

Figure 3. Rectangular diagram of a system for temperature control with feedback as modified from Barak and Waks (1992). A main control system functions throughout most hours of the day and night, especially in the early morning hours and in the late afternoon, when the ambient temperature outside the nest is relatively low (less than 26-27 °C), and the main role of the system is to warm the pupae and maintain them at a steady temperature of about 30°C. From this diagram one can gather that the antennae of the adult hornet gauge the temperature upon the silk as an indicator of the pupal temperature. The measuring of the temperature here is either direct or through measurement of the change in voltage (induced spontaneously from any change in the temperature) upon the silk. By use of its neural system, the adult hornet is able to compare the measured temperature B to the desired temperature R, which amounts to 30°C, and this when the error signal $E>0$, i.e., the pupal temperature, is below the desired value; the correction activity (M) entails, on the part of the hornet, the resting of its underside upon the pupal stretch, following which the hornet executes enhanced respirations. The correction is achieved *via* the thermic center located in the dorsal part of its thorax (see Fig. 1C). In short, the hornet warms the pupa resting beneath her and thereby raises its temperature to existing temperature C. Again, the hornet now measures the pupal temperature by means of its antennae and if it finds the error signal to be $E<0$, i.e., that the pupal temperature is higher than the desired value, the correction activity (M) goes into effect, whereafter the hornet reduces its respiration rate and departs the site of warming. Subsequently, however, the pupal temperature drops again and the entire process needs to be repeated.

Acknowledgments

We wish to thank the Israeli Ministry of Immigrant Absorption for the generous support of Dr Stanislav Volynchik. Thanks are due also to the Y. H. Mirzoeff & Sons Foundation Inc. for their generous financial support which enabled us to perform this study.

References

Anderson, P. W. (1997). *Basic notions of condensed matter physics.* Westview Press, Boulder CO.

Barak, M. & Waks, S. (1992). *Introduction to control systems.* Department of Education in Technology and Science, Technion, Haifa Israel (in Hebrew).

Benshalom-Shimony, T. & Ishay, J. (1990). Luminescence and thermoconductive properties of the Oriental hornet cocoon: evidence for photoinduced electron transfer system. *Comp. Biochem. Physiol.* 95A, 349–358.

Brokate, M., Kenmochi, N., Müller, I., Rodriguez, J. F. & Verdi, C. (1993). *Phase transitions and hysteresis*. Springer-Verlag, Berlin.

Golodnitsky, D., Ulus, A. & Ishay, J. S. (2003). Thermal characterization of social vespid silk. *J. Therm. Analysis and Calorimetry* 73, 85–96.

Drummy, L. F., Phillips, D. M., Stone, M. O., Farmer, B. L. & Naik, R. R. (2005). Thermally induced α-helix to β-sheet transition in regenerated silk fibers and films. *Biomacromolecules* 6, 3328-3333.

Edwards, R. (1980). *Social Wasps*. Rentokil Ltd, Felcourt England.

Grodzicki, P. & Caputa, M. (2005). Social versus individual behaviour: a comparative approach to thermal behaviour of the honeybee (*Apis mellifera* L.) and the American cockroach (*Periplaneta americana* L.). *J. Insect. Physiol.* 51, 315-322.

Heinrich, B. (1993). *The Hot-Blooded Insects: strategies and mechanisms of thermoregulation.* Harvard University Press, Massachusetts.

Heinrich, B. (1996). *The Thermal Warriors: strategies of insect survival.* Harvard University Press, Massachusetts.

Himmer, A. (1927) Ein Beitrag zur Kenntnis des Wärmehaushaltes im Nestbau sozialer Hautflügler. *Z vergl Physiol* 5, 375-389.

Ishay, J. (1964). Observations sur la biologie de la Guêpe orientale *Vespa orientalis* en Israël. *Insectes Sociaux* XI, 193-206.

Ishay, J. S., Plotkin, M., Ermakov, N. Y., Volynchik, S., Barkay, Z. & Bergman, D. J. (2006). The thermogenic center in social wasps. *JEM* 55, 41-49.

Ishay, J. & Ruttner, F. (1971). Die thermoregulation im hornisennest. *Z. Vergl. Physiol.* 72, 423-434.

Kleinhenz, M., Bujok, B., Fuchs, S. & Tautz, J. (2003). Hot bees in empty broodnest cells: heating from within. *J. Exp. Biol.* 206, 4217-4231.

Klingner, R., Richter, K., Schmolz, E. & Keller, B. (2005) The role of moisture in the nest thermoregulation of social wasps. *Naturwissenschaften* 92, 427-430.

Plotkin, M., Ermakov, N. Y., Volynchik, S., Bergman, D. J. & Ishay, J. S. (2007). Prevention of hyperthermia with silk of the Oriental hornet, *Vespa orientalis*: a hypothesis. *Microsc. Res. & Tech.* 70, 69-75.

Plotkin, M., Ermakov, N. Y., Volynchik, S., Barkay, Z., Bergman, D. J. & Ishay, J. S. (2005a). A thermoregulatory center in hornets: IR photography. *Microsc. Res. & Tech.* 68, 321-328.

Plotkin, M., Ermakov, N. Y., Bergman, D. J. & Ishay, J. S. (2005b). Electrical properties of hornet silk: temperature dependence. *Physiol. Chem. Physics. & Med. NMR* 37, 71-84.

Riabinin, K., Kozhevnikov, M. & Ishay, J. S. (2004). Ventilating activity at the hornet nest entrance. *J. Ethol.* 22, 49–53.

Sogl B., Gellissen, G. & Wiesner, R. J. (2000). Biogenesis of giant mitochondria during insect flight muscle development in the locust, *Locusta migratoria* (L.). Transcription, translation and copy number of mitochondrial DNA. *Eur. J. Biochem.* 267, 11-17.

Spradbery J. P. (1973). *Wasps*. Sidgwick and Jackson, London England.

Stabentheiner, A., Pressl, H., Papst, T., Hrassnigg, N. & Crailsheim, K. (2003). Endothermic heat production in honeybee winter clusters. *J. Exp. Biol.* 206, 353-358.

Verdú, J. R., Arellano, L. & Numa, C. (2006). Thermoregulation in endothermic dung beetles (Coleoptera: Scarabaeidae): effect of body size and ecophysiological constraints in flight. *J. Insect. Physiol.* 52, 854-860.

Volynchik, S., Plotkin, M., Ermakov, N. Y., Bergman, D. J. & Ishay, J. S. (2006). Presence of a thermoregulatory hot spot in the prothorax of the large Carpenter bee and the Bumble bee. *Microsc. Res. & Tech.* 69, 903-12.

Xu Y. (1991). *Ferroelectric materials and their applications*. Elsevier Science Pub Co, Amsterdam North-Holland.

In: Body Temperature Regulation
Editor: A. B. Cisneros and B. L. Goins
ISBN: 1-60741-282-3

Chapter XV

Characteristics, Mediators and Modulators of the Fever Response in Birds

***David A. Gray*[1]*, Manette Marais*[1]*, Peter R. Kamerman*[1] *and Shane K. Maloney*[1,2]**

[1]School of Physiology, University of the Witwatersrand, Johannesburg, South Africa;
[2]Physiology, School of Biomedical, Biomolecular and Chemical Sciences, University of Western Australia, Perth, Australia

Abstract

The phylogenetic development of the febrile mechanism is not well understood because few studies have investigated the physiology of fever in species other than mammals. Considerable insight could be gained from studies in the other homeothermic group, the birds. Like mammals, birds develop fevers when exposed to bacterial endotoxin, and like in mammals, lipopolysaccharide-induced fevers are mediated by prostaglandin and nitric oxide. However, the signaling pathways that lead to the synthesis of prostaglandin and nitric oxide in birds have not been elucidated. A role for pro-inflammatory cytokines has been suggested but, as yet, there is limited evidence for the existence of such a pathway. Likewise, avian toll-like receptors have been identified and have been shown to initiate innate immune responses on pathogen exposure. It is still unknown whether such immune reactions mediate fever. Hitherto glucocorticoids have been identified as the only likely modulators of the avian febrile response. Taken together, evidence for the involvement of prostaglandin, nitric oxide and glucocorticoids in the mediation and modulation of avian fevers support the idea that the febrile mechanism in birds and mammals has a common origin. Future experiments will need to examine the physiology of fever induction and the precise roles of mediators and modulators in the avian febrile response to broaden our scientific understanding of the evolutionary and phylogenetic development of the febrile mechanism.

Introduction

Fever is considered to have an ancient phylogenetic origin. Febrile responses have been observed in invertebrates [1-4], reptiles [5-8], amphibians [9, 10], fishes [11] and all endothermic animals [12-14]. The prevalence of fever through such a long evolutionary timeline is often cited as proof that fever has survival value and is beneficial to the host [14, 15]. The most generally accepted advantage of fever to the host is thought to be an inherent improvement in immune function and a reduction in pathogen proliferation (especially bacteria) [14]. While the occurrence of fever is well documented in groups other than mammals, there is a large bias in mechanistic studies to the mammalian lineage. Consequently the phylogenetic development of the febrile response is poorly understood.

In particular, a great deal could be learned from studies in the other homeothermic group, birds. The purpose of this report is to survey our present understanding of the physiology of avian fever, in particular its mediators and modulators.

Characteristics of the Avian Febrile Response

Experimental fevers in birds can be evoked with bacterial endotoxin [16-19]. Most birds tested thus far, including chickens [20], pigeons [19] and Pekin ducks [12], require a dose of 10 – 100μg/kg of LPS injected peripherally to stimulate a fever response. A notable exception is the Japanese quail *(Coturnix Coturnix Japonica)* which seems to be insensitive to LPS, as it requires in excess of 1mg/kg to induce a febrile response in these birds [21]. Japanese quail also rapidly develop tolerance to LPS, as evidenced by a lack of thermal response to 6.5mg/kg administered 48 hours after a priming dose [21]. In chickens a repeat dose of LPS (1mg/kg), two or seven days after the priming dose, resulted in fevers that were shorter in duration, whilst only the repeat injection of LPS two days after the priming dose, attenuated the peak temperature of the febrile response [22].

Studies in Pekin ducks have shown the LPS-induced fever response to be a monophasic elevation of body temperature which starts within two hours of LPS administration. The animals had a similar degree of sensitivity to LPS when injected intravenously or intramuscularly. Both the duration and peak of the fever were shown to be dose responsive, higher dosages caused significantly longer fevers with higher peak temperatures. The maximum dose of 100μg/kg produced a peak temperature of about 1.5°C higher than normal body temperature [12] (see Figure 1). It is unclear whether the monophasic profile seen in ducks and sometimes chickens [23] is the "typical" avian febrile response, or the biphasic response sometimes reported in pigeons [19, 24] and chickens [18, 22] treated with pyrogens. It is probable that, similarly to mammals [25], the specific avian febrile response is dependent on several factors including type, route, and dose of pyrogen, as well as species. Mammals generally develop monophasic fevers when treated with low intravenous (IV) dosages

(<10ug/kg) of LPS and polyphasic fevers when given moderate IV dosages ($10 - 10^4$ug/kg), whilst high dosages (>10^4ug/kg) cause hypothermia due to septic shock [13].

Other pathogens, such as cellular components of Gram-positive bacteria and viruses, which are known pyrogens in mammals [26], have rarely been investigated in birds. Thus far, one study found that muramyl dipeptide (MDP), a cellular component of Gram-positive bacteria, did not initiate a febrile response in Japanese quail [21].

The handling of birds causes a stress-induced hyperthermia. In mammals this elevation of body temperature has been categorized as a fever by some researchers [27]. However, studies with Pekin ducks demonstrated that the handling-induced rise in body temperature was not changed by any of the inhibitors of an LPS-induced fever. These results suggest that the stress-induced hyperthermia in at least some birds is not a fever, at least not one that works via the same mechanisms as LPS [17].

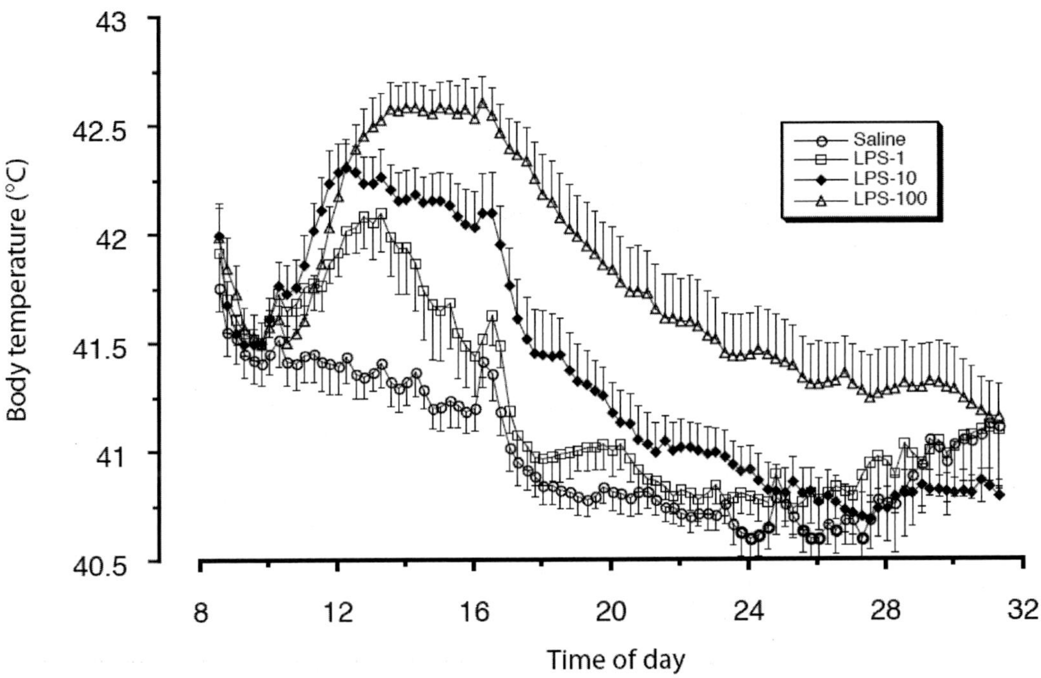

Figure 1. Fifteen minute averages of rectal temperatures in seven Pekin ducks given intravascular injections of saline and three different dosages of lipopolysaccharide LPS (indicated by key on figure) at 10:00 hours (indicated by arrows). Each point shows 15-min average of 1 min measurements ±SEM (after Maloney & Gray, 1998).

Induction of Fever

Fever is part of the innate immune system's acute phase response and fever mediation thus requires immune system activation. Immune responses are thought to be triggered when Toll-like receptors (TLRs), expressed on immune cells, recognize the conserved microbial structures of invading pathogens. The TLR family is considered a highly conserved component of the immune system which originated in pre-historic animals before the divergence of the major vertebrate lineages [28]. Thus far genomic studies have identified genes for ten toll-like receptors in chicken cDNA [29, 30]. The different TLRs identified were shown to recognize different evolutionarily conserved molecular patterns of infectious microbes [29, 30]. *In vitro* studies with chicken macrophages and heterophiles (the primary polymorphonuclear leukocytes in birds) showed LPS to serve as a ligand for both chicken (ch)TLR4 [31] and (ch)TLR2 type 2 [32]. Moreover, Kogut *et al.* (2001) and Fukui *et al.* (2001) identified (ch)TLR2 type 2 binding both LPS and lipoteichoic acid, a Gram-positive bacterial product. As such (ch)TLR2 type 2 potentially signals the presence of both Gram-negative and Gram-positive bacterial invasions. Unfortunately it is not yet known whether Gram-positive bacteria evoke fevers in birds. The activation of (ch)TLR4 (expressed on chicken heterophiles) with LPS was shown to initiate a signaling cascade, resulting in the differential expression of inducible nitric oxide synthase (iNOS), a significant upregulation of IL-1β, IL-6 and IL-8 as well as an oxidative burst and degranulation [29]. Of these only NO has thus far been implicated in the mediation of LPS-induced fevers in birds.

Mediators of the Avian Febrile Response

Prostaglandins are enzymatically derived eicosanoids that are synthesized from the cell membrane component arachidonic acid (AA) by the enzyme cyclo-oxygenase (COX). Prostaglandins are distributed ubiquitously throughout the body [33]. The upregulated synthesis of PGE_2, converted from AA by inducible COX_2, is known to fulfill a vital role in fever mediation in mammals [13, 15, 34]. PGE_2 is thought to exert its action during the fever response by binding to prostaglandin E3 receptors (EP3) expressed on temperature sensitive neurons [35], an action which is thought to inhibit warm sensitive neurons and lead to an activation of heat production and heat retention mechanisms [36, 37]. There is evidence from studies in chickens [20, 38] and ducks [17], that prostaglandins ultimately mediate the fever response in birds too.

Like mammals, birds also express constitutive [39] and inducible [40] isoforms of COX [41] and PGE_2 receptors were found in the spinal cord of chickens [42]. However, a study in chickens found no change in hypothalamic PGE_2 production after LPS inoculation [43], suggesting that PGs might not be the common final mediator of fever in all homeotherms and/or that the hypothalamus is not the site of mediation in birds. Studies in chickens [38] and Pekin ducks [17] did however find a significant reduction in the febrile response to LPS when PG synthesis was inhibited with the peripheral administration of a non-selective COX inhibitor, namely diclofenac sodium (see Figures 2A and 2B). Because diclofenac sodium inhibits the prostaglandin synthesis pathway before the differentiation of the different classes of prostaglandin [44], it is not possible to determine, from these studies, whether inducible or constitutive COX synthesized the series of prostaglandins responsible for driving, at least partially, the avian febrile response.

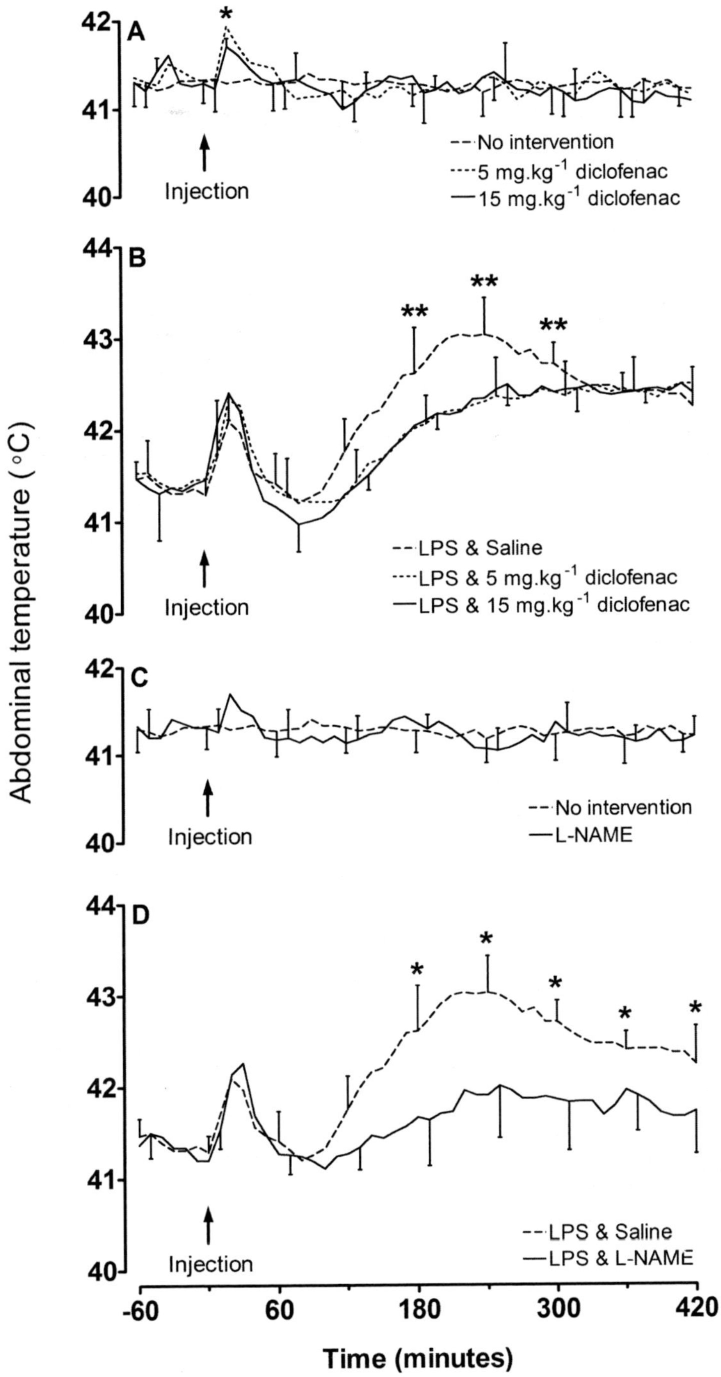

Figure 2. Mean (SD) body temperature as a function of time in ducks (n = 7); (A) injected with 5 mg/kg diclofenac or 15 mg/kg diclofenac, and when no intervention was made; (B) the same ducks injected with LPS and saline, LPS and 5 mg/kg diclofenac, and LPS and 15 mg/kg diclofenac; (C) injected with N-nitro-L-arginine methyl ester (L-NAME; 200 mg/kg), and when no intervention was made, and (D) the same ducks injected with LPS and saline, and LPS and L-NAME (after Gray *et al.* 2005).

Interestingly, in pigeons there was no evidence for a PG mediated febrile response after LPS administration [19]. These results support the idea of a PG-independent mechanism for the induction of LPS-fever in some birds. It is also possible that two pyrogenic mechanisms exist in birds, and that the relative importance of these mechanisms varies in different avian species. A PG-independent febrile mechanism might explain why in ducks there was no difference in the degree of inhibition produced by high and low dosages of diclofenac, suggesting that there is a maximum achievable inhibition. These investigations furthermore found that PG inhibition ceased after about five hours [17]. It is therefore attractive to hypothesise that other molecules maintain the latter part of the febrile response.

Evidence also suggests that the avian febrile response is mediated by nitric oxide (NO). In ducks inhibition of NO synthesis with L-NAME markedly attenuated the febrile response (see Figure 2D) [17]. (L-NAME inhibits NO synthase, the enzyme responsible for the synthesis of NO from oxygen and arginine [33]). Whether the role of NO in avian fever is the same as in mammalian fever is equivocal. Unfortunately, even in mammals, the mechanisms by which NO influences fever is a matter of speculation. Inhibition of NO synthesis may attenuate fever development by disrupting afferent fever signaling pathways, or central processing of pyrogenic signals, or efferent output to thermoeffector organs, or the functioning of those thermoeffectors. Within the central nervous system, NO has consistently been shown to promote thermolysis in mammals [45-48]. Thus, if the gas has similar actions in birds, a reduction in NOS activity should result in heat storage and higher body temperatures not an attenuation of fever and reduced body temperatures. Since L-NAME has no affect on normal body temperatures in ducks (see Figure 2C), it is possible that an inhibition of NO decreases the activation of thermoeffectors or decreases the capacity of thermoeffectors to increase heat storage in ducks treated with LPS [49-51]. Another possibility is that L-NAME influences avian fever by disrupting pyrogenic signaling across the blood-brain barrier. NO was shown to be important for transduction of pyrogenic signals across the blood-brain barrier in rabbits [52-54].

The signaling pathway that leads to the synthesis of either PG and/or NO during the avian febrile response has not yet been elucidated. A role for pro-inflammatory cytokines, especially interleukin (IL)-1 and IL-6, has been suggested [22, 23, 38, 41]. Pro-inflammatory cytokines are considered the first endogenous mediators of mammalian fever [13, 55, 56], but there is no concrete evidence that pro-inflammatory cytokines are involved in the avian febrile response. But birds, like mammals, do possess well developed cytokine systems [57-60]. Support for an avian cytokine scheme includes the identification of bioactive homologues for mammalian IL-6 and mammalian IL-1β in chickens [58, 61-63]. *In vitro* and *in vivo* studies have confirmed that chicken macrophages synthesize these cytokines [62, 64, 65], and secrete them into the circulation after an immune challenge [65, 66]. Avian pro-inflammatory cytokines are thought to have similar immune system regulating functions to their mammalian counterparts even though they have very different structures [57]. Thus far, one study, completed before the identification of chicken IL-1β (chIL-1β), showed that intra-cerebroventricular (i.c.v.) administration of recombinant human IL-1β (rhuIL-1β) caused fever in broiler chickens [38]. De Boevers *et al.* [22] and Xie *et al.* [41] reported significantly elevated plasma IL-6 levels after LPS administration [41]. Evidence for IL-6 as mediator of the febrile response is, however, equivocal. Xie *et al.* [41] found LPS (5mg/kg) to cause a febrile response which peaked three hours post injection, whilst plasma IL-6 concentrations were elevated 12 and 24 hours after LPS treatment. In contrast De Boevers *et al.* [22] found plasma IL-6 concentrations to peak after three hours and to return to baseline nine hours post LPS (1mg/kg) inoculation, whilst body temperatures reached a peak about eight hours and resolved 24 hours after LPS treatment. It is also not clear how the febrile response to i.c.v rhuIL-1β was mediated, because the species cross-reactivity between avian- and mammalian IL-1β is now known to be limited [57, 67, 68].

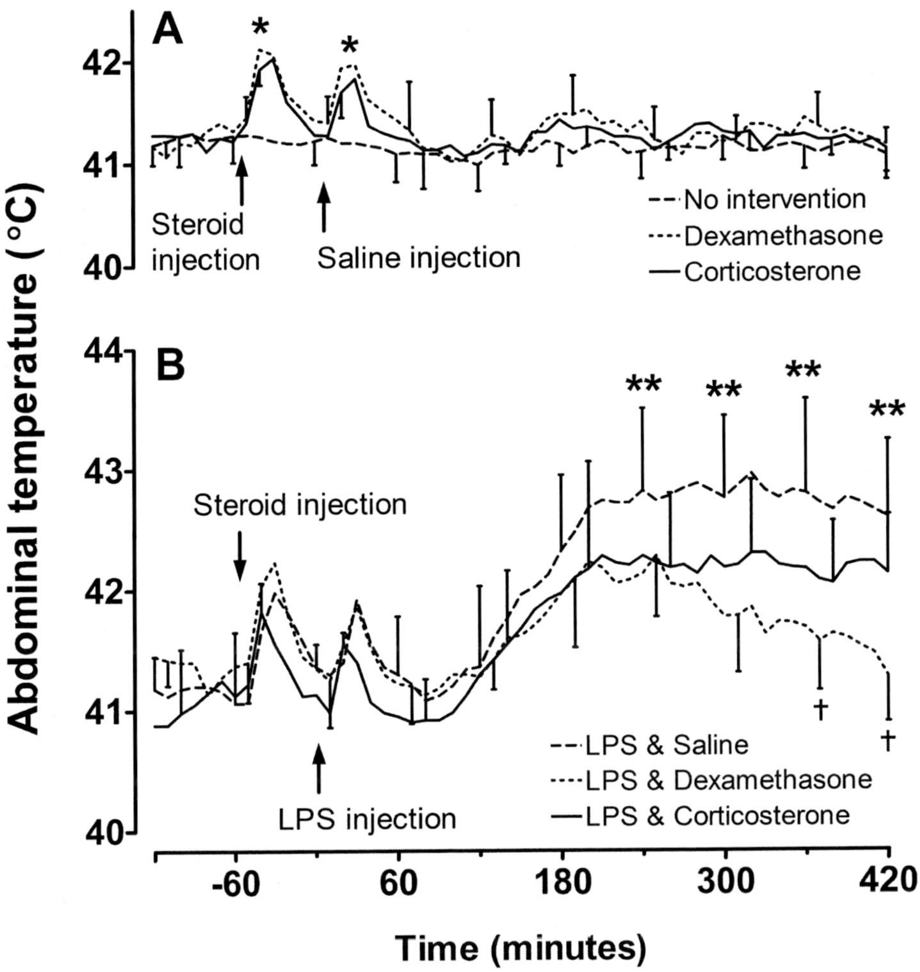

Figure 3. Mean (SD) body temperature as a function of time of ducks injected (A) dexamethasone (n = 6), corticosterone (n = 6), and when no intervention was made (n = 6), and (B) the same ducks injected with lipopolysaccharide (LPS) and saline (n = 6), LPS and dexamethasone (n = 7), and LPS and cortocosterone (n = 6) (after Gray *et al.*, 2005).

Modulators of Avian Fevers

Adrenocortical hormones are the only modulators of the avian febrile response identified thus far. Circulating concentrations of corticosterone are elevated during fever in chickens [20, 69], and in ducks, LPS-induced fever is lowered when birds were treated with exogenous corticosterone or a synthetic glucocorticoid, namely dexamethasone [17] (see Figures 3A and 3B). Interestingly, experimental restraint also modulates LPS-induced fevers in ducks and it is thought that restraint induces a stress reaction which, in turn, stimulates the secretion of glucocorticoids. Restraint modulated the fever response in ducks in such a way that fevers resolved quicker and peak temperatures were lower, when compared to fever responses evoked in unrestrained animals [70] (see Figure 4). The precise role that glucocorticoids play in avian fever is difficult to assess because plasma levels of the autocoid were shown to be elevated during the early part of the fever, whereas exogenously administered steroid only affected the later phase of the febrile temperature response. It could well be that two pyrogenic mechanisms are involved in fever mediation in birds, and glucocorticoids have differential effects and/or interactions with the two mechanisms. However, this is extremely speculative and unlikely to be resolved until our understanding of the identity and physiological actions of regulatory mediators (especially cytokines) in birds is enhanced.

The mammalian febrile response is modulated by a number of endogenous antipyretic substances including steroid hormones, neuropeptides and anti-inflammatory cytokines [13, 34]. Thus far, the fever modulating ability of two of these, namely arginine vasotocin (the avian equivalent of mammalian antidiuretic hormone) [71] and natriuretic peptide [72] have been tested in birds and found not to affect the avian febrile response.

Conclusion

Some similarities between avian and mammalian fevers provide important information about the evolution of fever. The apparent involvement of NO and PGs as mediators of fever in both birds and mammals, together with an analogous modulatory role for glucocorticoids in both groups of animals, is evidence supporting the idea that the febrile mechanism has a common origin, even if the relative roles of mediators such as PGE_2 may differ. Future experiments will need to examine the precise roles of PGs, NO, and adrenocorticoids, especially since as yet unidentified modulators of avian fever may be important at different stages of the fever profile in birds. In addition, if the physiology of avian cytokines can be characterized, then our understanding of the similarities and differences between mammalian and avian fever will be markedly enhanced.

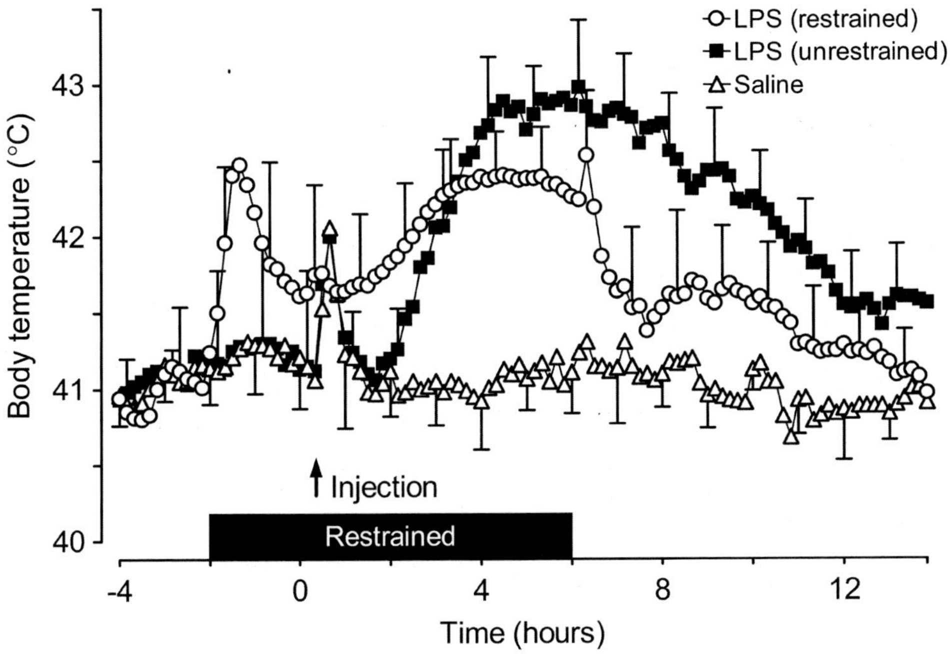

Figure 4. Tb measured every 10 minutes in Pekin ducks (n = 6). Ducks were restrained for 8 h (indicated by black bar on x-axis from -2 to 6 h) and injected (indicated by arrow at time 0) with lipopolysaccharide (LPS, restrained), remained unrestrained in their home pen and were injected with LPS at time 0 (LPS, unrestrained), or remained unrestrained in their home pen and were injected with saline at time zero (saline, unrestrained) (Gray *et al.*, 2008).

References

[1] Boorstein, S.M. and P.W. Ewald, *Costs and benefits of behavioral fever in Melanoplus sanuinipes infected by Nosema acridophagus.* Physiol Zool, 1987. 60: p. 586-595.

[2] Bronstein, S.M. and W.E. Conner, *Endotoxin-induced behavioural fever in the Madagascar cockroach, Gromphadorhina portentosa.* J Insect Physiol, 1984. 30: p. 327-330.

[3] Cabanac, M. and L.L. Guelte, *Temperature regulation and prostaglandin E1 fever in scorpions.* J Physiol, 1980. 303: p. 365-370.

[4] Casterlin, M.E. and W.W. Reynolds, *Fever and antipyresis in the crayfish Cambarus bartoni.* J Physiol, 1980. 303: p. 417-421.

[5] Monagas, W.R. and R.E. Gatten(Jr), *Behavioural fever in the turtles Terrapene carolina and Chrysemys picta.* J Therm Biol, 1983. 8: p. 285-288.

[6] Hallman, G.M., C.E. Ortega, and M.C. Towner, *Effect of bacterial pyrogen on three lizard species.* Comp Biochem Physiol A, 1990. 96: p. 383-386.

[7] Firth, B.T., C.L. Ralph, and T.J. Boardman, *Independent effects of the pineal and a bacterial pyrogen in behavioural thermoregulation in lizards.* Nature, 1980. 285: p. 399-400.

[8] Muchlinski, A.E., R.J. Stoutenburgh, and J.M. Hogan, *Fever response in laboratory-maintained and free-ranging chuckwallas (Sauromalus obesus).* Am J Physiol-Reg I, 1989. 257: p. R150-R155.

[9] Casterlin, M.E. and W.W. Reynolds, *Behavioral fever in anuran amphibian larvae.* Life Sci, 1977. 20: p. 593-596.

[10] Sherman, E., L. Baldwin, and G. Fernandez, *Fever and thermal tolerance in the toad Bufo marinus.* J Therm Biol, 1991. 16: p. 297-301.

[11] Reynolds, W.W., M.E. Casterlin, and J.B. Covert, *Behavioural fever in teleost fishes.* Nature, 1976. 259: p. 41-42.

[12] Maloney, S.K. and D.A. Gray, *Characteristics of the febrile response in Pekin ducks.* J Comp Physiol B, 1998. 168: p. 177-182.

[13] Romanovsky, A.A., et al., *Fever and hypothermia in systemic inflammation: Recent discoveries and revisions.* Front Biosci, 2005. 10: p. 2193-2216.

[14] Kluger, M.J., et al., *The adaptive value of fever.* Infect Dis Clin N Am, 1996. 10(1): p. 1-20.

[15] Blatteis, C.M., *The cytokine-prostaglandin cascade in fever production: fact of fancy?* J Therm Biol, 2004. 29(359-368).

[16] Pittman, Q.J., et al., *Changes in body temperature produced by prostaglandins and pyrogens in the chicken.* Am J Physiol, 1976. 230(5): p. 1284-1287.

[17] Gray, D.A., S.K. Maloney, and P.R. Kamerman, *Lipopolysaccharide-induced fever in Pekin ducks is mediated by prostaglandins and nitric oxide and modulated by adrenocortical hormones.* Am J Physiol-Reg I, 2005. 289: p. R1258-R1264.

[18] D'Alecy, L.G. and M. Kluger, *Avian febrile response. 253:223-232.* J Physiol, 1975. 253: p. 223-232.

[19] Nomoto, S., *Role of prostaglandin E2 and indomethacin in the febrile response of pigeons.* Jpn J Physiol, 2003. 53: p. 253-258.

[20] Johnson, R.W., et al., *Central and peripheral prostaglandins are involved in sickness behaviour in birds.* Physiol Behav, 1993. 53: p. 127-131.

[21] Koutsos, E.A. and K.C. Klasing, *The acute phase response in Japanese quail (Coturnix coturnix japonica).* Comp Biochem Phys C, 2001. 128: p. 255-263.

[22] DeBoever, S., et al., *The influence of age and repeated lipopolysaccharide administration on body temperature and the concentration of interleukin-6 and IgM antibodies against lipopolysaccharide in broiler chickens.* Avian Pathol, 2008. 37(1): p. 39-44.

[23] Leshchinsky, T.V. and K.C. Klasing, *Divergence of the inflammatory response in two types of chickens.* Dev Comp Immunol, 2001. 25: p. 629-638.

[24] Nomoto, S., *Diurnal variations in fever induced by intravenous LPS injection in pigeons.* Pflugers Arch, 1996. 431: p. 987-989.

[25] Kozak, W., et al., *Attenuation of lipopolysaccharide fever in rats by protein kinase C inhibitors.* Am J Physiol-Reg I, 1997. 273: p. R873-879.

[26] Denyer, S.P., et al., *Hugo and Russel Pharmaceutical Microbiology*. 2004: Blackwell Publishing. 481.

[27] Cabanac, M. and E. Briese, *Handling elevates the colonic temperature of mice.* Physiol Behav, 1991. 51: p. 95-98.

[28] Temperley, N.D., et al., *Evolution of the chicken Toll-like receptor gene family: a story of gene gain and gene loss.* BMC Genomics, 2008. 1(9): p. 62.

[29] Kogut, M.H., et al., *Expression and function of Toll-like receptors in chicken heterophils.* Dev Comp Immunol, 2005. 29: p. 791-807.

[30] Boyd, A., V.J. Philbin, and A.L. Smith, *Conserved and distinct aspects of the avian Toll-like receptor (TLR) system: implications for transmission and control of bird-borne zoonoses.* Biochem Soc T, 2007. 35(6): p. 1504-1507.

[31] Dil, N. and M.A. Qureshi, *Involvement of lipopolysaccharide related receptors and nuclear factor kB in differential expression of inducible nitric oxide synthase in chicken macrophages from different genetic backgrounds.* Vet Immunol Immunopat, 2002. 88: p. 149-161.

[32] Fukui, A., et al., *Molecular Cloning and Functional Characterization of Chicken Toll-like Receptors: A single chicken toll covers multiple molecular patterns.* J Biol Chem, 2001. 276(50): p. 47143-47149.

[33] Voet, D. and J.G. Voet, *Biochemistry.* Vol. 3. 2004, NJ, USA: Wiley International.

[34] Roth, J., *Endogenous antipyretics.* Clin Chim Acta, 2006. 371: p. 13-24.

[35] Vasilache, A.M., J. Andersson, and C. Nilsberth, *Expression of PGE2 EP3 receptor subtypes in the mouse preoptic region.* Neurosci Lett, 2007. 423(3): p. 179-183.

[36] Cartmell, T. and D. Mitchell, *The molecular basis of fever.* Handbook of Stress and the Brain, ed. T. Steckler, N.H. Kahn, and J.M. Reul. Vol. 15. 2005, Amsterdam: Elsevier.

[37] Scott, I.M., R.H. Fertel, and J.A. Boulant, *Leucocytic pyrogen effects on prostaglandins in hypothalamic tissue slices.* Am J Physiol-Reg I, 1987. 253: p. 71-76.

[38] Macari, M., et al., *Effects of endotoxin, interleukin 1-beta and prostaglandin injections on fever response in broilers.* Br Poult Sci, 1993. 34(5): p. 1035-1042.

[39] Reed, D.W., et al., *In vivo and in vitro expression of a nonmammalian cyclooxygenase-1.* Prostaglandins, 1996. 52: p. 269-284.

[40] Laurent, F., et al., *Analysis of chicken mucosal immune response to Eimeria tenella and Eimeria maxima infection by quantitative reverse transcription-PCR.* Infect Immun, 69. 69: p. 2527-2534.

[41] Xie, W., et al., *Structural determination and promoter analysis of the chicken mitogen-inducible prostaglandin G/H synthase gene and genetic mapping of the murine homolog.* Arch Biochem Biophys, 1992. 300(1): p. 247-252.

[42] Pralong, E., M.-F. Vesin, and B. Droz, *Prostaglandin E2 receptors in the chicken spinal cord: biochemical characterization.* Eur J Neurosc, 1990. 2(11): p. 897-903.

[43] Fraifeld, V., et al., *Is hypothalamic prostaglandin E2 involved in avian fever?* Life Sci, 1995. 56: p. 1343-1346.

[44] Samad, T.A., A. Sapirstein, and C.J. Woolf, *Prostanoids and pain: unraveling mechanism and revealing therapeutic targets.* Trends Mol Med, 2002. 8: p. 390-396.

[45] Gourine, A.V., *Does central nitric oxide play a role in thermoregulation?*, in *Thermal Balance in Health and Disease: Recent Basic Research and Clinical Progress*, E. Zeisberger, E. Schonbaum, and P. Lomax, Editors. 1994, Birkhaquser Verlag: Basel. p. 491-495.

[46] Gourine, A.V., *Pharmacological evidence that nitric oxide can act as an endogenous antipyretic factor in endotoxin-induced fever in rabbits.* Gen Pharmacol, 1995. 26: p. 835-841.

[47] Monda, M., et al., *Nitric oxide reduces body temperature and sympathetic input to brown adipose tissue during PGE1- hyperthermia.* Brain Res Bull, 1995. 38: p. 489-493.

[48] Steiner, A.A., et al., *Antipyretic role of the NO-cGMP pathway in the anteroventral preoptic region of the rat brain.* Am J Physiol-Reg I, 2002. 282: p. R584-R593.

[49] Nagashima, T., H. Ohinata, and A. Kuroshima, *Involvement of nitric oxide in noradrenaline-induced increase in blood flow through brown adipose tissue.* Life Sci, 1993. 54: p. 17-25.

[50] Kamerman, P.R., H.P. Laburn, and D. Mitchell, *Inhibitors of nitric oxide synthesis block cold-induced thermogenesis in rats.* Can J Physiol Pharmacol, 2003. 81: p. 834-838.

[51] Rowsey, P.J. and C.J. Gordon, *A peripheral mechanism of fever: differential sensitivity to the antipyretic action of methyl scopolamine.* Auton Neurosci Lett, 2000. 85: p. 148-155.

[52] Huang, W.T., M.T. Lin, and S.J. Won, *Mechanisms and sites of pyrogenic action exerted by staphylococcal enterotoxin A in rabbits.* Neurosci Lett, 1997. 236: p. 53-56.

[53] Lin, M.T., J.H. Lin, and Y.L. Yang, *Dexamethasone administered into organum vasculosum laminae terminalis of rabbits induced antipyresis via inhibiting nitric oxide pathway in situ.* Neurosci Lett, 1997. 230: p. 53-56.

[54] Lin, J.H. and M.T. Lin, *Nitric oxide synthase-cyclooxygenase pathways in organum vasculosum laminae terminalis: possible role in pyrogenic fever in rabbits.* Br J Pharmacol, 1996. 118: p. 179-185.

[55] Rothwell, N.J., G.N. Luheshi, and S. Toulmond, *Cytokines and their receptors in the central nervous system: Physiology, pharmacology and pathology.* Pharmacol Ther, 1996. 69: p. 85-95.

[56] Kluger, M.J., *Fever: Role of pyrogens and cryogens.* Physiol Rev, 1991(71): p. 93-127.

[57] Staeheli, P., et al., *Cytokines of birds: conserved functions-A largely different look.* J Interf Cytok Res, 2001. 21: p. 993-1010.

[58] Klasing, K.C., *Avian macrophages: Regulators of local and systemic immune response.* Poult Sci, 1998. 77: p. 983-989.

[59] Erf, G.F., *Cell-Mediated immunity in Poultry.* Brit Poultry Sci, 2004. 83: p. 580-590.

[60] Wigley, P. and P. Kaiser, *Avian cytokines in health and disease.* Rev. Bras. Cienc Avic, 2003. 5.

[61] Weining, K.C., et al., *A chicken homolog of mammalian interleukin-1 beta: cDNA cloning and purification of active recombinant protein.* Eur J Biochem, 1998. 258: p. 994-1000.

[62] Scott, T.R. and H.S. Lillehoj, *Monoclonal antibodies against chicken interleukin-6.* Vet Immunol Immunop, 2006. 114: p. 173-177.

[63] Gyorfy, Z., et al., *Truncated chicken interleukin-1b with increased biologic activity.* Journal Interf Cytok Res, 2003. 23: p. 223-228.

[64] Sijben, J.W.C., et al., *Early in vivo cytokine genes expression in chickens after challenge with Salmonella typhimurium lipopolysaccharide and modulation by dietary n-3 polyunsaturated fatty acids.* Dev Comp Immunol, 2003. 27: p. 611-619.

[65] Schneider, K., et al., *Chicken interleukin-6 cDNA structure and biological properties.* Eur J Biochem, 2001. 268: p. 4200-4206.

[66] Nakamura, K., et al., *Serum levels of interleukin-6, a1-acid glycoprotein, and corticosterone in two-week-old chickens inoculated with escherichia coli lipopolysaccharide.* Poultry Sci, 1998. 77: p. 908-911.

[67] Higgins, D.A., et al., *An examination of the immune system of the duck (Anas platyrhynchos) for factors resembling some defined mammalian cytokines.* Dev Comp Immunol, 1993. 17: p. 341-355.

[68] Guida, S., A. Heguy, and M. Melli, *The chicken IL-1 receptor: differential evolution of the cytoplasmic and extracellular domains.* Gene, 1992. 111: p. 239-243.

[69] Baert, K., et al., *Antipyretic effect of oral sodium salicylate after an intreavenous E. coli LPS injection in broiler chickens.* Br Poultry Sci, 2005. 46: p. 137-143.

[70] Gray, D.A., S.K. Maloney, and P.R. Kamerman, *Restraint increases afebrile body temperature but attenuates fever in Pekin ducks (Anas platyrhynchos).* Am J Physiol-Reg I, 2008. 294: p. R1666-1671.

[71] Gray, D.A., C.S. Ironside, and S.K. Maloney, *Modulation of plasma antidiuretic hormone levels does not change the magnitude of the LPS-induced febrile response in Pekin ducks.* J Comp Physiol B, 2000. 170(4): p. 315-320.

[72] Marais, M. and D.A. Gray, *A role for natriuretic peptide in lipopolysaccharide-induced fever in Pekin ducks (Anas platyrhynchos): is natriuretic peptide an endogenous antipyretic in birds?* J Comp Physiol B, 2008. in press.

In: Body Temperature Regulation
Editor: A. B. Cisneros and B. L. Goins
ISBN: 1-60741-282-3

Chapter XVI

Decreased Thermoregulation with Age in Mice Exhibits Significant Genetic Variation Early in the Aging Process

Patrick Gonzales, Brad A. Rikke
Institute for Behavioral Genetics, University of Colorado, Boulder, Colorado 80309, USA

Abstract

Aging leads to a decline in homeostasis; therefore, a decline in homeostasis might serve as a useful biomarker of aging. A homeostatic system that shows a marked decline with age in mammals is thermoregulation, which is particularly easy to monitor in mice. In this study, we measured the drop in mean body temperature of 34 female mice from 22 LSXSS recombinant inbred strains; the strains were divided into two cohorts that were four months apart in age. Both cohorts were monitored at the same time and showed a significant drop in mean body temperature over a period of 5 months: from 8 to 13 months of age, Cohort 1 exhibited a mean drop of 1 °C; from 4 to 9 months of age, Cohort 2 dropped 0.6 °C. Interestingly, the temperature drop in Cohort 1 (the older cohort), but not 2, exhibited significant genetic variation ($P = 0.004$). We discuss how his variation, occurring relatively early in the aging process, is potentially a powerful tool for elucidating the mechanisms underlying loss of thermoregulation with age.

Introduction

Thermoregulation is an essential homeostatic system in mammals, responsible for maintaining body temperature (T_b) within a narrow range (36-38 °C) for optimal biochemical activity. This homeostasis is achieved by integrating neuronal processes for temperature sensing; hormonal processes for signaling; metabolic processes in muscle, fat, and liver for adjusting heat production; and glandular, pulmonary, and vascular processes for controlling heat loss (Annianson et al., 1986; Lauria et al., 1999; Passlick-Deetjen et al., 2005; Van Someren et al., 2002). This intricate system is undoubtedly responsible for the great success of mammalian adaptability; however, such complexity makes it difficult to disentangle the root causes of impaired thermoregulation with age.

A decline in thermoregulation with age is one of the hallmarks of senescence in mammals (Reynolds et al., 1985; Florez-Duquet and McDonald, 1998; Moran and Mendal, 2002; Degroot and Kenney, 2006; Van Someren, 2007). The consequences can be quite serious: human epidemiological studies indicate that most deaths due to hyperthermia and hypothermia occur in elderly subjects (Bai et al., 1995). Even in the absence of extreme temperatures, however, the elderly have an attenuated ability to thermoregulate (Reynolds et al. 1985; Degroot and Kenney 2006). An unresolved question is whether mild impairments prior to the onset of pathology reflect differences in the rate of aging (Ingram et al., 1982). To address this question, a potentially powerful approach is to identify naturally occurring genetic variation in these impairments. In this study we tested whether the loss of thermoregulation with age in mice exhibits significant genetic variation early in the aging process.

Materials and Methods

The mice used in this study were female, LSXSS recombinant inbreds (RIs). The LSXSS were originally derived from two outbred strains, LS (long sleep) and SS (short sleep), which were created by selecting for differences in sedation sensitivity to ethanol; this selection started from a heterogeneous stock of eight classical inbred strains (Markel et al., 1996).

The ages of the mice were closely matched within cohorts: typically differing by less than a week (Rikke et al., 2004). The mice in Cohort 1 comprised all of the longitudinally studied, *ad libitum*-fed, control mice from Cohort 2a of Rikke et al. (2004), which only reported on the T_bs of food restricted mice. Similarly, the mice in Cohort 2 were the longitudinally studied, *ad libitum*-fed, control mice from Cohort 2b of Rikke et al. (2004). The 17 mice in Cohort 1 are representatives of 11 strains, with 6 strains represented twice. Cohort 2 is comprised of 17 mice from 11 different strains with 6 strains represented twice. The mice in Cohort 2 were 4 months younger than those in Cohort 1.

The husbandry conditions were the same as previously described (Rikke et al., 2003, 2004). In brief, the mice were singly housed in shoe-box style cages under specific-pathogen-free conditions with a 12-h light/dark cycle (lights on 7 am). The mice were given *ad libitum* access to food (Harlan-Teklad 7912 mouse chow) and water.

The T_bs were measured by passive, radio telemetry using rice-grain-size transponders (Biomedic Data Systems, Seaford, DE) implanted subdermally (Rikke et al., 2004). The transponders permit the measurement of T_b within seconds after disturbing the cage, thereby avoiding the potential confound of thermal arousal. Each transponder was calibrated against rectal T_b immediately after each temperature trial.

The T_bs were measured in Cohorts 1 and 2 at the same time during two temperature trials spaced five months apart. In the first trial, we measured T_bs every 4 hours for 7 days (July, 2001); the average room temperature was 23.3 °C (SD = 0.3 °C). In the second temperature trial, T_bs were measured every 6 hours (0330, 0930, 1530, 2130) for a total of 10 days (December, 2001); the average room temperature was 22.7 °C (SD = 0.1 °C).

All procedures were approved by the University of Colorado's Institutional Animal Care and Use Committee.

Results

A decrease in mean T_b over five months was tested for simultaneously in two cohorts of mice that differed in age by four months. All 17 mice in the older cohort (Cohort 1), assessed at 8 and 13 months of age, exhibited a decrease in mean T_b. Averaged across all 17 mice, mean T_b dropped from 36.7 °C to 35.7 °C. The difference of 0.99 °C (SE = 0.13 °C) was highly significant ($P = 1.0 \times 10^{-6}$, paired-sample T-test, 2-tailed).

In the younger cohort of 17 mice (Cohort 2), assessed at 4 and 9 months of age, all but one mouse exhibited a decrease in mean T_b. Averaged across all 17 mice, mean T_b dropped from 36.9 °C to 36.3 °C. This decrease of 0.63 °C was also statistically significant (SE 0.14 °C, $P = 0.003$, paired-sample T-test, 2-tailed).

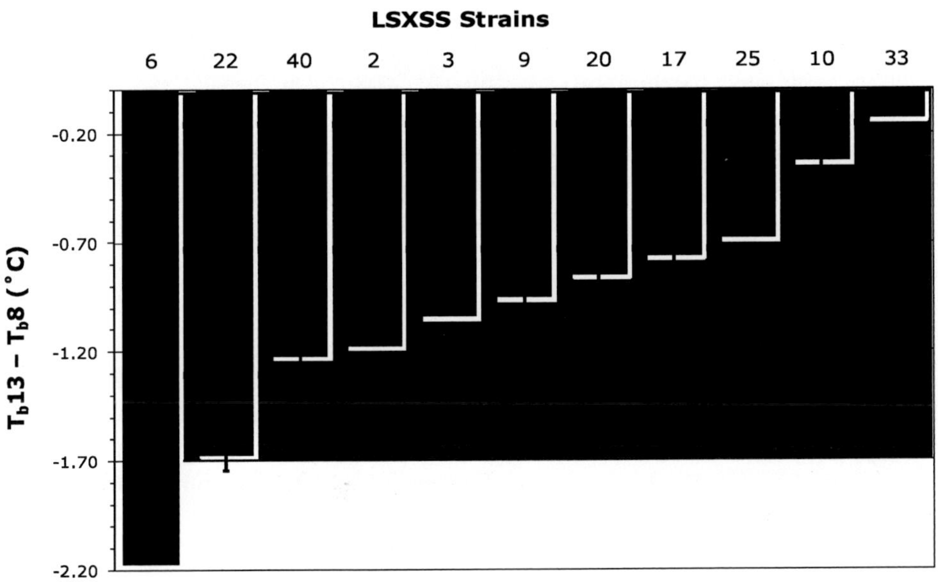

Figure 1. The drop in mean body temperature (T_b) with age exhibits significant genetic variation. Mean T_b was measured at 8 mos and again at 13 mos of age on 17 mice from 11 strains (LSXSS recombinant inbreds) shown on the X-axis. Each bar represents the average decrease in mean T_b (°C) for that strain. The error bars are standard errors based on measurements from two mice; strains represented by one mouse have no error bar. On average, the standard deviations in this cohort were only 0.15 °C, suggesting that even a single mouse estimates the strain mean for the decrease in mean T_b within 0.3 °C.

As shown in Figure 1, the drop in mean T_b for the 11 strains in Cohort 1 exhibited considerable variation—ranging continuously from 0.13 °C to 2.2 °C. Consistent with this variation being related to aging, the drops did not correlate with the initial T_b means measured at 8 months of age (R = -0.35, P = 0.30, 2-tailed), and they were modestly correlated with the T_b means measured at 13 months (R = 0.58, P = 0.03, 1-tailed). Importantly, the variation *between* strains for the drop in mean T_b was markedly greater than the variation *within* strains (P = 0.004, ANOVA), indicating significant genetic variation. The variance of mean T_b increased from 0.15 °C^2 at 8 months to 0.20 °C^2 at 13 months, which is consistent with an increase in variation due to aging.

In Cohort 2, the drop in mean T_b among the 11 strains ranged from 0 °C to 1.5 °C; however, the variation between strains was not significantly different from the variation within strains (P = 0.60, ANOVA). The drops in mean T_b did not correlate with mean T_b at 4 months of age (R = -0.31, P = 0.35, 2-tailed), and they were highly correlated with mean T_b at 9 months of age (R = 0.89, P = 0.0001, 1-tailed). The variance of mean T_b increased from 0.04 °C^2 at 4 months of age to 0.15 °C^2 at 9 months. These results are consistent with a loss of thermoregulation due to aging that was not appreciably affected by genetic variation.

Conclusion

Our major finding is that loss of thermoregulation with age in mice exhibits significant genetic variation relatively early in the aging process. This variation developed between 8 and 13 months of age, which is early mid-life for a laboratory mouse and much earlier than the onset of pathology. Concurrently, we also detected a significant but smaller loss of thermoregulation in our younger cohort, assessed at 4 and 9 months of age; this loss was not associated with genetic variation. Therefore, the genetic variation in Cohort 1 appears to be specific to the older age of the mice. This interpretation is tempered somewhat by the fact that the strains in the two cohorts were not the same; nevertheless, they were very closely related (the genetic equivalent of siblings) and each represents a random sampling from the LSXSS RI panel. Another limitation is that we sampled only one or two mice per strain; however, the average standard deviation was only 0.3 °C (n =12), which is consistent with mean T_b being highly regulated and conserved. It is also unlikely that the early onset of lower mean T_b in the LSXSS is due to unusually short lifespans: we have recently measured the lifespans of strains very closely related to the LSXSS—ILS, ISS, and more than forty of the LXS RIs (Williams et al., 2004)—and all of the means fall within the normal range for inbred strains of 20 to 33 months (unpublished results). In other studies, we have also found that there is no appreciable seasonal variation in T_b (unpublished results).

It's intriguing that thermoregulation declines so rapidly in the mouse. This decline is possibly a consequence of the highly integrated nature of temperature regulation and the inherent difficulty of maintaining such a complex homeostatic system, perhaps similar in principal to the declines in homeostatic complexity that characterize old-age frailty in humans (Lipsitz, 2004). In addition, the mean T_b of mice may be extremely sensitive to small impairments in thermoregulation because mice lose heat and equilibrate very rapidly due to their small size (Rikke and Johnson 2007). The decline of mean T_b may thus be a particularly sensitive and robust biomarker of aging for mice. Interestingly, a study conducted on C57BL/6J, A/J, C57BL/6 x A/J F_1s, and a pen-bred strain of wild *Mus Musculus* has previously suggested a genetic correlation between loss of thermoregulation and lifespan in mice (Talan and Ingram, 1986), with greater loss of thermoregulation being associated with a shorter lifespan. Greater age-associated loss of thermoregulation has also been shown to be a predictor of lifespan within strain (Reynolds et al., 1985). The results of Conti et al. (2006), however, also indicate that lower mean T_b *per se* can be associated with life-extension benefits.

Separating the effects associated with a decline in T_b from those associated with lower T_b *per se* poses a difficult challenge. Nevertheless, as we have previously shown, genetic variation can be a powerful tool for uncoupling traits that are tightly linked (Rikke and Johnson, 2007). In addition, we found in this study that the lower T_b at the older age in Cohort 1 was only modestly correlated with the drop in mean T_b (R = 0.56), suggesting that one could distinguish between factors that correlate more strongly with T_b drop versus T_b *per se*.

In conclusion, the results of this study suggest that a large panel of recombinant inbreds could be used to genetically dissect loss of thermoregulation due to aging in mice. Such studies could potentially address whether loss of thermoregulation correlates temporarily with changes in temperature sensing, hormone signaling, heat production, or heat loss. In addition, one could map quantitative trait loci (QTLs) and test whether the individual loci affect one or more of these processes and also lifespan. QTLs of particular interest can be pursued using genomic and bioinformatic approaches to identify the underlying molecular pathways and genes (DiPetrillo et al., 2005; Drake et al., 2006). This systems approach is still being perfected, but it promises to be a powerful tool for elucidating the mechanisms underlying complex traits and their interrelationships. Genetic variation provides an important starting point.

Acknowledgments

We're grateful to Christine Martin and John E. Yerg for technical support. Funding was provided by the Ellison Medical Foundation and the National Institute on Aging (R01 AG024354).

References

Aniansson, A., M. Hedberg, G.-B. Henning, & Grimby, G. (1986). Muscle morphology, enzymatic activity, and muscle strength in elderly men: a follow-up study. *Muscle Nerve, 9*, 585-591.

Bai H., Islam M.N., Kuroki H., Honda K., & Wakasugi C. (1995). Deaths due to heat waves during the summer of 1994 in Osaka Prefecture, Japan. *Nippon Hoigaku Zasshi, 49*, 265–274.

Conti, B., Sanchez-Alverez, M., Winsky Sommerer, R., Morale, M.C., Lucero, J., Brownell, S., Fabre, V., Huitron-Resindez, S., Henriksen, S., Zorilla, E.P., De Lecea, L., & Bartfai, T. (2006). Transgenic mice with a reduced core body temperature have an increased life span. *Science, 314*, 825-828

Degroot, D.W., & Kenney, L.W. (2006). Impaired defense of core temperature in aged humans during mild cold stress. *Am J Physiol Regul Integr Comp Physiol, 292*, R103-R108.

DiPetrillo K., Wang X., Stylianou I.M., & Paigen B. (2005). Bioinformatics toolbox for narrowing rodent quantitative trait loci. *Trends Genet, 21*, 683-692.

Drake T.A., Schadt E.E., & Lusis A.J. (2006). Integrating genetic and gene expression data: application to cardiovascular and metabolic traits in mice. *Mamm Genome, 17*, 466-479.

Florez-Duquet, M., & McDonald, R.B. (1998). Cold induced thermoregulation and biological aging. *Physiol Rev, 78*, 339-358.

Ingram, D.K., Archer, J.A., Harrison, D.E., & Reynolds, M.A. (1982). Physiological and Behavioral correlates of lifespan in aged C57BL/6J mice. *Exp Gerontol, 17*, 295-303.

Lauria, G., Holland, N., Hauer, P., Cornblath, D.R., Griffin, J.W., & Mcarthur, J.C. (1999). Epidermal innervation: changes with aging, topographic location, and in sensory neuropathy. *J. Neurol Science, 164,* 172-178.

Lipsitz L.A. (2004). Physiological complexity, aging, and the path to frailty. *Sci Aging Knowledge Environ, 2004(16),* pe16.

Markel P.D., Fulker D.W., Bennett B., Corley R.P., DeFries J.C., Erwin V.G., & Johnson T.E. (1996). Quantitative trait loci for ethanol sensitivity in the LS x SS recombinant inbred strains: interval mapping. *Behav Genet, 26*, 447-458.

Moran D.S., & Mendal L. (2002). Core temperature measurement: methods and current insights. *Sports Med, 32*, 879-85

Passlick-Deetjen, J., & Bedenbender-Stoll, E. (2005). Why Thermosensing? A Primer. *Nephrol Dial Transplant, 20*, 1784-1789.

Reynolds, M.A., Ingram, D.K., & Talan, M.I. (1985). Relationship of body temperature stability to mortality in aging mice. *Mech Ageing Dev, 30*, 143-152.

Rikke B.A., Yerg J.E. 3rd, Battaglia M.E., Nagy T.R., Allison D.B., & Johnson T.E. (2003). Strain variation in the response of body temperature to dietary restriction. *Mech Ageing Dev, 124,* 663-678.

Rikke B.A., Yerg J.E. 3rd, Battaglia M.E., Nagy T.R., Allison D.B., & Johnson T.E. (2004). Quantitative trait loci specifying the response of body temperature to dietary restriction. *J Gerontol A Biol Sci Med Sci, 59,* 118-125.

Rikke B.A., & Johnson T.E. (2007). Physiological genetics of dietary restriction: uncoupling the body temperature and body weight responses. *Am J Physiol Regul Integr Comp Physiol, 293,* R1522-527.

Talan M.I., & Ingram D.K. (1986). Age comparisons of body temperature and cold tolerance among different strains of *Mus musculus*. *Mech Ageing Dev, 33*, 247-256.

Van Someren, E.J. (2007). Thermoregulation and aging. *Am J Physiol Regul Integr Comp Physiol*, 292, R99-R102.

Van Someren, E.J., Raymann, R.J., Scherder, E. J., Daanan, H.A., & Swaab, D.F. (2002). Circadian and age related modulation of thermoreception and temperature regulation: mechanisms and functional implications. *Ageing Res Rev, 1*, 721-728.

Williams R.W., Bennett B., Lu L., Gu J., DeFries J.C., Carosone-Link P.J., Rikke B.A.,Belknap J.K., & Johnson T.E. (2004). Genetic structure of the LXS panel of recombinant inbred mouse strains: a powerful resource for complex trait analysis. *Mamm Genome, 15*, 637-647.

Reviewed by Donald K Ingram, PhD; Pennington Biomedical Research Center, Baton Rouge, LA

In: Body Temperature Regulation
Editor: A. B. Cisneros and B. L. Goins

ISBN: 1-60741-282-3

Chapter XVII

Role of the Nitric Oxide Pathway in Body Temperature Regulatory Mechanisms in the Central Nervous System

Hiromi Tsushima[1], Yoshihiro Hotta[1], Kumiko Mitsui-Saitoh[1], Hiroaki Ando[1], Tunemasa Nonogaki[1] and Mayumi Mori[2]

[1]College of Pharmacy, Kinjo Gakuin University, Nagoya 463-8521, Japan
[2]Department of cellular and Molecular Pharmacology, Nagoya City University Graduate School of Medical Sciences, Nagoya 467-8601, Japan

Abstract

Body temperature (BT) is strictly regulated to a constant level in human, after temperature-sensitive sensory neurons in the skin usually detect any changes of environmental temperature. Pathological conditions sometimes produce fever that results from the releases of endogenous pyretic substances after attacks such as those from bacteria and viruses. The mechanisms of BT regulation underlying these two cases are naturally different; however, in both cases the final step involves the thermoregulatory center in the central nervous system (CNS). We discuss whether the NO pathway is a pyretic, antipyretic or an unrelated one for BT regulation of the CNS. In conclusion, NO likely contributes to both hyperthermia and hypothermia under pathological conditions, but not under physiological conditions. The role of NO in BT regulation differs depending on when, where, and how NO is generated under pathological conditions, which is likely determined by cytokines biosynthesized after attacks of bacteria and virus.

Abbreviation

LPS:	lipopolysaccharide
PG:	prostaglandin
IL:	interleukin
L-NA:	L-nitro-arginine: a nNOS and eNOS inhibitor
L-NNA:	NG-nitro-L-arginine: a nNOS and eNOS inhibitor
L-NAME:	NG-nitro-L-arginine methylester: a potent cNOS inhibitor
L-NMMA:	NG-methyl-L-arginine acetate: a nonselective NOS inhibitor
L-NIO:	L-N5-(1-iminoethyl)ornithine: an irreversible NOS inhibitor
vL-NIO:	vinyl-N5-(1-imino-3-butrnyl)ornithine: a nNOS inhibitor
7NiNa:	7-nitroindazole sodium salt: a nNOS inhibitor
AG:	aminoguanidine: a specific iNOS inhibitor
SNP:	sodium nitropurusside: a NO donor
SNAP:	S-nitroso-actylpenisillamine: a NO donor
SIN-1:	3-morpholinosynonimine hydrocholide: a NO donor
carboxy-PTIO:	2-[4-carboxyphenyl]-4,4,5,5-tetra-methylimidazoline-1-oxy-3-oxide, pottasium salt: a NO scavenger
AVPO:	anteroventral preoptic region
OVLT:	the organum vasculosum laminae terminalis
POA:	preoptic area of the hypothalamus
AV3V:	the anterioventral third ventricular region

Introduction

The strict regulation of body temperature (BT) at a constant level is an essential homeostatic process required for the survival of homoeothermic animals. Under physiological conditions, the sensory neurons in the skin monitor changes in BT resulting from the external environment temperature. This information is then transferred to the brain. On the other hand, there are temperature-sensitive neurons in the hypothalamus, and which also receive the brain/blood temperature changes. The existence of these neurons in the hypothalamus has been demonstrated in electrophysiological studies. After modulating such information, various signals are sent from the brain to the peripheral tissues for maintenance of constant BT; for instance, vasodilation and sweat secretion for decreasing BT, and vasoconstriction, metabolism facilitation, and skeletal muscle shivering to increase BT (Figure 1). It was recently demonstrated that some members of the transient receptor potential ion channels play important roles in the temperature sensitivity in both the peripheral sensory neurons and the hypothalamus [4].

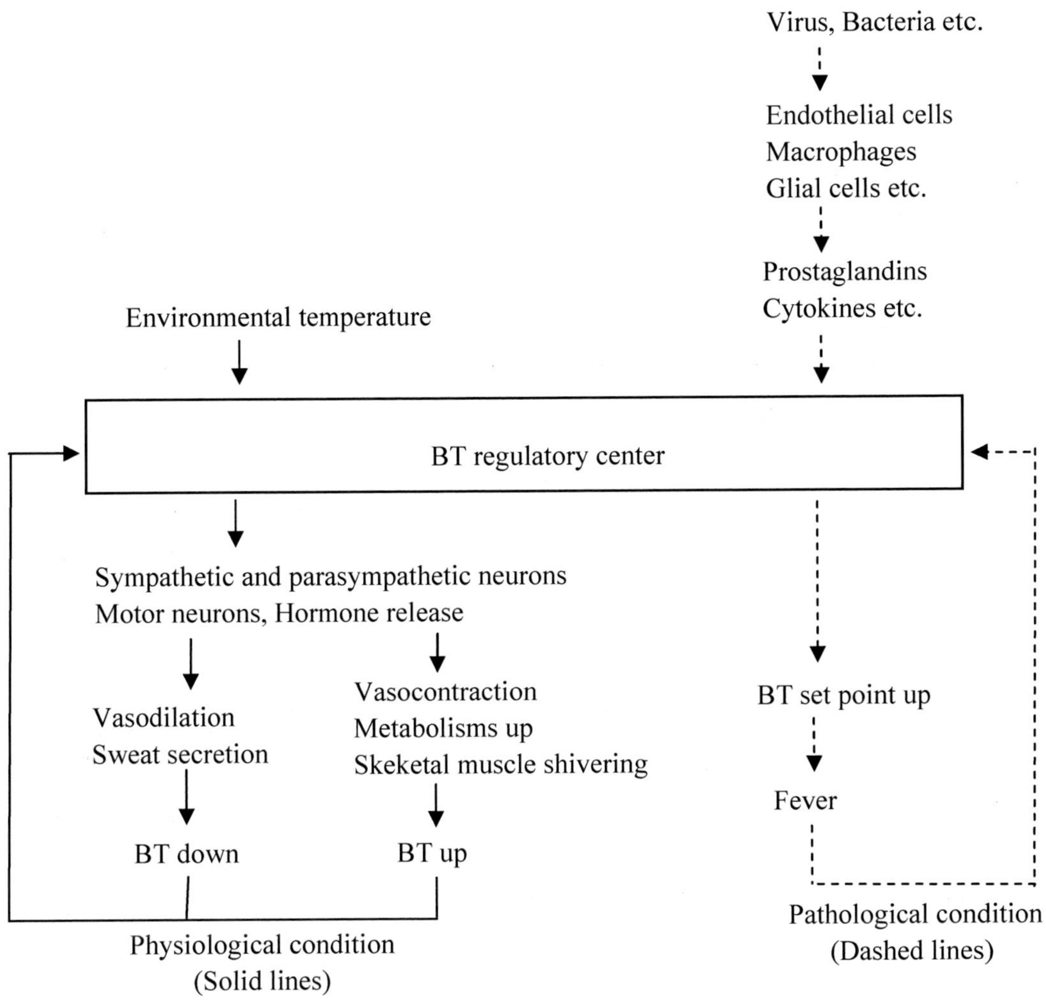

Figure 1. Schematic regulatory system for body temperature under physiological and pathological conditions.

Despite these regulatory mechanisms of BT, pathological conditions can often produce fever, resulting from the release of endogenous pyretic substances after attacks of bacteria and viruses, for instance [19]. Typical endogenous pyretic substances include prostaglandin E_2 (PGE_2) and interleukin-1beta (IL-1beta), which are biosynthesized in endothelial cells of blood vessels, glial cells, and macrophages, and can affect the BT regulatory center in the brain. As these substances typically shift up the BT set-point, fever is produced. Although there are a number of different mechanisms underlying under the both the physiological and pathological regulation of BT, in both cases the final step is through the CNS center of BT regulation (Fig. 1). The general consensus is that this center resides in the hypothalamus, and is likely the organum vasculosum laminae terminalis (OVLT) or the preoptic area (POA). However, detailed mechanisms for neural BT regulation are still largely unknown.

Nitric oxide (NO) is produced from L-arginine by NO synthase (NOS), which is divided in constitutional (c) NOS (comprise of neuronal (n) and endothelial (e) NOS), and inducible (i) NOS. As NO is oxidized as soon as it is generated, it only functions in restricted sites where it is synthesized. NO influences the activities of various enzymes, including activation of guanylyl cyclase (GC) and cyclooxygenase (COX), and inhibition of lipoxygenase, cytochrome P450, and protein kinase (PK) C. Therefore, it plays an important role in numerous physiological and the pathological signaling pathways. One of the most well-known downstream pathways resulting from NO production is the NO-cGMP-PKG pathway, which has been repeatedly demonstrated to play an important role in regulation of tension of vascular smooth muscle. NO also functions as a neuromodulator in the CNS, with many studies demonstrating that NOS inhibitors and NO donors can influence long-term potentiation and neurotransmitter release. Although there are some studies suggesting that NO can contribute to BT regulation, its role remains controversial; NO has been suggested to exert both pyretic and antipyretic roles, and has also been shown to be unrelated to BT regulation. Herein, we discuss the potential role of NO in the BT regulatory system in the CNS.

The NO Pathway in the Central Nervous System

As NO exerts its functions in the limited sites where it is synthesized, for NO to play a role in BT regulatory mechanisms in the hypothalamus, the precursors (L-arginine and O_2), enzymes (NOS) and coenzymes (NADPH, FAD/FM, and calmodulin[1]) required for NO production must be present. These substances have been shown to be widely distributed in the CNS. Primary cultured endothelial cells, glial cells, and cerebellar granule cells show iNOS activity even under the standard culture conditions, and increase NO production after stimulation. In addition, many studies have shown the presence of cNOS (nNOS and eNOS) and iNOS in the CNS including the hypothalamus, cerebrum, amygdala, dorsal motor nucleus of the vagus, reticular nucleus of medulla, and the area postrema [1, 3, 6, 10, 15, 23, 25]. Further, following injection of IL-1beta/LPS in living animals, expression of iNOS protein and mRNA, NOS activity and NO production are increased in the region surrounding the cerebroventricle and/or the hypothalamus [12, 16]. Immunohistochemical studies have also demonstrated expression of iNOS and cNOS in hypothalamic nuclei, including the OVLT and POA, and in other sites [3, 10].

The likely downstream signaling pathway following NO production involves GC activation, resulting in increased cGMP and PKG activation. Soluble GC and PKG are abundant in the cerebellar cortex, neostriatum, olfactory tubercle, and hypothalamus [3, 6, 8]. In the hypothalamic supraoptic nucleus, NOS and cGMP have been shown to co-localize in neurons using double-labeling immunofluorescence [23]. Unfortunately, there are no reports concerning the co-localization of NOS and GC/cGMP in OVLT or POA where the BT center is thought to exist. These data suggest that the hypothalamus contains NOS, GC, and PKG, although they do not express intense activity compared with the other brain sites, and it is unclear whether they exist in one cell.

---

[1] Note that calmodulin is not necessary for iNOS activation.

The NO Pathway in Pyretic Signaling

The OVLT is one of the strongest candidates for the BT regulatory center in the hypothalamus. Lin et al. [13-15] examined role of NO in BT regulation in the OVLT using rabbits, and demonstrated that pre-injection of some NOS inhibitors into the OVLT inhibited lipopolysaccharide (LPS)-induced hyperthermia, while a NO donor increased BT from the basal level (Table 1). In those studies, they also examined the roles of the different NOS isoforms, and demonstrated that the fever was inhibited by pretreatment with a specific iNOS inhibitor, aminoguanidine, but not by the cNOS inhibitor NG-nitro-L-arginine methyl ester (L-NAME), while increased iNOS immunoreactivity was found after LPS administration. Thus NO likely participates in the mechanisms of fever, with iNOS potentially more important than cNOS. In support of this, in studies using rats, administration of an iNOS inhibitor blocked LPS-induced hyperthermia more effectively than a cNOS inhibitor [20], and expression of iNOS mRNA increased in circumventricular organs after LPS administration [12]. In addition, LPS/cytokines are well known to stimulate iNOS transcription in various types of cultured cells. However, in another study, the cNOS inhibitors L-NAME and L-nitro-arginine (L-NA) inhibited LPS/IL-1beta-induced hyperthermia, although the authors did not discuss NOS isoforms [17, 24]. Overall, these data suggest a role of NO in fever. NO in fever is likely biosynthesized by iNOS. However, it is difficult to determine which type of three NOS isoforms is most important in fever in living animals. Many isoform-selective or non-selective NOS inhibitors such as L-NA, L-NAME, aminoguanidine, and NG-methyl-L-arginine acetate (L-NMMA), have been used, but the isoform selectivity of these drugs in living animals is largely unknown.

In LPS-induced hyperthermia, tyrosine kinase (TK) is included upstream of iNOS expression [15]. Further, we previously demonstrated that TK activation is necessary to produce fever [24]. This may show TK activity in the intracellular domain of cytokine receptors, as various kinds of cytokines are biosynthesized after LPS administration. The signaling pathway for LPS-induced fever after NO production is also thought to involve activation of PKG and generation of PG. The role of PKG activation following cGMP generation in fever has been described in detail [13]; as fever was inhibited by some GC inhibitors and was produced by a PKG activator itself, NO produces fever after activation of GC [3]; suggesting that the LPS-cytokines-TK-iNOS-GC-cGMP-PKG pathway is important for fever. However, NO activates several enzymes, not only GC, one of which is COX. As such, it is possible that NO directly activates COX, increases PGE_2 production, and then increases BT. Another pathway was proposed by Palmi et al. [18], who demonstrated that increased cGMP activates ADP-ribosyl cyclase followed by activation of cyclic ADP ribose (cADPR), then phospholipase (PL)A_2 is activated, and finally PGE_2 is synthesized. This pathway does not involve PKG activation. cADPR activates ryanodine receptors, and then increases intracellular Ca^{2+} concentration ($[Ca^{2+}]_i$). This increase in $[Ca^{2+}]_i$ is important for PLA_2 activation. Increased Ca^{2+} in the cerebrospinal fluid was shown after administration of IL-1beta and NO, and which correlated to fever. As NO production may contribute to several downstream, further studies are required.

Table 1. The NO pathway is pyretic one for BT regulation

Drug	Dose	Site	Effect	Experimental condition	Ref.
L-NA	110 μg	icv	inhibition	NMDA (100 nmol, icv)-induced hyperthermia	9
L-NAME	6 μg	icv	inhibition	LPS (3 μg, icv)-induced hyperthermia	24
vL-NIO	10 μg	icv	inhibition	LPS (50 μg/kg, ip)-induced hyperthermia	20
AG	10 μg	icv	inhibition	LPS (50 μg/kg, ip)-induced hyperthermia	
L-NAME	5 mg/kg	iv	inhibition	IL-1beta (100 ng, icv)-induced hyperthermia	17
L-NIO	100 μg	OVLT	inhibition	LPS (0.2 μg, OVLT)-induced hyperthermia	13-15
L-NIO	100 μg	OVLT	inhibition	IL-1beta (5 ng, icv)-induced hyperthermia	
L-NMMA	100 μg	OVLT	inhibition	LPS (0.2 μg, OVLT)-induced hyperthermia	
AG	1000 μg	OVLT	inhibition	LPS (0.2 μg, OVLT)-induced hyperthermia	
AG	1000 μg	OVLT	inhibition	IL-1beta (5 ng, icv)-induced hyperthermia	
methylene blue	100 μg	OVLT	inhibition	LPS (0.2 μg, OVLT)-induced hyperthermia	
SNAP, SNP	1-10 μg	OVLT	hyperthermia	basal condition	

The NO Pathway in Antipyretic Signaling

Peripheral injections, as well as central injections, of LPS can produce hyperthermia, in which NO is considered to function as an antipyretic signaling molecule in the anteroventral preoptic region (AVPO), a candidate BT center in the CNS ([7, 22]; Table 2). This hyperthermia was facilitated by pretreatment with a NOS inhibitor, L-NAME, or a NO scavenger, carboxy-PTIO; and was inhibited by a NO donor, SIN-1, injected into the CNS. In addition, in the AVPO, L-NAME itself produces hyperthermia, and an NO donor produced hypothermia from basal BT levels. During hyperthermia induced by intraperitoneal (i.p.) injections of LPS, the NO concentration actually decreased in the anteroventral third ventral region (AV3V) which contains the AVPO [22]. Signaling after NO production likely involves the GC-cGMP-PKG pathway in AVPO as: 1) 8-Br-cGMP, a phosphodiesterase-resistant cGMP analog, also produced hypothermia by itself under basal conditions, 2) 8-Br-cGMP inhibited LPS-induced hyperthermia, 3) neither of these effects were observed after peri-AVPO injections, and 4) a PKG inhibitor caused hyperthermia in the AVPO [21, 22]. On the other hand, central injections of PGE_2 into the OVLT induced hyperthermia, which was potentiated by L-NIO, aminoguanidine, and L-NMMA [13]. These data suggest that NO synthesized by both cNOS and iNOS decreases BT. Feleder et al. [7] also demonstrated a role of norepinephrine (NE) and PGE_2 release in the POA for induction of hyperthermia after intravenous (i.v.) injection of LPS. In that study, release of NE and PGE_2 were inhibited by an NO donor, and were potentiated by an NO scavenger. Further, injection of an alpha-2 adrenergic receptor agonist into the POA induced hypothermia followed by hyperthermia, which were accompanied by a respective decrease and increase in PGE_2. These effects were altered by an NO donor and an NO scavenger. The authors hypothesized that peripheral LPS administration decreases NO production, increases NE release, activates COX2, increases PGE_2 production, and then induces fever. This mechanism may not involve the cGMP-PKG pathway.

Intraperitoneal and i.v. injections of LPS can sometimes produce hypothermia, in which the NO pathway is also involved. L-NNA injected i.c.v inhibited LPS-induced hypothermia [26]. However, a nNOS inhibitor 7-nitroindazole sodium salt (7NiNa) or an iNOS inhibitor aminoguanidine had no effect on the LPS-induced hypothermia. Therefore, the authors concluded that NO is biosynthesized by eNOS in the CNS after i.v. injection of LPS, and suggested that hypothermia resulted from inhibition of PVN neuronal activity and IL-1beta

Table 2. The NO pathway is antipyretic one for BT regulation

Drug	Dose	Site	Effect	Experimental condition	Ref.
L-NAME	12.5 μg	AVPO	potentiation	LPS (100 μg/kg, ip)-induced hyperthermia	22
L-NAME		AVPO	hyperthermia	basal condition	
SNP	5 μg	AVPO	hypothermia	basal condition	
SIN-1	20 μg/μl perfusio n	POA	inhibition	LPS (2 μg/kg, iv)-induced hyperthermia	7
carboxy-PTIO	20 μg/μl perfusio n	POA	potentiation		
L-NMMA	280 μg	Icv	inhibition (hyperthermia)	LPS (50 μg/kg, ip)-induced hypothermia	2
L-NMMA	280 μg	Icv	hyperthermia	basal condition	
L-NNA	100 nmol	Icv	inhibition	LPS (100 μg/kg, iv)-induced hypothermia	26
L-NIO	200 μg	OVLT	potentiation	PGE_2 (0.2 μg, OVLT)-induced hyperthermia	13
AG	1000 μg	OVLT	potentiation	PGE_2 (0.2 μg, OVLT)-induced hyperthermia	
L-NMMA	100μg	OVLT	potentiation	PGE_2 (0.2 μg, OVLT)-induced hyperthermia	

mRNA expression. In another study using near-term pregnant rats [2], LPS-induced hypothermia was also inhibited by L-NMMA, while the inhibitor itself did not change BT under basal conditions. However, in virgin or mid-term pregnant rats, LPS produced hyperthermia, and L-NMMA had no effect. These data suggest that the role of NO in BT regulation is limited to near-term pregnancy. Electrophysiological studies have also shown an increased firing rate of heat-sensitive neurons in non-POA, but not in POA, following cGMP analog administration [25]. As this phenomenon is considered to induce decreased BT, this suggests that the NO-cGMP pathway contributes in hypothermia.

In a recent study, nNOS knockout (KO) mice were shown to have no changes in BT after i.p. injection of Δ^9-tetrahydrocannabinol (THC), while wild-type (WT) mice developed hypothermia [1]. THC is considered to bind brain-type cannabinoid receptor 1 (CB1) in the hypothalamus, resulting in hypothermia. CB1 and nNOS are immunohistochemically co-expressed in the ventromedial hypothalamus. Thus nNOS in the CNS participates in this effect of WT mice.

The NO Pathway is Unrelated to BT Regulation

In contrast to the above data, a number of studies have demonstrated that the NO pathway is uninvolved in BT regulation (Table 3). Most of the results were collected from controls in experiments examining the effects of various NO-related drugs on hyperthermia or hypothermia. Under basal conditions without any treatment, the NO pathway is suggested to have no function in physiological BT regulation as NOS inhibitors, NO scavengers, and PKG inhibitors did not modulate basal BT levels. In addition, basal BT levels were not different between iNOS KO mice [5], nNOS KO mice [1], and WT mice; there are no studies examining BT regulation in eNOS KO mice. Further studies have demonstrated no effects of an NO donor or 8-Br-cGMP on BT, while NOS inhibitors did not influence LPS, IL-1beta, or ghrelin-induced hyperthermia, which is mediated through release of PGs [11, 14, 24, 26]. These results suggest that the NO pathway is not involved in BT regulation under physiological conditions, and may demonstrate that the PG pathway and the NO pathway are independent of each other in CNS fever mechanisms.

Conclusion

In conclusion, NO likely contributes to both hyperthermia and hypothermia under pathological conditions, but not under physiological conditions. In the case of hyperthermia, iNOS in the OVLT plays an important role in NO production. iNOS expression requires activation of TK in the CNS, and the downstream pathway involves activation of GC and PKG. In addition, an increased CNS Ca^{2+} concentration may participate in this signaling pathway. In contrast, in the case of NO as an antipyretic molecule, eNOS in the POA plays an important role in NO production, and GC is involved in the downstream pathway. Decreased NE and PGE_2 in the hypothalamus may also be connected to this pathway. Taken together, the role of NO in BT regulation differs depending on when, where, and how NO is generated under pathological conditions, which is likely determined by cytokines biosynthesized after attacks of bacteria and virus.

Table 3. The NO pathway is unrelated one for BT regulation

Drug	Dose	Site	Effect	Experimental condition	Ref.
L-NAME	20 μg	Icv	no change	Ghrelin (20 μg, icv)-induced hyperthermia	11
L-NAME	20 μg	Icv	no change	basal condition	
L-NAME	6 μg	Icv	no change	IL-1beta (50 ng, icv)-induced hyperthermia	24
L-NAME	6 μg	icv	no change	basal condition	
L-NIO	10 μg	icv	no change	basal condition	20
L-NIO	100 μg	OVLT	no change	basal condition	13
L-NMMA	100 μg	OVLT	no change	basal condition	
L-NAME	1000 μg	OVLT	no change	basal condition	
L-NAME	1000 μg	OVLT	no change	LPS (0.2 μg, OVLT)-induced hyperthermia	
L-NAME	1000 μg	OVLT	no change	IL-1beta (5 ng, OVLT)-induced hyperthermia	
AG	1000 μg	OVLT	no change	basal condition	
methylene blue	100 μg	OVLT	no change	basal condition	
7NiNa	16.3 μg	icv	no change	LPS (100 μg/kg, iv)-induced hypothermia	26
AG	2.2 μg	icv	no change	LPS (100 μg/kg, iv)-induced hypothermia	
L-NMMA L-NMMA	280 μg 280 μg	Icv icv	no change no change	basal condition, LPS (50 μg/kg, ip)-induced hyperthermia	2
SIN-1	20 μg/μl perfusion	POA	no change	basal condition	7
carboxy-PTIO	20 μg/μl perfusion	POA	no change	basal condition	
L-NAME	12.5 μg	AVPO	no change	basal condition	22
L-NAME	1000 μg	OVLT	no change	IL-1beta(5 μg, OVLT)-induced hyperthermia	14

References

[1] Azad, S. C., Marsicano, G., Eberlein, I., Zeiglgansberger, W., Spanagel, R. & Lutz, B. (2001) Differential role of the nitric oxide pathway on delta9-THC-induced central nervous system effects in the mouse. *Eur. J. Neurosci.*, 13, 561-568.

[2] Begg, D. P., Kent, S., McKinley, M. J. & Mathai, M. L. (2007) Suppression of endotoxin-induced fever in near-term pregnant rats is mediated by brain nitric oxide. *Am. J. Physiol.*, 292, R2174-R2178.

[3] Bhat, G., Mahesh, V. B., Aguan, K. & Brann, D. W. (1996) Evidence that brain nitric oxide synthase is the major nitric oxide synthase isoform in the hypothalamus of the adult female rat and that nitric oxide potently regulates hypothalamic cGMP levels. *Neuroendocrinology*, 64, 93-102.

[4] Caterina, M. J. (2007) Transient receptor potential ion channels as participants in thermosensation and thermoregulation. *Am. J. Physiol.*, 292, R64-76.

[5] Dias, M. B., Almeida, M. C., Carnio, E. C. & Branco, L. G. S. (2005) Role of nitric oxide in tolerance to lipopolysaccharide in mice. *J. Appl. Physiol.*, 98, 1322-1327.

[6] Ding, L.-D., Burette, A., Nedvetsky, P. I., Schmidt, H. H. H. W. & Weinberg, R. J. (2004) Distribution of soluble guanylyl cyclase in the rat brain. *J. Comp. Neurol.*, 472, 437-448.

[7] Feleder, C., Perlik, V. & Blatteis, C. M. (2007) Preoptic nitric oxide attenuates endotoxic fever in guiea pigs by inhibiting the POA release of norepinephrine. *Am. J. Physiol.*, 293, R1144-R1151.

[8] Feil, S., Zimmermann, P., Knorn, A., Brummer, S., Schlossmann, J., Hofmann, F. & Feil, R. (2005) Distribution of cGMP-dependent protein kinase type 1 and its isoforms in the mouse brain and retina. *Neuroscience*, 135, 863-868.

[9] Hara, S., Mukai, T., Kuriiwa, F., Iwata, N., Yanase, T., Kano, S. &, T. (1997 Endo) Inhibition of NMDA-induced increase in brain temperature by N-w-nitro-L-arginine and indomethacin in rats. *Brain Res.*, 756, 301-304.

[10] Ishihara, T., Orikasa, C., Araki, T. & Sakuma, Y. (2002) Sex difference in the expression and regulation of nitric oxide synthase gene in the rat preoptic area. *Neurosci. Res.*, 43, 147-154.

[11] Jaszberenyi M., Bujdoso, E., Bagosi, Z. & Telegdy, G, (2006) Mediation of the behavioral, endocrine and thermoregulatory actions of ghrelin. *Hormones and Behavior*, 50, 266-273.

[12] Konsman, J. P., Kelley, K. & Dantzer, R. (1999) Temporal and spatial relationships between lipopolysaccharide-induced expression of Fos, interleukin-1beta and inducible nitric oxide synthase in rat brain. *Neuroscience*, 89, 535-548.

[13] Lin, J.-H. & Lin M.-T. (1996) Nitric oxide synthase-cyclo-oxygenase pathways in organum vasculosum laminae terminalis: possible role in pyrogenic fever in rabbit. *Br. J. Pharmacol.*, 118, 179-185.

[14] Lin, J.-H. & Lin M.-T. (1996) Inhibition of nitric oxide synthase or cyclo-oxygenase pathways in organum vasculosum laminae terminalis attenuates interleukin-1beta in rabbits. *Neurosci. Lett.*, 208, 155-158.

[15] Lin, M. T. & Lin J. H. (2000) Involvement of tyrosine kinase in the pyrogenic fever exerted by NOS pathways in organum vasculosum laminae terminalis. *Neuropharmacology*, 39, 347-352.

[16] Lopez-Figueroa, M. O., Day, H. E., Lee, S., Rivier, C., Akil, H. & Watson, S. J. (2000) Temporal and anatomical distribution of nitric oxide synthase mRNA expression and nitric oxide production during central nervous system inflammation. *Brain Res.*, 852, 239-246.

[17] Monroy, M., Kuluz, J. W., He, D., Dietrich, W. D. & Schleien, C. L. (2001) Role of nitric oxide in the cerebrovascular and thermoregulatory response to interleukin-1beta. *Am. J. Physiol. Heart Cire. Physiol.*, 280, H1448-H1453.

[18] Palmi, M. & Meini, A. (2002) Role of the nitric oxide/cyclic GMP/Ca2+ signaling pathway in the pyrogenic effect of interleukin-1beta. *Mol. Neurol.*, 25, 133-147.

[19] Rivest, S., Lacroix, S., Vallieres, L., Nadeau, S., Zhang, J. & Laflamme, N. (2000) How the blood talks to the brain parenchyma and the paraventricular nucleus of the hypothalamus during systemic inflammatory and infectious stimuli. *Proc. Soc. Exp. Biol. Med.*, 223, 22-38.

[20] Soszynski, D. & Chelminiak, M. (2007) Intracerebroventricular injection of neuronal and inducible nitric oxide synthase inhibitors attenuates fever due to LPS in rats. *J. Physiol Pharmacol.*, 58, 351-561.

[21] Steiner, A. A., Antunes-Rodrigues, J. & Branco, L.G. S. (2002) Role of preoptic second messenger systems (cAMP and cGMP) in the fenrile response. *Brain Res.*, 944, 135-145.

[22] Steiner, A. A., Antunes-Rodrigues, J., McCann, S. M. & Branco, L.G. S. (2002) Antipyretic role of the NO-cGMP pathway in the anteroventral preoptic region of the rat brain. *Am. J. Physiol.*, 282, R584-R593.

[23] Stern, J. & Zhang, W. (2005) Cellular sources, targets and actions of constitutive nitric oxide in the magnocellular neurosecretiry system of the rat. *J. Physiol.*, 562, 725-744.

[24] Tsushima, H. & Mori, M. (2000) In vivo evidence that activation of tyrosine kinase is a trigger for lipopolysaccharide-induced fever in rats. *Brain Res.*, 852, 367-373.

[25] Wright, C. L., Burgoon, P. W., Bishop, G. A. & Boulant, J. A. (2008) Cyclic GMP alters the firing rate and thermosensitivity of hypothalamic neurons. *Am. J. Physiol.*, 294, R1704-R1715.

[26] Yang, W. W. & Krukoff, T. L. (2008) Nitric oxide regulates body temperature, neuronal activation and interleukin-1beta gene expression in the hypothalamic paraventricular nucleus in response to immune stress. *Neuropharmacol.*, 39, 2075-2089.

In: Body Temperature Regulation
Editor: A. B. Cisneros and B. L. Goins
ISBN: 1-60741-282-3

Short Communication

Thermogenic Effects of Orexin A are Modified by Neuroleptic Drugs

*Marcellino Monda**, *Alessandro Viggiano, Giovanni Messina, Teresa Iannaccone, Eleonora Andreozzi, Emanuela Viggiano, Vincenzo De Luca*

Department of Experimental Medicine, Section of Human Physiology, and Clinical Dietetic Service, Second University of Naples, Via Costantinopoli 16, 80138 Naples, Italy

Abstract

This paper describes the effects of many neuroleptic drugs on the hyperthermia due to orexin A, a neuropeptide with influences on body temperature and food intake. Haloperidol reduces, while clozapine and olanzapine block the hyperthermia induced by orexin A. Risperidone enhances the elevation of body temperature due to orexin A. Quetiapine delays these hyperthermic effects. The implications on human therapeutic strategies are discussed, including the role of orexinergic system in the development of body weight gain, induced by these neuroleptic substances.

Introduction

The hypothalamic neuropeptide "hypocretin 1/orexin A" [8] is a neuropeptide so named for its influence on food intake [47]. An intracerebroventricular (icv) injection of orexin A does not merely affects eating behavior, but it also induces an increase in heart rate [35],

* Corresponding author: Prof. Marcellino Monda, MD, Dipartimento di Medicina Sperimentale, Sezione di Fisiologia Umana, Seconda Università di Napoli, Via Costantinopoli 16, 80138 Napoli, Italy, Tel. +39 +81 566 5804, Fax +39 +81 5667500. E-mail: marcellino.monda@unina2.it

blood pressure [41] and metabolic rate [21]. These autonomic modifications indicate that this neuropeptide plays a role in the control of vegetative functions. Orexin A also influences body temperature. In fact, an icv administration of orexin A induces an increase in firing rate of the sympathetic nerves to interscapular brown adipose tissue (IBAT), accompanied with a rise in IBAT and colonic temperatures [35]. The simultaneous increase in heart rate and body temperature after an icv injection of orexin A indicates a generalized activation of the sympathetic nervous system. In addition, the presence of orexin receptors in many cerebral areas suggests that additional functions are played by orexin A [17]. A role for the orexins in sleep regulation has been demonstrated [1].

Neuroleptic drugs, as haloperidol, clozapine, olanzapine, risperidone and quetiapine, can modify body temperature.

Intraperitonel administration of haloperidol produces hypothermic effects in the rats [20], and in the rabbits [46]. Oral administration of haloperidol in normal volunteers causes significant attenuation of the normal daytime increase of body temperature [7].

Clozapine is able to induce modification on body temperature. An undesirable effect of clozapine is the hypothermia in patients treated with this drug [15]. Intraperitonel administration of clozapine produces also hypothermic effects in the rats [26, 38]. This hypothermia appears to be unrelated to effects of the drug on movement. Indeed, intraperitoneal administration of clozapine lowers core temperature, having minimal motor side effects [16].

Olanzapine reduces the metamphetamine-induced hyperthermia [13]. This drug can induce a reduction of body temperature during the neuroleptic malignant syndrome [11].

Risperidone can influence body temperature. There is evidence showing that both hypothermia [40] and hyperthermia [12] are also affected by risperidone.

Quetiapine significantly attenuates the hyperthermia due to amphetamine [14], suggesting an influence of this drug on the thermogenic reactions. A sudden loss of body temperature control can be seen during the neuroleptic malignant syndrome [5] induced by administration of quetapine.

The aim of the present experiments was to test the influence of various neuroleptic substances on sympathetic and hyperthermic reactions due to an icv injection of orexin A, so to reveal influences of these neuroleptic drugs on the control of body temperature, exerted by orexinergic system.

Method

The firing rates of the sympathetic nerves to interscapular brown adipose tissue (IBAT), along with IBAT, colonic temperatures were monitored in urethane-anesthetized male Sprague-Dawley rats before an injection of orexin A (1.5 nmol) into the lateral cerebral ventricle and over a period of 180 min after the injection. The same variables were monitored in rats with an intraperitoneal administration of a neuroleptic substance: haloperidol (1 mg/kg bw), clozapine (8 mg/kg bw), olanzapine (10 mg/kg bw), risperidone (50 mg/kg bw), quetiapine (5 mg/kg bw).

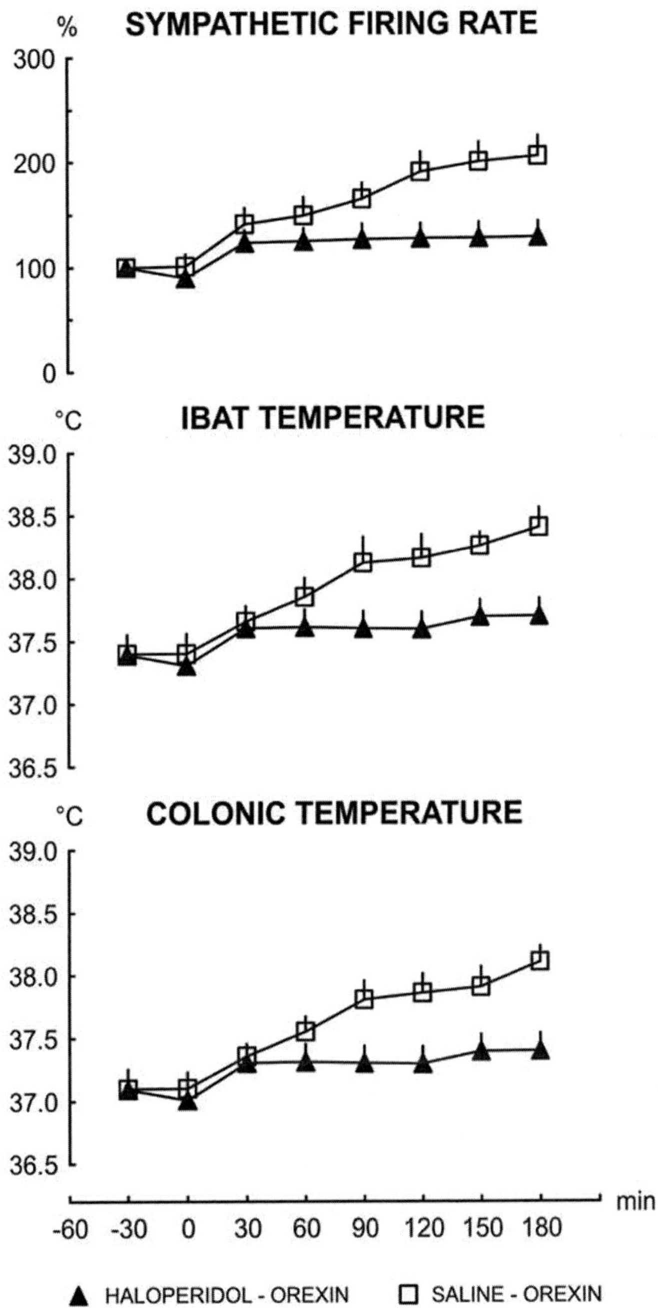

Figure 1. Means ± S.E.M. of sympathetic firing rate of nerves to interscapular brown adipose tissue (IBAT), IBAT and colonic temperature in rats with injection of haloperidol plus orexin A or saline.

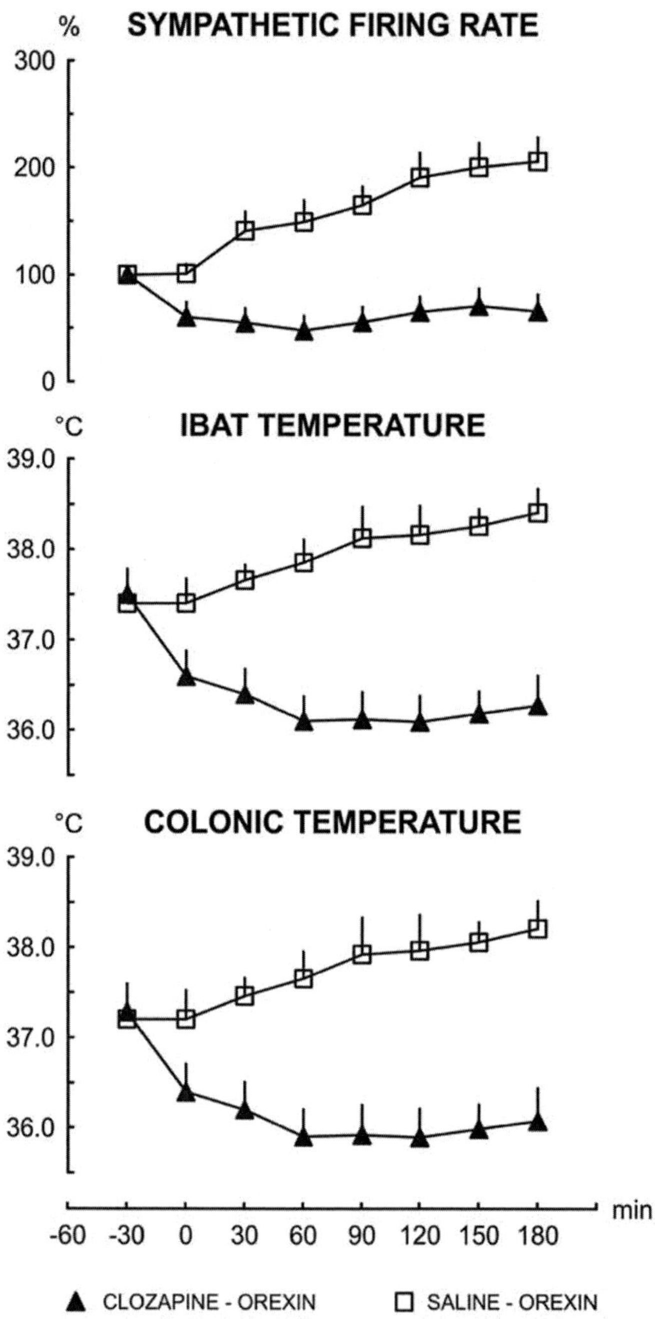

Figure 2. Means ± S.E.M. of sympathetic firing rate of nerves to interscapular brown adipose tissue (IBAT), IBAT and colonic temperature in rats with injection of clozapine plus orexin A or saline.

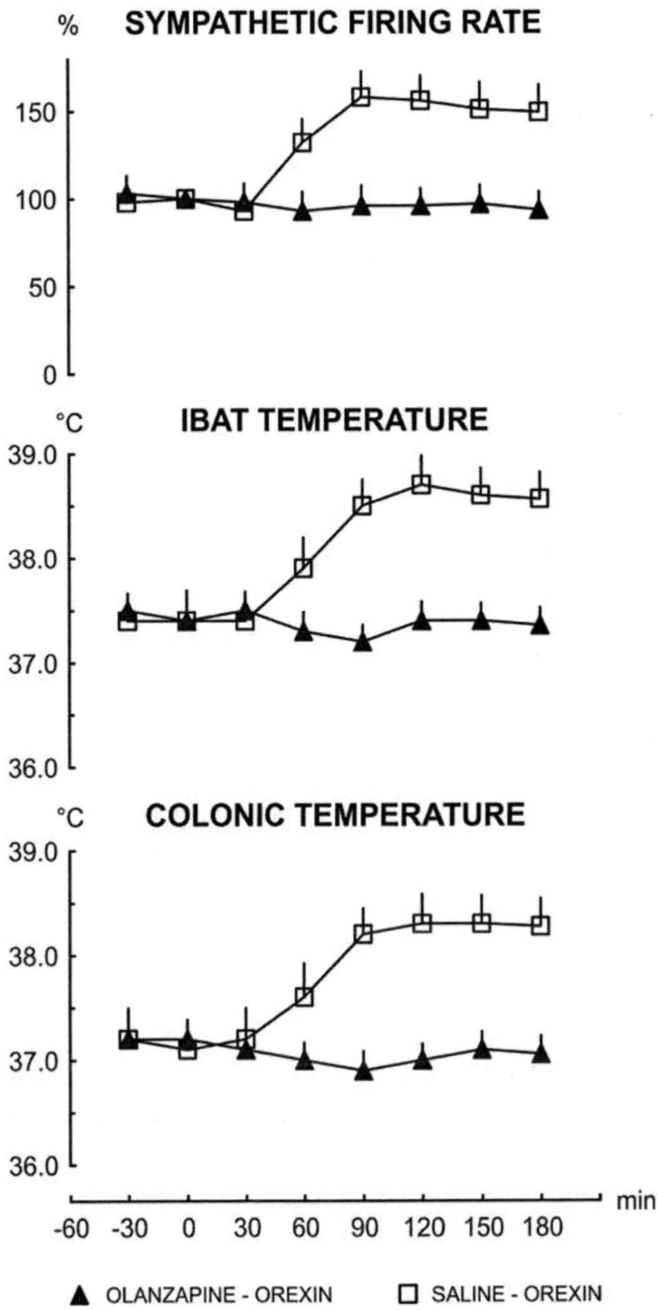

Figure 3. Means ± S.E.M. of sympathetic firing rate of nerves to interscapular brown adipose tissue (IBAT), IBAT and colonic temperature in rats with injection of olanzapine plus orexin A or saline.

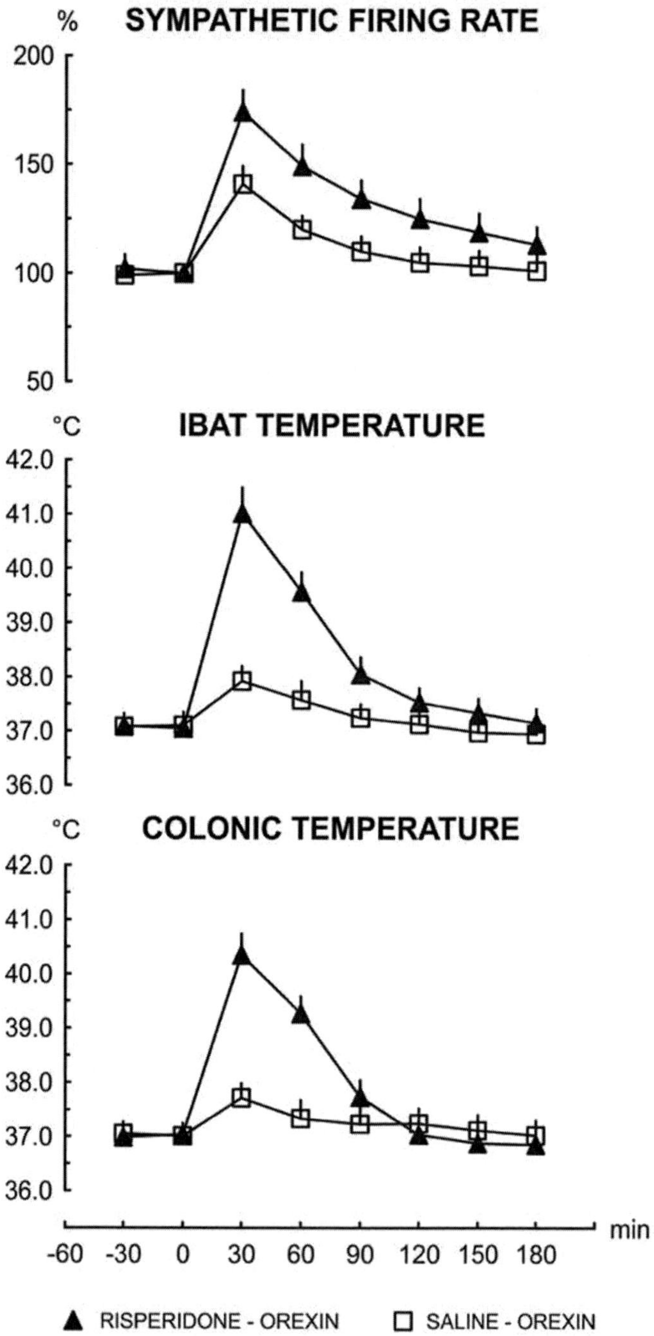

Figure 4. Means ± S.E.M. of sympathetic firing rate of nerves to interscapular brown adipose tissue (IBAT), IBAT and colonic temperature in rats with injection of risperidone plus orexin A or saline.

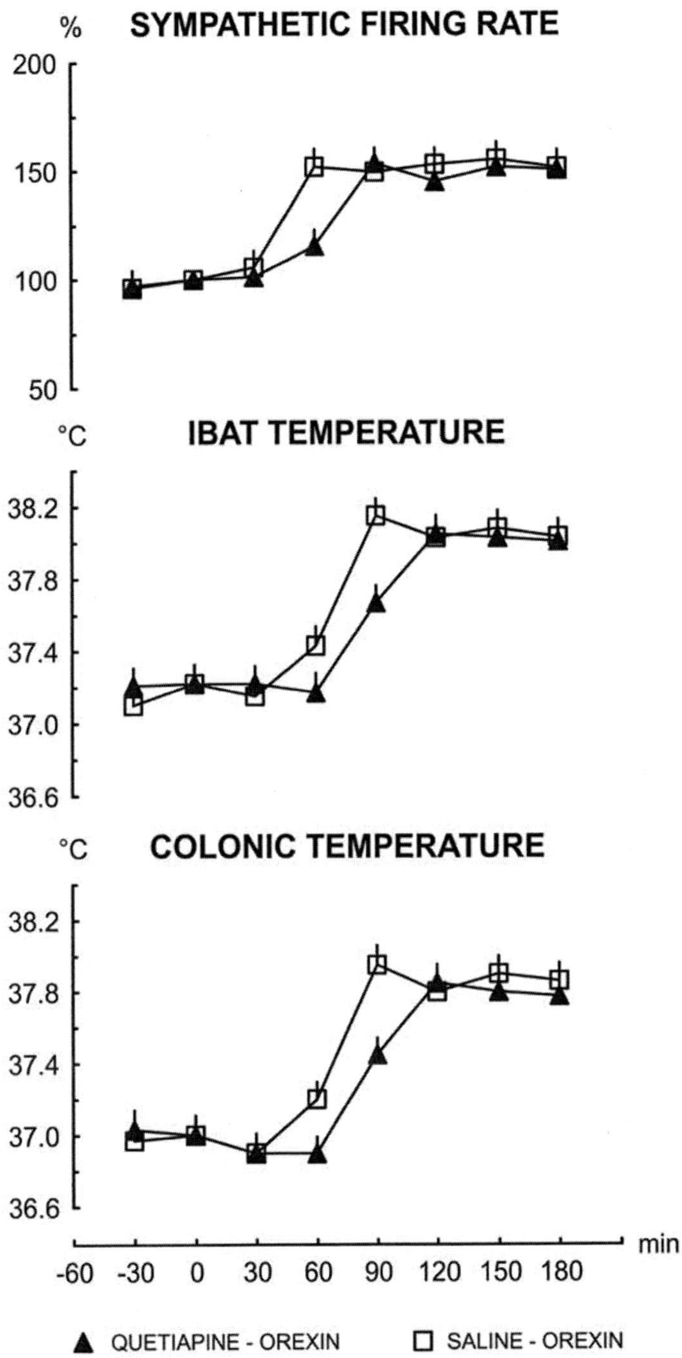

Figure 5. Means ± S.E.M. of sympathetic firing rate of nerves to interscapular brown adipose tissue (IBAT), IBAT and colonic temperature in rats with injection of quetiapine plus orexin A or saline.

Results

The results show that : a) haloperidol reduces the effects of orexin A on the sympathetic firing activity and body temperature. These effects are illustrated in the Figure 1; b) clozapine and olanzapine block the increases in the sympathetic firing rate, IBAT and colonic temperatures induced by orexin A, as reported in the Figure 2 and Figure 3; c) risperidone enhances the effects of orexin A. The Figure 3 shows that the thermic elevation due to orexin A are increased by risperidone; d) orexin A increases the sympathetic firing rate, IBAT, colonic temperatures. This increase is delayed by the injection of quetiapine, as shown in the Figure 5.

Discussion

These findings demonstrate that an ip injection of haloperidol reduces the sympathetic and thermogenic effects due to an icv administration of orexin A [28]. Since the increase in the sympathetic activity induced by orexin A is reduced by haloperidol (antagonist of D_2 receptors of the dopaminergic system), an involvement of the D_2 receptors can be hypothesized. On the other hand, an injection of quinpirole, a D_2 receptor agonist, induces an increase in the sympathetic discharge [6]. This indicates that an activation of dopaminergic system could be a possible mechanism of sympathetic and thermogenic stimulation induced by orexin A.

Since clozapine has high affinity for dopamine receptors [18], an involvement of the dopaminergic system can be hypothesized in the block of hyperthermia induced by orexin A [31]. There are evidences showing that dopamine D_3 receptors participate in clozapine-induced hypothermia [26]. On the other hand, an involvement of serotoninergic pathways in the orexin-induced hyperthermia cannot be excluded. A review of the psychiatric cases revealed that nine of ten hypothermic patients were treated with drugs that are potent antagonists of 5-HT_2 receptors [42]. Since clozapine is an antagonist of 5-HT_2 receptors [9], the block of these receptors could be important in the lack of the hyperthermia induced by orexin A. In other words, orexin A can produce an elevation in body temperature by acting on 5-HT_2 receptors. Furthermore, several laboratories found an even higher affinity for α_1-adrenergic receptors [3, 23]. Indeed, it was shown that central blockade of α_1-adrenergic receptors produces hypothermia [43], and that α_1-agonists block the hypothermic effects of clozapine [24]. Then, the central blockade of α_1-adrenergic receptors due to clozapine could be involved in the lack of the hyperthermia induced by orexin A. Clozapine blocks completely the hyperthermic action of orexin A. This indicate that patients treated with clozapine are non responsive to physiological variations of orexin A. This non-responsiveness to orexin A could induce not only hypothermia, but also body weight gain [22]. Indeed, orexin A is involved in the activation of thermogenesis related to eating behaviour. The thermogenesis due to food intake is an important factor in the regulation of body weight and the lack of this thermogenesis causes the obesity [2, 27]. In this experiment, the complete lack of hyperthermic effects due to injection of orexin A makes reasonable this supposition. Clozapine influences strongly the thermogenic role of orexin A, which is ineffective to modify the clozapine-induced hypothermia. These data are useful to address the choice of antipsychotic drugs for the therapy of mental disorders.

Furthermore, the treatment of rare but perilous hypothermia induced by clozapine should exclude substances, which utilize the orexinergic pathway to elevate body temperature. This pathway is completely blocked by clozapine, as demonstrated by the findings of these experiments.

Since olanzapine is an antagonist of 5-HT_2 and D_1-D_2 receptors [44], the block of these receptors could be involved in the block of the hyperthermia induced by orexin A [33]. An involvement of histaminergic, adrenergic and cholinergic systems cannot be excluded, because olanzapine affects also these systems [45]. Because this substance is largely used in the treatment of schizophrenia and manic disorders, these data are useful to address the choice of antipsychotic drugs for the therapy of mental disorders. The treatment of the rare but perilous hypothermia induced by olanzapine [19], for instance, could not include drugs acting on the orexinergic pathway to elevate body temperature: this pathway is blocked by olanzapine. Although the discussion on the body weight gain due to olanzapine [13] is not justified by the present experiment, we cannot exclude that the weight gain is due, at least in part, to the reduction in the thermogenic effects of orexin A. Since weight gain induced by antipsychotics is the second most frequently given reason for noncompliance with pharmacologic therapy [25], investigations on mechanisms inducing obesity assume a relevant importance in the therapeutic strategy of psychiatric disorders. These studies are the base for further investigations regarding the role of orexinergic system in the control of body weight that is affected also by thermogenic changes induced by neuroleptic drugs.

Risperidone is an atypical antipsychotic drug that possesses 5-HT_2 receptor antagonism combined with milder dopamine D_2 receptor antagonism [10, 39]. A risperidone injection potentiates hyperthermic effects of orexin A [30], indicating that these thermic reactions involve serotoninergic and dopaminergic systems. This experiment indicates that a 5-HT_2 receptor antagonism is stronger than a D_2 receptor antagonism in the induction of the amplified hyperthermia due to orexin A. The results of this experiment suggest that the mechanism inducing hypothermia, as occasional side effect of psychiatric therapy with risperidone, is independent of the involvement of orexinergic system. On the other hand, rare cases of high fever have been described in the patients treated with risperidone. Thus, the findings of the present experiment can suggest an involvement of orexinergic system in the fever due to risperidone. These data are useful for addressing the choice of antipsychotic drugs for the therapy of mental disorders. Furthermore, the treatment of rare but perilous hypothermia induced by risperidone could include substances, which utilize the orexinergic pathway to elevate body temperature. This pathway is not blocked by risperidone, as demonstrated by our experiments.

Quetiapine delays (but it does not reduce) the effects of orexin A, but the sympathetic discharge and body temperature still reach the same elevation obtained injected with orexin A alone [34]. Since quetiapine is an antagonist of 5-HT_{2c} and D_2 receptors [38], the block of these receptors could be involved in the delay of the hyperthermia induced by orexin A. An involvement of histaminergic and adrenergic systems cannot be excluded, because quetiapine affects these systems [38]. These data are useful to address the choice of antipsychotic drugs for the therapy of mental disorders. The treatment of the rare but perilous hypothermia induced by quetiapine, for instance, could include drugs acting on utilize the orexinergic pathway to elevate body temperature, because this pathway is not blocked by quetiapine, as demonstrated by the findings of the present experiment. The control of body temperature is also related to the control of body weight, in accord to the thermoregulatory hypothesis of food intake. We cannot exclude that the weight gain due to quetiapine is due, at least in part, to the reduction in the thermogenic effects of orexin A.

In general, these experiments demonstrate that an icv injection of orexin A increases the temperature of IBAT, which is the most important effector of non shivering thermogenesis in the rat [4], illustrating that the rise in heat production is also due to the activation of thermogenesis unrelated to muscle activity. IBAT activity is controlled by the sympathetic nervous system, and factors, which influence thermogenesis, appear to act centrally to modify the sympathetic outflow to IBAT [36].

The present experiment confirms the strong influence of orexin A on body temperature, independently of eating behavior. The name assigned to orexin A, called primarily hypocretin-1 [8], was due to the strong increase in food intake after administration of this peptide, assigning a fundamental role in the induction of food intake [47]. The present findings confirm that orexin A induces strong elevations of body temperature in anesthetized rats, thus independently on food intake. For this reason, we can insist to suggest the term "hyperthermine A" as an additional denomination of "orexin A" [29, 32].

References

[1] Beuckmann, CT; Yanagisawa, M. Orexins: from neuropeptides to energy homeostasis and sleep/wake regulation. *J. Mol. Med.* 2002 80, 329-342.

[2] Bray, GA. Reciprocal relation of food intake and sympathetic activity: experimental observations and clinical implication. *Int. J. Obesity.* 2000 24, S8-S17.

[3] Brunello, N; Masotto, C; Steardo, L; Markstein, R; Racagni. G. New insights into the biology of schizophrenia through the mechanism of action of clozapine. *Neuropsychopharmacol.* 1995 13, 177-213.

[4] Cannon, B; Nedergaard, J. Brown adipose tissue: function and physiological significance. *Physiol. Rev.* 2004 84, 277-359.

[5] Chungh, DS; Kim, BN; Cho, SC. Neuroleptic malignant syndrome due to three atypical antipsychotics in a child. *J. Psychopharmacol.* 2005.19, 422-425.

[6] Damase-Michel, C; Montastruc, JL; Gharib, C; Geelen, G; De Saint-Blanquat, G; Tran, MA. Effect of quinpirole, a specific dopamine DA2 receptor agonist on the sympathoadrenal system in dogs. *J. Pharmacol. Exp. Ther.* 1990 252, 770-777.

[7] De Koning, P; De Vries, MH. A comparisons of the neuro-endocrinological and temperature effects of DU 29894, flesinoxan, sulpiride and haloperidol in normal volunteers. *Br. J. Clin. Pharmacol.* 1995 39, 7-14.

[8] De Lecea, L; Kilduff, TS; Peyron, C; Gao, XB; Foye, PE; Danielson, PE; Fukuhara, C; Battenberg, ELF; Gautvik, VT; BartlettIi, FS; Frankel, WN; Van Den Pol, AN; Bloom, FE; Gautvik, KM; Sutcliffe, JG. The hypocretins: two hypothalamic peptides with neuroexcitatory activity. *Proc. Natl. Acad. Sci.* USA 1998 95, 322-327.

[9] Gobbi, G; Janiri, L. Clozapine blocks dopamine, 5-HT_2 and 5-HT_3 responses in the medial prefrontal cortex: an in vivo microiontophoretic study. *Eur. Neuropsychopharmacol.* 1999 10, 43-49.

[10] Grant, S, Fitton, A.. Risperidone. A review of its pharmacology and therapeutic potential in the treatment of schizophrenia. *Drugs.* 1994 48, 253-273.

[11] Hall, KL; Taylor, WH; Ware, MR. Neuroleptic malignant syndrome due to olanzapine. Psychopharmacol. *Bull.* 2001 35, 49-54.

[12] Halloran, Ll; Bernard, DW. Management of drug-induced hyperthermia. *Curr. Opin. Pediatr.* 2004 16, 211-215.

[13] He, J; Xu, H; Yang, Y; Zhang, X; Li, XM. Neuroprotective effects of olanzapine on methamphetamine-induced neurotoxicity are associated with an inhibition of hyperthermia and prevention of Bcl-2 decrease in rats. *Brain Res.* 2004 1018,186-192.

[14] He, J; Xu, H; Yang, Y; Zhang, X; Li, XM. Chronic administration of quetiapine alleviates the anxiety-like behavioural changes induced by a neurotoxic regimen of dl-amphetamine in rats. *Behav. Brain Res.* 2005 160, 178-187.

[15] 15 Heh, CW; Herrera, J; Demet, E; Potkin, S; Costa, J; Sramek, J; Hazlett, E; Buchsbaum, MS. Neuroleptic-induced hypothermia associated with amelioration of psychosis in schizophrenia. *Neuropsychopharmacol.* 1988 1, 149-156.

[16] Huberman, A; Turek, VF; Carlisle, HJ. Clozapine does not induce a motor impairment in operant responding for heat reinforcement. *Pharmacol. Biochem. Behav.* 2000 67, 307-312.

[17] Kukkonen, JP; Holmqvist, T; Ammoun, S; Akerman, KE. Functions of the orexinergic/hypocretinergic system. *Am. J. Physiol.* 2002 283, C1567-1591.

[18] Kulkarni, SK; Ninan, I. Dopamine D_4 receptors and development of newer antipsychotic drugs. *Fundam. Clin. Pharmacol.* 2000 14, 529-539.

[19] Lee, TW; Tsai, SJ; Hwang, JP. Severe cardiovascular side effects of olanzapine in an elderly patient: case report. *Int. J. Psychiatry. Med.* 2003 33, 399-401.

[20] Lin, MT; Wang, HS; Wang, Z; Chern, YF. Haloperidol produces hypothermic effects in the rats. *Experientia* 1979 35, 1469-1470.

[21] Lubkin, A; Stricker-Krongrad, A. Independent feeding and metabolic action of orexins in mice. Biochem. *Biophys. Res. Commun.* 1998 253, 241-245.

[22] Mcintyre, RS; Mccann, SM; Kennedy, SH. Antipsychotic metabolic effects: weight gain, diabetes mellitus, and lipid abnormalities. *Can. J. Psychiatry.* 2001 46, 273-281.

[23] Menon, MK; Gordon, LI; Fitten, J. Interaction between clozapine and a lipophilic alpha 1-adrenergic agonist. *Life Sci.* 1988 43, 1791-1804.

[24] Menon, MK; Lloyd, LR; Fitten, LJ. Antagonism of the hypothermic effect of clozapine in mice by centrally-active alpha 2-adrenergic antagonists and alpha 1-adrenergic agonists. *Psychopharmacol.* 1990 101, 67-72.

[25] Milano, W; Grillo, F; Del Mastro, A; De Rosa, M; Sanseverino, B; Petrella, C; Capasso, A. Appropriate intervention strategies for weight gain induced by olanzapine: a randomized controlled study. *Adv. Ther.* 2007 24, 123-134.

[26] Millan, MJ; Audinot, V; Melon, C; Newman-Tancredi, A. Evidence that dopamine D3 receptors partecipate in clozapine-induced hypothermia. *Eur. J. Pharmacol.* 1995 280, 225-229.

[27] Monda, M; Sullo, A; De Luca, B. Lesions of ventromedial hypothalamus reduce postingestional thermogenesis. *Physiol. Behav.* 1997 61, 687-691.

[28] Monda, M; Viggiano, A; De Luca V. Haloperidol reduces the sympathetic and thermogenic activation induced by orexin A. *Neurosci. Res.* 2003 45, 17-23.

[29] Monda, M; Viggiano, A; De Luca, V. A paradoxical effect of orexin A: the hypophagia induced by hyperthermia. *Brain Res*. 2003 961, 220-228.

[30] Monda, M; Viggiano, A; Viggiano, A; Viggiano, E; De Luca V. Risperidone potentiates the sympathetic and hyperthermic reactions induced by orexin A in the rat. *Physiol. Res.* 2006 55, 73-78.

[31] Monda, M; Viggiano, An; Viggiano, Al; Fuccio, F; De Luca, V. Clozapine blocks the hyperthermia induced by orexin A in the rat. *Physiol. Res.* 2004 53, 507-513.

[32] Monda, M; Viggiano, An; Viggiano, Al; Fuccio, F; De Luca, V. Cortical spreading depression blocks the hyperthermic reaction induced by orexin A. *Neuroscience.* 2004 123, 567-574.

[33] Monda, M; Viggiano, An; Viggiano, An; Mondola, R; Viggiano, E; Messina, G; Tafuri, D; De Luca, V. Olanzapine blocks the sympathetic and hyperthermic reactions due to cerebral injection of orexin A. *Peptides* 2008 29, 120-126.

[34] Monda, M; Viggiano, An; Viggiano, Al; Viggiano, E; Messina, G; Tafuri, D; De Luca, V. Quetiapine delays sympathetic and hyperthermic reactions due to cerebral injection of orexin A.*Neuropeptides* 2006 40, 357-363.

[35] Monda, M; Viggiano, An; Viggiano, Al; Viggiano, E; Messina, G; Tafuri, D; De Luca, V. Sympathetic and hyperthermic reactions by orexin A: role of cerebral catecholaminergic neurons. *Regul. Pept.* 2007 139, 39-44.

[36] Morrison, SF. Central pathways controlling brown adipose tissue thermogenesis. News Physiol. Sci. 2004 19, 67-74.

[37] Nemeroff, CB; Kinkead, B; Goldstein, J. Quetiapine: Preclinical studies, phatmacokinetics, drug interactions, and dosing. *J. Clin. Psychiatry* 2002 63, 5-11.

[38] Oerther, S; Ahlenius, S. Atypical antipsychotics and dopamine D1 receptor agonism: an in vivo experimental study using core temperature measurements in the rat. *J. Pharmacol. Exp. Ther.* 2000 292, 731-736.

[39] Ota, M; Mori, K; Nakashima, A; Kaneko, YS; Fujiwara, K; Itoh, M, Nagasaka, A, Ota, A. Peripheral injection of risperidone, an atypical antipsychotic, alters the body weight gain of rats. *Clin. Exp. Pharmacol. Physiol.* 2002 29, 980-989.

[40] Razaq, M; Samma, M. A case of risperidone-induced hypothermia. Am. J. Ther. 2004 11, 229-230.

[41] Samson, WK; Gosnell, B; Chang, JK; Resch, ZT; Murphy,.C. Cardiovascular regulatory actions of the hypocretins in brain. *Brain Res.* 1999 831, 248-253.

[42] Schwaninger, M; Weisbrod, M; Schroder, M, Hacke, W. Hypothermia induced by atypical neuroleptics. *Clin. Neuropharmacol.* 1998 21, 344-346.

[43] Stone, EA; Zhang, Y; Rosergaten, H; Yeretsian, J; Quartermain, D. Brain alpha 1-adrenergic neurotransmission is necessary for behavioral activation to enviromental changes in mice. *Neurosci.* 1999 94, 1245-1252.

[44] Tauscher, J, Hussain, T; Agid, O; Verhoeff, NP; Wilson, AA; Houle, S, Remington, G; Zipursky, RB; Kapur, S. Equivalent occupancy of dopamine D1 and D2 receptors with

clozapine: differentiation from other atypical antipsychotics. *Am. J. Psychiatry.* 2004 161, 1620-1625.

[45] Uchida, S; Kato, Y, Hirano, K, Kagawa, Y; Yamada, S. Brain neurotransmitter receptor-binding characteristics in rats after oral administration of haloperidol, risperidone and olanzapine. *Life. Sci.* 2007 80, 1635-1640.

[46] Vybiral, S; Vesela, R. Can haloperidol disguise fever. *Physiol. Res.* 1994 43, 359-363.

[47] Wolf, G. Orexins: a newly discovered family of hypothalamic regulators of food intake. *Nutr. Rev.* 1998 56, 172-173.

In: Body Temperature Regulation
Editor: A. B. Cisneros and B. L. Goins

ISBN: 1-60741-282-3

Expert Commentary A

Thermoregulation: Role of Cerebral Cortex

Marcellino Monda*[*]*, Giovanni Messina, Andrea Viggiano*[1]*, Domenico Tafuri*[1]*, Alessandro Viggiano, Sergio Chieffi, Eleonora Andreozzi, Gennaro Izzo, Bruno De Luca

Department of Experimental Medicine, Section of Human Physiology, and Clinical Dietetic Service, Second University of Naples, Via Costantinopoli 16, 80138 Naples, Italy;

[1]Faculty of Motor Sciences, University of Naples "Parthenope", Naples, Italy

Abstract

This expert commentary reports the role exerted by cerebral cortex in the thermoregulation. Generally, the thermoregulation is described almost exclusively with the role of the hypothalamus, while the control exerted by other cerebral structures is not sufficiently underlined. Also the influences of cerebral cortex on body temperature are neglected. Indeed, a search on "PubMed" with the words "thermoregulation and cerebral cortex" reports only 261 articles, while the search of the words "thermoregulation and hypothalamus" produces 2485 papers. This commentary underlines especially the experimental evidences regarding the effects of functional decortication (obtained with cortical spreading depression) on thermogenic changes due to various conditions in different experimental models.

[*] Corresponding author: Prof. Marcellino Monda, MD, Dipartimento di Medicina Sperimentale, Sezione di Fisiologia Umana, Seconda Università di Napoli, Via Costantinopoli 16, 80138 Napoli, Italy, Tel. +39 +81 566 5804, Fax +39 +81 5667500. E-mail: marcellino.monda@unina2.it

The hypothalamus controls body temperature. Although the hypothalamic areas exert relevant influences, the role played by the cerebral cortex on thermoregulation is not negligible. The aim of this commentary is to underline the scarcity of investigations about the control exerted by cerebral cortex on the regulation of body temperature. The preoptic-anterior and posterior hypothalamic areas are known as the playmakers of thermoregulatory mechanisms. Intracerebroventricular injection of prostaglandin E_1 acts on preoptic/anterior hypothalamic area, inducing the elevation of body temperature [11]. Although the preoptic-anterior hypothalamic area is considered the “thermostat” of body temperature regulation, other hypothalamic areas are involved in the thermoregulation.

Stimulations of the ventromedial hypothalamus causes an increase in temperature of interscapular brown adipose tissue [15, 16], that is the main effector of non-shivering thermogenesis [1]. Lesions of the ventromedial hypothalamus reduce the hyperthermia due to various stimuli. Indeed elevation of body temperature induced by an intracerebroventricular injection of prostaglandin E_1 is reduced by ibotenate damage of the ventromedial hypothalamus [9], such as post-ingestional thermogenesis is lowered by the ventromedial hypothalamic lesion [8]. Lesion of the lateral hypothalamus induces hyperthermia [14], while lysine acetylsalicylate modifies thermogenic changes induced by this lesion [10]. These experiments demonstrate a key role played by the hypothalamus in the control of body temperature.

On the other hand, the cerebral cortex exerts considerable influence on the hypothalamic areas. The functional ablation of cerebral cortex reduces the hyperthermia due to lateral hypothalamic lesions [4]. Stimulations of the frontal neocortex induces an elevation of body temperature [3], whereas cortical spreading depression blocks prostaglandin E_1 and endotoxin fever [7]. The cerebral cortex exerts significant influences on hyperthermic reaction induced by orexin A [13], which is a hypothalamic neuropeptide controlling body temperature and food intake [12].

In general, cerebral cortex controls not only voluntary behavior, but also autonomic functions. The ablation of the cerebral cortex reduces the increase of the sympathetic discharge induced by various stimuli [2, 6, 13]. Since the sympathetic activation is associated with the rise of body temperature, the influence of the cerebral cortex on the sympathetic discharge affects also body temperature. Although experimental investigations have demonstrated an evident influence of the cerebral cortex on thermoregulation, the researchers investigate scarcely on these aspects of cortical functions. Indeed, scientific papers regarding the hypothalamic influences on body temperature are much more abundant than articles about the cortical control of thermoregulation. Using a search performed on July 2008, PubMed (U.S. National Library of Medicine) cites 2485 papers on “thermoregulation and hypothalamus”, while it reports only 261 articles on “thermoregulation and cerebral cortex”.

In the future, this difference should be reduced by higher investments on research about the role played by the cerebral cortex on vegetative functions, as the thermoregulation. Particular should be addressed to study regarding the influences exerted by cerebral cortex on “thermostatic hypothesis” of food intake [5]. These experimental investigations could shed further light on the controversial issue regarding the relationship between thermoregulation and control of food intake.

References

[1] Cannon, B; Nedergaard, J. Brown adipose tissue: function and physiological significance. *Physiol. Rev.* 2004 84, 277-359.

[2] De Luca, B; Cerciello, A; Monda, M. Cortical control of neurally-mediated arrhythmogenic properties of desacetyl lanatoside C: The role of the posterior hypothalamus. *Neuropharmacology.* 1982 21, 1211-1214.

[3] De Luca, B; Monda, M; Amaro, S; Pellicano, MP; Cioffi, LA. Thermogenic changes following frontal neocortex stimulation. *Brain Res. Bull.* 1989 22, 1003-1007.

[4] De Luca, B; Monda, M; Pellicano, MP; Zenga, A. Cortical control of thermogenesis induced by lateral hypothalamic lesion and overeating. *Am. J. Physiol.* 1987 253, R626-R633.

[5] Himms-Hagen, J. Role of brown adipose tissue thermogenesis in control of thermoregulatory feeding in rats: a new hypothesis that links thermostatic and glucostatic hypotheses for control of food intake. *Proc. Soc. Exp. Biol. Med.* 1995 208, 159-169.

[6] Monda, M; Amaro, S; De Luca, B. Non-shivering thermogenesis during prostaglandin E1 fever in rats: role of the cerebral cortex. *Brain Res.* 1994 651, 148-154.

[7] Monda, M; Pittman, QJ. Cortical spreading depression blocks prostaglandin E_1 and endotoxin fever in rats. *Am. J. Physiol.* 1993 264, R456-R459.

[8] Monda, M; Sullo, A; De Luca, B. Lesions of ventromedial hypothalamus reduce post-ingestional thermogenesis. *Physiol. Behav.* 1997 61, 687-691.

[9] Monda, M; Sullo, A; De Luca, V; Viggiano, A. Ibotenate lesion of the ventromedial hypothalamus lowers hyperthermic effects of prostaglandin E1. *Physiol. Res.* 50, 321-326, 2001.

[10] Monda, M; Sullo, A; De Luca, E; Pellicano, MP. Lysine acetylsalicylate modifies aphagia and thermogenic changes induced by lateral hypothalamic lesion. *Am. J. Physiol.* 1996 271, R 1638-1642.

[11] Monda, M; Sullo, A; Viggiano, A; De Luca, V. Eating behavior is related to sympathetic activation induced by icv injection of prostaglandin E1 in the rat. *Nutr. Neurosci.* 1999 2, 69-74.

[12] Monda, M; Viggiano, A; Mondola, P; De Luca, V. Inhibition of prostaglandin synthesis reduces hyperthermic reactions induced by hypocretin-1/orexin A. *Brain Res.* 2001 909, 68-74.

[13] Monda, M;, Viggiano, An; Viggiano, Al; Fuccio, F; De Luca, V. Cortical spreading depression blocks the hyperthermic reaction induced by orexin A. *Neurosci.* 2004 123, 567-574.

[14] Refinetti, R; Carlisle, HJ. Effects of lateral hypothalamic lesions on thermoregulation in the rat. *Physiol. Behav.* 1986 38, 219-228.

[15] Thornhill, JA; Halvorson, I. Electrical stimulation of the posterior and ventromedial hypothalamic nuclei causes specific activation of shivering and nonshivering thermogenesis. *Can. J. Physiol. Pharmacol.* 1994 72, 89-96.

[16] Yoshimatsu, H; Egawa, M; Bray, GA. Sympathetic nerve activity after discrete hypothalamic injections of L-glutamate. *Brain Res.* 1993 601, 121-128.

In: Body Temperature Regulation
Editor: A. B. Cisneros and B. L. Goins

ISBN: 1-60741-282-3

Expert Commentary B

Aspects of the Case Evidence of Antipsychotics-associated Hypothermias

***Stefan Löffler*[*]*, Ansgar Klimke, Mignon Löffler-Ensgraber*[1]**

Department of Psychiatry and Psychotherapy, Klinikum Offenbach GmbH, Germany
[1]European Management School, Mainz, Germany

Abstract

For the most part, publications especially of severe antipsychotics-associated hypothermias meet criteria — in form — of uncontrolled observations and — in content — of adverse drug reactions of type B or idiosyncratic drug reactions. Both attributes may have an inhibitory effect on publication procedures in journals setting a high value on data integrity and pathomechanism. Whether publication bias is already true or not, the restrictive handling of (single) cases or uncontrolled observations hampers the constructive clinical as well as scientific discussion of this hardly known side effect.

Antipsychotics-associated hypothermias (AAH) of moderate and especially severe degree, i.e. with a core temperature of ≤ 32°C rectal, dimming of consciousness and bradyarrhythmias, respectively, turned out to be mainly published as case studies with a lower rank (IP ≤ 3). The same seems to go for reviews of this theme. In case reports clinicians often point out that on medication a hypothermia b occurred and that after discontinuation of medication a hypothermia b declined. Persons affected are often aged, rather multimorbid and polymedicated (Poncini et al., 2004).

* Corresponding author: Stefan Löffler[*], M.D., Department of Psychiatry and Psychotherapy, Klinikum Offenbach GmbH, Starkenburgring 66, 63069 Offenbach, Germany. E-mail: SLoeffler@klinikum-offenbach.de

In contrast in (prospective) studies — often higher-ranking published — rather titratable, therefore limited dose-dependent AAH of rather low degree in mammals and not multimorbid subjects, respectively, often under exposition of one substance were investigated (e.g. Van Marum et al., 2007). It is worth mentioning that antipsychotic monopharmacy and mono- / oligomorbidity correspond for example to the German guideline (Gaebel and Falkai, 2006), but not often reflects the clinical reality (Messer et al., 2006).

It is arguable to what extent the severe case reported AAH resembling type B-reactions* are etiopathologically related to the milder study-AAH, which rather correspond to type A-reactions. It seems that the rather severe forms of AAH tend to be influenced by host factors whereby environmental co-factors like changing (co-/poly-)medication play a pivotal role.

The text of a case rejection for publication in a journal with a higher impact factor could be:

"Evaluation of case reports always presents a difficult challenge. In general, such reports are intended to raise the possibility (though certainly do not prove) that Drug A is causally related to Event B. However the public media and even some aspects of the medical/scientific community do not always take it this way-not uncommonly, the case report is misinterpreted as clear evidence of cause-and-effect. Therefore we need to exercise caution as we proceed with editorial assessments of case reports. If there are plausible reasons to connect Drug A to Event B (and the connection is not already established through more rigorous study), we would be more positively inclined to accept the report for publication. On the other hand, if the clinical picture is so complex and muddy that the connection is improbable and/or uncertain, then we are less inclined to acceptance. It may actually do more harm than good to the medical community to publish uncertain reports of this type, since they can unnecessarily raise alarm and concern among patients and physicians. The case that you describe falls more in the latter category. Temperature dysregulation associated with neuroleptics has been recognized for decades. It can be variable and unpredictable* in terms of prevalence and time-course. The patient you describe had been exposed to multiple neuroleptics, and it is not clear that exposure or non-exposure to Drug Z was plausibly related to the occurrence or non-occurrence of temperature dysregulation in either direction" (author known).

In fact it seems among others that in a single uncontrolled observation the temporal relationship between a sudden stop of the putative inducing antipsychotic (mono- or polymedication) and the recovery of temperature regulation within hours or a few days is too uncertain for assumption of a close, if not causal, connexion. In the course of this some points cause astonishment because (i) alternative hypotheses (no or other connexion) may be unlikely or absurd, (ii) cumulative evidences of many case reports of different authors as well as studies exist and (iii) the causal connexion between antipsychotics and hypothermia may be plausible for the following reasons: The relation reveals more (type A-reactions) or less (type B-reactions*) (i) dose-concentration-response properties and (ii) transient and reversible effects, respectively, i.e. basic pharmacological principles, whose closer characterization is still to be found.

Moreover, antipsychotic polypharmacy (successively, simultaneously and/ or overlapping) as well as the impossibility of reonset of the putative adverse drug for scientific reasons remain clinical requirements.

"There can be no discussion that pharmacodiligence requires prospective studies, access to regulatory filings for controlled and monitored use of drugs, and some sense of relative potency for comparison between different drug entities. However, large cohorts are needed for the detection and analysis of type B-reactions*. Since these data are lacking, case and spontaneous reports by alert clinicians must be analyzed. The outcome of such an analysis cannot be seen as solid evidence, but can help to get more insight in the adverse drug reaction. The use of measures like Reporting Odds Ratios and disproportionality can help to detect drugs with an increased association with an adverse drug reaction. A statistically significant Reporting Odds Ratio may be indicative of a higher risk for that particular event during the use of a specific medication, but is never conclusive for the actual existence of a causal relation" (Van Marum et al., 2007).

Generally speaking case reports of AAH broaden our knowledge of new constellations concerning exposition factors (medication), host factors and circumstances (life style). At least case reports remind us of the possibility as well as conditions of appearance of AAH in a way that is taken from clinical everyday life (Vandenbroucke, 2001). As always we should critically meet the fast direct naturalistic descriptive and preliminary evidence of case studies, but not strictly with refusal. This should be true for clinicians as well as scientists.

"Casual observations strike us when they are unexpected. They are therefore refutations of our previously held beliefs and will lead to new conjectures — new ideas, and new theories" (Karl Popper in Vandenbroucke, 2001).

**Explanation of type A- vs. type B-reactions:* Adverse drug reactions can be divided into at least two forms (Rawlins and Thompson, 1977; Park et al., 1992): Type A-reactions are extreme effects based on the pharmacology of a drug and therefore mostly dose-dependent and predictable. Type B-reactions or idiosyncratic drug reactions cannot be derived from the pharmacology and occasionally result often in fatalities particularly because they are hardly detectable in drug development.

References

Gaebel, W & Falkai, P. *S3 Praxisleitlinien in Psychiatrie und Psychotherapie; Vol. 1: Behandlungsleitlinie Schizophrenie*. Edition. Darmstadt: Steinkopff; 2006.

Messer, T; Tilscher, C; Schmauss, M. Polypharmacy in the treatment of schizophrenia. *Fortschr Neurol Psychiatr,* 2006, 74, 377-391.

Park, BK; Pirmohamed, M; Kitteringham, NR. Idiosyncratic drug reactions: a mechanistic evaluation of risk factors. *Br J Clin Pharmacol*, 1992, 34, 377-395.

Poncini, M; Stricker, H; Senn, V. Sindrome da assideramento fra le mure domestiche. *Tribuna Medica Ticinese,* 2004, 69, 129–131.

Rawlins, MD & Thompson, JW. Pathogenesis of adverse drug reactions. In: Davies, DM editor. *Textbook of Adverse Drug Reactions*. Oxford: Oxford University Press, 1977, 44.

Vandenbroucke, JP. In defense of case reports and case series. *Ann Intern Med,* 2001, 134, 330–334.

Van Marum, RJ; Wegewijs, MA; Loonen, AJM; Beers, E. Hypothermia following antipsychotic drug use. *Eur J Clin Pharmacology,* 2007, 63, 627-631.

In: Body Temperature Regulation
Editor: A. B. Cisneros and B. L. Goins

ISBN: 1-60741-282-3

Expert Commentary C

Expert Commentary: Thermoregulation during upper Body Exercise

Michael J. Price
Department of Biomolecular and Sports Sciences, Faculty of Health and Life Sciences, Coventry University, Priory Street, Coventry, UK
Lindsay M. Bottoms
Centre for Applied Sport and Exercise Science, School of Psychology, University of Central Lancashire, Preston, UK

Thermoregulation during upper Body Exercise

A large amount of research regarding body temperature regulation during exercise has been undertaken using lower body exercise modes such as cycling and running. However, although it may be convenient to consider arm exercise (predominantly arm cranking or arm cycling) to elicit similar physiological responses to lower body cycling, but of a lower magnitude, there are important factors to take into account which may result in different thermoregulatory responses. Indeed, upper body exercise has been suggested to pose unique thermoregulatory problems (Sawka et al, 1984). This commentary will address a number of factors that may be considered 'unique' to thermoregulation during upper body exercise.

Influence of the Lower Body during upper Body Exercise

Within a vocational and sporting context upper body exercise rarely involves just the upper body musculature. The lower body is often used to stabilise the upper body or to contribute to force production. For example, the set-up of both hand grip width and foot bar distance has been shown to be related to performance in elite kayakers (Ong et al., 2006). During incremental arm crank ergometry the range of motion of lower body joints have been shown to increase as power output requirement increases (Price et al., 2007). Similarly, during synchronous and asynchronous arm crank ergometry the leg muscle have been suggested to aid torso stability and contribute to the effective transmission of force (Smith et al., 2008). Where the lower body is unable to produce force such as in a braced position or in individuals with a spinal cord injury peak power output is significantly reduced when compared to the able-bodied (Price and Campbell, 1997, 1999, 2003). Furthermore, popular sports and fitness activities such as rowing, where both the lower body and upper body contribute more obviously to force production, require a well-balanced distribution of muscle mass throughout the body (Tachibana et al., 2007). Therefore, the lower body should not be overlooked for studies involving upper body exercise from either a physiological or thermoregulatory perspective.

From a local metabolic or heat production point of view the lower body may only contribute a small proportion of total heat production during arm crank exercise. The potential for heat loss and heat gain from external sources though may be of greater interest. For example, while resting in hot conditions significant amounts of heat may be gained from the environment in able-bodied persons due to whole body cutaneous vasodilation when compared to individuals with a spinal cord injury who demonstrate reduced vasomotor control and sweating capacity in the lower body (Price and Goosey-Tolfrey, 2008). This data, along with core temperature responses, allowed the proportional contribution of heat gain from the environment and heat production from metabolism for able-bodied and athletes with a spinal cord injury to be estimated. It was suggested that during exercise the contribution to heat storage from the environment would be greater for the spinal cord injured than able-bodied group due to reduced heat dissipation even though absolute metabolic rates would be lower.

When considering heat loss, although the lower body does not produce large amounts of heat during arm crank exercise and there is a large surface area for heat gain there is also a large potential surface area for heat loss. This has been illustrated during cooling of the lower body during upper body exercise in the heat (Price and Mather, 2004) and during upper body exercise heat acclimation where the thigh as well as the forehead demonstrate greater sweat production after a seven day heat acclimation programme (Bottoms, 2008). Furthermore, other measures of alleviating heat strain utilising the lower body during upper body exercise, such as foot cooling, have been examined (Hagobian, et al., 2001). However, foot cooling during 30 minutes of arm crank ergometry in the heat was only effective in a group of persons with spinal cord injury when compared to the able bodied as increases in tympanic temperature were much greater in the no cooling control trial (1.6 vs. 0.5°C, spinal cord injured and able-bodied, respectively) and cooling trial (1.0 vs. 0.3°C, respectively). As the lower body has been suggested to be a potent site for heat storage in both able-bodied and persons with spinal cord injury during upper body exercise in the heat (Price and Campbell, 1997, 2002; Price, 2006) further research into this aspect of thermoregulation may be of interest.

Regional Blood Flow during upper Body Exercise

During arm crank exercise skin blood flow of the active arm has been demonstrated to increase linearly with exercise intensity (Volianitis and Secher, 2002; Ahlborg and Jensen-Urstad, 1991). Such data suggests that upper body exercise elicits similar skin blood flow responses in relation to exercise intensity as lower body exercise. However, data from a number of studies examining thermoregulation during upper body exercise have observed differing and unique localised responses within the relatively inactive leg when compared to those of the relatively inactive arm during lower body exercise (Sawka et al., 1984; Price and Campbell, 1997, 1999, 2002). For example, a decrease in calf volume as a result of sympathetic activity and vasoconstriction occurs during submaximal upper body exercise (Hopman et al.,1993) which appears to be accompanied by a decrease in calf skin temperature (Sawka et al., 1984; Price and Campbell, 1997, 1999, 2001). The changes in calf volume noted here were measured using strain gauge plethysmography, and represent whole limb blood flow. It has therefore been suggested that the decrease in calf skin temperature during exercise is a result of blood flow being redirected away from the lower body to the active upper body similar to how visceral blood flow is redistributed during lower body exercise (Price and Campbell, 1997). However, in the absence of specific blood flow data, this response presently can only be speculated.

Although calf volume decreases during submaximal upper body exercise (Hopman, et al., 1993) skin blood flow at the calf as measured by laser doppler techniques, has been observed to increase. Theisen et al., (2001b) observed an increase in skin blood flow of ~150% during 3 min of arm cranking at 80% peak power (W_{peak}) and ~436% during 60 min of arm cranking at 50%W_{peak} (Theisen et al. 2001a). However, laser doppler techniques only measure the surface layer of the skin to a depth of ~1mm and therefore reflect blood flow to only one component of the limb when compared to plethysmography which conversely measures the whole limb response. Studies have yet to report simultaneous measurements of these specific components of leg blood flow during upper body exercise and their effects on thermoregulation. Data from our laboratory combining these techniques (Bottoms, 2008) supports previous studies in that cutaneous blood flow increases even though calf volume decreases. In order for the increase in skin blood flow to occur there must be a more substantial decrease in other tissues of the leg to a greater extent, namely muscle and bone. Indeed, Taylor et al (1989) demonstrated an increase in skin blood flow with a concurrent decrease in muscle blood flow in the inactive calf during one legged cycling. Further studies of muscle, skin and whole limb flow would be useful to determine how regional circulations within the lower limb integrate during upper body exercise.

Leg Blood Flow during Recovery from upper Body Exercise

A consistent thermoregulatory response for upper body exercise is that of a significant reduction in calf skin temperature during recovery from prolonged arm exercise. For example, Price and Campbell (1997, 2002) demonstrated calf skin temperature to continue to decrease below resting values during passive recovery. During 20 min of passive recovery following arm cranking at 80% maximal heart rate (HR_{max}) for 90 min calf skin temperature was observed to decrease a further 2.6°C below resting calf skin temperature in addition to the decrease in skin temperature during exercise (Price and Campbell, 1997). Price and Campbell (2002) also demonstrated a 2.5 (±0.8)°C decrease in calf skin temperature during 30 min of passive recovery following 60 min of arm exercise at 60% peak oxygen uptake ($\dot{V}O_{2peak}$). This is supported by Bottoms et al. (2008) where calf skin temperature decreased to various degrees during passive recovery from upper body exercise after a range of exercise intensities and durations. During the initial five minutes of recovery calf volume increased and skin blood flow decreased, both returning to baseline values. Consequently, the decreased calf skin temperature throughout passive recovery may most likely or partly be due to counter-current heat exchange related mechanisms. This may be suggested based on an increase in blood flow of the whole leg post upper body exercise, possibly as a hyperaemic response, with concurrent decreases in skin temperature (decreased skin blood flow). If arterial inflow increased during recovery, heat could be transferred to the veins within vascular bundles draining the limb and returning to the central circulation. This may allow blood flow to return to pre-exercise levels whilst preventing heat loss via a cooler lower limb. However, whether this mechanism occurs and how the reduction in skin blood flow during recovery from laser Doppler measurement (Bottoms, 2008) contributes to this requires further study.

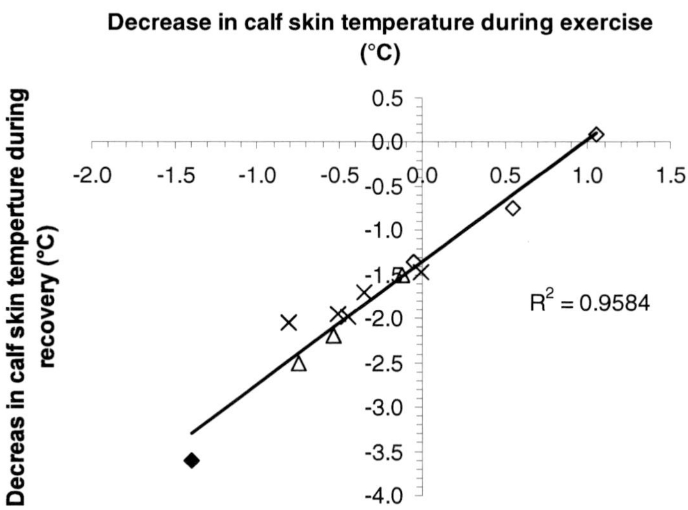

Figure 1. Relationship between the change in calf skin temperature during upper body exercise and the change in calf skin temperature from the end of exercise to the end of passive recovery. All data is from ~21°C ambient temperature and from arm crank exercise at 60% peak power or 60% $\dot{V}O_{2peak}$ followed by 30 minutes passive recovery unless otherwise indicated. Points represent mean values of data taken from Bottoms et al. (2008) and Price and Campbell (2002) as follows; Open diamonds = exercise of 30 min at 21, 27 and 33°C; Crosses = 30 minutes exercise pre, during and post 8 weeks training; Open triangles = exercise of 15, 30 and 45 minutes duration; Closed diamond = data from Price and Campbell (2002), exercise of 60 minutes.

Whatever the mechanism at work within the lower limb during exercise and recovery the calf skin temperature responses are consistent and repeatable for durations of exercise from 15 to 60 minutes. Plotting mean values for the change in calf skin temperature during exercise (15 to 45 min, 60% peak power) against the subsequent change in calf skin temperature from the end of exercise to the end of 30 minutes recovery (Bottoms 2008) results in a strong positive correlation ($r = 0.985$, $P<0.05$; Figure 1). This correlation is maintained when data from Price and Campbell (2002) is added in which calf skin temperature measurements were recorded after 60 min of arm cranking exercise and 30 min of passive recovery ($r = 0.979$). Therefore, it appears that the change in calf skin temperature occurring during passive recovery is directly related to the change in calf skin temperature during arm exercise. This proportional response may strengthen the argument for counter current heat exchange occurring post exercise, as the cooler the lower limb appears to be at the end of exercise, the cooler it appears to be after a period of recovery. As noted previously, studies examining the integration of local circulations would help us to understand this response more fully.

Wheelchair Exercise and Disability

A final point for discussion is that upper body exercise, in the form of wheelchair exercise, represents the habitual mode of locomotion for many individuals. Body temperature regulation within the range of disabilities to which this applies has yet to be examined in any great detail when compared to lower body exercise modes in able-bodied individuals. This is especially true when considering the range of exercise intensities examined. Most often these intensities reflect those undertaken by athletes rather than those during activities of daily living in non athletic populations (Price, 2006). In light of the fact that wheelchair exercise is less efficient than arm crank ergometry (Brown et al., 1990) greater thermal strain may be observed for a given exercise intensity. However, Price and Campbell (1999) observed lower heat storage during 60 min of wheelchair roller exercise in cool conditions when compared to arm crank ergometry at the same absolute and relative exercise intensity (~1.2 – 1.3 $l.min^{-1}$; 60 – 65% $\dot{V}O_{2peak}$). Lower heat storage may have been contributed to by differences in convective cooling currents between exercise modes. This would most likely be from the arm moving relative to the body during wheelchair exercise rather than direct cooling from the wheel itself. Although differences in thermoregulation may occur due to exercise mode the main concern regarding temperature regulation and disability would most likely relate to any physiological differences as a result of a given disability or condition.

Common medical problems associated with wheelchair athletes are generally considered to include intolerance of environmental extremes (Shephard, 1988) or difficulty with thermoregulation (Martinez, 1988; Klenck and Gebke, 2007). Although this is predominantly in consideration of athletes with a spinal cord injury where a significant reduction in vasodilation and sweating capacity exists below the level of injury (Price, 2006) other disabilities may also be wheelchair dependent and have specific thermoregulatory needs. One example is cerebral palsy (CP), a condition which may require the use of either or both upper and lower body exercise modes as a result of a wide range of functional capacities. Children with CP have been shown to elicit greater oxygen consumption than non-CP children during treadmill walking at the same speed (Rose et al., 1990) and have also demonstrated greater rectal temperature (~0.4°C) when undertaking three ten-minute treadmill walking bouts (0.9 $m.s^{-1}$, 3.3% gradient) in the heat (35°C, 50% relative humidity). However, no differences were observed for sweat rates between the CP and non CP groups with the authors suggesting that future research was needed to determine whether those with CP rely more on dry heat loss (Maltais et al., 2004b).

Maltais et al., (2004a) also examined the responses of CP children during arm exercise in the heat. In a similar design to the treadmill walking study children undertook arm crank exercise for three ten-minute bouts in the heat. As an exercise intensity relative to body mass was utilised (~0.55 $W.kg^{-1}$) and body mass was similar between groups the exercise intensity and metabolic rates were similar between groups (~21W, oxygen consumption; ~0.49 $l.min^{-1}$). The results showed that thermal responses were similar between the CP and non CP children during arm exrcise, although skin temperatures were slightly, but significantly, lower in those with CP. The difference in skin temperature of ~0.6°C was considered not to be physiologically or clinically significant. Furthermore, any differences in skin blood flow were not considered sufficient to evoke differences in cardiovascular strain.

It was noted by Maltais et al. (2004a) that the participants of the study were not able to undertake longer durations of arm crank exercise which may be due to their level of disability not requiring habitual wheelchair use. Participants were classed as having mild (Level I and II; n=7) to moderate CP (Level III; n=4) on the Gross Motor Function Classification System which would suggest predominantly lower body habitual exercise with some assistance. Children in Level III need assistive mobility devices, and frequently orthoses, to walk, while children in Level II do not require assistive mobility devices after age four (Palisano et al., 2008). The authors noted that their participants had relatively mild upper limb motor impairment. Consequently, further research regarding those CP individuals using predominantly wheelchair exercise as habitual locomotion, those with greater upper limb impairment and in those with more severe CP is required.

The area within disability that has received most attention regarding thermoregulation is that of individuals with a spinal cord injury (SCI). In brief, spinal cord injury results in a loss of vasodilation and sweating below the level of injury. If the injury is sufficiently high and above the sympathetic outflow, such as injuries in the neck or cervical region of the spinal cord (quadriplegcs; aka tetraplegics), then sweating is either absent or significantly reduced. For injuries in the thoracic and lumbar regions thermoregulatory responses are usually proportional to the level of lesion. A recent review (Price, 2006) summarised the research to date in relation to exercise and thermoregulation in this population and suggested research should concentrate upon the following areas. Firstly, from a thermoregulatory control and heat balance perspective the specific roles of absolute and relative workloads between and within athletes with spinal cord injuries is of interest. This stems from the fact that although athletes with SCI can be matched quite well with their able-bodied counterparts in terms of the number and type of training sessions and training status, $\dot{V}O_{2peak}$ and peak power output are significantly reduced as a result of SCI (Price and Campbell, 1997, 2003). As a consequence, although athletes may be exercising at the same relative intensity, the absolute intensity is significantly lower. This aspect has been addressed in a paper within this volume and suggests that for a given level of oxygen consumption athletes with SCI do appear to have greater increases in core temperature in relation to metabolic rate in cool conditions. This increase however, is relatively small and may not be physiologically important. Specifically designed studies to examine this further rather than from a retrospective approach should therefore be undertaken. It will also be of interest to consider training status and environmental conditions.

Secondly, specific exercise scenarios should be considered in relation to both sports performance and vocational capacities. Such examinations would ideally cover the full range of metabolic and environmental conditions likely to be experienced in athletes and non athletes. For example, some studies have examined high intensity intermittent sprint based protocols in tetraplegic athletes (Webborn et al., 2008) whereas others have examined exercise at lower power outputs (~40W) in paraplegics with a range of lesion levels (Yamasaki et al, 2001). Determining body temperature responses in a greater range of conditions would be useful from both performance and health perspectives.

In relation to exercise a third area for future work focuses upon the enhancement of performance. This may consider aspects of exercise and thermoregulatory physiology already known for able-bodied athletes but not yet reported for athletes with a disability namely; fluid replacement strategies and cooling techniques. Of the latter, the effects of ice vests on repeated sprint performance have been examined in athletes with tetraplegia (Webborn et al., 2005, 2008), those athletes with the greatest thermal dysfunction. Foot cooling has been examined via the use of a negative pressure device (Hagobian et al., 2004) and hand cooling from more conventional water cooling (Goosey-Tolfrey et al., 2008). Both these studies used a mixed group of wheelchair users or athletes with SCI, in terms of the level of spinal injury with both high and low level injuries combined in one group. The latter study also compared able-bodied individuals during a 3 km cycling performance trial whereas the wheelchair athletes were examined over a 1 km wheelchair ergometry time trial. Both time trials occurred after five 10 minute exercise bouts at 50% peak power output in the heat (~31°C). However, although the modes of exercise and performance distance differed between the two groups of athletes, performance was enhanced in both. More specifically designed studies should be undertaken to further our understanding of the mechanisms underlying performance in both the spinal cord injured and able-bodied athlete populations.

A final area of interest and development within the remit of spinal cord injury and thermoregulation is the development and use of functional-electrical stimulation (FES) devices (Davoodi et al., 2002). Recently FES rowing has become important not only in the rehabilitation of spinal cord injury but also for competition. This produces an interesting scenario for thermoregulation. Firstly, the otherwise paralysed muscle is able to contract and produce force. As a consequence this increases metabolism compared to exercise without FES (for a review of physiological responses to FES see Davis et al., 2008). Secondly, the heat produced from muscle stimulation will have to be dissipated along with the heat produced from the non paralysed upper body. Consequently, the ability to produce and dissipate heat may become imbalanced, a factor which has been considered important for athletes with SCI during wheelchair exercise in the heat (Price and Campbell, 2003). It is possible that the sensate areas above the lesion could adapt in order to dissipate the extra heat produced. Conversely, as the stimulated muscle demonstrates increased cross sectional area and capillary density (Chillibeck et al., 1999) heat transfer from the muscle may be enhanced. However, it should be noted that although cross sectional area and capillary density both increased the capillary to fibre area ratio was unchanged. Nevertheless, research into this area may produce some fascinating thermoregulatory findings

In summary, upper body exercise represents an important component of sporting, recreational and vocational activities in a number of population groups. Pertinent areas for future research have been suggested to expand our understanding of thermoregulation within them.

References

Ahlborg, G. and Jensen-Urstad, M. (1991). Metabolism in Exercising Arm Vs. Leg Muscle. *Clin. Physiol.* 11,(5) 459-468.

Bottoms, L.M. (2008). *Themoregulaotry responses during upper body exercise, thermal stress, training and heat acclimation.* PhD thesis, Coventry University.

Brown DD, Knowlton RG, Hamill J, Schneider TL, Hetzler RK. (1990).

Physiological and biomechanical differences between wheelchair-dependent and able-bodied subjects during wheelchair ergometry. *Eur J Appl Physiol Occup Physiol.* 60(3):179-82.

Chilibeck PD, Jeon J, Weiss C, Bell G, Burnham R. (1999). Histochemical changes in muscle of individuals with spinal cord injury following functional electrical stimulated exercise training. *Spinal Cord.* 37(4):264-8.

Davis GM, Hamzaid NA, Fornusek C. (2008) Cardiorespiratory, metabolic, and biomechanical responses during functional electrical stimulation leg exercise: health and fitness benefits. *Artif Organs.* 32(8):625-9.

Davoodi R, Andrews BJ, Wheeler GD, Lederer R (2002). Development of an indoor rowing machine with manual FES controller for total body exercise in paraplegia. *IEEE Trans Neural Syst Rehabil Eng.* 10(3):197-203.

Goosey-Tolfrey V, Swainson M, Boyd C, Atkinson G, Tolfrey K. (2008). The effectiveness of hand cooling at reducing exercise-induced hyperthermia and improving distance-race performance in wheelchair and able-bodied athletes. *J Appl Physiol.* 105(1):37-43.

Hagobian TA, Jacobs KA, Kiratli BJ, Friedlander AL. (2004). Foot cooling reduces exercise-induced hyperthermia in men with spinal cord injury. Med Sci Sports Exerc. 36(3):411-7.

Hopman, M. T., Verheijen, P. H., and Binkhorst, R. A. (1993). Volume Changes in the Legs of Paraplegic Subjects During Arm Exercise. *J. Appl. Physiol.* 75, (5) 2079-2083.

Klenck, C.,Gebke, K., (2007). Practical management: common medical problems in disabled athletes. *Clin J Sport Med.* 17(1):55-60.

Maltais D, Unnithan V, Wilk B, Bar-Or O. (2004a). Responses of children with cerebral palsy to arm-crank exercise in the heat. *Med Sci Sports Exerc.* 2004a Feb;36(2):191-7.

Maltais D, Wilk B, Unnithan V, Bar-Or O. (2004b). Responses of children with cerebral palsy to treadmill walking exercise in the heat. *Med Sci Sports Exerc.* 36(10):1674-81.

Martinez, S.F. (1988) Medical concerns amongst wheelchair road racers, *Physician and Sports Medicine.* 17, 63-68.

Ong, K., Elliott, B., Ackland, T., Lyttle, A. (2006) Performance tolerance and boat set-up in elite sprint kayaking. *Sports Biomech.* 5(1):77-94.

Palisano, R., Rosenbaum, P., Walter, S., Russell, D., Wood, E., Galuppi, B. (2008) Gross Motor Function Classification System for Cerebral Palsy. http://www.canchild.ca/Portals/0/outcomes/pdf/GMFCS.pdf Accessed 18/09/08.

Price, M.J., Collins, L., Smith, P.M., Goss-Sampson, M. (2007). The effects of cadence and power output upon physiological and biomechanical responses to incremental arm-crank ergometry. *Appl. Physiol. Nutr. Metab.* 32(4):686-92.

Price, M.J. (2006). Thermoregulation during exercise in individuals with spinal cord injuries. *Sports Med.* 36(10):863-79.

Price, M.J., Campbell, I.G. (1999). Thermoregulatory and physiological responses of wheelchair athletes to prolonged arm crank and wheelchair exercise. *Int J Sports Med.* 20(7):457-63.

Price, M. J. and Campbell, I. G. (1997)'Thermoregulatory Responses of Paraplegic and Able-Bodied Athletes at Rest and During Prolonged Upper Body Exercise and Passive Recovery.'*Eur. J. Appl. Physiol Occup. Physiol* 76, (6) 552-560.

Price, M. J. and Campbell, I. G. (2002) 'Thermoregulatory Responses During Prolonged Upper-Body Exercise in Cool and Warm Conditions.'*J. Sports Sci.* 20, (7) 519-527.

Price MJ, Campbell IG. (2003). Effects of spinal cord lesion level upon thermoregulation during exercise in the heat. *Med Sci Sports Exerc.* 35(7):1100-7.

Price, M.J. Goosey-Tolfrey, V.L. (2008) Heat Flow of the paraplegic and Able-bodied lower limb during resting heat exposure. *Journal of Thermal Biology,* 33; 255-260

Price MJ, Mather MI. (2004). Comparison of lower- vs. upper-body cooling during arm exercise in hot conditions. *Aviat Space Environ Med.* 75(3):220-6.

Rose J, Gamble JG, Burgos A, Medeiros J, Haskell WL. (1990). Energy expenditure index of walking for normal children and for children with cerebral palsy. *Dev Med Child Neurol.* 32(4):333-40.

Sawka MN, Pimental NA, Pandolf KB. (1984). Thermoregulatory responses to upper body exercise. *Eur J Appl Physiol Occup Physiol.* 52(2):230-4.

Shephard, R.J (1988). Sports medicine and the wheelchair athlete. *Sports Med.* 5(4):226-47.

Smith, P.M., Chapman, M.L., Hazlehurst, K.E., Goss-Sampson, M.A. (2008). The influence of crank configuration on muscle activity and torque production during arm crank ergometry. *Journal of electromyography and Kinesiology.* 18(4), 598-605.

Tachibana K, Yashiro K, Miyazaki J, Ikegami Y, Higuchi M. (2007). Muscle cross-sectional areas and performance power of limbs and trunk in the rowing motion. *Sports Biomech.* 6(1):44-58.

Taylor, J. A., Joyner, M. J., Chase, P. B., and Seals, D. R. (1989) 'Differential Control of Forearm and Calf Vascular Resistance During One-Leg Exercise.'*J. Appl. Physiol* 67, (5) 1791-1800.

Theisen, D., Vanlandewijck, Y., Sturbois, X., and Francaux, M. (2001a). Cutaneous vascular response and thermoregulation in individuals with paraplegia during sustained arm-cranking exercise. *Int. J. Sports Med.* 22, (2) 97-102.

Theisen, D., Vanlandewijck, Y., Sturbois, X., and Francaux, M. (2001b). Central and peripheral haemodynamics in individuals with paraplegia during light and heavy exercise. *J. Rehabil. Med.* 33, (1) 16-20.

Volianitis, S. and Secher, N. H. (2002). Arm blood flow and metabolism during arm and combined arm and leg exercise in humans. *J. Physiol* 544, (Pt 3) 977-984.

Webborn. N., Price, M.J., Castle, P., Goosey-Tolfrey, V. (2008). Cooling strategies improve intermittent sprint performance in the heat of athletes with tetraplegia. *Br J Sports Med.* [Epub ahead of print].

Webborn, N., Price, M.J., Castle, P.C., Goosey-Tolfrey, V.L. (2005). Effects of two cooling strategies on thermoregulatory responses of tetraplegic athletes during repeated intermittent exercise in the heat. *J Appl Physiol.* 98(6):2101-7.

Yamasaki, M., Kim, K.T., Choi, S.W., Muraki, S., Shiokawa, M., Kurokawa, T. (2001). Characteristics of body heat balance of paraplegics during exercise in a hot environment. *J Physiol Anthropol Appl Human Sci*. 20:227-32.

Index

A

B

C

D

E

F

G

H

I

J

K

L

M

N

O

P

Q

R

T

U

V

W

X

Y